# Competition for Water Resources

# Competition for Water Resources

## Experiences and Management Approaches in the US and Europe

*Edited by*

**Jadwiga R. Ziolkowska**
The University of Oklahoma, Norman, OK, United States

**Jeffrey M. Peterson**
University of Minnesota, Saint Paul, MN, United States

ELSEVIER

AMSTERDAM • BOSTON • HEIDELBERG • LONDON • NEW YORK • OXFORD
PARIS • SAN DIEGO • SAN FRANCISCO • SINGAPORE • SYDNEY • TOKYO

Elsevier
Radarweg 29, PO Box 211, 1000 AE Amsterdam, Netherlands
The Boulevard, Langford Lane, Kidlington, Oxford OX5 1GB, United Kingdom
50 Hampshire Street, 5th Floor, Cambridge, MA 02139, United States



**Library of Congress Cataloging-in-Publication Data**
A catalog record for this book is available from the Library of Congress

**British Library Cataloguing-in-Publication Data**
A catalogue record for this book is available from the British Library

ISBN: 978-0-12-803237-4



*Publisher:* Candice Janco
*Acquisition Editor:* Louisa Hutchins
*Editorial Project Manager:* Tasha Frank
*Production Project Manager:* Anitha Sivaraj
*Designer:* Greg Harris

Typeset by TNQ Books and Journals

# Dedication

***To Our Families***

# Contents

# List of Contributors

**M.P. Aillery**, Economic Research Service, USDA, Washington, DC, United States

**E. Assoumou**, MINES ParisTech, PSL Research University, Sophia Antipolis, France

**J.W. Balay**, Susquehanna River Basin Commission, Harrisburg, PA, United States

**J.S. Bergtold**, Kansas State University, Manhattan, KS, United States

**S. Bouckaert**, MINES ParisTech, PSL Research University, Sophia Antipolis, France

**T.A. Boyer**, Oklahoma State University, Stillwater, OK, United States

**N. Brozović**, University of Nebraska-Lincoln, Lincoln, NE, United States

**K.M. Cobourn**, Virginia Polytechnic Institute and State University, Blacksburg, VA, United States

**B. Colby**, University of Arizona, Tucson, AZ, United States

**D. Drabik**, Wageningen University, Wageningen, The Netherlands

**L. Elbakidze**, West Virginia University, Morgantown, WV, United States

**J.E. Fewell**, Lake Region State College, Devils Lake, ND, United States

**S. Ghosh**, Northeastern State University, Tahlequah, OK, United States

**P. Grundmann**, Leibniz-Institute for Agricultural Engineering Potsdam-Bornim e.V., Potsdam, Germany; Humboldt University of Berlin, Berlin, Germany

**K. Hagedorn**, Humboldt University of Berlin, Berlin, Germany

**K. Hansen**, University of Wyoming, Laramie, WY, United States

**M. Hopkins**, Oklahoma State University, Stillwater, OK, United States

**R. Huffaker**, University of Florida, Gainesville, FL, United States

**K. Knapp**, University of California-Riverside, Riverside, CA, United States

**Y. Kuwayama**, Resources for the Future, Washington, DC, United States

**I. Luviano**, Michoacana San Nicolas de Hidalgo University, Morelia, Michoacán, Mexico

**O. Maaß**, Leibniz-Institute for Agricultural Engineering Potsdam-Bornim e.V., Potsdam, Germany; Humboldt University of Berlin, Berlin, Germany

**N. Maïzi**, MINES ParisTech, PSL Research University, Sophia Antipolis, France

**N. Miller**, Kansas State University, Manhattan, KS, United States

**J.Q. Moss**, Oklahoma State University, Stillwater, OK, United States

**J.M. Peterson**, University of Minnesota, Saint Paul, MN, United States

**M. Pulido-Velazquez**, Research Institute of Water and Environmental Engineering (IIAMA), Universitat Politecnica de València, Valencia, Spain

**R. Puls**, Oklahoma Water Survey, The University of Oklahoma, Norman, OK, United States

**S. Ramsey**, Kansas State University, Manhattan, KS, United States

**L. Reins**, KU Leuven University, Leuven, Belgium

**R. Reyes**, The University of Oklahoma, Norman, OK, United States

**J. Richenderfer**, Susquehanna River Basin Commission, Harrisburg, PA, United States

**A. Sahuquillo**, Polytechnic University of Valencia, Valencia, Spain; Royal Academy of Sciences, Madrid, Spain

**L.D. Sanders**, Oklahoma State University, Stillwater, OK, United States

**A.C. Sant'Anna**, Kansas State University, Manhattan, KS, United States

**G.D. Schaible**, Economic Research Service, USDA, Washington, DC, United States

**C. Schleyer**, Alpen-Adria University Klagenfurt, Vienna, Austria

**K. Schwabe**, University of California-Riverside, Riverside, CA, United States

**M.K. Shank**, Susquehanna River Basin Commission, Harrisburg, PA, United States

**A. Tarhule**, The University of Oklahoma, Norman, OK, United States

**T.J. Venus**, Wageningen University, Wageningen, The Netherlands

**S. Villamayor-Tomas**, Thaer Institute, IRI THESys; Humboldt University of Berlin, Berlin, Germany

**F.A. Ward**, New Mexico State University, Las Cruces, NM, United States

**M.E. Webber**, The University of Texas at Austin, Austin, TX, United States

**R. Young**, Mammoth Trading, Lincoln, NE, United States

**D. Zikos**, Humboldt University of Berlin, Berlin, Germany

**J.R. Ziolkowska**, The University of Oklahoma, Norman, OK, United States

**B. Ziolkowski**, Rzeszow University of Technology, Rzeszow, Poland

# Preface

In recent decades, growing population, depletion of aquifers, and extreme drought have exacerbated long-term water scarcity problems in many countries of the world, especially in the United States and Europe. Nowadays, agriculture, the energy sector, municipalities, feedstocks for biofuels, as well as the oil and gas industry are competing for water, with very few instances of well-functioning markets to regulate water allocation. Moreover, water is generally underpriced, because water rates do not reflect the actual economic value of water. As water scarcity emerges as a global problem, strategies for sustainable and cost-effective ways of dealing with water shortages are an urgent and highly relevant topic. This book offers answers to those issues by following two main goals:

1. It evaluates sectoral water use and competition for water resources on both sides of the Atlantic, and provides case study examples from the United States and Europe.
2. It discusses water management strategies and approaches applied in the United States and Europe to optimize water use and allocation, mitigate water scarcity, and adapt to water scarcity.

The comparative analysis between the United States and Europe points out national and regional perspectives on water problems and management strategies, and more importantly lessons learned from applying specific strategies and approaches. The condensed knowledge can be valuable for scientists, practitioners, stakeholders, and other policymakers evaluating different water management strategies for their potential effectiveness. This knowledge can also help with designing and establishing effective strategies and policies subject to regional natural conditions, regional socioeconomic and environmental needs, and weather patterns. By learning from experiences in different countries and regions, potential successful ways of dealing with water scarcity could be learned while mistakes in water management policy could also be avoided in the future.

Chapter 1 provides an overview of global and regional problems related to water scarcity based on the example of the western United States, Europe, as well as institutional and policy deliberations on water scarcity issues. In Chapter 2, a country-specific analysis with case study examples is provided for different sectors both in the United States and Europe to evaluate similarities and differences in the existing water issues on both continents. The section is summarized by a discussion on the water–energy–food nexus that combines the presented sector-specific examples. Chapter 3 is focused on management approaches and strategies to mitigate and/or adapt to water scarcity based on experiences from both continents, through public- and policy-driven approaches as well as new water innovations and technologies. The section is summarized with a discussion detailing experiences from different countries, their transferability, and accessibility to other countries at a larger scale. Chapter 4 gives an outlook toward challenges for water management in the 21st century based on the past and recent developments in the water sector.

*Competition for Water Resources—Experiences and Management Approaches in the US and Europe* will be of a great interest to scientists, practitioners, stakeholders with research/work fields related to water scarcity, natural resource management, environmental economics, and water economics, as well as students and any person interested in water issues and water management.

We hope you will enjoy reading and working with this book.

**Jadwiga R. Ziolkowska**
**Jeffrey M. Peterson**
Norman, Oklahoma
St. Paul, Minnesota
September 2016

# Acknowledgment

The editors wish to acknowledge the National Science Foundation under Grant No. OIA-1301789 for supporting work on some of the material included in this book.

The authors also wish to thank Candice Janco (Elsevier Publisher in the Earth & Environmental Science Program), Rowena Prasad (Elsevier Editorial Project Manager with the Earth & Environmental Science Program), Louisa Hutchins (Acquisitions Editor in the Earth & Environmental Science Program), Anitha Sivaraj (Elsevier Production Manager), and Tasha Frank (Elsevier Editorial Project Manager with the Earth & Environmental Science Program) for their professional help and continuous support with this book project.

PART 1

# Regional Water Scarcity Problems

Chapter 1.1

# Meeting the Challenge of Water Scarcity in the Western United States

**K. Hansen**
*University of Wyoming, Laramie, WY, United States*

## 1. INTRODUCTION

Over the past 100 years, all regions of the United States have experienced increased temperatures and marked changes in average annual precipitation. A range of climate models project more dramatic changes to come (IPCC, 2013). Existing water supply infrastructure was built with outdated population estimates and environmental constraints in mind and, it now turns out, often optimistic estimates of future water supply availability. Water managers are working to find creative solutions to the resulting imbalance between supply and demand.

This is especially true in the western United States, where recent and ongoing droughts have already challenged existing water infrastructure and management institutions. California is in its fourth year of a severe drought, which is having profound impacts on humans and ecosystems in the state (CDWR, 2015; Howitt et al., 2015). The Colorado River Basin is also in the midst of a multiyear drought (16 years and counting), prompting significant regional discussion on how to adapt to a drier future. The higher temperatures projected for the western United States will lead to earlier runoff and increased variability in the intensity and timing of flows (Stewart et al., 2005; Shinker et al., 2010; IPCC, 2014). These additional challenges will increase the difficulties water managers face in maintaining a reliable water supply. They will also increase the frequency of conflicts among water users. This is especially true in light of continued population growth and increasing environmental constraints in many parts of the western United States.

The good news to consider against this rather bleak backdrop is that water managers and policymakers at the federal, state, and local levels are working within existing water management institutions to resolve conflicts. In many cases they are also seeking new and innovative solutions when and where the old institutions no longer generate satisfactory outcomes. For example, the Western Governors Association (WGA), an organization comprised of the governors of 19 states and three territories in the western

**Competition for Water Resources. http://dx.doi.org/10.1016/B978-0-12-803237-4.00001-X**

United States, recognizes drought as a foremost concern.[1] In 2015, the WGA implemented the Western Governors' Drought Forum, which creates a framework for states to share information and best practices so that they can better anticipate and manage drought impacts.

Themes raised by water managers and policymakers through the Forum fall broadly into three categories. First is finding mechanisms for addressing present and future imbalance between demands and available supplies in any particular location, whether these mechanisms be maintenance and expansion of existing infrastructure, new water sources, or conservation strategies. Second is working effectively within water management institutions through increased communication and collaboration between state, local, and federal agencies, water providers, agricultural users, and citizens. This can be particularly challenging when existing legal frameworks and regulations slow response to drought. The final theme is increased recognition of the interconnectedness of water users and the ecosystems upon which they depend. This is simultaneously a call for improved data collection and analysis and better land management practices for forests and farmland.

This chapter describes existing mechanisms for increasing—or reallocating—water supplies to meet changing demands. Such mechanisms are most effective when they can resolve water crises before they occur and reflect changes in societal priorities regarding water resource allocation and management. The next section provides an overview of water resources and uses in the United States. It very quickly identifies the western United States as the location where water management institutions have been stressed the most to date and where they likely require the most attention moving forward. Section 3 describes methods used in the past to resolve conflict between water users and some of the challenges currently facing water managers in the western United States. Section 4 describes the complex regulatory relationships between state, federal, and local water managers that must be part of water management solutions moving forward. Section 5 concludes.

## 2. OVERVIEW OF WATER IN THE WESTERN UNITED STATES

The majority of water conflicts in the United States have occurred in the West, where average precipitation levels and high inter- and intraannual variability create challenges to water management. Average annual precipitation levels have decreased over the past 100 years in this region (IPCC, 2013), further straining resource availability. Some comparison with conditions in the eastern region of the country is useful, but the remainder of the chapter will focus on the western region.

### 2.1 Resource Availability

The westernmost 17 states form the western region of the contiguous United States and represent approximately 59% of its landmass and 35% of its population. Average annual

1. The WGA is an association of the governors of the 19 westernmost states (the 17 discussed later plus Alaska and Hawaii) and the three western US territories of American Samoa, Guam, and Northern Mariana Islands. The mission of the WGA is to "be an instrument… for bipartisan policy development, information exchange, and collective action on issues of critical importance to the Western U.S. (WGA, 2016)."

state-level precipitation in this western region varies between a low of 10 in. in Nevada and a high of 39 in. in Washington. The regional average is just 21 in., significantly below the national average of 37.[2] The region is characterized by significant variability in average precipitation, within as well as between states. Precipitation often falls as high-elevation snowpack, which acts as a natural reservoir, storing water until temperatures rise in the spring, causing the snow to melt (Svoboda et al., 2002; Pierce et al., 2008).

More of the western United States is currently experiencing drought conditions than is the eastern United States (Fig. 1). This is no surprise given the average precipitation levels mentioned previously. But also note that the concept of drought encompasses resource availability for demands rather than just absolute precipitation levels. The National Drought Mitigation Center defines drought to be a moisture deficit bad enough to have social, environmental, or economic effects (NDMC, 2016).

## 2.2 Water Use

Public supply (deliveries to households, commercial, and industrial customers over public supply systems) comprises 12% of water withdrawals in both the western and eastern

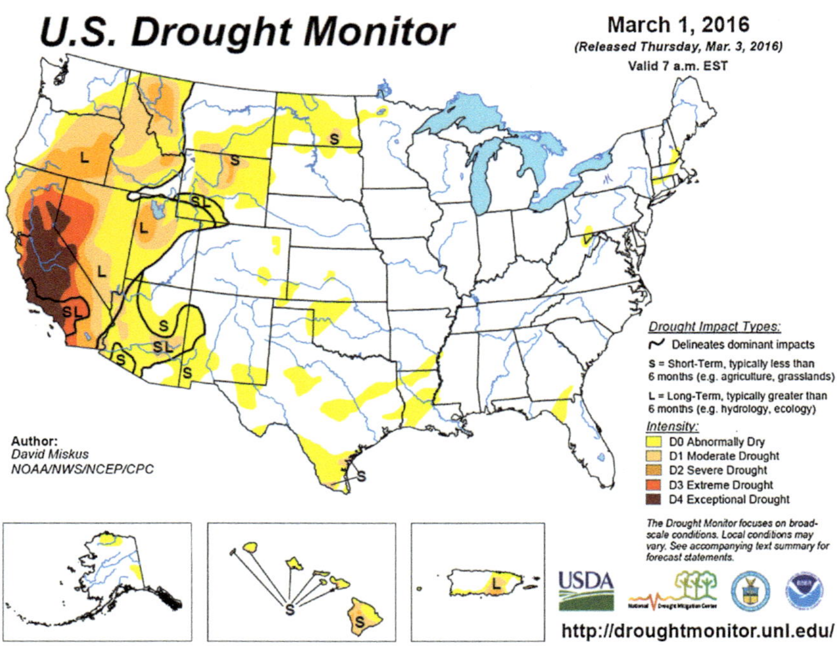


FIGURE 1 Snapshot of current drought conditions across the United States.

2. By contrast, average annual precipitation in the eastern region is 46 in. The eastern states with the lowest and highest annual precipitation levels are Minnesota (27 in.) and Louisiana (60 in.), respectively. Averages, compiled by NOAA (2016), are based on historical data (1971–2000).

regions (Fig. 2).[3] However, in stark contrast to the eastern region, 69% of withdrawals in the western region are for agriculture (primarily irrigation but also including livestock watering and aquaculture).[4] The remaining 15% and 4% of water withdrawals are for thermoelectric power and industrial uses (self-supplied rather than from a public supply system and including withdrawals for mining activity), respectively. Approximately 83% of withdrawals for irrigation and 74% of irrigated acres were in the western region in 2010. Surface water supplied approximately 57% of total irrigation withdrawals nationwide. Surface water accounts for 65% of withdrawals in the western region (compared to only 31% in the eastern region).

Greater variability in states' reliance on water for different uses from different sources exists between western states than between eastern states because of the higher variability in climate and precipitation conditions. Irrigated agriculture accounts for 80% of consumptive water use in the 17 westernmost states and is as high as 90% in some states (Schaible and Aillery, 2012). Water improves farm production value; nationwide, the average value of production for an irrigated farm is more than three times that of a dryland farm (Schaible and Aillery, 2012).

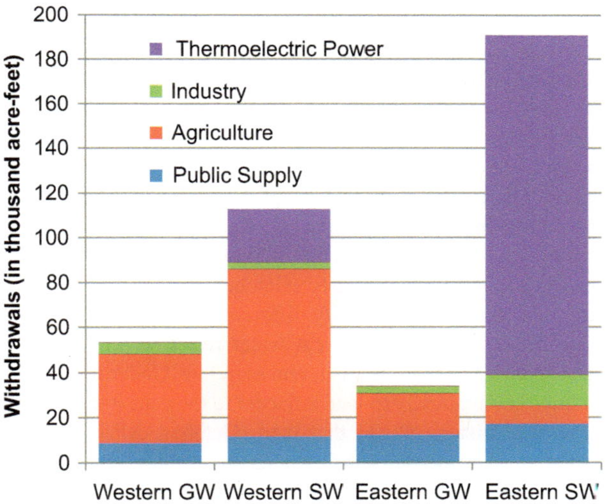


**FIGURE 2** Water withdrawals by region, use, and source for the contiguous United States. *GW*, groundwater; *SW*, surface water.

3. The most comprehensive, recurring source of water withdrawals in the United States is the estimates released every 5 years by the US Geological Survey. The information presented in this section is compiled from the most recent report (Maupin et al., 2014). The public supply total includes domestic use by individuals who self-supply, primarily from groundwater wells in rural areas. Approximately 14% of the US population self-supplies their water rather than receiving it through a public system.
4. By contrast, the largest portion of withdrawals in the eastern region is for thermoelectric power, which uses 69% of water withdrawn in the eastern region. Thermoelectric power use is by and large sourced from surface water. Only 12% of withdrawals in the eastern region are for agriculture, though irrigation there has been expanding (Schaible and Aillery, 2012).

Water use is measured in one of two ways: by quantifying the amount of water withdrawn or the amount of water consumed. Withdrawals far exceed consumption in many uses because much of the water returns to the streams, wetlands, and aquifers from which the water was taken. The US Geological Survey (USGS) numbers reported earlier refer to water withdrawn rather than consumptively used. For water resource management and policy purposes, consumptive use is defined as water removed from an available supply (both surface and groundwater) without return to the system. Thus some withdrawn water is returned to rivers, wetlands, or aquifers and is available for use by others downstream. The percentage of water withdrawn that is returned to the system varies by use.

The uses described previously are consumptive in nature. Some water uses—hydropower, recreation on streams and lakes, and instream flows for fisheries—are entirely nonconsumptive; the water remains entirely available for use downstream. Nonconsumptive uses do not reduce the quantity of water in a river but they may change the quality, location, and timing of flows in ways that affect downstream users. Instream flows are waters that are retained in the river, rather than diverted, to support fish populations.

## 2.3 Water Management

States have primary authority to administer waters within their borders. Western states rely in part or entirely on the doctrine of prior appropriation. Under prior appropriation, the first to divert water from a waterway has the more senior right, giving rise to the phrase, "first in time is first in right." Thus in dry years a more senior right must be entirely satisfied before a more junior right receives any water. To retain a prior appropriation water right, diverted water must be put to a use defined under relevant state law as "beneficial." Uses defined as beneficial in all western states are, for example, municipal, agricultural, industrial, and mining. Recreational and/or instream flows are also identified as beneficial in an increasing number of states. Water is also subject to abandonment if it is not used continuously (Getches, 2009).

Water law has developed fundamentally differently in the western region than it did in the eastern region because of their different water resource profiles. The eastern states inherited the riparian doctrine from English common law. Under the riparian doctrine, the right to water accrues automatically to land adjacent to the waterway, regardless of whether water has been applied historically to that land (Getches, 2009). Settlers moving west found that the riparian doctrine did not fit the harsher, more arid climate of the western United States. For example, gold miners in California wanted to use water in hydraulic mining operations located on public lands far removed from waterways. They would not have had a right to the water under the riparian doctrine. But under prior appropriation, they were able to lay claim to the water by diverting it and putting it to beneficial use. Prior appropriation further helped to ensure that a mining or irrigation project with a senior right would likely have sufficient access to water to justify their often significant capital investment, even in dry years (Hundley, 2001).

# 3. ADDRESSING WATER SCARCITY

Inhabitants of western North America have managed and grappled with scarce water resources since the earliest human settlements (Hundley, 2001). In more recent history, settlers of European descent have set to the task of harnessing land and water resources to human purpose in a way that has had a profound impact on the environment. This is particularly true for water.

## 3.1 Water Development

The first solution to the water scarcity problem implemented on a wide scale was construction of storage and conveyance infrastructure to move water from where it arose naturally to places where people wanted to live. The US Bureau of Reclamation (USBR) played a significant role in this effort. Originally established in 1902 as the US Reclamation Service and housed within the USGS, its original mission was to develop water projects in the 17 western US states with the goal of "reclaiming" arid lands for irrigated agriculture.[5] Today, USBR is the largest wholesale supplier of water in the United States. It operates 337 reservoirs, with a total storage capacity of 245 million acre-feet. One in five farmers receives irrigation water from USBR. USBR water irrigates 10 million acres, which produce 60% of vegetables and 25% of fresh fruit and nut crops in the United States. USBR is also responsible for over 8000 miles of irrigation canals. Approximately 31 million people rely on USBR water deliveries for municipal, residential, and/or industrial use (USBR, 2016).

Many individual states have also constructed significant infrastructure to store and convey water from where it falls to population centers and locations most conducive to agricultural production. One early example (and the inspiration for the classic 1974 movie *Chinatown*) is the purchase of land in Owens Valley, in eastern California, by the City of Los Angeles. Many credit this purchase, controversial though it was, with Los Angeles' subsequent population boom and current economic importance to the California state economy (Libecap, 2005). More recently, the state of California constructed the State Water Project (SWP) to deliver water to 29 municipal and agricultural water suppliers across the state through a system of reservoirs, aqueducts, and pumping plants. Unlike USBR projects, well over half (70%) of SWP water is delivered to municipal rather than agricultural users. SWP includes 34 storage facilities and over 700 miles of open canals, through which it delivers supplemental water to 25 million Californians and over 700,000 acres of irrigated farmland (CDWR, 2016).

The agricultural landscape created by water projects such as these is now a fundamental part of the western identity. Many environmental services, most notably wildlife habitat, have come to depend on the pastoral landscape created by agriculture. We see water removed from agriculture and transferred to other uses, but not without resistance, for both of these reasons.

Some of this infrastructure has been constructed for the purpose of transferring water from one basin to another. For example, the Colorado-Big Thompson project was built in the mid-twentieth century by USBR to transport water from the western slopes of the Rocky Mountains under the Continental Divide to population centers in northeastern Colorado, including Denver. The project consists of 12 reservoirs and 130 miles of tunnels and canals. It provides over 200,000 acre-feet of water to municipal, agricultural, and industrial users (NCWCD, 2016).

Interbasin transfers are not without controversy (Howe and Easter, 1971). Basins with plentiful water supplies and low populations that might be easily tapped for imports are increasingly rare. The potential for ecological harm to native species in the basin of origin

5. Many earlier private and state efforts to establish irrigation projects had failed, either because of lack of funds or lack of engineering capability. USBR projects were originally funded from revenues received by the federal government on federal lands. Beginning in the 1920s, projects were funded directly from Congressional appropriations.

is often an issue, as is the risk of importing exotic and nonnative species to the importing basin. Economic harm in the basin of origin is not always fully compensated financially to everybody's satisfaction, which can contribute to political tension. For all of these reasons, new interbasin transfers, and even new water development projects located within a single basin, are not common. Siting such projects in places where the water supply benefits outweigh construction costs and environmental concerns is difficult. Instead, water managers are increasingly finding ways to use existing water resources more efficiently, either through conservation or transfers.

## 3.2 Conservation

USGS reports that per capita withdrawals for domestic use decreased by nearly 10% nationwide between 2005 and 2010, from 98 to 88 gallons per day.[6] Cities in particular have made a concerted effort to reduce reliance on what are often imported water supplies with complicated relationships with the basin of origin. For example, Phoenix, Arizona, has reduced water use by 35% between 1980 and 2014. Urban southern California imports less water now than it did 20 years ago, even though the region's population has grown by 4 million (approximately 20%) over the same timeframe. Denver, Colorado, and Las Vegas, Nevada, have reduced overall water use by 20% and 30%, respectively, since 2002, even though the populations of both communities have been growing steadily (USBR, 2014).

Economists like prices as a form of water management because they send a clear signal of water's relative value among competing uses. Prices, however, are not always the preferred instrument in a residential setting because of concerns about universal and affordable access to water. What water managers tend to do instead is to ask for voluntary reductions in water use during drought, impose mandatory restrictions on landscape watering, and fund incentive programs to replace old, inefficient appliances. Nonetheless, water managers are increasingly turning to price to regulate demand. Between 2000 and 2012, the number of public and private water systems in the United States with an increasing block rate structure (so that the rate per unit of water consumed increases as consumption increases) rose from 29% to 52% (Smith and Zhao, 2015). One challenge water managers often face is maintaining affordable rates in light of decreasing consumption. As more communities adopt a rate structure based on volume consumed or otherwise encourage conservation, some water supply systems are faced with the prospect of increasing rates to cover the costs of maintaining their supply infrastructure.

Total irrigated acres in the western region increased by 2.1 million acres between 1984 and 2008, though total agricultural water applied declined by nearly 100,000 acre-feet. The portion of all crop agricultural water in the western region using inefficient gravity irrigation systems such as flood and furrow irrigation decreased from 71% to 48% over this time period (Schaible and Aillery, 2012). The pressure irrigation systems brought on line included drip, low-pressure sprinkler, and low-energy precision application systems.

There is room for more efficiency gains, as more than half of irrigated cropland acreage in the United States is irrigated with less efficient irrigation systems such as

6. Withdrawals for most uses (including public supply systems, self-supplied domestic, livestock, irrigation, thermoelectric power, and industrial) decreased between 2005 and 2010. (Withdrawals for mining and aquaculture, though small in absolute terms, did increase.) Population increased from 300.7 to 313 million over the same period.

flood irrigation (Schaible and Aillery, 2012). Pressures to increase agricultural efficiency come from diminished supplies (for example, aquifer drawdown in areas that rely on groundwater) and, in states where water users are able to market their water savings, the lure of selling water to higher-value municipal, industrial, and environmental uses. Agricultural efficiency improvements can be an important part of integrated watershed-level planning, when used in conjunction with other tools, for example, drought water banks, contingent water markets through which water is only transferred in dry years, reservoir management, irrigated acreage and groundwater pumping restrictions, and irrigated acreage retirement (Schaible and Aillery, 2012).

A few words of caution are in order. First, efficiency measures may have unintended, negative effects at the watershed level. Flooding fields may indeed be inefficient and result in significant losses through evaporation and runoff. However, much of the runoff often reenters streams and aquifers and is available downstream for other users. In some systems, where return flows are de minimis, it may make sense to increase efficiency, in the interest of decreasing withdrawals and evaporation rates. In other systems, flood irrigation may provide artificial wetlands for wildlife habitat and generate return flows that provide benefits to downstream users (Peck and Lovvorn, 2001; Conner et al., 2012; Mount et al., 2016).

Even in systems where return flow is relatively small, agricultural conservation measures may not have the intended effects. For example, in areas that rely on declining groundwater aquifers for irrigation supplies, the recharge rate is so low as to be negligible. In such locations, water managers discuss strategies for managed depletion rather than sustainable use. Pfeiffer and Lin (2014) find that irrigators in western Kansas, drawing water from the Ogallala Aquifer, increase water use after installing drop nozzles on their sprinkler systems. Increased irrigation efficiency may increase yields by reducing water losses from runoff and evaporation and allow irrigators to plant more valuable crops. It does not, however, necessarily reduce water use if irrigators are able to use the saved water intensively on the same fields (thus increasing yields or allowing irrigators to grow more valuable crops) or apply the water extensively on fields that would otherwise be fallowed. To be truly effective, efficiency measures must be linked to measurable reductions in withdrawals.

## 3.3 Water Transfers

Water transfers facilitate reallocation of water to higher-value uses dynamically in response to relative changes in supply and demand conditions (Howe et al., 1986; Easter et al., 1998). They allow water managers to adapt quickly to shortfalls in water availability, whether such shortfalls result from increased demands or decreases in supply related to a changing climate or, as is increasingly common in the western United States, a combination of both. Water transfers provide a cost-effective alternative to the development and construction of expensive infrastructure and may be of particular use when additional conservation measures are relatively more expensive or politically infeasible.

Transfers are increasingly common in the western United States. They generally fall into two categories: leases, in which the seller retains the water right but simply allows another party to use the water that flows from the right for a specified period of time; and rights transfers, where the water permit actually changes hands. Leasing activity is more responsive to annual fluctuations in precipitation than sales activity; lease volume is higher in dry years than in normal/wet years. The most common sellers of water and water rights are from the agricultural sector. The user group that has acquired the most water through rights transfers is municipalities (Hansen et al., 2015).[7]

One would expect a certain price differential between a 1-year lease of water and a water right, reflecting the expected stream of benefits a rights holder would accrue in perpetuity from the right. Even given this expected price differential, water rights prices tend to be high relative to 1-year leases. One reason is the high transaction costs associated with a rights transfer; a water right once sold would be expensive to reacquire. High transaction costs notwithstanding, many rights holders prefer to retain a water right even when a strict cost–benefit analysis would suggest that the water user would be better off with a cash payment. Water is life, as the saying goes, and westerners are reluctant to part with it. This feature makes water different than standard commodities, more easily bought and sold in the marketplace, a fact that is reflected in the price of water rights.

Another trend in water transfer activity is that environmental purchasing (both leases and rights) has increased over the past 25 years. In fact, more leasing has occurred to environmental use than to agricultural or municipal use. One oft-cited example is the Environmental Water Account (EWA) in the San Francisco Bay Sacramento Delta. The EWA is a fund established by the CALFED Bay-Delta Program through which state and federal fishery managers could purchase water in real time to help fisheries during critical periods (Hanak, 2003). In 2003, USBR also implemented a temporary water bank in the Klamath River Basin of southern Oregon and northern California to protect three endangered fish species. USBR purchased the water from irrigators, who idled land and pumped groundwater to make the water available (Burke et al., 2004).

Regnacq et al. (2016) observe that in Australia's Murray-Darling Basin, one-third or more of total water available has been traded, whereas in California, trading volumes in the late 2000s were only roughly 3–5% of total water use in the urban and agricultural sectors. Similarly, Hansen et al. (2013) show percentages for all western US states ranging between roughly 0.25% and 6% of total water for 1990 through 2008. These numbers are a lower bound on trading activity, as the data source utilized by these researchers is not comprehensive of trading activity. Further, short-term leases may often happen on a more informal basis, for example, between agricultural producers located within the same irrigation district. Despite increasing political acceptance of water transfers, however, trading volume is still low compared to what it might be. Why are there not more transfers?

Young (1986) detailed the reasons why water transfers were not more common in the western United States 30 years ago. The situation and the underlying reasons remain unchanged. Water is mobile and in cases where consumptive use is something less than the quantity diverted from the waterway, downstream users come to depend on the flow patterns created by the upstream use. These physical challenges and costs associated with measuring and harnessing water have contributed to transfer proceedings at the state level that are complex, time-consuming, and expensive. One feature of transfer proceedings

7. See also Howitt and Hansen (2005), Brown (2006), Brewer et al. (2008), and Doherty and Smith (2014) for detailed discussion on various aspects of water trading in the western United States. These articles all rely on data from the *Water Strategist*, a trade journal that reported water transaction details (price, quantity, buyer/seller identity, and some additional terms) in western US states through 2010. Transactions reported in the *Water Strategist* are not comprehensive of all trading activity and results should thus be taken with a grain of salt. However, the *Water Strategist* is the most comprehensive source of western US water trading activity during the time period. See Hanak and Stryjewski (2012) for a more comprehensive database of transfers in California.

(though specifics vary by state) is a determination that no harm will accrue to other water users as a result of the transfer. As a consequence, a water transfer is generally limited to the quantity of water that has historically been consumptively used (rather than the full amount a user has a right to), to prevent harm to downstream water users who have come to depend on return flows (Getches, 2009). Transfer proceedings also increasingly recognize environmental impacts of changes in the timing or location of flows, largely because of the influence of federal environmental legislation, such as the Endangered Species Act and the Clean Water Act.

Even in the absence of harm to downstream users and the environment, transfers can have negative economic impacts on the exporting community. When fields are fallowed, the exporting community can experience unemployment and income loss (Howitt, 1994; Howe and Goemans, 2003). Water purchasers are not generally required to compensate exporting communities for these costs, though some transfers (most notably the purchase of up to 111,000 acre-feet annually for 35 years by Metropolitan Water District of Southern California from the Palo Verde Irrigation District) have included mitigation funds to assist the exporting communities adapt to a reduced resource base (O'Donnell and Colby, 2009).

A recent report commissioned by the WGA documents the increase in water transfer activity that has occurred in recent decades and suggests ways to make water transfers more efficient and equitable (Doherty and Smith, 2012). This report demonstrates significant political will in the western United States to use water markets, where appropriate, to increase allocative efficiency between competing sectors by redirecting water to its highest-value use. Creatively structured water transfers such as dry-year options, interruptible leases, and water banks[8] can avoid the sale of water rights out of rural communities, a phenomenon known as "buy and dry." This type of arrangement avoids many of the political pitfalls historically associated with water rights transfers from agriculture to urban areas but also provides nearby cities with a way to access water in dry years when they need it most.

State laws have been changing in response to the need for increased flexibility to address drought and protect the environment (Schempp, 2009; Doherty and Smith, 2012; Hansen et al., 2015). One important example is instream flow laws. Prior appropriation has often historically failed to take into account public interests. Nondiversionary uses such as instream flows for fish and wildlife habitat were not originally identified as beneficial use under most state laws. However, when state laws do allow water to be re-allocated to instream flows, the rights are generally assigned a more junior priority date. The question is whether incremental change within the framework of prior appropriation will be sufficient to keep pace with the changing needs of water users and the environment. Chapter 3 of this book provides a thorough discussion of the shortcomings of prior appropriation in the current climate of water scarcity. However, if water transfers can be structured so that they reallocate water more efficiently, without undue harm to existing water users, the environment, or the exporting basin, the answer may be yes.

8. These transfer mechanisms facilitate water sharing during dry years. Under a dry-year option, the buyer pays a premium for the right to lease water should the year turn out to be dry. Under an interruptible lease, irrigation is temporarily suspended and the water transferred to some other user during dry years. Through a water bank, users make water available for future use by storing it in an aquifer or reservoir. If correctly structured, water banks can prevent users from losing a water right from lack of use, thereby facilitating efficient use of water over time and between users.

# 4. STATE, FEDERAL, AND LOCAL GOVERNANCE OF WATER RESOURCES

## 4.1 State Authority

States have primary authority to administer waters within their borders. Western states' constitutions and statutes generally claim ownership to all the water within their boundaries and reserve the right to administer water consistent with the public interest (Getches, 2009). It is certainly the case that all of the transfer activity described earlier has taken place with the approval and oversight of the relevant state, and that virtually all such transfers have been intrastate. States are also primarily responsible for water supply planning that takes place within their borders.

## 4.2 Federal Influence

Nonetheless, the federal government has historically exercised considerable influence over water allocation in the western United States. In addition to the water development already discussed, the United States also owns 32.7% of the land in the 17 westernmost states.[9] In the early 1970s, the federal government extended its reach even further into water management through the passage of the Clean Water Act of 1972 and the Endangered Species Act of 1973. These environmental protection laws, designed to protect and improve water quality, ecosystem health, and biodiversity, placed limits on new and existing water users and somewhat further constrained states' ability to administer land and water resources (Getches, 2001; Mount et al., 2016).

Getches (2001) argues persuasively that federal influence on western US water policies has always been significant. Getches refers to the "myth of state control" of water, given the power of the federal government to supersede that control through court rulings or legislative action. He notes that although the federal government has repeatedly deferred to the states in control over water resources, it has always done so with reference to an early Supreme Court case involving the Rio Grande Irrigation Company, in which the Court found that "state-authorized water use must not interfere with federal rights to protect the flow of the stream and can be superseded by the exercise of federal powers over commerce and public land."[10] In practice, this has meant that whenever a conflict arises between states' exercise of authority over water allocation and federal programs (namely, construction of water projects and enforcement of environmental regulations), the federal purpose has prevailed.

This federal authority, though not always exercised, is an opportunity for improved water management. The Public Policy Institute of California issued a report describing the

9. Federal ownership is even higher in the 11 westernmost US states (46.9%) and is highest of all in Nevada (84.9%). By comparison, federal land ownership at the national level is only 27.4% of the US land surface. These percentages are based on management by five federal agencies (Bureau of Land Management, Fish and Wildlife Service, and National Park Service in the Department of Interior, Forest Service in the Department of Agriculture, and the Department of Defense). These percentages exclude lands managed by other federal agencies, for example, USBR and Department of Energy (Vincent et al., 2014).

10. The quote is from page 6 of Getches (2001). The court case is United States v. Rio Grande Irrigation Co., 174 US 690 (1899).

ways in which federal water policy could change to improve drought resilience in the western United States (Mount et al., 2016). The report is based on interviews with water resource managers and policymakers in western US states and Washington DC. Those interviewed recognized efforts by the federal government to improve coordination between federal agencies and align funding with needs but also suggested additional federal actions that could help western states ready themselves for drought, drought emergencies, and general water scarcity.

The report's first suggestion is to align federal farm program activity—primarily subsidies to irrigators that have historically focused on farm efficiency and easement programs—with local watershed and river basin conservation objectives. Second is to improve the health of headwaters forests (often federally owned and managed) by taking actions to reduce wildfire risk. Third is to improve coordination in the collection and dissemination of water information. Given the multiple roles that the federal government plays in water management, it has great potential to be a positive influence on water management in the western United States.

Although there may well be a larger role for the federal government in resolving conflicts over water allocation, tensions remain regarding the proper extent of federal authority over water. In 2015, the Environmental Protection Agency (EPA) issued an order designed to improve protection for public health and aquatic resources and to clarify the scope of waters of the United States protected under the Clean Water Act. Within days, more than half of US states (along with industrial and agricultural groups) had filed suit against the EPA, claiming that the clarifications extended federal regulatory reach into smaller, previously unregulated waterways and ditches in a way that would increase uncertainty and liability for farmers and ranchers. The case awaits resolution by the courts (Copeland, 2016). USFS also recently proposed a directive, this one on groundwater resource management, intended to establish a clear approach to evaluating and monitoring the effects of actions on USFS groundwater resources. USFS subsequently withdrew the directive amid complaints from states (as well as industry and agriculture) that the directive was an infringement on states' authority over groundwater management (WGA, 2015).

These tensions between state and federal authority over water management will continue. They are likely ultimately useful as well, given, as Huffaker suggests in Chapter 3, relying on traditional state policy to allocate water among competing uses without continued federal intervention might not be enough to protect environmental uses of water that society increasingly finds important.

## 4.3 Tribal Water Claims

The federal government has also played a significant role in allocation of tribal water rights. Most Indian reservations were established by treaty with the federal government before the turn of the 20th century, and without reference to water. The coexistence of these implicitly granted federal rights with the more conventional, state-administered prior appropriation rights led to conflicts that could not be resolved within the rubric of prior appropriation. The 1908 Winters Doctrine clarified the situation somewhat by stating that Indian reservation establishment carried with it sufficient water for the purposes of the reservation.[11] The priority date of these "reserved rights" is reservation establishment,

11. See Winters v. United States, 207 US 564 (1908).

which in most cases predates the general allocation of water in a region through prior appropriation (Doherty and Smith, 2012; Wilkinson, 2015). Reserved water quantities are litigated and quantified through a general stream adjudication, to determine the nature, extent, and relative seniority of all water claimants in a river basin. The first of these occurred in 1978. Since then, general stream adjudications have been completed or are now under way in at least 12 western states (Thorson, 2015).

Wilkinson (2015) describes the tension inherent between state-granted prior appropriative rights and tribal water rights, which are senior to most non-Indian rights and, unlike rights granted under prior appropriation, do not require diversion or actual use to remain valid. States have historically taken umbrage at tribal and federal lawyers' claims to a superior right to water over the generations-old diversions for irrigated farmlands. Prior appropriation is, Wilkinson notes, "infused with history, myth, emotion, politics, economics, and public acceptance." He also notes that state and federal court judges in general stream adjudications have not been unfair but that the proceedings "do not reflect the normal supremacy of valid federal laws over contradictory state provisions."

General stream adjudications tend to be lengthy, contentious court proceedings. Tribes tend to fair better in negotiated settlement. Settlement is more likely to result in "wet" rather than "paper" water for tribes. It also promotes flexibility in finding solutions that involve conservation and wise water management and a spirit of cooperation between tribes and states (Thorson et al., 2006). A combination of litigation and settlement is most common, as tribes can leverage court cases to negotiate settlements outside of court (Thorson, 2015).

Some of these observations are borne out by the experience of the Eastern Shoshone and Northern Arapaho tribes. These two tribes, state and federal agencies, and countless water claimants in the Big Horn Basin of central Wyoming recently completed a 37-year general stream adjudication. As one former Wyoming State Engineer involved in the proceeding has noted, the parties got off on the wrong foot and would likely have achieved a better outcome through settlement rather than litigation (Wilkinson, 2015). In particular, the tribes received rights to 500,000 acre-feet of water, though approximately half of these adjudicated rights so far remain paper rights because the tribes have been unable to obtain funding to develop them (Wilkinson, 2015). Also of note is the failure of the courts in the Big Horn Adjudication to allow the tribes to use their water for instream flows to enhance the fishery, as the tribes did not historically rely on water for this purpose. As Wilkinson (2015) notes, "The weight of classic prior appropriation surely played a role here, for these uses were wholly unrecognized under the consumptive, out-of-stream imperative that drives western water law."

## 4.4 Local Collaboration

As competition between water uses has become more pronounced, basin- and watershed-level planning processes have become increasingly important to water conflict management. Such processes incorporate input from multiple stakeholder groups, thus increasing the probability of successful resolution. Getches (2001) argues that the state agencies established by state constitutions and statutes to administer prior appropriation did not deal well with the pressing water management issues of the 1990s: efficiency and conservation, conjunctive use of groundwater, protection of instream flows, more comprehensive planning, and inclusive public participation at the local level. Rather, it was locally based problem-solving efforts motivated by federal regulatory pressure that implemented needed water reforms.

One such example is the Dungeness Water Exchange located in western Washington state, which arose in response to recent ESA regulations for salmon and pressure on land and water resources from agricultural and residential development. Washington Water Trust developed a mitigation and voluntary leasing program, in coordination with local water users, funded by the State Department of Ecology.[12] New groundwater appropriators must mitigate their impacts to water resources through the program. The mitigation portion of the exchange consists of water rights purchases from Dungeness irrigators used to support instream flows and aquifer recharge projects. The voluntary leasing portion of the exchange serves restoration needs in the watershed. Local irrigators sign forbearance agreements to cease late-season irrigation in exchange for payment. Leasing activity has occurred in 2009 and 2015 and is expected to continue in 2016. The leasing program is flexible and consistent with state law. The watershed-level objectives of the program and the active involvement of local entities in the watershed and state agencies with regulatory oversight responsibilities have been critical to the program's success (Amanda Cronin, personal communication, March 3, 2016).

Given the hydrologic interconnectivity and the fact that many water users are affected by changes in water use patterns, there are many examples moving forward of ways in which local communities can work together to incorporate water transfers into broader, more integrated strategies for addressing water scarcity. Another example is the case of Deschutes County, Oregon. In response to increasing municipal demands, higher environmental standards, and no obvious ways of acquiring new supplies, stakeholders are implementing a combination of agricultural conservation (lining ditches and installing pipes), water transfers, a water bank for irrigators, and improved reservoir management to meet local needs. Part of the result is increased instream flows to meet requirements for endangered fish species (steelhead and salmon). Important to the resolution was a USBR study that provided detailed projections on supply and demand. This information served as a catalyst for action on the part of local stakeholders (Doherty and Smith, 2012).

Local stakeholders can also come to agreement on how to address water scarcity even in the absence of a federal regulatory driver over endangered species. Groundwater Management Districts were formed in Kansas in the early 1970s to establish local control over groundwater rights. Irrigators in Sheridan County, Kansas, wanted even more local control and so pushed for state legislation to create a new kind of management institution called Local Enhanced Management Areas (LEMAs). Under a LEMA, the Kansas Chief Engineer approves locally generated management plans and corrective controls. Based on an economic study, stakeholders in Sheridan County decided to voluntarily reduce present groundwater pumping, which was estimated to produce a slightly lower gross profit in the present but a proportionately higher gross profit in the future, as a result of the greater groundwater reserves generated in the present (Golden et al., 2008). Local stakeholders raised no objections to the plan, so the Chief Engineer approved it (Kansas, 2013). Farmers have altered cropping patterns and implemented deficit irrigation in response to the reductions (Bill Golden, personal communication, February 16, 2015).

12. Also involved in the program is Washington Water Trust, an environmental nonprofit organization that works to use voluntary, market-based transactions and cooperative partnerships to improve water management.

## 5. CONCLUSIONS

Wide-scale water development is no longer the option for addressing water scarcity it once was in the western United States. The relatively low marginal benefits of such new projects do not outweigh the costs, especially given increased awareness of the resulting environmental damage. Water transfers can do some of the work needed to reallocate water in response to changing circumstances. However, prior appropriation has not so far proven flexible enough to meet all of these needs. The question is how to increase transfer activity that improves allocative efficiency while at the same time discouraging transfers that harm water users and the environment downstream.

Local, stakeholder-driven, collaborative processes driven by federal regulations may hold the answer, as evidenced by some of the examples presented here. Such collaborative processes facilitate common understanding and can implement solutions not strictly envisioned by existing water management institutions. Support from federal and state agencies with information on the consequences of alternative courses of action can also aid stakeholders in the decision-making process. And in such examples, water transfers are often one part of the solution.

Water managers must find ways to adapt to emerging needs and accommodate changing circumstances. This is no small challenge, given increased water scarcity and the complexity and interconnectedness of water systems. Strong, flexible water management institutions can tip the balance from protracted conflict to rational and orderly competition. Local, collaborative processes guided by watershed-level needs can help communities and regulators meet their objectives as effectively as possible within the prior appropriation framework.

## REFERENCES

Brewer, J., Glennon, R., Ker, A., Libecap, G.D., 2008. Water markets in the West: prices, trading, and contractual forms. Economic Inquiry 46 (2), 91–112.

Brown, T.C., 2006. Trends in water market activity and price in the Western United States. Water Resources Research 42, W09402. http://dx.doi.org/10.1029/2005WR004180.

Burke, S.M., Adams, R.M., Wallender, W.W., 2004. Water banks and environmental water demands: case of the Klamath project. Water Resources Research 40, W09S02. http://dx.doi.org/10.1029/2003WR002832.

California Department of Water Resources (CDWR), 2015. Drought in California, 15 pp. http://www.cdwr.gov/drought.

California Department of Water Resources (CDWR), 2016. California State Water Project Overview. Available at: http://www.water.ca.gov/swp/.

Conner, J.D., Schwabe, K., King, D., Knapp, K., 2012. Irrigated agriculture and climate change: the influence of water supply variability and salinity on adaptation. Ecological Economics 77, 149–157.

Copeland, C., 2016. EPA and the Army Corps' Rule to Define 'Waters of the United States', p. R43455. Congressional Research Service. 7-5700.

Doherty, T., Smith, R., 2012. Water Transfers in the West: Projects, Trends, and Leading Practices in Voluntary Water Trading, Report of the Western Governors. Association and Western States Water Council.

Easter, K.W., Rosegrant, M.W., Dinar, A. (Eds.), 1998. Markets for Water: Potential and Performance. Kluwer Academic Publishers, Boston, MA.

Getches, D.H., 2001. The metamorphosis of Western water policy: have federal laws and local decisions eclipsed the states' role? Stanford Environmental Law Journal 20, 3–72.

Getches, D.H., 2009. Water Law in a Nutshell. Thomson/Reuters, St Paul, MN.

Golden, B., Peterson, J., O'Brien, D. Potential Economic Impact of Water Use Changes in Northwest Kansas. Kansas State University Agricultural Experiment Station and Cooperative Extension Service. Staff Paper No. February 08–02, 2008.

Hanak, E., Stryjewski, E., 2012. California's Water Market, by the Numbers, Update 2012. Public Policy Institute of California, San Francisco, CA.

Hanak, E., 2003. Who Should Be Allowed to Sell Water in California? Third-Party Issues and the Water Market. Public Policy Institute of California, San Francisco, CA.

Hansen, K., Howitt, R., Williams, J., 2013. Water trades in the western United States: risk, speculation, and property rights. In: Maestu, J. (Ed.), Water Trading and Global Water Scarcity: International Perspectives. RFF Press Water Policy Series, New York, pp. 55–67.

Hansen, K.M., Howitt, R.E., Williams, J., 2015. An econometric test of water market institutions. Natural Resource Journal 55 (1), 127–152.

Howe, C.W., Easter, K.W., 1971. Interbasin Transfers of Water: Economic Issues and Impacts. Johns Hopkins University Press, Baltimore, MD.

Howe, C.W., Goemans, C., 2003. Water transfers and their impacts: lessons from three Colorado water markets. Journal of the American Water Resources Association 39 (5), 1055–1065.

Howe, C., Schurmeier, D., Shaw Jr., W., 1986. Innovative approaches to water allocation: the potential for water markets. Water Resources Research 22 (4), 439–445.

Howitt, R., Hansen, K., 2005. The evolving western water markets. American Agricultural Economics Association Choices 20 (1), 59–63.

Howitt, R.E., Medellin-Azuara, J., MacEwan, D., Lund, J.R., Sumner, D.A., 2015. Economic Analysis of the 2015 Drought for California Agriculture. Center for Watershed Sciences. University of California, Davis, California. Available at: http://watershed.ucdavis.edu.

Howitt, R.E., 1994. Empirical analysis of water market institutions: the 1991 California water market. Resource and Energy Economics 16 (4), 357–371.

Hundley, N., 2001. The Great Thirst. University of California Press, Berkeley, CA.

IPCC, 2013. Summary for policymakers. In: Stocker, T.F., Qin, D., Plattner, G.-K., Tignor, M., Allen, S.K., Boschung, J., Nauels, A., Xia, Y., Bex, V., Midgley, P.M. (Eds.), Climate Change 2013: The Physical Science Basis. Contribution of Working Group I to the Fifth Assessment Report of the Intergovernmental Panel on Climate Change. Cambridge University Press, Cambridge, United Kingdom and New York, NY, USA.

IPCC, 2014. Summary for policymakers. In: Field, C.B., Barros, V.R., Dokken, D.J., Mach, K.J., Mastrandrea, M.D., Bilir, T.E., Chatterjee, M., Ebi, K.L., Estrada, Y.O., Genova, R.C., Girma, B., Kissel, E.S., Levy, A.N., MacCracken, S., Mastrandrea, P.R., White, L.L. (Eds.), Climate Change 2014: Impacts, Adaptation, and Vulnerability. Part A: Global and Sectoral Aspects. Contribution of Working Group II to the Fifth Assessment Report of the Intergovernmental Panel on Climate Change. Cambridge University Press, Cambridge, United Kingdom and New York, NY, USA, pp. 1–32.

Kansas Chief Engineer (Kansas), 2013. April 17, 2013 Order of Designation.

Libecap, G.D., 2005. Rescuing Water Markets: Lessons from Owens Valley. PERC Policy Series Issue Number PS-33, Bozeman, MT. Available at: http://perc.org/sites/default/files/ps33.pdf.

Maupin, M.A., Kenny, J.F., Hutson, S.S., Lovelace, J.K., Barber, N.L., Linsey, K.S., 2014. Estimated Use of Water in the United States in 2010: U.S. Geological Survey Circular 1405, 56 p. http://dx.doi.org/10.3133/cir1405.

Mount, J., Hanak, E., Chappelle, C., Colby, B., Frank, R., Gartrell, G., Gray, B., Kenney, D., Lund, J., Moyle, P., Szeptycki, L., 2016. Improving the Federal Response to Drought: Five Areas for Reform. Public Policy Institute of California, San Francisco, CA, 35 pp.

National Drought Mitigation Center (NDMC), 2016. U.S. Drought Monitor. Available at: http://droughtmonitor.unl.edu/.

Northern Colorado Water Conservancy District (NCWCD), 2016. Colorado-Big Thompson Project. http://www.northernwater.org.

O'Donnell, M., Colby, B., 2009. Dry-Year Water Supply Reliability Contracts: A Tool for Water Managers. The University of Arizona, Department of Agricultural and Resource Economics, Water Transaction Guidebooks for Water Professionals and Stakeholders. Available at: http://cals.arizona.edu/arec/pubs/facultypubs/ewsr-dyo-Final-5-12-10.pdf.

Peck, D., Lovvorn, J.R., 2001. The importance of flood irrigation in water supply to wetlands in the Laramie basin, Wyoming, USA. Wetlands 21 (3), 370–378.

Pfeiffer, L., Lin, C., 2014. Does efficient irrigation technology lead to reduced groundwater extraction? Empirical evidence. Journal of Environmental Economics and Management 67, 189–208.

Pierce, D.W., Barnett, T.P., Hidalgo, H.G., Das, T., Bonfils, C., Santer, B.D., Bala, G., Dettinger, M.D., Cayan, D.R., Mirin, A., Wood, A.W., Nozawa, T., 2008. Attribution of declining western U.S. Snowpack to human effects. Journal of Climate 21, 6425–6644.

Regnacq, C., Dinar, A., Hanak, E., 2016. The gravity of water: water trade friction in California. In: Proceedings of the Meetings of the Allied Social Science Association, San Francisco, CA, USA, January 3–5, 2016.

Schaible, G.D., Aillery, M.P., September 2012. Water Conservation in Irrigated Agriculture: Trends and Challenges in the Face of Emerging Demands. EIB-99. U.S. Department of Agriculture, Economic Research Service.

Schempp, A., 2009. Western Water in the 21st Century: Policies and Program that Stretch Supplies in a Prior Appropriation World. Environmental Law Institute. http://www.elistore.org/reports_detail.

Shinker, J.J., Shuman, B.N., Minckley, T.A., Henderson, A.K., 2010. Climatic shifts in the availability of contested waters: a long-term perspective from the headwaters of the North Platt river. Annals of the Association of American Geographers 100 (4), 1–14.

Smith, V.K., Zhao, M., 2015. Residential water management: an economic perspective on policy instruments. In: Dinar, A., Schwabe, K. (Eds.), Handbook of Water Economics. Edward Elgar Publishing, pp. 103–125.

Stewart, I.T., Cayan, D.R., Dettinger, M.D., April 2005. Changes toward earlier streamflow timing across western North America. Journal of Climate 18, 1136–1155.

Svoboda, M., LeComte, D., Hayes, M., Heim, R., Gleason, K., Angel, J., Rippey, B., Tinker, R., Palecki, M., Stooksbury, D., Miskus, D., Stephens, S., 2002. The drought monitor. Bulletin of the American Meteorological Society 83 (8), 1181.

Thorson, J., Britton, S., Colby, B. (Eds.), 2006. Tribal Water Rights Essays in Contemporary Law, Policy and Economics. University of Arizona Press.

Thorson, J.E., 2015. Reflections on western general stream adjudications upon the signing of Wyoming's Big Horn river adjudication final decree. Wyoming Law Review 15 (2), 383–411. Available at: http://repository.uwyo.edu/wlr/vol15/iss2/8.

U.S. Department of Commerce National Oceanic and Atmospheric Administration (NOAA), 2016. NOAA website. Available at: http://www.esrl.noaa.gov/psd/data/usclimate/pcp.state.19712000.climo (accessed 20.02.2016).

U.S. Department of Interior Bureau of Reclamation (USBR), 2014. Agreement Among the United States of America and [Various Urban Water Agencies] for a Pilot Program for Funding the Creation of Colorado River System Water through Voluntary Water Conservation and Reductions in Use. Agreement No. 14-XX-30-W0574. Available at: http://www.usbr.gov/newsroom/docs/2014-07-30-Executed-Pilot-SCP-Funding-Agreement.pdf.

U.S. Department of Interior Bureau of Reclamation (USBR), 2016. Bureau of Reclamation Website. Available at: http://www.usbr.gov.

Vincent, C.H., Hanson, L.A., Bjelopera, J.P., 2014. Federal Land Ownership: Overview and Data. Congressional Research Service 7-5700, p. R42346.

Western Governors Association (WGA), 2015. Western Governors Association 2015 Annual Report. Available at: http://www.westgov.org/reports.

Western Governors Association, 2016. The WGA Mission. Available at: http://www.westgov.org/about.

Wilkinson, C., 2015. Introduction to Big Horn general stream adjudication symposium. Wyoming Law Review 15 (2), 233–242. Available at: http://repository.uwyo.edu/wlr/vol15/iss2/2.

Young, R.A., 1986. Why are there so few transactions between water users? American Journal of Agricultural Economics 68 (5), 1143–1151.

Chapter 1.2

# Competition for Water Resources From the European Perspective

**D. Zikos, K. Hagedorn**
*Humboldt University of Berlin, Berlin, Germany*

## 1. INTRODUCTION

There are mounting pressures on water resources at the global level, as increasingly variable weather conditions resulting in extreme hydrological events (floods and droughts) create additional stress on water supplies essential both for human demand and for ecosystem health. These pressures arise from the natural variability in water availability and climatic changes, but they are also linked to national and international social, environmental, and economic policies (Estrela et al., 2007). In Europe, the problem of water scarcity is particularly intense in the Mediterranean region, where some semiarid zones are located (like in Spain, Greece, and Italy but most importantly in Malta and Cyprus). However, other European countries are increasingly exposed to similar dangers impacting their water resources, making water conservation strategies crucial to ensure water availability in the long term (EEA, 2007a). In qualitative terms, humans' perceptions and actions regarding water resources are changing, which is reflected with the southern EU Member States gradually recognizing the importance of water quality issues (traditionally a concern rather typical for western Member States) in the general water availability discourse. This makes the call for Europe-wide water strategies, "effective" water policies, and "good" management practices crucial for ensuring both water availability and quality across the continent. The EU Water Framework Directive (WFD) is arguably one of the most significant manifestations of this logic, as it incorporates scattered elements referring to qualitative and quantitative targets of previous EU water legislations and regulations in an integrative way.

However, the impact of climate change on water resources further contributes to Europe's vulnerability to extreme weather phenomena and is expected to magnify existing regional differences of Europe's natural resources and assets, where southern countries will become even drier in comparison to northern EU Member States (EEA, 2012a). Growing water stress and the risk of more people in the future living in river basins under high water stress raise further societal concerns (EEA, 2007b; Alcamo et al., 2007). The combination of these factors will intensify competition between single users or whole sectors of the economy

**Competition for Water Resources. http://dx.doi.org/10.1016/B978-0-12-803237-4.00002-1**

(eg, agriculture and tourism) and might even spark regional conflicts within or across Member States because of the unequal distribution and allocation of water resources (Zikos, 2010). Moreover, it is argued with high confidence (Alcamo et al., 2007) that numerous economic sectors will be challenged and this might lead to a redistribution of economic activities. The European Environmental Agency (EEA) summarizes the situation by identifying a growing imbalance between water demand and availability, reaching critical levels in some parts of Europe (EEA, 2012b). Anthropogenic factors like overabstraction, coupled with natural phenomena like extensive periods of low rainfall or drought are deemed responsible for the situation (EEA, 2012a,b). The recognized economic, social, and environmental impacts of the identified imbalance might contribute to the final breakdown of the fragile economies in the Mediterranean Member States that have already been devastated by the financial crisis of 2008/2009. Under these anticipated severe changes, even the basis of the founding principles[1] upon which the European "dream" has been built might be shaken because of water scarcity and its contribution to the increasing inequalities between the Member States.

The aim of this contribution is twofold: (1) provide a brief overview of water resources in Europe[2] and the associated anthropogenic and natural pressures, and (2) introduce the main European instrument (the WFD) to enable a response to the expected water crisis. The authors argue that despite WFD's careful design allowing flexibility in the implementation of the Directive in the respective Member States, the regional characteristics of Europe, both natural and socioeconomic, urges for a rather differentiated approach. Such an approach should explicitly take into account the major differences between Member States and thus also explicitly deal with existing, underlying, or future conflicts caused by the increased competition for water. It might further deal with the very political economy of water, the process of how public policy related to water is created and implemented, how institutions develop in different social and economic conditions, and how actors from different fields interplay. An in-depth analysis associating the water crisis and its manifestations with the financial and consecutive economic and social crisis in the European Union might sound particularly interesting in the described context. However, the authors decided to address a more integrative approach aligning this chapter with the broader scope of the book, and providing an analysis of a regional water scarcity problem.

For this chapter, mostly secondary data sources from the European Environmental Agency were used. A second data and information source was a metaanalysis of the research conducted by one of the authors between 2005 and 2015 on a multitude of issues concerning European waters and water governance with a particular focus on European institutions and the WFD. Finally, four interviews were conducted with water administrators in Athens and the Cyclades Island complex in Greece in early September 2015 that provided a number of expert opinions on water issues in relation to the ongoing European financial crisis.

The next section presents the "great divide" of Europe in terms of water availability, use, and the main sectors competing for water resources. Further, the WFD is presented as

1. Since December 2009, the Treaty of Lisbon (EC, 2007) sets the fundamental EU objectives, aiming at promoting the Union's main values (rights, freedom, solidarity, and security) and highlights the importance of solidarity between Member States: the Union and its Member States act jointly in a spirit of solidarity if a Member State is the subject of a terrorist attack or the victim of a natural or man-made disaster. Solidarity in the area of energy is also emphasized.
2. It should be noticed that the terms "Europe" and "European Union" are used interchangeably in this chapter, unless specifically noted.

the main instrument addressing water-related problems across the European Union and, in a rather implicit way, facilitating the settlement of disputes. In the following section we argue that the limited success of the WFD, especially in terms of competition for water, is largely because of the great European divide in terms of water availability and use and thus stability in Europe. The authors conclude that the WFD fails as the main European instrument to deal with water issues with respect to conflict resolution and mitigating competition for water resources between users, sectors, and whole European regions.

## 2. THE "GREAT DIVIDE" OF THE EUROPEAN UNION

Water is perhaps the most emblematic natural resource when viewed from the perspective of direct linkages and interfaces of nature and society (EEA, 2012a; Zikos, 2010). Managing the whole spectrum of water functions and uses presents a fundamental example of how ecological, physical, social, economic, political, and even cultural processes can fuse together in the modes of organizing, regulating, controlling, and/or accessing natural resources. Water bodies, be it surface waters or groundwater, provide an extremely variable multitude of functions crucial to the human population. They are a source of drinking water, providers of relaxation and recreation, as well as a transportation route. They receive treated wastewater, provide water for irrigation, and are used for industrial cooling or for energy generation. Water is also closely connected to traditions, cultural, or historical events. Furthermore, water sustains life and as such it is absolutely essential for a healthy ecosystem to fulfill its ecological functions. Therefore water, conceived as a hydrosocial cycle, constitutes an "encompassing vector" (Swyngedouw et al., 2002) to such a degree that the ecological processes of water, the natural hydrocycle, can no longer meaningfully be abstracted from its twin social hydrocycle of sociopolitical, economic, and cultural embeddedness.

Because of this multitude of functions, it is self-evident that water as a resource cannot be managed like other natural resources such as minerals, either fulfilling very specific functions or treated simply as a "commodity" that can be easily evaluated in monetary terms, distributed and allocated according to clearly defined economic and social needs in a direct way, either in a centralized or in a free-market economy. Because of the limited space in this chapter, we choose not to discuss how water was managed historically and the contemporary discourses on the typology of goods, property rights and regimes, "environmental wars" and "environmental peace," and the unique role that water maintains in such debates. Instead, and from the perspective described previously, we employ water as the lens through which different paths to development have been manifested within Europe.

The European Environmental Agency (EEA, 2012c) identified several driving forces behind the existing pressures on European[3] waters. These can be broadly categorized as pressures both in water quality and quantity arising from agriculture, public water

3. The European Environmental Agency divides Europe into three regional blocks of countries: (1) eastern Europe (which includes several central and southeastern European countries) consisting of Bulgaria, Czech Republic, Estonia, Latvia, Lithuania, Hungary, Poland, Romania, Slovenia, Slovakia; (2) western Europe (including the several central European and all north or Nordic countries) consisting of Belgium, Denmark, Germany, Ireland, France, Luxembourg, Netherlands, Austria, Finland, Sweden, England, and Wales, including the non-EU countries of Iceland, Norway, and Switzerland; and (3)southern or Mediterranean Europe consisting of Greece, Spain, Italy, Cyprus, Malta, and Portugal (Fig. 1).

FIGURE 1 Map of the European continent, with EU Member States highlighted. *Own map.*

provision, production of energy, and industry. Another driving force having a direct impact on the hydrocycle is the increasing land use change mainly caused by urbanization and technological and structural change in agriculture and land conversion giving way to the tourism industry (especially in the Mediterranean Member States). The effects of climatic changes multiply the impact of pressures arising from increased sectoral demand for water and from land use change.

According to the European Environmental Agency, agriculture accounts for 33% of total water use in Europe (EEA, 2012a). However, this is an average not reflecting the great diversity of agricultural water demand in different EU Member States. In some parts of Europe, agricultural water use can reach or even exceed 80% (especially in arid and semiarid areas of the Mediterranean Member States including much of southern France, Greece, Italy, Portugal, Cyprus, and Spain). In all those regions, water shortage would be a limiting factor to crop production (EEA, 2012b). The dependency of some countries on agriculture comes at a price. In summer 2005, Portugal faced a severe drought that destroyed large amounts of crops (60% loss of wheat and 80% loss of maize) (Isendahl and

Schmidt, 2006). The cost of this damage is estimated at half a billion EUR[4] (EEA, 2012a). Contrary to the high water demand for irrigated agriculture in the European south, the western and northern Member States only use a fraction of this amount: barely 1% in some cases (like Finland and Ireland) and for a total of 5% (Table 1). There are several reasons for this startling difference. For thousands of years most irrigation has been practiced in southern Europe and was historically and overwhelmingly associated with huge numbers of very small farms. Higher temperatures in summer and lower precipitation make the irrigation vital for the agricultural sector in southern Europe, resulting in an average use of 7000 $m^3$/ha water compared to less than 2000 $m^3$/ha in western and northern Europe (EEA, 2003). The Common Agricultural Policy (CAP) also plays a role in the continuation of the current trend as crops like maize, rice, tobacco, and olives receiving support under CAP are typically produced in the European south.

Supplying public water systems (including households, small businesses, hotels, and small industries connected to the public network) accounts for 25% of water abstraction across Europe although some variation exists between the European countries, with the eastern Member States accounting for an average of 18% water for public supply, with 27% in the western Member States and 29% in the southern Member States (EEA, 2012b). Differences between countries, but also within countries, reflect the greater degree of urbanization in the west, but also the different patterns of domestic water use between countries (household use vs. tourism, for example).

Water scarcity may result in severe impacts on the supply of drinking water. For example, the 2008 extreme drought in Spain left some reservoirs in Catalonia (supplying almost 6 million inhabitants) with only 20% of their capacity, which ultimately caused restrictions on domestic water use (Collins, 2009). In recent years water has become a scarce commodity in Cyprus and it is regularly rationed, even in Nicosia, the capital city and the largest human settlement on the island (Zikos and Roggero, 2012). The division between southern European states and the rest of Europe in terms of how water is distributed between different categories (agriculture, industry, energy, and public supply) becomes even more acute when differences within each category are also examined (Fig. 2). In this respect, water for tourism is very often included in public water, which might change the big picture of water consumption and the data accuracy. For example, per capita daily water use in Spain is twice as high as in Germany (265 L/day and 122 L/day, respectively). However, a large share of this amount is actually water consumed by tourists or tourism-related infrastructure. While the economic impact of tourism is well documented, the true costs of tourism in terms of water usage have not been fully determined, despite the fact that tourism facilities are usually much more water demanding than households (cf. EEA, 2007b).

In the European Union, a significant amount of water is used for energy production (hydropower) and for cooling (especially in nuclear plants), while some industries are particularly water demanding (like the textile industry). The picture here is also very diverse. The average water use for energy generation in western and eastern Member States accounts for about 50% of total water use. In contrast, the average for southern Member States is around 30% (EEA, 2012d). What greatly differentiates the countries is the average for the industrial use of water, reflecting the stereotypical division between the industrialized north and the agricultural south. Western European industry accounts for almost

4. 1 EUR = 1.08 USD as of January 31, 2016.

**TABLE 1** Water Abstractions (mi $m^3$/year) Distributed per Region and Sector

| Region | Year Range | Energy | % | Industry | % | Agriculture | % | Public | % | Total |
|---|---|---|---|---|---|---|---|---|---|---|
| Southern Europe | 1990s | 6,635 | 11.4 | 2,010 | 3.5 | 35,542 | 61.3 | 13,828 | 23.8 | 58,015 |
| | 2009–2011 | 7,018 | 12.2 | 666 | 1.2 | 33,175 | 57.6 | 16,738 | 29.1 | 57,597 |
| Western Europe | 1990s | 67,088 | 51.3 | 22,548 | 17.2 | 7,570 | 5.8 | 33,682 | 25.7 | 130,888 |
| | 2009–2011 | 54,787 | 50.9 | 17,787 | 16.5 | 5,797 | 5.4 | 29,439 | 27.3 | 107,810 |
| Eastern Europe | 1990s | 21,901 | 36.8 | 12,573 | 21.1 | 13,945 | 23.5 | 11,058 | 18.6 | 59,477 |
| | 2009–2011 | 18,538 | 56.3 | 4,882 | 14.8 | 3,545 | 10.8 | 5,990 | 18.2 | 32,955 |

Adapted from EEA (2015). The European Environment—State and Outlook 2015: Synthesis Report. European Environment Agency, Office for Official Publications of the EU, Luxembourg.

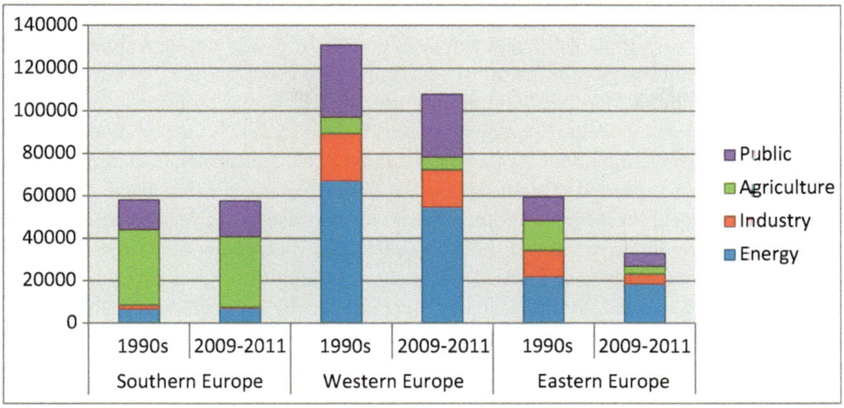


**FIGURE 2** Abstraction of water with respect to sectoral uses. *Own graph based on data from EEA, 2015. The European Environment—State and Outlook 2015: Synthesis Report. European Environment Agency, Office for Official Publications of the EU, Luxembourg.*

20% of all water use while the southern Member States consume barely above 1% of water for industrial use (EEA, 2012d). An impressive, but not surprising, fact is that after the rapid deindustrialization of the 1990s, eastern European Member States largely reduced water use for industrial purposes from more than 20% in the early 1990s down to 10% in the early 2000s, while consumption rose to around 15% as of today (EEA, 2012d). Water scarcity also largely impacts industrial/energy water use. In Portugal, during the 2005 drought, hydropower generation dropped 54% below the average, and 37% below the 2004 average (EEA, 2012a). Similarly, for nine consecutive summers between 1979 and 2007, Germany reduced production of nuclear power because of a combination of high surface water temperatures and low water flow rates (Müller et al., 2007).

The European Environmental Agency has acknowledged land use change as a factor seriously influencing both water flows and water availability, especially when combined with sealing of soils, for instance, when an agricultural area is transformed to an industrial zone (EEA, 2012a). The total area of land use change from agriculture to artificial surfaces greatly varies across Europe and does not follow the great divide although it can be generally said that southern and western Europe are more prone to such changes than eastern and northern Europe (EEA, 2012a). However, the highest share of land use change of this kind occurred in Cyprus (1.7% conversion from agriculture to other uses). It is suspected that rapid urbanization in combination with a booming tourism since 2000 caused this phenomenon (EEA, 2012a).

The diverse picture just described has direct implications for competition for water resources between users, sectors, regions, and even countries. What is more the interdependencies between resources caused by the characteristics of water, often creating a water–food–energy nexus, contain synergies, tradeoffs, and potential conflicts between its component parts (EEA, 2012d). Indeed, the impacts of water shortages are not equally distributed across Europe and can be a source of conflict, again differentiated depending on the European region and its specific characteristics. In Mediterranean Europe, for instance, the periods of peak demand for irrigation come during the summer, when rainfall is already low and when regions are already suffering from drought. This creates an additional pressure for tourism with its water demand peak occurring also in summer. Domestic water use also increases in hot months, creating a difficult situation of limited water resources. In Greece, for example, a Member State almost entirely depending on tourism and services and in some areas on agriculture, water users are affected by serious water shortage during the irrigation season using about 87% of the total available freshwater resources (Isendahl and Schmidt, 2006). Increased competition for water resources often leads to conflicts occurring at the local level between users, but it can also rise to regional importance when water transfers are employed as a response to shortages. The Tagus-Segura water transfer in Spain raised conflicts between the autonomous communities of Castilla-La Mancha and Murcia, and even created tensions at the binational level between Spain and Portugal regarding flow regimes (Isendahl and Schmidt, 2006). In eastern and western Europe such conflicts might take a completely different course. For example, many infrastructure projects promoting the development of rivers for inland navigation, financed through the EU cohesion policy as a freight mode that can reduce greenhouse gas (GHG) emissions, can have negative effects on the hydromorphology of water bodies, destroy natural habitats, and/or decrease their recreational value (EEA, 2012d). On the other hand, in some countries, for example, Germany, the Netherlands, Poland, the Baltic, and Scandinavian countries water scarcity is not of a particular concern. Beyond an intensified effort to ensure high water quality (especially surface waters), large infrastructure projects serve to get the land dry and regulate water flows, as flooding occurrences are on the rise (EEA, 2012a). In northern Germany, for example, the dykes at the North Sea are increased by 1 m, although skeptics argue that with the current rate of climatic changes the existing dams might not be sufficient.

Attempting to summarize the comprehensive materials from the European Environmental Agency for cases across Europe (EEA, 2007b, 2012a,b,c,d,e, 2015) we can distinguish several key characteristics of the great divide between the European south and western Europe:

- Mainly, southern Europe faces limitations in terms of water quantity while western Europe is generally challenged by qualitative problems and increasingly by floods. However, this gap is gradually closing.
- Agriculture is by far the main water user in southern Europe, with industry consuming a relatively tiny fraction of water, while in northern Europe the situation is the opposite.
- Patterns of water use, reflecting the actual paths to economic development, also determine the form of competition: in southern Europe, agriculture, tourism, and domestic users compete for the scarce resource, while water scarce regions and countries are often in conflict with their privileged neighbors. In western Europe, industry, especially power plants, and other users of water like navigation and recreation compete not only for water quantity but also for the overall high water quality and status of water bodies.
- Land use changes caused by urbanization, tourism, and the decrease of agricultural land use play an equally important role in competition for water resources, although the specificities differ, with tourism playing a much more important role in southern Europe.
- The effects of climatic changes are also affecting both southern Europe and western Europe but might take a much more dramatic turn when applied in southern Europe.

## 3. THE WATER FRAMEWORK DIRECTIVE

Historically, many European countries faced barriers such as weak and sector-specific legislation, fragmented administrative structures, exclusive decision-making procedures, inadequate financial resources, and entrenched organizational cultures resulting to a large extent in a separation of water resource management from other societal and economic sectors (Zikos et al., 2006). Following the Earth Summit in 1992, there have been numerous, rather universal, calls for governments and their agencies to reform their water institutions and to adopt more integrated, holistic, and/or ecosystem-based approaches for sustainable water management. The European Council, following the global discourses, gradually recognized the emerging water crisis as one of governance and introduced the WFD (2000/60/EC) in an attempt to introduce a sound basis to meet the principles of effective water governance. This Directive offers a new perspective on water governance and presents an opportunity to enhance interactions between water stakeholders. Since the early 2000s, the WFD and issues related to its implementation were the main focus of several interesting dialogs across Europe. This was especially true in cases where some of the introduced innovations were almost completely unknown in practice (for instance, the concept of public participation in decision making and consultation of nongovernmental actors in east and southeastern Member States).

The European WFD (EC, 2000) requires significant changes in the procedures and performance of water management in all EU countries. It replaced older pieces of European water legislation, such as Directives 76/160/EEC (EEC, 1976a) (quality of bathing water), 76/464/EEC (EEC, 1976b) (water pollution by discharges of certain dangerous substances), 80/68/EEC (EEC, 1980) (groundwater protection against dangerous substances), 91/676/EEC (EEC, 1991a) (nitrates directive), and 91/271/EEC (EEC, 1991b) (urban wastewater treatment). At the time, it reflected the changing sociopolitical and economic context of the 1990s: (1) the increasing internationalization and complexity of water resource management; (2) the rising number of actors and institutions involved; (3) the newly vested economic interests in water supply; and (4) the growing concern and sensitivity toward environmental protection (Kaika, 2003). The Directive promotes an integrated and holistic water management approach, targeting all water bodies and pursuing a sustainable use of water resources, both from a quantitative and a qualitative perspective. Economic, environmental, and ethical issues are incorporated in the overall aim of achieving good water status by 2015 (Giannoccaro et al., 2011; Demetropoulou et al., 2010).

The Directive's objective of "good ecological and chemical status of all European waters" introduces an indirect incentive for Member States to address functional, temporal, and spatial interrelations between multiple water uses, various water systems, and institutions to mediate their interdependencies. Petersen et al. (2009) therefore argued that a fundamental shift in the public mandate to regulate water use has occurred. The WFD has significantly enlarged the scope of responsibilities of the respective EU countries in water management by demanding provision of a good environmental status, a task that the authors associate with greater country interference.

Specifically, the WFD introduced a series of key innovations, including organization of water management around river basins and widening of participation in water policymaking (Page and Kaika, 2003). For river basins, several measures have been devised: identification of key river-basin natural, social, and economic characteristics, prediction on how human activities impact the quality and quantity of basin waters, and a program of measures to achieve good ecological and chemical status of waters. The corresponding processes often require significant changes in national water management and planning practices (cf. Limburg et al., 2002).

The elaboration of such plans is informed by the operationalization of the Directive in the so-called pilot river basins. Measures to improve waters are to be founded on in-depth data. The WFD contains further institutional prerequisites regarding water management, such as "increasing public participation and balancing interest of various groups," as well as ensuring that the "price charged to water users integrates the true costs" (Bressers and Kuks, 2004). Broad integration with other policies and coordinated procedures for data gathering throughout stock-taking exercises are also required. The process by which the WFD was elaborated (the so-called Common Implementation Process) is a top-down approach, which gained local, bottom-up input and legitimization only in a very abstract fashion (EC, 2000). Only later, participation and consultation exercises throughout the elaboration of River Basin Plans allowed for more bottom-up involvement, necessitating a case-by-case analysis of the construction of legitimacy. Several authors have concluded that participation exercises have varied greatly (Borowski et al., 2008). It has also been argued that gathering information about water management problems, bringing actors together to develop and legitimize measures and potentially changing water use and management culture, and, at the same time, researching such processes of transformation is a lengthy, time, and resource consuming process. Such a process often goes well beyond the limited scope of bureaucratic assessments casually conducted by national authorities with respect to River Basin Plans (Zikos and Thiel, 2013). From such a perspective the processes of implementing these various aspects of the Directive become a veritable field of social science studies. Early warning signs have been given, for example, with assessments highlighting the need for collaboration across sectors and levels, which also set agendas for research to support the implementation of the WFD (Borja, 2005; Mohaupt et al., 2007; Ravesteijn and Kroesen, 2007; Liefferink et al., 2011; Dieperink et al., 2012). Another set of studies highlighted a potential role of social learning as a necessary component to meet the integrative management challenges and uncertainties involved in achieving the WFD objectives (Watson and Howe, 2006; Ison et al., 2007; Wright and Fritsch, 2011). These challenges in designing processes that contribute to achieving the integrative aims of the WFD and establishing mechanisms for resolving crises and responding to potential conflicts are yet to be addressed.

The WFD has substantive, performance-related, and procedural goals with regard to water management. In all European Member States, changes are required in water management practices, and therefore extensive institutional change, formal and informal, is implied. Thus despite key provisions for participation, consultation, and deliberation, the WFD constitutes a typical top-down policy, aiming at changing water institutions toward a "desired" direction. To reach its goals it relies on a variety of mechanisms to foster and implement interrelated, multifaceted institutional change. For example, certain prerequisites concerning organization of water planning and management in river basin districts aim at the change of institutions external to people's practices. They require water managers and users to adapt to WFD or face possible European sanctions for noncompliance (Zikos and Thiel, 2013).

Similarly, requirements regarding the achievement of a good ecological and chemical water status and the introduction of water pricing policies presume changes of rational actors' practices, as a result of changes in formal requirements. In contrast, extensive data and knowledge gathered as an input for water planning and goals of WFD can be considered a prerequisite for endogenous mechanisms triggering institutional change through learning about challenges for water management. They are necessary to adapt to the substantive aims of the WFD and adaptation of practices to achieve these aims. Similarly, the Common Implementation Strategy of the WFD aims at an institutional change by bringing together European high-level water managers and fostering a collective learning process. Furthermore, the WFD's rules on participation and consultation often require changes in formal institutions to avoid EU sanctions. They finally aim at unleashing information exchange and learning processes among local water users and managers to draw up cohesive and legitimate measures for attaining the substantive objectives of the WFD (Zikos and Thiel, 2013).

Thus, in the context of the WFD, the intended institutional change with regard to water management unquestionably relies on a clear understanding of such a change as being multifaceted with strong reliance on certain presumed conditions. Above all though, it relies on a very rational assumption that the European integration process will continue as expected and all changes will occur in a rather stable political and economic setting, as set and anticipated in the EU's founding principles from 1958.

## 4. WISHFUL THINKING OR A DANGEROUS MISMATCH?

As early as of 2002 experts claimed that institutionalizing river basin management in accordance with the WFD will require substantial changes to the established modes of water governance (Water Directors, 2002; Heinelt et al., 2002). It has been further argued that in the future water governance in the European Union will become more open and transparent, inclusive and communicative, coherent and integrative, accountable, equitable and ethical, and thus efficient (GWP, 2002). A parallel synthesis process, with continuous coordination and integration of top-down and bottom-up approaches, was acknowledged as a vital requirement to ensure that the implementation of any water-related policies and plans can satisfy the objectives of the WFD (Heinelt et al., 2002). Moreover, many examples from different countries around the world showed that to enhance the democratic mechanisms in the water sector, particular attention needs to be paid to putting ideas into practice and learning from experience through networks and partnerships (Water Directors, 2002).

It is our position though that water governance implemented in a form of organization that is almost entirely based on physical boundaries resulting from the spatial extent of the main transactions (in the given case the water flows in river basins defined by the WFD) is a rather poor unit of analysis for complex social–ecological systems, especially when technology plays an influential role. In a largely diverse setting like on the European continent there might be other physical or cultural boundaries that have to be taken into account. Hagedorn (2008) suggested to consider transaction as a unit of analyses, since "(…) the properties of the transactions are strongly influenced by attributes that are typical of natural systems (…)" (p. 358). As additional properties of transactions, Hagedorn (2008) identified complexity, irreversibility, time lags between a transaction and the consequences, and jointness of production, meaning that the production of one desired good cannot be separated from another, maybe undesired, outcome. From this perspective, institutional diversity and tailored governance structures are in fact seen as an advantage, since they account for the diversity of the natural system (Hagedorn, 2008), which are typical on the European continent. An additional element pointing at an inherited weakness of the river basin as the main unit for the implementation of the WFD is pooling of resources, in our case of water. Whereas in man-made systems pooling of resources is achieved artificially (for example, in agricultural cooperatives based on capital shares provided by the members), in nature-related or social–ecological systems the resources used as commons are often naturally pooled already (for example, a lake used by waterside communities for diverse purposes going beyond the use of its water to irrigate their fields) (Hagedorn, 2013). In such cases, tailored governance structures (often already existing in cases of common pool resources) might present certain advantages over the river basin approach of the WFD as they account for the complexity of the social–ecological system in question.

Such a superficial, rather mechanistic approach to water indicates that certain failures were to be expected, and more importantly give rise to new types of competition between sectors, thus struggling to reach often contradicting European and national policy objectives. Indeed, recent assessments of river basin management plans under the WFD indicate that to reach the objectives of the WFD a reduction of inputs in agriculture are urgently needed (EEA, 2012b). However, this is a huge burden that Mediterranean Member States would be affected by most. Given the current financial crisis and its crippling effects in economic development and social cohesion, it remains unclear how such a reduction could be implemented, especially as reviving agriculture by unemployed youth recently emerges as a still largely unstudied response. Indeed, the ongoing financial and economic crisis has triggered an uncoordinated reruralization process, especially in Greece (Kasimis and Zografakis, 2013). This process underlines a growing momentum of young people seeking alternative development paths relying on agriculture as a response to the crisis (Koutsouris, 2013). In parallel, a much stronger driver for agricultural intensification has been emerging throughout Europe, namely, the structural and technological change in agriculture. It includes an extensive use of crops for bioenergy production, actually intensifying the production processes, which fosters the objectives and targets regarding biofuel production and GHG emissions in the European Union.

Similarly, reports from the European Environmental Agency reveal that more than half of the surface water bodies in Europe are currently below a good ecological status and will need mitigation and/or restoration measures to meet the WFD's objectives (EEA, 2012d). Northern Germany, the Netherlands, and Belgium are the worst areas of Europe concerning the ecological status and pressures on freshwater bodies (EEA, 2012d). Pollution from urban waste and agricultural effluents constitutes a major pressure on water ecosystems and is largely diffused across Europe (EEA, 2012d). This pollution denigrates water quality making it unusable for human purposes, thus creating scarcity of pure water, without meteorological drought. In quantitative terms, water scarcity is reported for nearly all river basin districts in the Mediterranean area (EEA, 2012d).

Since late 1990s, demand-side strategies emerged across the European Union as a promising solution to most of the described (and expected) water problems. Such strategies focused on water demand management instead of the most traditional approaches of managing water supply. Accordingly, the liberalization of the water sector and the introduction of demand-side measures like differentiated water tariffs, water metering, cost-reflective pricing, and increasing the efficiency of the water system gained momentum over costly infrastructure that would aim at increasing water supply (Zikos, 2008). The European Environmental Agency, however, recognized as soon as in 2007 and increasingly since then (EEA, 2015; EEA, 2012a; EEA, 2007a,b) that the reliance on demand-side strategies have the potential to create conflicts between competing demands from economic sectors or entire regions. Furthermore, the European Environmental Agency indirectly acknowledged that the WFD is not a panacea. Although cooperation among water users is a primary goal that requires appropriate institutional frameworks to guarantee that water users "play by the rules," potential conflicts are to be expected in the face of decreased water availability (EEA, 2012a, p. 68). Going beyond the requirements of the WFD and increasing awareness and participation is a necessary precondition not only for success but also for "even greater priorities" (EEA, 2012a, p. 68).

This is better understood given the increasing pressures for cuts in public spending, threatening the provision of water in rural areas with large water deficits. This may lead to intensified competition between water users and entire sectors, which has already occurred in several countries in Europe and further afield. In Greece, for example, according to water administrators interviewed by the authors, in summer 2015 intense conflicts have been observed in the Aegean Archipelago between farmers, domestic users, and tourism businesses about the allocation of the diminishing water reserves. An attempt undertaken by the municipalities in this region to use local water resources to relieve water scarce areas within their administrative boundaries has caused additional conflicts between communities, sometimes even violent.

In such a chaotic situation, the WFD provisions are often not only forgotten, but even officially breached in an attempt to find new equilibria in terms of reallocating water. Italy and Greece, but most notably Spain and Cyprus, were discussing significant increases in desalination plants, in an attempt to ease the pressure on their existing water bodies and support the implementation of the WFD's requirements. The alternatives were discussed despite potential environmental concerns, fears about high dependency on energy supplies, rapid increase in energy demand for operating desalination plants (in Cyprus energy demand was expected to rise from 4% to 40%), and concerns about land use change (Zikos et al., 2015; EEA, 2012b). The inherent risk involved in this approach was proven when an explosion of stored ammunition at a naval base destroyed a major power plant in the country on July 11, 2011, shutting down half of Cyprus's power supply. This tragedy forced desalination plants to operate at one-third of their capacity during the driest month of the year. It needs to be mentioned though that accidents like this do not happen frequently, and current energy availability in Europe does not present any technological or economic impediment to desalination.

In a study by Gikas and Tchobanoglous (2009), comparing seawater desalination, imports of water, and wastewater reclamation on the Aegean islands, the latter was found as the solution with lowest costs and energy requirements. This approach could, however, generate another type of conflict between local communities and water administrators, as the use of wastewater has not gained public acceptance compared to desalination. According to an interview conducted by the authors with the water directorate of the Aegean, desalination has been acknowledged by the public as an acceptable solution to additional sources for water supply.

What really casts a dark shadow on any Europe-wide solutions and blueprints is the increasing frustration of administrators, questioning the very rationale of European policymaking, especially in the face of the ongoing economic crisis. Interviewees in Greece stated that the water divide in Europe is nothing more than one of the facets of other existing divides. They could be summarized as follows: increasing economic inequalities, the role of southern Europe as a buffer to immigrants, dump area for waste, potential special economic zone for foreign investments, source of raw materials, and others. Without a lengthy discussion, if such views are substantiated with real evidence, we judge that when such opinions are expressed by ministerial officials in a country with long-standing experience in water management and who are responsible for the implementation of the WFD's requirements in Greek regions, the emerging political divide gains special weight in this debate. Even though analyzing such a possible political divide and further examining its potential role in European water systems and management is very interesting, this topic goes well beyond the scope of this chapter.

## 5. CONCLUSION

Based on the analysis of secondary data sources and our empirical research findings from several research projects in the past, we claim that the WFD does not meet the challenge of rescoping water management in Europe. Further, it does not foster determining the appropriate process of institutional change, conflict resolution, and softening competition between users, sectors, and all European regions.

We argue that the great divide of Europe makes the implementation of the ambitious targets of the WFD a particularly challenging task, not only because of different socio-economic stages of development and the subsequent uses of water, but also because of hugely diversified cultural and physical settings in the respective EU Member States. Moreover, a governance approach relying almost entirely on the physical boundary of transactions, the river basins, and the assorted River Basin Plans has failed to take into account complexity, irreversibility, time lags between a transaction and the consequences, the jointness of production, and the natural and artificial pooling of water during millennia of European history. We argue that in the case of the European water management system, institutional diversity and tailored governance structures can and should be seen as an advantage, since they account for the diversity of natural systems.

The authors suggest that this failure might not be blamed on the Directive as an instrument of European water legislation as such, but rather an issue related to the political economy of water systems that have not been explicitly dealt with in the short time of European water legislature. Water, in every context but especially in the European diverse setting as described in this chapter, retains high exchange value, gives economic and political power, and allows nations to flourish, while its lack can doom entire economic sectors. The European Union as a political entity has largely failed to address issues related to the political economy of water through the WFD, and has instead treated water as another natural resource requiring EU regulation. Despite ambitious targets and important steps forward, the WFD is challenged by the same obstacles, consisting of elements that Europe constantly faces: an amalgamation of diverse nation-states, with competing and often conflicting interests, entangled in a power struggle in the EU's political arena.

It is an educated guess of the authors informed by changes in European water governance in the last 15 years that the competition for European waters, be it between users, regions, sectors, or countries, is related to the governance style and the ability for institutional change in the first place. Also, vested interests of big international actors and uncoordinated competing political decisions such as the European bioenergy policy (often negating any positive steps resulting from the WFD) play an important role as well.

## REFERENCES

Alcamo, J., Moreno, J.M., Nováky, B., Bindi, M., Corobov, R., Devoy, R.J.N., Giannakopoulos, C., Martin, E., Olesen, J.E., Shvidenko, A., 2007. Europe. In: Parry, M.L., Canziani, O.F., Palutikof, J.P., van der Linden, P.J., Hanson, C.E. (Eds.), Climate Change 2007: Impacts, Adaptation and Vulnerability. Contribution of Working Group II to the Fourth Assessment Report of the Intergovernmental Panel on Climate Change. Cambridge University Press, Cambridge, pp. 541–580.

Borja, A., 2005. The European Water Framework Directive: a challenge for nearshore, coastal and continental shelf research. Continental Shelf Research 25, 1768–1783.

Borowski, I., Le Bourhis, J.-P., Pahl-Wostl, C., Barraque, B., 2008. Spatial misfit in participatory river basin management: effects on social learning, a comparative analysis of German and French case studies. Ecology and Society 13. Article 7.

Bressers, H., Kuks, S., 2004. Integrated Governance and Water Basin Management: Conditions for Regime Change and Sustainability. Kluwer Academic Publ., Dordrecht, The Netherlands.

Collins, R., 2009. Water scarcity and drought in the Mediterranean. EU Change 5 (3), 58–62.

Demetropoulou, L., Nikolaidis, N., Papadoulakis, V., Tsakiris, K., Koussouris, T., Kalogerakis, N., Koukaras, K., Chatzinikolaou, A., Theodoropoulos, K., 2010. Water Framework Directive implementation in Greece: introducing participation in water governance—the case of the Evrotas river basin management plan. Environmental Policy and Governance 20, 336–349.

Dieperink, C., Raadgever, G.T., Driessen, P.P.J., Smit, A.A.H., van Rijswick, H.F.M.W., 2012. Ecological ambitions and complications in the regional implementation of the Water Framework Directive in the Netherlands. Water Policy 14, 160–173.

EC, December 22, 2000. Directive 2000/60/EC of the European Parliament and of the Council of 23 October 2000 Establishing a Framework for Community Action in the Field of Water Policy. Volume 43/Official Journal of the European Communities, OJ L 327. Publications Office of the European Union, Luxembourg.

EC, December 17, 2007. Treaty of Lisbon Amending the Treaty on European Union and the Treaty Establishing the European Community. Volume 50/Official Journal of the European Communities, OJ C 306. Publications Office of the European Union, Luxembourg.

EEA, 2003. Indicator Factsheet: Mean Water Allocation for Irrigation in Europe. European Environmental Agency. Available at: http://www.eea.europa.eu/data-and-maps/indicators/mean-water-allocation-for-irrigation/irrig-water_revoct03.pdf/at_download/file.

EEA, 2007a. Climate Change and Water Adaptation Issues. European Environmental Agency Technical Report No. 2/2007, Office for Official Publications of the EU, Luxembourg.

EEA, 2007b. Europe's Environment: The Fourth Assessment. European Environmental Agency. Office for Official Publications of the EU, Luxembourg.

EEA, 2012a. Water Resources in Europe in the Context of Vulnerability. EEA 2012 State of Water Assessment. European Environmental Agency Report No. 11/2012, Office for Official Publications of the EU, Luxembourg.

EEA, 2012b. Towards Efficient Use of Water Resources in Europe. European Environmental Agency Report No. 01/2012, Office for Official Publications of the EU, Luxembourg.

EEA, 2012c. European Waters—Assessment of Status and Pressures. European Environmental Agency Report No. 08/2012, Office for Official Publications of the EU, Luxembourg.

EEA, 2012d. European Waters—Current Status and Future Challenges. European Environmental Agency Report No. 09/2012, Office for Official Publications of the EU, Luxembourg.

EEA, 2012e. Territorial Cohesion and Water Management in Europe: The Spatial Perspective. European Environmental Agency Report No. 04/2012, Office for Official Publications of the EU, Luxembourg.

EEA, 2015. The European Environment—State and Outlook 2015: Synthesis Report. European Environment Agency, Office for Official Publications of the EU, Luxembourg.

EEC, February 5, 1976a. Council Directive 76/160/EEC of 8 December 1975 Concerning the Quality of Bathing Water; Official Journal of the European Communities, OJ L 31. Publications Office of the European Union, Luxembourg.

EEC, May 18, 1976b. Council Directive 76/464/EEC of 4 May 1976 on Pollution Caused by Certain Dangerous Substances Discharged into the Aquatic Environment of the Community; Official Journal of the European Communities, OJ L 129. Publications Office of the European Union, Luxembourg.

EEC, January 26, 1980. Council Directive 80/68/EEC of 17 December 1979 on the Protection of Groundwater Against Pollution Caused by Certain Dangerous Substances; Official Journal of the European Communities, OJ L 020. Publications Office of the European Union, Luxembourg.

EEC, December 31, 1991a. Council Directive 91/676/EEC of 12 December 1991 Concerning the Protection of Waters against Pollution Caused by Nitrates from Agricultural Sources; Official Journal of the European Communities, OJ L 375. Publications Office of the European Union, Luxembourg.

EEC, May 30, 1991b. Council Directive 91/271/EEC of 21 May 1991 Concerning Urban Waste-Water Treatment; Official Journal of the European Communities, OJ L 135. Publications Office of the European Union, Luxembourg.

Estrela, T., et al., 2007. Sustainable Water Use in Europe. Part 3: Extreme Hydrological Events: Floods and Droughts. Environmental issue report No. 21, European Environmental Agency, Office for Official Publications of the EU, Luxembourg.

Giannoccaro, G., Prosperi, M., Zanni, G., 2011. Economic effects of legislative framework changes in groundwater use rights for irrigation. Water 3, 906–922.

Gikas, P., Tchobanoglous, G., 2009. Sustainable use of water in the Aegean Islands. Journal of Environmental Management 90, 2601–2611.

GWP, 2002. Dialogue on Effective Water Governance. Global Water Partnership.

Hagedorn, K., 2008. Particular requirements for institutional analysis in nature-related sectors. European Review of Agricultural Economics 35 (3), 357–384.

Hagedorn, K., 2013. Natural resource management: the role of cooperative institutions and governance. Journal of Entrepreneurial and Organizational Diversity 2 (1), 101–121.

Heinelt, H., Getimis, P., Kafkalas, G., Smith, R., Swyngedouw, E., 2002. Participatory Governance in a Multi-level Context: Concepts and Experience. Leseke & Budrich, Oplanden.

Isendahl, N., Schmidt, G., 2006. Drought in the Mediterranean: WWF Policy Proposals. WWF-World Wide Fund for Nature.

Ison, R., Roeling, N., Watson, D., 2007. Challenges to science and society in the sustainable management and use of water: investigating the role of social learning. Environmental Science & Policy 10, 499–511.

Kaika, M., 2003. The Water Framework Directive: a new directive for a changing social, political and economic European framework. European Planning Studies 11, 299–316.

Kasimis, C., Zografakis, S., 2013. Η ύπαιθρος, η γεωργία και η κρίση. In: Kasimis, C., Zografakis, S. (Eds.), Αφιέρωμα: Επιστροφή στη γεωργία. Athens, pp. 17–32 (in Greek).

Koutsouris, A., 2013. Ο ρόλος της κατάρτισης και της τεχνικής βοήθειας στην εγκατάσταση νεοεισερχόμενων στη γεωργία. Διεθνής Ευρωπαϊκή και Οικονομική Πολιτική 27, 60–68 (in Greek).

Liefferink, D., Wiering, M., Uitenboogaart, Y., 2011. The EU Water Framework Directive: a multi-dimensional analysis of implementation and domestic impact. Land Use Policy 28, 712–722.

Limburg, K.E., O'Neill, R.V., Costanza, R., Farber, S., 2002. Complex systems and valuation. Ecological Economics 41, 409–420.

Mohaupt, V., Crosnier, G., Todd, R., Petersen, P., Dworak, T., 2007. WFD and agriculture activity of the EU: first linkages between the CAP and the WFD at EU level. Water Science and Technology 56, 163–170.

Müller, U., Greis, S., Rothstein, B., 2007. Impacts on Water Temperatures of Selected German Rivers and on Electricity Production of Thermal Power Plants Due to Climate Change. DKKV/CEDIM, Karlsruhe University.

Page, B., Kaika, M., 2003. The EU Water Framework Directive: part 2—policy innovation and the shifting choreography of governance. European Environment 13, 299–316.

Petersen, T., Klauer, B., Manstetten, R., 2009. The environment as a challenge for governmental responsibility—the case of the European Water Framework Directive. Ecological Economics 68, 2058–2065.

Ravesteijn, W., Kroesen, O., 2007. Tensions in water management: Dutch tradition and European policy. Water Science and Technology 56, 105–111.

Swyngedouw, E., Page, B., Kaika, M., 2002. Sustainability and policy innovation in a multi-level context: Crosscutting Issues in the Water Sector. In: Heinelt, H., Getimis, P., Kafkalas, G., Smith, R., Swyngedouw, E. (Eds.), Participatory Governance in Multi-Level Context: Concepts and Experience. Leske & Budrich, Opladen, pp. 107–131.

Water Directors, 2002. Guidance on Public Participation in Relation to the Water Framework Directive. Final Draft 24/10/2002.

Watson, N., Howe, J., 2006. Implementing the EU Water Framework Directive: experiences of participatory planning in the ribble basin, North West England. Water International 31, 472–487.

Wright, S.A.L., Fritsch, O., 2011. Operationalising active involvement in the EU Water Framework Directive: why, when and how? Ecological Economics 70, 2268–2274.

Zikos, D., Beveridge, R., Medd, W., 2006. In pursuit of the effective water governance. Transactions on Business and Economics 3 (1), 17–25.

Zikos, D., 2008. Urban water dilemmas under the multi-dimensional prism of sustainability. Transactions on Business and Economics 8 (5), 413–422.

Zikos, D., 2010. Community involvement in the implementation of the WFD in Greece. In: Ambientalia Special Issue Series (II): Ten Years of the Water Framework Directive – An Overview from Multiple Disciplines, pp. 1–20.

Zikos, D., Roggero, M., 2012. The patronage of thirst: exploring institutional fit on a divided Cyprus. Ecology and Society 18 (2), 25.

Zikos, D., Thiel, A., 2013. Action research's potential to foster institutional change for urban water management. Water 5, 356–378.

Zikos, D., Sorman, A., Lau, M., 2015. Beyond water security: asecuritisation and identity in Cyprus. International Environmental Agreements: Politics, Law and Economics 15 (3), 309–326.

Chapter 1.3

# Institutional Aspects and Policy Background of Water Scarcity Problems in the United States

**R. Huffaker**
*University of Florida, Gainesville, FL, United States*

## 1. INTRODUCTION

Interest in overhauling water management institutions spikes in response to devastating water shortages. As the Murray–Darling basin endured an extended drought during the 2000s, Australia's prime minister urged that "the only hope of restoring the river to health lies in a complete overhaul of how it is managed" (Unnamed, 2007). When California suffered its third worst drought in 106 years during the the 2010s, the California governor observed that "it's a different world...we have to act differently" (Fear, 2015). Editorials in the *Washington Post* advised that the California drought was "a crisis worth exploiting" by making changes needed "to shift [away from] an arcane and antiquated system" (Bittman, 2014), and "finally creat[ing] a transparent and efficient water market" (Editorial, 2015).

Economists have long recommended water markets (Milliman, 1959). Water markets are attractive because they rely on decentralized voluntary transactions, not political consensus, to maximize social benefits by equalizing marginal water values across competitive uses.[1] Howitt contended that a free water market would "leave everyone better off" in drought-stricken California, and that "small engineering changes [could] move the water from east to west, from the $20 region to the $2000 region" (Charles, 2014). Hamilton et al. (1989) estimated that water transfers from irrigation to hydroelectric power in the Columbia Basin would generate potential benefits 10 times greater than lost farm income.

Despite large prospective benefits, water markets have not developed substantially as a mitigation measure in the drought-stricken western United States (Charles, 2014; Perala and Benson, 1995; Howitt and Lund, 2014). This chapter investigates a key contributing factor: the prior appropriative doctrine governing water use in the western United States provides poor soil for growing markets, similar to relying on hardpan to grow

1. The equi-marginal principle holds that the social benefits of water allocation are maximized when competitive water users each earn the same marginal benefits per unit of water. Otherwise, social benefits are increased by allocating an available unit to the water user who earns the greatest marginal benefits.

**Competition for Water Resources. http://dx.doi.org/10.1016/B978-0-12-803237-4.00003-3**

award-winning roses. How can the prior appropriation doctrine be amended to grow better markets?

## 2. INSTREAM FLOWS

The prior appropriation doctrine grants a user's right to the quantity of publicly owned water first diverted to a beneficial use on a fixed tract of appurtenant land. The priority of the right is determined by the date of first diversion. During water shortages, appropriators with senior priority receive their full water duty until no water remains at the source. Appropriators with more junior priority receive no water at all. If senior appropriators wish to expand their water duties, they must execute a new appropriation with the most junior priority.

The requirement that water be diverted to obtain an appropriative right secured 80–90% of dependable river flows to irrigated agriculture in the late 19th and 20th centuries, and excluded modern instream uses such as hydropower production and aquatic ecosystem protection. The result is that traditional appropriative rights can "with impunity, flood deep canyons and literally dry up streams, as has happened with some regularity" (Wilkinson, 1992, p. 21).

Water markets cannot equalize marginal water values across competitive uses that are excluded from participation. Reallocating water from irrigation to hydropower production in the Pacific Northwest has been demonstrated to generate benefits 10 times greater than lost farm income (Hamilton et al., 1989), and two times greater if the flows are shaped specifically to meet the migratory needs of endangered fish species (Hamilton and Whittlesey, 1992). Without markets, advocates for nonappropriative uses must lobby state water agencies to establish instream flow rights that typically have junior priority to existing senior rights already allocated to the bulk of dependable river flows (see, eg, Instream Flows). Consequently, instream flow rights are liable to be among the first curtailed during water shortages.

Do western states have the legal authority to extend their instream flow policies to allow private parties to compete in water markets for appropriative rights that could be converted to instream uses? In sum, states would be relying on water markets to recondition private water rights in the public interest. Can states recondition appropriative rights?

There is reason to believe that states have this authority. Appropriative water rights grant a private right to use *publicly owned* water. Representing the public owners of water, states retain the authority to condition private water use so that it is consistent with public values, and they do this to varying degrees in their water statutes (Wilkinson, 1992). The public trust doctrine, recognizing the government's obligation to manage some types of natural resources in trust for the public benefit, provides an additional font of legal authority. States are empowered to condition appropriative water rights to the degree needed to protect public resources and uses under the trust. The US. Supreme Court has determined that the trust applies to navigable freshwater bodies and tidelands. Individual states have extended public trust obligations to rural parklands; wetlands associated with navigable water bodies; nonnavigable tributaries; and waters usable for fish and wildlife habitat and recreational purposes (Stevens, 1980).

## 3. SPECIFICATION OF APPROPRIATIVE RIGHTS

The quantity of water conveyed by a prior appropriative right (the "water duty") is the diversion deemed sufficient to irrigate an average mix of crops on the appurtenant land with the irrigation technology prevailing when the water right was granted. Water that is not beneficially used reverts back to the public, and can be reappropriated by another person ("use it or lose it").

Limiting the definition of the water duty to diversionary quantities gives an incomplete accounting of water use in a river basin to the detriment of appropriators and instream flows (Allen et al., 1997). Water users are linked in a complicated hydrologic web because of the fugitive nature of water. When crops consume less water than is diverted from the stream, as is almost always the case, the unconsumed water may: (1) return to the stream as surface runoff or as underground spring flow after deep percolation to an underlying aquifer (return flows); or (2) escape by the same means to a second water course (escape flow). Return and escape flows often supply part of the water required to satisfy the water rights of other appropriators, and constitute an essential component of instream flows in western US rivers (Pulver, 1988). Consequently, a farm that increases consumptive use (ie, the volume of water consumed by crops and lost in evaporation) and reduces return flows can impair the water rights of other appropriators.

Consumptive water use has not remained static in irrigated agriculture. Although water duties are tied to the irrigation technology prevailing when the right was issued, farms have been allowed to gradually improve on-farm irrigation technology, which has tended to increase consumptive use at the expense of basin-wide water supplies (Ward, 2008). Improved irrigation technology may increase consumptive water use at the intensive margin of production because water is applied to crops more uniformly in space and time on appurtenant land. Consumptive use may also increase at the extensive margin if efficiency-improving farmers are allowed to spread irrigation water to land that is not appurtenant to the water right ("nonappurtenant land") in further contravention of the prior appropriation doctrine. Efficiency improvements may require less of the original water duty to achieve increased levels of consumptive use in crop production on appurtenant land (Huffaker et al., 2000). Efficiency-improving irrigators reasonably fear that the unused portion of the water duty will be forfeited subject to the "use it or lose it" criterion of the prior appropriation doctrine unless they can use it to irrigate nonappurtenant land (water spreading). Several western states have been convinced by this argument (see, eg, Chapter 537 Oregon Revised Statutes; Water Code, Revised Code of Washington, 2003).

The hydrologic impact of turning a blind eye to noncompliant water spreading is that instream flows and aquifer levels decrease while concealing the true culprit—increased consumptive use. For example, improvements in irrigation technology in the upper Snake River Plain aquifer were eventually identified as the cause of an "invisible drought" in the lower aquifer (Johnson et al., 1999). In another example, water spreading pursuant to shifts from flood to center-pivot irrigation contributed to water shortages in the Columbia Basin Project in Washington State (Huffaker et al., 2000).

The legal impact of noncompliant water spreading is to allow senior appropriators to expand their water use to nonappurtenant land. Compliant senior appropriators would request a new and junior appropriative right to irrigate nonappurtenant land, and this would not be granted to the detriment of preexisting water rights. Allowing water spreading allows efficiency-improving irrigators to increase water consumption to the detriment of other water rights.

Markets require well-specified property rights so that buyers can be secure that what they purchase will be delivered. Water markets cannot provide this security so long as appropriative rights conceal the true hydrologic impacts of water use. Buyers cannot reasonably expect that their rights will not be jeopardized as noncompliant senior appropriators are allowed to enlarge their water use without requesting a new junior appropriation. To provide this security in water markets, appropriative rights must be more fully specified with the "use of terms and definitions that clearly describe the effects of various water uses, both consumptive and non-consumptive, within a hydrologic system" (Allen et al., 1997, p. 72). Fully specified water rights must include consumptive use, and return/escape flow parameters. Satellite remote-sensing Earth-observing systems (eg, Landsat) offer increased capacity to monitor agricultural water consumption (NASA).

## 4. TRANSFERABILITY OF APPROPRIATIVE RIGHTS

States impose moderate to severe restrictions on water-right transfers to prevent changes in the timing, quantity, and quality of return/escape flows detrimental to third-party rights (Gould, 1988; Young, 1986). For example, some states place the burden of proof on transfer proponents to prove with "clear and convincing" evidence that third-party rights will not be injured (eg, Wyoming), or ban transfers of agricultural water to nonagricultural uses (eg, Nebraska) (Huffaker et al., 2000). These restrictions have been identified as the principal reason why water markets have failed to develop substantially within the framework of the prior appropriation doctrine (Perala and Benson, 1995; Young, 1986).

Economists have long recommended that transferred water rights be restricted to the seller's consumptive water use to protect irrigation return/escape flows relied upon by other water users (Milliman, 1959). Unfortunately (as discussed earlier), the water duty that would be transferred under appropriative right is quantified in terms of diversion, not consumption. Moreover, restricting transfers to consumptive use does not completely resolve quantity-related impairments to use-dependent rights (Anderson and Johnson, 1986), and does not reach timing- or quality-related impairments (Gould, 1988). In response, economists have designed specialized transfers to limit the extent and duration of third-party impairment. For example, "trial transfers" could be modified or revoked if actual impairment occurred, "one-time-temporary transfers" would reduce the duration of an injury, and "contingent transfers" would be triggered by some predetermined, often drought related, contingency (Huffaker et al., 2000).

Another barrier to transferability is that appropriative rights holders may be reticent to participate in water markets because of the fear that marketed water may be lost to the "use it or lose it" requirement (Howitt and Lund, 2014). This was the root cause of why Idaho potato growers continued irrigating in a poor market year rather than lease their water rights to hydroelectric power generators (Unnamed, 2001).

A third barrier is that rural agriculturally dependent communities may actively resist water exports that reduce local economic activity and tax receipts. As a farmer in Turlock, California, stated, "If we sold our water off, the jobs would go away here, too. There would be less commerce going on in our county" (Charles, 2014). Some westerns states require that projected economic impacts on the area of origin be considered in transfer applications (Texas), while other states require an export fee on transferred water and possibly creation if a mitigation fund (Nevada) (Unnamed, 2012).

## 5. ENFORCEMENT OF APPROPRIATIVE WATER RIGHTS

Water users whose rights are impaired by expanded consumptive use of efficiency-improving irrigators or negative transfer externalities can seek redress from state courts or administrative water agencies. However, judicial actions are costly, lengthy, and possibly futile. Courts have not consistently recognized the complex hydrologic web of use dependencies linking water users. For example, in *Estate of Steed* v. *New Escalante Irrigation Co.* [846 P.2d 1223 (Utah 1992)], Steed (plaintiff) contended that improved irrigation technology adopted by the New Escalante (defendant) substantially reduced escape flows providing water for Steed's water right. The court refused to enforce Steed's water right on the basis that it was based on a previously wasteful irrigation practice. The ruling allowed New Escalante to enlarge its water use at the expense of Steed's water right.

Enforcement through administrative channels may be even less promising. For example, a Washington State court ruled that constitutional due process requires a judicial basin-wide adjudication of water rights before the state water agency is authorized to evaluate and enforce the priorities of water rights [*Rettkowski* v. *Department of Ecology* 1983 (Sinking Creek)]. A renowned water lawyer in Washington concluded: "Since most of the state's waters remain unadjudicated, the most obvious effect of Sinking Creek is that for most water users, priority—the keystone of Western water law—is now meaningless" (Dufford, 1994).

Extreme and prolonged drought can motivate states to take extraordinary enforcement measures. For example, in 2015, California water regulators proposed a $1.5 million fine against an irrigation district for illegally diverting water in violation of a state-imposed cutbacks. Similar to the Sinking Creek case, the district claimed a breach of constitutional due process in curtailment of its senior water rights (Nagourney, 2015).

## 6. CONCLUSION

The prior appropriation doctrine promoted the rapid economic development of the western United States by providing irrigated agriculture with a secure water supply when it was the major economic sector. However, the doctrine's past success works against current pressures to reallocate water to emerging nonappropriative uses whose contribution to social welfare increases as irrigated agriculture's historic contribution declines.

The doctrine provides poor soil for growing water markets that could work toward equilibrating substantial differentials in the marginal value products of water across competing uses. This chapter investigated several reasons for this, and what might be done about it. First, prior appropriative rights require water diversion and thus exclude nondiversionary instream-flow uses from owning rights that could be acquired in market transactions. As managers of water in trust for the public owners, states could recondition traditional appropriative rights to include modern instream-flow uses. Second, prior appropriative rights are incompletely specified as diversionary quantities so that the hydrologic and legal implications of water use under right are concealed from water market participants. Satellite remote-sensing systems offer prospects for monitoring key parameters of water use such as agricultural water consumption. Third, water-right transfers may impose negative water quantity/quality and timing externalities on third-party rights that decrease social benefits from water marketing. Externalities can be mitigated by restricting water transfers to consumptive use and promoting specialized transfer mechanisms designed to limit the extent and duration of third-party impairment. Finally, water market activity is frustrated by the difficulty of enforcing the priority of appropriative rights. The judicial branch may mistake agricultural runoff as waste instead of return/escape flow essential to supply another's appropriative right, and also strip away the authority that state agencies require to enforce water-right priorities. Unfortunately, the most effective remedy for poor enforcement may be a drought so devastating that state governors and legislatures order water agencies to police water diversions.

## REFERENCES


Allen, R., Willardson, L., Frederiksen, H., 1997. Water use definitions and their use for assessing the impacts of water conservation. In: de Jager, J., Vermes, P., Ragab, R. (Eds.), ICID Workshop on Sustainable Irrigation in Areas of Water Scarcity and Drought. International Committee on Irrigation and Drainage, New Delhi, pp. 72–82.

Anderson, T., Johnson, R., 1986. The problem of instream flows. Economic Inquiry 24, 535–554.

Bittman, M., 2014. Exploiting California's Drought. The New York Times.

Charles, D., 2014. California Farmers Ask: Hey Buddy, Can You Spare Some Water? NPR. http://www.npr.org/blogs/thesalt/2014/06/09/317011698/california-farmers-ask-hey-buddy-can-you-sparesome-water.

Dufford, W., 1994. Water law after Sinking Creek. In: The Sinking Creek Decision: Water Rights in the 21st Century. University of Washington.

Editorial, 2015. An Unrelenting Drought Should Spur California to Overhaul Its Water System. The Washington Post.

Fear, D., 2015. Calif. Governor Orders Statewide Mandatory Water Restrictions. The Washington Post.

Gould, G., 1988. Water rights transfers and third-party effects. Land and Water Law Review 23, 1–41.

Hamilton, J., Whittlesey, N., 1992. Contingent Water Markets for Salmon Recovery. University of Idaho.

Hamilton, J., Whittlesey, N., Halverson, P., 1989. Interruptible water markets in the Pacific Northwest. American Journal of Agricultural Economics 71, 63–73.

Howitt, R., Lund, J., 2014. Five Myths About California's Drought. New York Times.

Huffaker, R., Whittlesey, N., Hamilton, J., 2000. The role of prior appropriation in allocating water resources into the 21st century. Water Resources Development 16, 265–272.

Instream Flows. Washington State Department of Ecology. Available: http://www.ecy.wa.gov/programs/wr/instream-flows/isfhm.html.

Johnson, G., Sullivan, W., Cosgrove, D., Schmidt, R., 1999. Recharge of the Snake River Plain aquifer: transitioning from incidental to managed. Journal of the American Water Resources Association 35, 123–131.

Milliman, J., 1959. Water law and private decision making: a critique. Journal of Law and Economics 2, 41–63.

Nagourney, A., July 20, 2015. California Farm District Accused of Diverting Water. The New York Times.

NASA. Landsat's Role in Water Resource Management. Available: http://landsat.gsfc.nasa.gov/?p=8770.

Perala, O., Benson, R., 1995. Water transfers: can we get there from here? Illahee 11, 16–18.

Pulver, R., 1988. Liability rules as a solution to the problem of waste in western water law: an economic analysis. California Law Review 76, 671–726.

Stevens, J., 1980. The public trust: a sovereign's ancient prerogative becomes the people's environmental right. University of California Davis Law Review 14, 195.

Unnamed, July 12, 2001. You Say Potato, I Say Electricity. The Economist.

Unnamed, 2007. Australia's Water Shortage: The Big Dry. The Economist.

Unnamed, 2012. Water Transfers in the West. The Western Governors' Association.

Ward, F., 2008. Pulido-Velazques, water conservatin in irrigation can increase water use. PNAS 105, 18215–18220.

Wilkinson, C., 1992. Crossing the Next Meridian: Land, Water, and the Future of the West. Island Press, Washington, DC.

Young, R., 1986. Why are there so few transactions among water users? American Journal of Agricultural Economics 68, 1143–1151.

PART 2

# Areas of Competition for Water Resources - Experiences from the US and Europe

SUBPART 2.1

# Water for Food Production

Chapter 2.1.1

# Challenges for US Irrigated Agriculture in the Face of Emerging Demands and Climate Change

G.D. Schaible, M.P. Aillery
*Economic Research Service, USDA, Washington, DC, United States*

## 1. INTRODUCTION

Across the United States, human and environmental demands for water resources have increased significantly over the last 50 years. Population and economic growth, changing social norms regarding the importance of water quality and ecosystems, and longstanding Native American water-right claims have increased pressures on available water supplies, particularly in the arid western states. Given that agriculture accounts for roughly 85% of US consumptive water use, growing water demands with relatively fixed water supplies have heightened conflicts over agricultural allocations in water-short years.

Water conflicts have required a variety of legislative and judicial remedies, generally involving reallocation of agricultural water supplies to meet increasing competing water demands (NRC, 1996; CBO, 1997; Schaible, 2000; Schaible et al., 2010). Historically, federal and state policy response has focused on agricultural water conservation and mandatory withdrawal restrictions, and more recently the use of water markets to meet the nation's various water needs. Expanding water demands for energy development and other uses, together with shifting regional water balances under projected climate change, have heightened awareness of the importance of water conservation for the long-term sustainability of irrigated agriculture. Knowledge about the status and the social and institutional dimensions of competing uses of water resources provides a better understanding of the supply and demand challenges facing irrigated agriculture.

## 2. WATER SUPPLY AND DEMAND CHALLENGES FOR US IRRIGATED AGRICULTURE

The US Geological Survey (USGS) has developed water use estimates for major water demand sectors of the United States, reported every 5 years since 1950 (Fig. 1). Water withdrawals across all sectors—including public use (largely municipal), rural/domestic use, livestock use, irrigation, thermoelectric power generation, and all other uses—

**Competition for Water Resources. http://dx.doi.org/10.1016/B978-0-12-803237-4.00004-5**

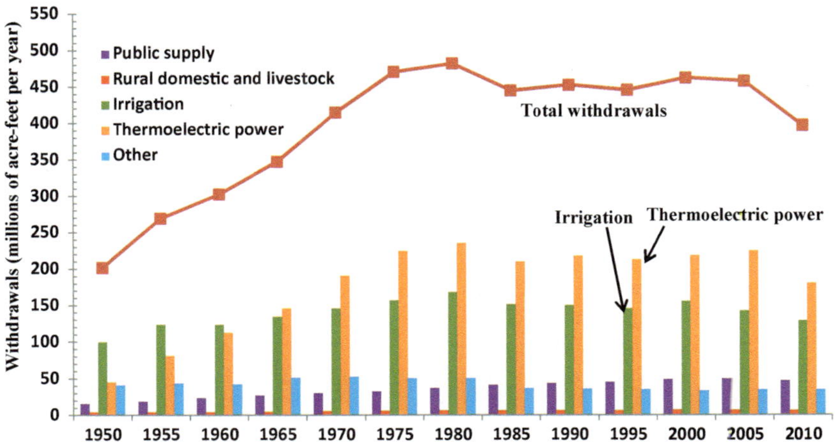


**FIGURE 1** Trends in US water demands by major sector, 1950–2010 (agriculture vs. nonagriculture withdrawals). "Other" category includes water use for the self-supplied industrial, mining, commercial, and aquaculture sectors. Note: US geological survey water use numbers were converted to million acre-feet units. *Adapted from Maupin, M.A., Kenny, J.F., Hutson, S.S., Lovelace, J.K., Barber, N.L., Linsey, K.S., 2014. Estimated use of water in the United States in 2010, US Geological Survey, Circular 1405, Table 14, p. 45, http://dx.doi.org/10.3133/cir1405.*

increased dramatically between 1950 and 2013.[1] Total water withdrawals peaked at about 482 million acre-feet (maf)[2] in 1980 before declining slightly after 1985 to about 458 maf in 2005 (126% higher than in 1950) and declining further to 398 maf by 2010 (still 97% higher than in 1950 but a 13% decline since 2005) (Maupin et al., 2014). Water withdrawals for irrigated agriculture and thermoelectric power are the dominant sources of water demand. Nationally, water withdrawals for thermoelectric power (primarily for cooling purposes) have increased threefold since 1950, accounting for 45% of total US withdrawals in 2010 (about 180 maf). However, efficiency gains in thermoelectric cooling have reduced water demand in recent decades, contributing to a decline in withdrawals of 23% from peak demand in 1980 and 20% since 2005. Nearly 98% of water withdrawals for thermoelectric cooling systems currently return to their source of origin, where the water can be reused for other purposes, including irrigation.

Irrigated agriculture, with withdrawals of about 129 maf, accounted for 32% of the nation's total in 2010. Irrigation withdrawals, while 29% greater than in 1950, have declined 23% from peak demand in 1980 and 9% below the level for 2005. For the 17 western states,[3] where much of the nation's irrigated production is concentrated, irrigated

1. Water withdrawals (one measure of water demand) refer to the quantity of water removed during a period of time from streams, rivers, lakes, reservoirs, and groundwater aquifers, for an intended use.
2. An acre-foot represents the water quantity required to flood 1 acre at 1 ft in depth, equivalent to 325,851 gal.
3. The 17 western states include Arizona, California, Colorado, Idaho, Kansas, Montana, Nebraska, Nevada, New Mexico, North Dakota, Oklahoma, Oregon, South Dakota, Texas, Utah, Washington State, and Wyoming. All other states within the contiguous United States are referred to in this chapter as the 31 eastern states (or eastern states).

agriculture continues to account for most water demand from both surface water and groundwater sources (Maupin et al., 2014). In 2010, irrigation water withdrawals in the West totaled approximately 107 maf, or 64% of total water withdrawals in the region; irrigated agriculture accounted for 61% of surface water withdrawals and 72% of groundwater withdrawals across the region.[4]

## 2.1 Challenges Facing Irrigated Agriculture

Competing demands for US water resources have continued to increase and are expected to intensify water resource conflicts over the foreseeable future. Important sources of expected growth and/or emerging water demands include Native American water rights, instream (environmental) flow requirements, and an expanding energy sector. In addition, climate change is projected to affect both the supply of and demand for freshwater.

### 2.1.1 Native American Water Rights

Native American reservation water rights were established by the US Supreme Court in its 1908 *Winters v. United States* decision. The ruling established reserved water rights based on the amount of water necessary for Native Americans to maintain and survive on the land granted to the reservation by the federal government, even if those rights were not explicitly stated in the reservation treaty. In subsequent decisions, the US Supreme Court quantified those water rights as the water needed to irrigate all "practicably irrigable acreage" on the reservation and made such rights generally superior to the rights of all other appropriators by vesting them with a "priority" date equivalent to the date the reservation was established (Gregory, 2008; Moore, 1989). In addition, while *Winters v. United States* applies to surface waters, in 1976 the US Supreme Court (in *Cappaert v. United States*) opened the door for Native American reserved water-right claims to apply to groundwater. No definitive decision on Native American reserved rights to groundwater has been made, but some states recognize these rights (Gregory, 2008).

Native American water-right claims have been estimated at nearly 46 maf annually (Western States Water Council, 1984). At present, the claims for many reservations are under negotiation or remain unresolved within settlement disputes or judicial proceedings. Future resolution of these water-right claims will undoubtedly affect the water resources available for competing uses, including off-reservation irrigated agriculture. However, settlement of Native American water-right claims may not necessarily result in less water for agriculture, but rather a reallocation of existing water rights. While water delivered to reservation lands generally originates from existing water-right allocations, tribes through settlement arrangements are generally allowed to assign, exchange, or lease their water-right allocation. Within existing negotiated settlements, some reallocated water supports

4. Water withdrawals as a measure of water demand are used here because they are the best and most recently available data by water demand sector. Some portion of withdrawals returns to the hydrologic system, is lost to the system, or is otherwise irrecoverable after its initial use. Consumptive use by sector would provide improved estimates of water demand; however, USGS estimates of consumptive water use were discontinued after 1995.

irrigation expansion on reservation lands, but Tribes may also agree to lease water to off-reservation agricultural users, to non-Indian lessees on reservation lands, and to nonagricultural users such as municipalities (Claims Resolution Act of 2010). To the extent that tribes accept compensation in lieu of wet water, the actual reallocation of water from existing agricultural users may be limited. However, because of the political and financial challenges in negotiating or adjudicating water-right claims and a lack of ability to finance irrigation projects and related storage, exercising reservation water rights have moved historically at a relatively slow pace. The reality is that, for many reservations, future development of these claims will likely continue to progress slowly barring an infusion of economic, legal, and technical assistance.[5]

### *2.1.2 Instream (Environmental) Flows*

Historically, water resources were managed to fulfill the needs of out-of-stream development, such as crop irrigation and municipal or industrial expansion. Water not withdrawn from a stream for economic development was generally considered wasted water. Until relatively recently, water-flow needs for fish and wildlife habitat and other ecosystem benefits were not a legally recognized water management priority. From the 1970s on, however, changing social values with respect to water quality and environmental/ecosystem services have had greater influence on federal and state water resource management institutions and policies. Changing environmental values initially led to the establishment of minimum streamflow requirements to meet legally recognized instream water needs. Subsequently, watershed/basin-level water management agencies were legally bound to manage water resources consistent with maintaining sustainable ecosystems.

Minimum streamflow management focused primarily on the need for a minimum amount of water to be left in a stream, generally to maintain fish habitat (Poff et al., 2003; Zellmer, 2009; MacDonnell, 2009). In basins with significant irrigation withdrawals, minimum flow provisions often reallocate water supplies from agriculture, particularly during low-flow (drought) years. More recently, the use of flow provisions designed to enhance ecosystem services has become more complex (eg, minimal flow requirements for seasonal time durations by stream node) and broader in scope. Often referred to as "environmental flows," these flow regimes are intended to provide multiple instream benefits, including enhanced filtration, dilution of sewage and other effluents, fish and wildlife habitat, recreation (fishing, hunting, boating, and environmental aesthetics), hydropower, navigation, groundwater recharge,[6] riparian wetlands, and migratory bird habitat, as well as exotic species control and local/regional economic development (Sophocleous, 2007; Zellmer, 2008; MacDonnell, 2009).

---

5. Under the Interior Department's Indian Water Rights Settlement Program the federal government has refocused settlement of tribal water-right claims, emphasizing negotiated settlements (with congressional approval) rather than litigation (US BoR, 2012b). Congressional hearings have revealed that in the last dozen or so years many more tribal claims were settled via negotiation than through litigation (US Senate Committee on Indian Affairs, 2012). While many tribal water rights remain unquantified, these settlements help to enhance certainty in water-right allocations, which may also contribute to new investment in improved irrigation systems.
6. Use of the hydrologic process to refill a groundwater aquifer by either pumping water back into wells or managing surface water to increase downward water percolation to the groundwater aquifer.

Environmental flows will likely play an increasingly important role in the ongoing competition among alternative water demands. Most western states have adopted some form of legislation establishing minimum instream flows, and provisions have evolved over time to reflect the complexities of hydrology and a range of instream uses.[7] Water demands for environmental flows very often exceed the historical "minimum instream flow" requirement, placing increasing pressures on limited water supplies. The following examples illustrate the rising importance of environmental water demands.

Stream and river restoration projects have become an important component of federal and state environmental management programs. Based on a review of these projects in the National River Restoration Science Synthesis database, the number of river restoration projects across the United States increased exponentially since 1990, costing more than $14 billion (1990–2004), averaging slightly more than $1 billion annually (Sophocleous, 2007; Bernhardt et al., 2005). These projects may be designed to achieve multiple objectives, including enhanced water quality, management of riparian zones, improved instream habitat for fish and other aquatic species, improved fish passage, bank stabilization, flood plain management, river/stream channel reconfiguration, and flow modification for fish, aesthetics, and recreation.

In many western states, water markets are increasingly being used to reallocate water from existing uses, particularly from agriculture, to enhance supplies for environmental flows within fully or overappropriated basins. Many state water laws now recognize environmental flows as a beneficial use and allow state and nongovernmental organizations, including conservation and environmental groups, to lease, purchase, or donate water or water rights to enhance river flows (Sophocleous, 2007; MacDonnell, 2009). In one of the few studies conducted, Landry (1998) reported that from 1990 to 1997 about 2.4 maf of water was "leased, purchased, or donated for purposes of enhancing river flows in the Western United States." This quantity represented about 5.2% of the surface water applied by irrigated agriculture in 1998.

Over the years, managing water supplies to enhance benefits for fisheries and ecosystem values has become an increasingly important focus for the Central Valley of California. The Federal Central Valley Project (CVP), initially authorized in 1933 and completed in the early 1970s, is comprised of 18 dams and reservoirs and over 500 miles of canals and aqueducts. The project has historically, in nondrought years, conveyed about 7.4 maf of water annually from the Sacramento, Trinity, American, Stanislaus, and San Joaquin Rivers to agricultural users (irrigating more than 3 million acres), municipal users, wildlife refuges, and for endangered fish species recovery in the Sacramento and San Joaquin Valleys and the San Francisco Bay/Delta Estuary. In 1992, the US Congress adopted the Central Valley Project Improvement Act, which formally identified fish and wildlife protection, restoration, and mitigation as project objectives of equal priority with irrigation and other domestic uses, as well as required the CVP to contribute to the state's efforts to protect the Bay/Delta Estuary (US BoR, 2009). The Act also reallocated 800,000 acre-feet of water from existing off-stream uses to fish and wildlife annually. Since 1992, and after nearly $1 billion has been spent on numerous restoration projects throughout the Central Valley, reallocating water supplies to meet environmental/

7. The evolution and status of state-specific minimum instream environmental flow programs, statutes, and policies, which vary widely across the western states, has been summarized by MacDonnell (2009).

ecosystem concerns within the Central Valley remains a high priority of the state/federal partnership (CALFED), an agreement by 25 state and federal agencies established in 2000 to "work collaboratively toward achieving balanced improvements" for the Bay/Delta Estuary (CALFED Bay-Delta Program, 2010). More recently, efforts of the state and CALFED have taken on a larger ecosystem sustainability focus. In 2006, California state agencies initiated the Bay Delta Conservation Plan, a collaborative effort by state, federal, and local water agencies, state and federal fish agencies, environmental organizations, and other interested parties to identify water flow and habitat restoration actions designed to recover endangered sensitive species and their habitats in the Bay/Delta area, while also providing for improved reliability of water supplies (US BoR, 2010).

### *2.1.3 Water for Energy Expansion*

US energy sector growth, for production of fossil fuels, biofuels, and other renewable energy sources, is also expected to place increasing demand on water resources.[8] In the western states, where surface water systems are already overappropriated and groundwater aquifer levels are declining in many areas, energy-related water demand is expected to affect local water supply/demand conditions with potential impacts on regional agricultural production.

An expanded biofuel sector requires water for both processing and feedstock production. Water demand for a biofuel plant with a given processing capacity is generally known (an engineering relationship), local (site specific), and typically managed through market-based permanent lease or purchase agreements between local farms and the biofuel firm. While total withdrawals for biofuel processing are comparatively low, local/regional impacts on water resources can be sizable. Water demand for irrigated feedstock production for biofuel production, however, is expected to be more significant. Chiu et al. (2009), in estimating the "embodied water in ethanol," revealed that: (1) more corn production for ethanol was taking place within highly irrigated regions, particularly in the northern High Plains (Ogallala Aquifer) region; and (2) consumptive water use for bioethanol production in the United States (including water for irrigated feedstock crops) (measured in acre-feet equivalent units) increased from 1.54 to 4.95 maf (246%) between 2005 and 2008. The National Research Council (2008) estimated that: (1) irrigated corn for ethanol (in Nebraska) required about 780 gal of freshwater withdrawals per gallon of ethanol; and (2) "while irrigation of native grass today would be unusual, this could easily change as cellulosic biofuel production gets underway." The US Government Accountability Office estimated that corn ethanol production (adjusting for irrigation return flows) for the Northern Plains states consumed 323.6 gal of water per gallon of ethanol, with nearly 88% of this requirement estimated to come from groundwater (US GAO, 2009). The full impact of biofuel expansion on agricultural land and water resources, however, is expected to be complex, involving the substitution of land and water among crops, cropland expansion, reduced use of idled cropland, expanded use of applied inputs, and increased double-cropping (producing two crops on the same land within the same year), depending on where biofuel development occurs (Wallander et al., 2011; Malcolm et al.,

8. For more information on the water–energy nexus, see the Resources of the Future infographic (Kuwayama, 2016) summarizing literature estimates of water intensity for various categories of fossil fuel sources at: http://www.rff.org/research/publications/infographic-exploring-water-energy-nexus-water-use-fossil-fuel-extraction-and.

2009). Expansion of corn acreage to meet biofuel feedstock demand would likely involve an increase in consumptive water use, particularly for the Plains states, both because of expanded irrigated corn acres and because water consumption by corn plants is greater than that for soybeans, placing additional pressure on groundwater resources where withdrawals have generally exceeded natural recharge.

Water demands are also expected to increase because of growth and technical innovation forecast in other energy-related uses, including thermoelectric generating capacity, development of utility-scale solar power across the southwestern United States, and development of a commercial oil shale industry in the Upper Colorado River Basin. In addition, expansion of hydraulic fracturing (fracking) for deep shale natural gas exploration is expected to continue to increase energy sector water demand in the eastern and central United States. Hydraulic fracking involves pumping water, sand, and chemicals under high pressure into a shale formation to generate fractures or cracks that allow oil or natural gas to flow out of the rock and into the well. Water demand for hydraulic fracking does not represent a long-term water resource commitment, as it occurs only during the drilling and completion phases of each well (Chesapeake Energy, 2012). However, the practice has raised public concern for groundwater use and quality.

Increased use of evaporative cooling technology for thermoelectric and solar power may also significantly increase consumptive water use requirements for the energy sector in areas where expansion occurs. Water demand for the oil shale industry could also be significant—ongoing studies by the US Department of Interior are intended to address the uncertainties of water resource impacts for this sector. In a study of hydraulic fracturing water use, the USGS indicates that based on information from 263,859 oil and gas wells drilled between 2000 and 2014, water use varies significantly across well types (horizontal, vertical, or slanted), well depths, regions and their hydrologic and geologic characteristics, type of water used (saline or freshwater), as well as the volume and composition of the produced water originating from the oil and gas well itself (Gallegos et al., 2015). The USGS study indicates that as oil and gas production associated with fracking has increased, the median annual water volume used to hydraulically fracture horizontal wells increased from less than 670 $m^3$ (176,995 gal) to nearly 15,275 $m^3$ (4 million gal) per oil well and 19,425 $m^3$ (5.1 million gal) per gas well. Because associated environmental problems are heavily linked to volumes of water used and produced by fracking wells, differences in local hydrologic, geologic, and fracking practices ultimately translate into significant differences in the potential for fracking-based wastewater environmental impacts.

For most new energy development, water quality and environmental impacts are potentially the more significant policy concern. Summarizing these demands is outside the scope of this chapter because of the unique needs by energy type, the complexities of energy forecasts, technological uncertainties, and the lack of aggregate water use estimates for projected energy expansion.[9]

### 2.1.4 Climate Change and Water Resources

Substantial evidence demonstrates that the global climate is changing, with important implications for agriculture and water resources (IPCC Report, 2007, 2014; US CCSP, 2008; Melillo et al., 2014; US BoR, 2012a). In much of the western United States, effective

9. For more specific information on these water-use demands, see NETL, 2008; GWPC and All Consulting, 2009; US DOE, 2010; US GAO, 2010; Bartis et al., 2005; US BLM, 2011.

precipitation available for crop uptake is projected to decline, particularly in the warmer summer months. Moreover, gradual temperature increases will shift the West's traditional source of freshwater supplies from winter snowpack to more frequent and intense early spring rain (IPCC Report, 2007, 2014; Knowles et al., 2006). These shifts are expected to alter both the quantity and timing of associated streamflow, with more flow in the early spring, and reduced late season reservoir storage amounts from precipitation and late spring and summer snowmelt. In many areas, streamflow and reservoir storage effects are expected to reduce water supplies for traditional peak irrigation water demands during the summer and fall growing seasons.

Studies conducted for the Intergovernmental Panel on Climate Change's (IPCC) *Fourth Assessment Report* (IPCC Report, 2007) revealed that: (1) the April 1 snow-water equivalent snow cover "has declined 15 to 30 percent since 1950 in the western mountains of North America" (Mote et al., 2003, 2005; Lemke et al., 2007); and (2) streamflow over the last century has "decreased by about 2 percent per decade" in the Central Rocky Mountain region (Rood et al., 2005). These studies indicated that these patterns were not uniform across the Mountain region and that, while there has been a general downward trend in snowpack levels across the western states, decreases have been relatively larger at lower elevations. In addition, results from various climate simulation models or analyses based on multicentury tree-ring reconstruction (1490–1998) indicate that expected warming temperatures and precipitation changes will reduce streamflow in the Upper Colorado River Basin. Streamflow could decline by 8–11% by the end of the 21st century, with declines as high as 25% by 2030 and 45% by 2060 (Christensen and Lettenmaier, 2007; Hoerling and Eischeid, 2007; McCabe and Wolock, 2007).

The US Climate Change Science Program's *Final Report of Synthesis and Assessment Product 4.3* (US CCSP, 2008), drawing on 2007 IPCC climate change assessments and other studies, projected that annual runoff would increase across the eastern United States, gradually transition to little change in the Missouri and Lower Mississippi basins, and substantially decrease (by up to 20%) in the western interior (particularly the Colorado and Great Basin areas). The Bureau of Reclamation (BoR) report to Congress (US BoR, 2011) further disaggregated climatic impact and hydrologic projections to eight reclamation river basins. For the Colorado Basin, this study indicates that the southern subbasins are expected to experience greater warming and a decrease in precipitation—while portions of the upper basin are expected to experience wetter conditions—but warming temperatures will dominate expected basin-wide effects. As a result, projected reductions in natural runoff and changes in runoff seasonality in the Colorado Basin are expected to reduce water supplies given current reservoir system capacity and operational regimes, with differences between northern and southern subbasins. In addition, because reservoir storage opportunities are limited by flood control considerations, increased winter runoff is not expected to translate into increased water storage for the spring season. However, reductions in runoff during the spring and early summer are expected to reduce reservoir levels and water supply deliveries during the irrigation season. In its 2012 Study Report, the US BoR projected that with warming temperatures and reduced snowpack, the mean annual natural flows for the lower Colorado River (at Lees Ferry) over the next 50 years could be reduced by nearly 9% (US BoR, 2012a).

The 2011 BoR report also indicates that warming temperatures are expected to be relatively uniform over the Columbia River Basin, with generally wetter conditions varying across subbasins (US BoR, 2011). Decreases in snowpack are expected to be more substantial over the western mountain ranges of the basin and the lower elevations of the basin's eastern mountain ranges, which "contribute significantly to runoff in headwater reaches of major Columbia River tributaries." Snowpack in northern and higher elevations of eastern portions of the basin, however, are projected to increase overall. These impacts are expected to result in varied annual runoff across subbasins. The BoR report recognized that, for the Columbia Basin, the impact on water supply and reservoir operations is less obvious because of the anticipated variability in climatic effects across subbasins. The report also notes, based on some studies, that general warming effects across the basin appear to have the most influence on runoff and ultimately on basin water supplies.[10]

Other climate change studies indicate that, as increasing temperatures thin snowpack and raise snowline elevations, mountain recharge rates will decline as recharge areas shrink, thereby reducing aquifer recharge and water table levels (Dettinger and Earman, 2007; Hall et al., 2008). For the Ogallala Aquifer region, groundwater recharge is expected to decrease by more than 20% if temperatures increase by 4.5°F (2.4°C) (IPCC Report, 2007). Aquifer recharge rates could decrease by as much as 25% in the Ellensburg Basin of the Columbia Basin Plateau (NWAG Report, 2000). While these studies provide some initial information on how climate change may affect groundwater resources, these processes are less well understood (USGS, 2009; Green et al., 2007). This uncertainty affects researchers' ability to isolate climate change influences on the subsurface hydrologic cycle and their effect on such factors as recharge, discharge, and groundwater storage. These factors are influenced significantly by groundwater residence time—the time it takes climate variability and long-run climate change to affect a groundwater resource—which can range from days to tens of thousands of years. The longer the groundwater residence time, the greater the challenge in detecting responses in groundwater supply caused by climate variability and change.

Climate-induced declines in snowpack and altered runoff also create uncertainties involving the interactions between evapotranspiration (ET), mountain recharge versus alluvial (fan) basin recharge, and their combined effect on lower-basin groundwater recharge (Dettinger and Earman, 2007). In addition, most groundwater systems have been altered substantially by human activities (Green et al., 2007). The USGS reports that improved groundwater monitoring systems and an expanded research focus are needed that go beyond concerns about groundwater-level fluctuations and also address groundwater uncertainties and processes occurring over multiple decades to improve our understanding of groundwater's response to climate change (USGS, 2009).

Moderate temperature increases are also expected to increase crop ET for the southern-tier western states, increasing irrigation water demands in the region, while enhancing ET efficiency for many crops in the northern-tier western states.[11] Even for northern-tier

10. For more information on how projected climate change affects water supplies for other river basins, see the Reclamation report (US BoR, 2011) at http://www.usbr.gov/climate/SECURE/docs/SECUREWaterReport.pdf.

11. Crop evapotranspiration (ET) is generally defined as the loss of water to the atmosphere through evaporation (from soil and plant leaf surfaces) and transpiration (water from inside the plant that vaporizes through plant stomata or microscopic pores on plant leaf surfaces). Crop ET efficiency, as used here, refers to the effect that rising temperatures have on crop yield per unit of water consumed in ET, alternatively recognized as crop water use efficiency (Izaurralde et al., 2003; Hatfield et al., 2008; Bates et al., 2008). Rising temperatures are expected to reduce crop yield per unit of ET in the southern-tier western states, while having a positive effect in the northern-tier western states.

states, however, moderate warming conditions will likely still impact irrigation water demands because, with less total water supply, the timing of irrigation becomes a more critical on-farm water management issue. Crop ET may also shift with projected changes in crop biomass because of temperature stress, carbon fertilization, and other factors, with expected yield declines for some crop/regions (eg, for corn) and positive $CO_2$ impacts for other crop/regions (eg, for wheat). In the eastern United States, where precipitation is generally sufficient to support rain-fed crop production, climate-induced changes in irrigation to meet water demands will depend on shifts in normal growing season rainfall, potential increases in the frequency and severity of drought, and relative returns to irrigated and dryland production.

## 2.2 The Challenge for Agricultural Water Conservation

New pressures on regional water budgets, particularly in the western states, have raised important questions concerning the sustainability of water resources for irrigated agriculture. Three critical questions include:

1. Can irrigated agriculture adapt to climate-adjusted water supplies and emerging water demands through conventional means alone (ie, the adoption of more efficient irrigation technologies, improved water management practices, and/or cropland allocation shifts)?
2. What changes in water institutions may be needed to complement and drive water conservation policy to more effectively manage increasingly scarce water supplies for agriculture?
3. How will these changes impact irrigated agriculture, land and water resource use, the environment, and rural economies?

### *2.2.1 Sustainability of US Western Irrigated Agriculture*

Reduced water supplies because of climate change will likely further constrain already overallocated water resources across much of the western United States, while increased water demand from alternative user groups, ecological requirements, and Native American claims will put additional pressure on water allocations. For agriculture, increased competition underscores the importance of managing irrigation applications effectively, that is, applying water at the time and in the amount needed to meet consumptive use requirements by crop growth stage. In addition, high-pressure sprinkler and traditional gravity irrigation systems will become even less efficient as application losses increase because of higher evaporation rates caused by rising temperatures.

The critical link between climate change vulnerability and sustainability is *adaptability* (Wall and Smit, 2005; Hall et al., 2008; IPCC Report, 2007, 2014; Brekke et al., 2009; Marshall et al., 2015).[12] Given growth in competing demands and projected climate

12. For purposes here and consistent with USDA reports, we define sustainable irrigation water use as a goal of conservation policy—ensuring a viable irrigated agriculture sector and adequate agricultural water availability for future generations, while also protecting offsite environmental services. Adaptation strategies involve various mechanisms for achieving agricultural water conservation and allocation goals.

changes, the adaptability of western irrigated agriculture to a more sustainable future could involve more widespread use of efficient gravity and pressurized irrigation systems, coupled with more intensive use of field-level water management practices to enhance irrigation efficiency and potential farm water savings. Such practices may include broader use of soil or plant moisture sensing devices, commercial irrigation-scheduling services, and computer-based crop growth simulation models that help producers decide when and how much to irrigate.

Practices that enhance gravity-flow systems through improved distributional uniformity of field water advance include field laser leveling, gated pipe systems with surge flow/cablegation applications, shortened furrow lengths, alternate row irrigations, reduced irrigation set times, and polyacrylamide (PAM) applications (a water-soluble soil amendment that stabilizes soil and waterborne sediment). Broader use of tailwater pits may also be used to enhance capture and reuse of irrigation drainage from fields. Pressurized system enhancements, including low-energy precision application/drop-tube systems, drip/trickle and low-flow microspray irrigation systems, and automated nozzle control systems, also improve the precision of applied water while reducing energy requirements for pressurization.

Under more efficient gravity and pressurized irrigation systems, intensive infield water management practices can enhance a producer's ability to apply water closer to a crop's consumptive use requirement. This is especially important when deficit irrigating a crop to maximize profits, particularly during drought years. Deficit irrigation is a water management strategy that concentrates the application of limited seasonal water supplies on moisture-sensitive crop growth stages to maximize the productivity of applied water. The quantity of water applied provides less than the full crop ET requirement, which inevitably results in plant moisture stress and reduced crop yield. With deficit irrigation, however, the farmer's goal is to maximize profits (net income) per unit of water used rather than per land unit used for production (Fereres and Soriano, 2007; Geerts and Raes, 2009). Thus appropriately integrating water management practices with efficient irrigation systems improve the adaptability of irrigated agriculture to water supply deficits, while enhancing long-run sustainability.

### 2.2.2 Sustainability of US Eastern Irrigated Agriculture

Conservation also ensures a more sustainable future for irrigated agriculture in the 31 eastern states. In the more humid East, irrigation generally complements growing season precipitation that normally provides sufficient water to meet crop consumptive use requirements in average rainfall years. When precipitation during the crop-growing season falls short, some producers supplement with irrigation to meet crop water use requirements.[13] Nearly 80% of crop water applied in the eastern states is pumped from shallow aquifers subject to annual recharge that also often serve as the primary source for downstream surface water flows for nonagricultural uses (USGS, 2011a). Less than 6% of the water for eastern irrigated agriculture comes from off-farm water sources (USDA/NASS, 2014b).

---

13. While all irrigation is supplemental to rain-fed crop production, irrigation in humid regions is often referred to as supplemental (or complementary) within the scientific literature (Evans and Sadler, 2008; Clemmens et al., 2008).

Historically, irrigated production has accounted for a small share of crop production in the eastern states. Since the mid-1990s, however, crop irrigation has expanded significantly across the East, increasing by nearly 42% from 1998 to 2013 and by 14% since 2008 (USDA/NASS, 2014b).[14] Irrigation has increased in the eastern states primarily because of increases in commodity prices and yields, increased risk avoidance because of recurring drought conditions, and access to available groundwater supplies at relatively low cost because of shallow aquifer pumping depths (Midwest Irrigation, 2010; Fischer Farm Services, 2011; Evett et al., 2003; Vories and Evett, 2010). At the same time, population growth has increased water demand to meet the needs of urban/industrial growth and recreation, while changing social values have increased pressure for improved water quality and ecosystem services. Expanded groundwater use for irrigated agriculture has contributed to declining aquifer water levels, rising pumping costs, and saltwater intrusion near coastal regions. The increasing importance of groundwater resources for nonagricultural uses, the lack of reliable surface water supplies because of limited reservoir storage capacity, rising irrigation pumping costs, and water quality concerns from irrigation system losses have all heightened concerns for on-farm water conservation as a critical component of a sustainable irrigated agriculture sector in the eastern states. As a result, advancing on-farm water conservation is as important throughout much of the 31 eastern states as it is in the 17 western states.

## 3. HOW IMPORTANT IS IRRIGATION TO US AGRICULTURE?

Nationwide, irrigated agriculture makes a significant contribution to the value of US agricultural production. In 2012, the market value of all agricultural products sold was $394.6 billion, with irrigated farms (farms with at least some irrigated cropland) accounting for roughly 39% of market sales, or $152.4 billion, and nonirrigated farms (farms not irrigating any cropland) accounting for the remainder (Table 1). While the average per-farm value of agricultural products sold by all farms in 2012 was $187,097, the average value for irrigated farms was nearly 2.7 times higher, at $514,412. The average value of farm products sold by irrigated farms was nearly 3.9 times the average value for nonirrigated (dryland) farms.

Irrigation also contributes to the value of livestock and poultry products via irrigated crop production used as animal forage and feed. In 2012, the total value of crop products sold (including nursery and greenhouse crops) by irrigated farms was $106.3 billion, representing 50.0% of the value of crop sales by all farms (Table 1). For irrigated farms only, the value of crop products sold accounted for nearly 70.0% of their agricultural sales in 2012, with livestock products accounting for the remainder.[15] In general, nonirrigated farms were more dependent upon livestock and poultry, with livestock/poultry sales accounting for 56.2% of agricultural product sales.

---

14. The largest irrigation increases in the East since 1998 (though 2013) have been in the Southeast (Georgia at 85% and Alabama at 118%), the Lower Mississippi Delta (Missouri at 48%, Arkansas at 22%, and Mississippi at 53%), and the Upper Midwest (Minnesota and Michigan, each at 60%).
15. The relative importance of irrigated forage and feed production varies across states. In California, irrigated forage acres (alfalfa and other hay, grass silage, and greenchop) account for 88% of acres devoted to irrigated forage and corn production. In the Plains states, however, irrigated corn for grain acres dominate production acres for irrigated forage and corn for grain, ranging from 62% in Texas to 87% and 93% in Kansas and Nebraska, respectively (USDA/NASS, 2014a).

**TABLE 1** Market Value of Agricultural Products Sold and Farm Production Expenses for Irrigated and Nonirrigated Farms, 2012

| | | Irrigated Farms | | |
|---|---|---|---|---|
| Farm Characteristic | All Farms | All Irrigated Farms (Mixed Irrigated and Dryland Cropland) | Farms With All Harvested Irrigated Cropland (no Dryland Cropland) | Dryland Farms (Farms With no Irrigated Cropland) |
| Market value of agricultural products sold ($1000) | 394,644,481 | 152,421,721 | 79,582,158 | 242,222,760 |
| **Average per farm ($)** | 187,097 | 514,412 | 471,687 | 133,603 |
| Crops, including nursery and greenhouse crops ($1000) | 212,397,074 | 106,281,346 | 57,540,345 | 106,115,728 |
| **Average per farm ($)** | 205,754 | 444,231 | 386,509 | 133,809 |
| Livestock, poultry, and their products ($1000) | 182,247,407 | 46,140,375 | 22,041,814 | 136,107,032 |
| **Average per farm ($)** | 181,419 | 433,614 | 502,504 | 151,541 |
| Total farm production expenses ($1000) | 328,939,354 | 123,022,726 | 64,792,431 | 205,916,628 |
| **Average per farm ($)** | 155,947 | 415,192 | 384,028 | 113,578 |
| Energy-related expenses (excluding customwork) ($1000) | 24,835,166 | 11,092,703 | 5,919,143 | 13,742,463 |
| **Average per farm ($)** | 7435 | 21,231 | 20,225 | 4876 |

USDA, National Agricultural Statistics Service, 2012 Census of Agriculture, 2014.

## 3.1 Where Does Irrigation Occur and What Does It Produce?

In 2012, 55.8 million farmland acres were irrigated across the United States (52.1 million acres of harvested cropland and 3.1 million acres of pastureland and other cropland), accounting for about 14.3% of all cropland, and 6.9% of all cropland, pastureland, and rangeland. About 16.5% of US harvested cropland acres were irrigated, while only 0.8% of pastureland acres were irrigated (USDA/NASS, 2014a). Nearly three-quarters of US irrigated agriculture occurred in the 17 western states, including 71.0% of harvested irrigated cropland and 92.3% of irrigated pastureland.

For 2012, 12 leading irrigation states accounted for 76.2% of all irrigated acres, including harvested cropland, pasture, and other lands (Fig. 2). Nebraska's 8.3 million irrigated acres led all other states (14.9% of the US total), followed by California with 7.9 million acres (14.1%), Arkansas with 4.8 million acres (8.6%), and Texas with 4.5 million acres (8.0%). Three eastern states—Arkansas, Mississippi, and Florida—were among the 12 leading irrigation states. Mississippi accounted for 1.7 million acres (3.0%) and Florida for 1.5 million acres (2.7%) of the total US irrigated area.

Irrigated agriculture and water use are not static; areas grow and decline over time, influencing regional demands for water, energy, and other inputs (Fig. 3). From 2002 to 2007, agricultural water use reflected a net increase of nearly 1.3 million irrigated acres across the United States. Nebraska accounted for nearly a million of those additional acres (72% of the increase), with lesser increases occurring in the Mississippi Delta and Southeast regions (Arkansas, Mississippi, Missouri, and Georgia). Irrigated acreage expansion in these states was attributed to availability of water supplies, improved irrigation economics [partly because of higher crop yields and reduced water costs associated with more efficient irrigation systems (USDA/NRCS, 2006)], increased biofuel demand for

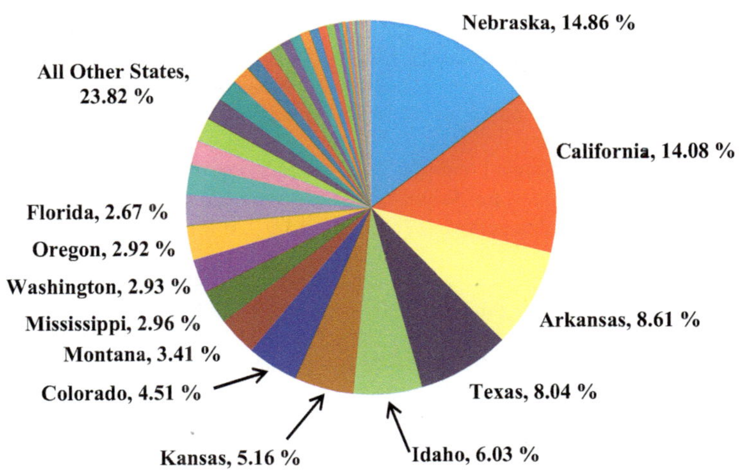


FIGURE 2 State shares of total US irrigated acres for 2012. Note: twelve leading irrigation states (nine from the West and Arkansas, Mississippi, and Florida from the Delta and Southeast) accounted for 76.2% of US irrigated acres, including harvested cropland, pasture, and other lands. *U.S. Department of Agriculture, National Agricultural Statistics Service (USDA/NASS). May 2014a. 2012 Census of Agriculture: U.S. Summary and State Data. Geographic Area Series, Part 51 (AC-12-A-51), vol. 1, p. 695, at: http://www.agcensus.usda.gov/Publications/2012/Full_Report/Volume_1,_Chapter_1_US/usv1.pdf.*

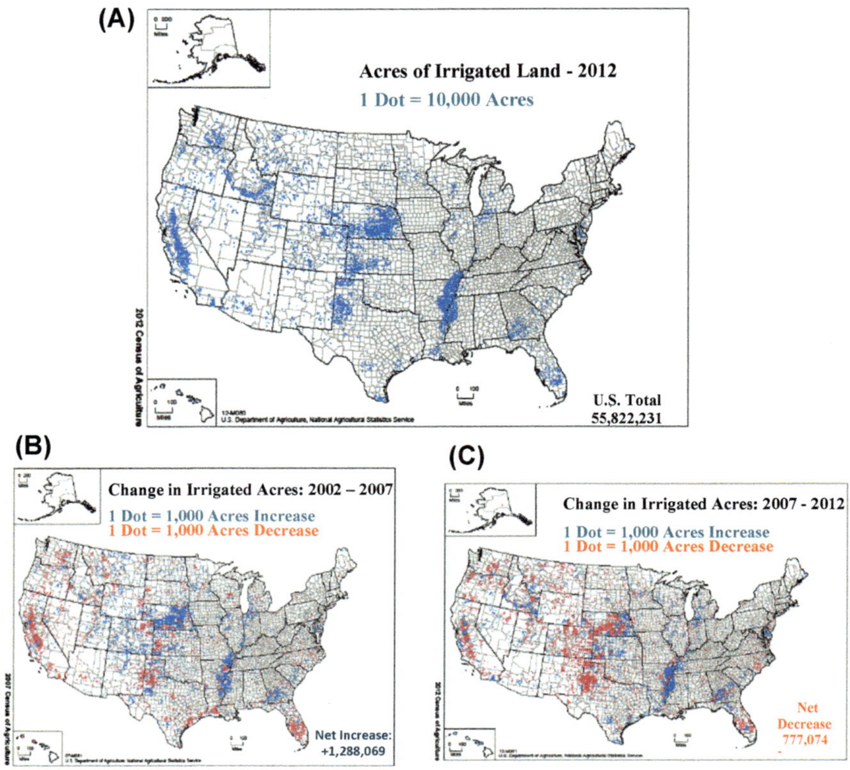


**FIGURE 3** US irrigated acres, 2012: how has it changed over time? *Adapted from US Department of Agriculture, National Agricultural Statistics Service, Map Atlases from the 2007 and 2012 Census of Agriculture (Maps 12-M080, 07-M081, and 12-M081, respectively).*

corn, recurring regional drought conditions, and the prospect of future restrictions on new irrigation development (at least for Nebraska).[16] California and Florida led the states where irrigated acres fell during this period (0.7 million and 0.3 million acres, respectively).[17,18]

16. Personal communication with Professor Raymond J. Supalla, University of Nebraska—Lincoln, Agricultural Economics Department.
17. Florida's irrigated acreage has been decreasing for a variety of reasons, including: (1) the reallocation of water supplies to restore the Everglades ecosystem; (2) declining groundwater aquifer levels and saltwater intrusion; (3) loss of competitive markets; (4) urbanization; and (5) crop diseases (Aillery et al., 2001; USGS, 2008; Florida DEP, 2010).
18. In California, irrigated acres have been declining because of: (1) increased use of pumping restrictions on water supplies from the San Francisco Bay–Delta Estuary to meet environmental regulations to protect endangered species; (2) continued urban growth (although more recently at a slower pace because of current economic conditions); and (3) reduced soil productivity because of increasing salinity (particularly in the Imperial and San Joaquin Valleys). Recurring droughts have heightened water supply pressures in California, resulting in significantly increased Delta pumping restrictions and subsequent reductions in irrigated cropland (Ayars, 2010; California Department of Conservation, 2011).

For the period 2007 to 2012, irrigated area had a net decline of 777,074 acres across the United States. The larger decreases occurred in Texas (521,000 acres), Colorado (351,000 acres), Nebraska (262,000 acres), and Oregon (215,000 acres), with smaller declines in California and New Mexico. The larger net gains in irrigated acres during this period occurred in Arkansas (343,000 acres) and Mississippi (283,000 acres), with smaller increases in Louisiana, Georgia, and Kansas. However, since about 1997, the dominant pattern of change across the United States reflects a shift in irrigated acreage from the 17 western states to the Delta and Southeast (with the exception of Florida).

Fig. 4 illustrates the longer-term changes that have taken place since the early 1980s in irrigated acres (Part A) and agricultural water applied (Part B) across United States Department of Agriculture (USDA) farm production regions. From 1982 to 1997, irrigated acres increased for most farm regions. Since 1997, however, most regions saw either a decline in irrigated acres or a slowing of irrigated expansion. The largest growth in irrigated acres since 1997 was concentrated in the Northern Plains, Delta, and Corn Belt regions, with more moderate expansion across the eastern United States (except Florida). Growth rates in the Northern Plains (primarily Nebraska) pushed irrigated acreage (11.9 million acres in 2007) above acreage irrigated in the Pacific region (11.6 million acres). While both regions had a net decline in irrigated acres from 2007 to 2012, the Pacific region's decline occurred at a slightly faster pace. Similarly, since 1997, irrigated acres in the Delta region surpassed acres irrigated in the Southern Plains. Since 1997, the largest contraction in irrigated acres has occurred in the more arid western Mountain, Pacific, and Southern Plains regions.

Agriculture in the Pacific region is the most dependent on irrigation, with about half (51%) of cropland acreage irrigated in 2012. Other arid western regions with sizable concentrations of irrigated cropland include the Mountain (30%), Northern Plains (12%), and Southern Plains (12%) regions. In the eastern states, irrigated acreage accounted for 44 and 25% of cropland in the warmer Delta and Southeast regions, respectively, but less than 5% of cropland acreage in the middle- and northern-tier regions.

Although more acres were irrigated in the Mountain states than in the Pacific or Northern Plains states, agriculture in the Pacific region uses significantly more water overall, in part because of higher application rates. Average per acre field-level water use for agriculture in the Pacific region was 2.8 acre-feet, compared with 2.0 acre-feet in the Mountain states. Differences reflect regional variation in crop consumptive use requirements associated with climate and cropping pattern choices, as well as variation in the contribution of natural precipitation. Applied water rates are also influenced by differences in irrigation efficiencies, water prices, and energy costs for irrigation pumping. Irrigated agriculture within the Pacific and Mountain states accounted for the largest share (62%) of total agricultural water applied across the continental United States.

*What does irrigated agriculture produce?* Irrigated agriculture accounts for a share of harvested acreage for most US crops. Vegetable, orchard, and rice crops had the dominant share of their harvested acres irrigated in 2012, with 82% for orchards and 100% for rice and vegetables (USDA/NASS, 2014a). For all other crops, irrigated acreage accounted for less than half of US harvested acreage by crop, with shares ranging from 41% for cotton to 5% for oats.

Irrigated cropping patterns differ regionally across the United States. For the West, the cliché that "if a crop is not irrigated it is not grown" is not universally true. Rice, vegetables, and orchard crops were the only crops in the West with more than 80% of their

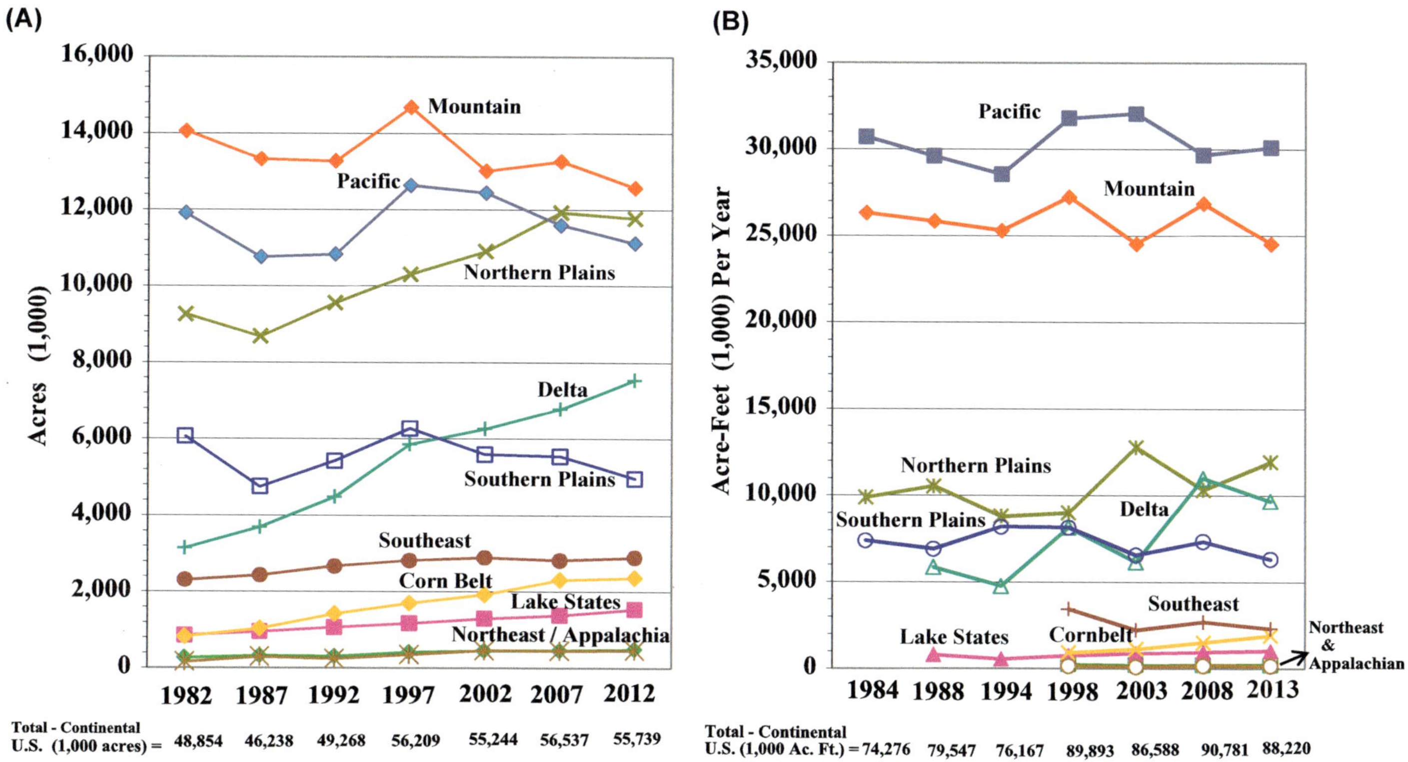


FIGURE 4 Irrigated acres and agricultural water applied by USDA farm production region. (A) Acres irrigated (1982–2012) *(Census of Agriculture Statistics, 1982–2012, National Agricultural Statistics Service, USDA. (Summarized by Economic Research Service, USDA))*. (B) Agricultural water applied (1984–2013) *(Farm & Ranch Irrigation Surveys, 1984–2013, National Agricultural Statistics Service, USDA. (Summarized by Economic Research Service, USDA))*. USDA farm production regions: **Pacific** (Washington, Oregon, and California); **Mountain** (Montana, Idaho, Wyoming, Nevada, Utah, Colorado, Arizona, and New Mexico); **Northern Plains** (North Dakota, South Dakota, Nebraska, and Kansas); **Southern Plains** (Oklahoma and Texas); **Lake states** (Minnesota, Michigan, and Wisconsin); **Corn Belt** (Ohio, Iowa, Missouri, Indiana, and Illinois); **Northeast** (New Hampshire, Pennsylvania, Maine, Maryland, Rhode Island, Massachusetts, Delaware, Connecticut, Vermont, New York, and New Jersey); **Appalachia** (West Virginia, Tennessee, North Carolina, Virginia, and Kentucky); **Delta** (Louisiana, Arkansas, and Mississippi); and **Southeast** (South Carolina, Alabama, Georgia, and Florida).

harvested cropland acres irrigated in 2012 (USDA/NASS, 2014a). Cotton, peanuts, and sugar beets grown in the West also relied heavily on irrigation, but only 50–64% of harvested cropland for these crops was irrigated. As much as 65–95% of the harvested cropland acres for corn for grain, sorghum, soybeans, wheat, oats, barley, and forage crops (hay, haylage, grass silage, and greenchop) in the West were farmed using dryland production systems.

Fig. 5 illustrates the relative distribution of 2012 harvested irrigated acres by major crop category for both the western and eastern United States. Corn for grain and forage crops accounted for about 49% of all harvested irrigated crop acres across the West (Part A). However, corn for grain, soybeans, rice, and vegetables accounted for nearly 80% of harvested irrigated crop acres across the eastern states (Part B). Cotton accounted for an additional 10% of irrigated harvested acreage in the East. Relative to the western states, the irrigated cropping pattern in the eastern states reflects a much smaller share of irrigated acres for forage crops and wheat, and a larger share of irrigated acres devoted to rice and soybeans.

## 3.2 How Much Water Is Applied, What Is Its Source, and What Does It Cost?

In 2013, irrigators across the western states applied about 72.9 maf of water for irrigated cropland production (for all "acres in the open," but excluding water applied to horticulture under protection), averaging about 1.8 acre-feet per acre (af/ac) overall (Table 2). Much of this water (51%) originated from surface water sources, with the remainder (49%) supplied from wells used to pump groundwater from local and regional aquifers (USDA/NASS, 2014b). Surface water is drawn from both on-farm and off-farm sources. On-farm surface water from ponds, lakes, or streams and rivers on the farm account for roughly 10% of total agricultural water applied in the West, while off-farm water sources accounted for nearly 41% of total water applied. Water from off-farm sources is generally supplied through local irrigation districts; mutual, private, cooperative or neighborhood water-delivery "ditch" companies; or from commercial or municipal water systems. Applied water from groundwater sources in the West averaged about 1.5 af/ac in 2013 (Table 2). In contrast, applied water averaged 1.7 af/ac for on-farm surface water and 2.2 af/ac for off-farm surface water over the same period. These application differences likely reflect the generally higher cost of groundwater and the fact that more off-farm surface water is applied to higher-valued, more water-intensive crops. In addition, more efficient systems are more likely to be used where groundwater is the primary water source. Center-pivot systems, for example, tend to be the more cost-effective system when drawing on groundwater. More than half (54%) of agricultural water for crop production in the western states was applied using pressure irrigation (sprinkler, drip/trickle, and/or low-flow microspray) systems, with most of the remainder (42%) applied with gravity irrigation systems.[19] Application rates using gravity systems, which are generally less water use efficient and more likely associated with lower-cost surface water, averaged about 2.3 af/ac

19. For more information on gravity and pressure (sprinkler) irrigation systems, see *Irrigation and Water Use Glossary* on the USDA/ERS website at: http://webarchives.cdlib.org/sw1rf5mh0k/ http://www.ers.usda.gov/Briefing/WaterUse/glossary.htm.

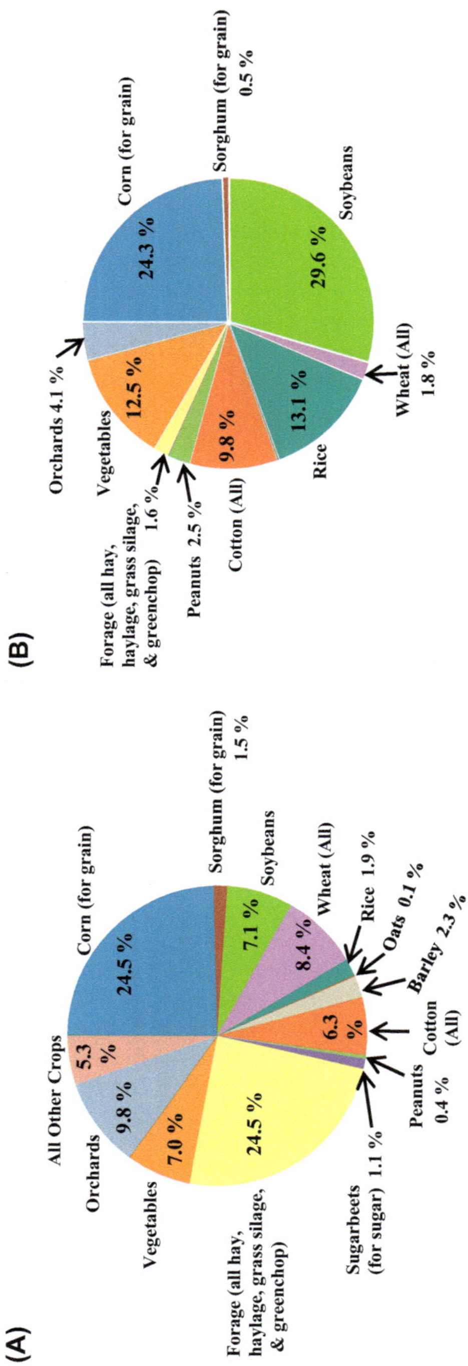


FIGURE 5 Distribution of harvested irrigated acres, by major crop category, 2012. (A) 17 Western States, (B) 31 Eastern States. *US Department of Agriculture, National Agricultural Statistics Service, 2012 Census of Agriculture.*

**TABLE 2** Water Application Statistics for the 17 Western States, for "Acres in the Open", by Type of Irrigation and Water Source, 2013

| Water Application | | Type of Irrigation[a] | | | | | | | |
|---|---|---|---|---|---|---|---|---|---|
| | Total Water Applied (Acre-Feet) | Gravity Systems (Acre-Feet) | Percent Gravity | Sprinkler Systems (Acre-Feet) | Percent Pressure | Drip/Trickle and/or Low-Flow Microsprinkler Systems (Acre-Feet) | Percent Other | Other Systems (Acre-Feet) | Percent Other |
| Water applied: | 72,896,810 | 30,970,461 | 42.0 | 36,144,815 | 50.0 | 3,276,408 | 4.0 | 2,505,126 | 3.0 |
| Average Application (acre-feet/ acre): | 1.83 | 2.29 | | 1.34 | | 0.90 | | | |
| | | Water Source[a] | | | | | | | |
| | | Wells | | Surface Water Sources | | | | | |
| | | Ground Water (Wells) (Acre-Feet) | Percent Ground Water | | | On-Farm Surface Water (Acre-Feet) | Percent On-Farm Surface | Off-Farm Surface Water (Acre-Feet) | Percent Off-Farm Surface |
| Water applied: | 72,896,810 | 36,023,636 | 49.0 | | | 6,984,525 | 10.0 | 29,888,649 | 41.0 |
| Average Application (acre-feet/ acre): | 1.83 | 1.49 | | | | 1.68 | | 2.18 | |

[a] *USDA, Economic Research Service calculations based on data from the USDA, National Agricultural Statistics Service, 2013 Farm and Ranch Irrigation Survey.*
USDA, National Agricultural Statistics Service, 2013 Farm and Ranch Irrigation Survey, 2014.

while rates for sprinkler systems averaged 1.3 af/ac, and drip/trickle and/or low-flow/ microspray systems averaged about 0.9 af/ac.

Irrigation water is generally pumped from groundwater wells, surface water sources, or from water-delivery ditches (canals), and may be conveyed under pressure to access irrigated fields. Pumps are also used to pressurize field-level sprinkler, drip/trickle, low-flow, and microspray systems for field application. As a result, producers typically incur significant energy expenses over and above typical crop production costs under nonirrigation production.[20] Both capital expenses (irrigation conveyance and distribution systems) and variable irrigation costs (depending on the quantity of water used) vary significantly by region and across irrigated crops. These cost differences impact irrigation profitability, which will fluctuate based on available water sources, type of irrigation system used, crops irrigated, energy source used to power irrigation pumps, and water costs charged for off-farm water supplies.

In 2013, irrigated agriculture in the western states incurred nearly $2.2 billion in energy expenses for on-farm pumping of irrigation water (Table 3). Costs for pumping water varied by type of irrigation [including water applied to "acres-in-the-open" (for field crops or horticulture crops) versus water applied to horticulture crops under protection (eg, greenhouse structures)], and whether the water was pumped from groundwater or surface-water sources. Water pumped from wells and for pressurizing irrigation systems used to irrigate field crops averaged about $65 per irrigated acre, compared with $33 per acre for water supplied from a surface water source. To pump water from wells to irrigate horticulture crops grown in the open cost about $92 per acre and nearly $101 per acre when using surface water. Pumping water from wells to irrigate horticulture crops grown under protection cost about $190 per 10,000 square feet of area under protection, while costing $57 per 10,000 square feet when pumping surface water for horticulture under protection. Expenses for scheduled irrigation replacement and maintenance and repairs for on-farm irrigation systems in the West totaled nearly $852 million (averaging $99 per affected irrigated acre). Irrigation labor costs across the western states in 2013 totaled about $814 million ($671 million for hired labor and $142 million for contract labor). Hired labor for irrigation averaged about $27,042 per irrigated farm while contract labor for irrigation averaged $28,535 per farm. In addition, irrigators using off-farm water supplies paid nearly $742 million to purchase water from irrigation districts and other off-farm water suppliers. Purchased water costs across the West averaged about $74 per affected irrigated acre, or $33 per acre-foot of water. However, total variable irrigation costs can vary significantly across states and water sources. In 2013, the sum of energy costs for pumping irrigation water, the cost of water purchased from off-farm suppliers, and scheduled replacement and maintenance and repair costs ranged from $55 per acre in Montana to $386 per acre in California for field crop acres irrigated in the open. For horticultural crops irrigated in the open, the sum of these costs ranged from $58 per irrigated acre in Oklahoma to nearly $550 per acre in Nevada. Costs for hired and contracted irrigation labor also varied significantly across states, influenced heavily by the crops irrigated and quantity of labor required. For 2013, average contract irrigation labor costs ranged from $2250 per farm in North Dakota to over $75,000 per farm in Arizona. Average hired irrigation labor cost ranged from $6000 per farm in South Dakota to $42,000 per farm in California.

20. In addition, irrigated production can often involve higher (nonenergy) input and harvest costs because of more intensive input use and higher yields relative to nonirrigated production.

**TABLE 3** Irrigation Cost Statistics for the 17 Western States, by Type of Irrigation and Irrigation Expense, 2013[a]

| | Energy Expenses for Onfarm Pumping of Irrigation Water, Total and by Irrigation Category and Water Source[b] | | | | | | Scheduled Irrigation Replacement and Maintenance/Repair Expenses | |
|---|---|---|---|---|---|---|---|---|
| | Expenses per Irrigated Acre | | | | Expenses per 10,000 sq.ft. | | | |
| | For Operations With Only Acres in the Open | | For Operations With Only Horticulture in the Open | | For Operations With Only Irrigated Horticulture Under Protection | | | |
| Total Pumping Expenditures | Water from Wells | Surface Water | Water from Wells | Surface Water | Water from Wells | Surface Water | Total Expenditures | Average Cost per Irrigated Acre |
| ($1,000 dollars) | Dollars per acre | | Dollars per acre | | Dollars per 10,000 Sq.Ft. | | ($1,000 dollars) | Dollars per acre |
| 2,147,696 | 64.56 | 32.70 | 91.58 | 100.87 | 190.00 | 57.00 | 851,265 | 99.39 |

| Irrigation Labor Costs by Type (Hired and Contract Labor) | | | |
|---|---|---|---|
| Total Expenses | | Average Cost per Irrigated Farm | |
| Hired Labor | Contract Labor | Hired Labor | Contract Labor |
| $1,000 dollars | | Dollars per farm | |
| 671,265 | 142,245 | 27,042 | 28,535 |

| Purchased Water Costs for Off-farm Water Supplies | | |
|---|---|---|
| | Average Cost | |
| Total Purchased Water Expenses | Per Acre | Per Acre-Foot |
| ($1,000 dollars) | Dollars | |
| 741,900 | 74.25 | 32.74 |

[a] *USDA, Economic Research Service calculations based on USDA*, National Agricultural Statistics Service, 2013 Farm and Ranch Irrigation Survey.

[b] Includes expenditures for all energy sources (electric, natural gas, LP gas, propane, butane, diesel fuel, gasoline and gasohol), except for solar.

USDA, National Agricultural Statistics Service, 2013 Farm and Ranch Irrigation Survey, 2014.

In the 31 eastern states, variable irrigation costs in 2013 were generally less than those for the western states. Average energy pumping costs for water pumped from wells ranged from $31 to $74 per acre for field crops and horticultural crops, respectively, to $120 per 10,000 square feet for horticultural crops irrigated under protection (USDA/NASS, 2014b). Energy costs for water pumped from surface sources ranged from $34 per acre for field crops irrigated in the open to $66 per 10,000 square feet for horticulture crops under protection. Pumping costs for water pumped from wells are lower in the eastern states because groundwater pumping depths are generally shallower. Purchased water costs averaged $36 per acre (or $41 per acre-foot). However, purchased water from off-farm sources in the eastern states account for less than 6% of water supplies for acres irrigated in the open, and about 14% of water supplies used to irrigate horticulture crops under protection. Irrigation labor costs averaged $12,687 per farm for hired labor and $16,095 per farm for contract labor (with an overall average of $23 per acre for affected irrigated acres, compared to $55 per acre in the western states). In the eastern states, costs for scheduled irrigation replacement and maintenance and repair costs were similar to those in the western states, averaging $107 per acre for affected irrigated acres.

## 4. HOW EFFICIENT IS IRRIGATED AGRICULTURE?

Prior to the 1970s, gravity-fed furrow and flood irrigation systems were the dominant production systems for irrigated crop agriculture. By 1978, sprinkler irrigation, including center-pivot systems, accounted for about 35% of crop irrigation in the western states. Virtually all of this transition involved adoption of high-pressure sprinkler irrigation.[21] While the center-pivot system improved field irrigation efficiency, water conservation was not the primary motivation for its widespread adoption. Other factors, such as yield enhancement from uniform water application and irrigation's expansion into productive lands that were not suitable for a gravity system because of topography, soils, or distance from traditional riparian boundaries, were the primary drivers behind the early transition from gravity-flow irrigation to center-pivot sprinkler irrigation.

The expansion of irrigated agriculture, along with increasing water demands from nonagricultural users, significantly intensified the competition for available water resources. Over time, federal and state resource conservation programs provided financial and technical assistance to promote adoption of more efficient irrigation systems, to improve irrigation returns and enhance the health and productivity of the resource base, and to help ensure a more sustainable future for small farm and rural livelihoods. Adoption of more efficient irrigation systems and water management practices has been examined extensively, particularly within the 17 western states (Schaible and Aillery, 2006; Schaible, 2013; Schaible et al., 2010). Fig. 6 illustrates that between 1984 and 2013 a substantial shift has occurred across the western states away from gravity irrigation to pressure-sprinkler irrigation systems. In 1984, for example, 71% of crop agricultural water in the West was applied using gravity irrigation systems. By 2013, operators used gravity systems to apply just 41% of water for crop production, while pressure irrigation systems accounted for 59%, or an increase of 31 percentage points from 1984. By 2013, much of the acreage in more efficient pressure irrigation systems included drip/trickle or low-flow microspray

21. Sprinkler irrigation systems operating with greater than 60 pounds per square inch (PSI) of pressure.

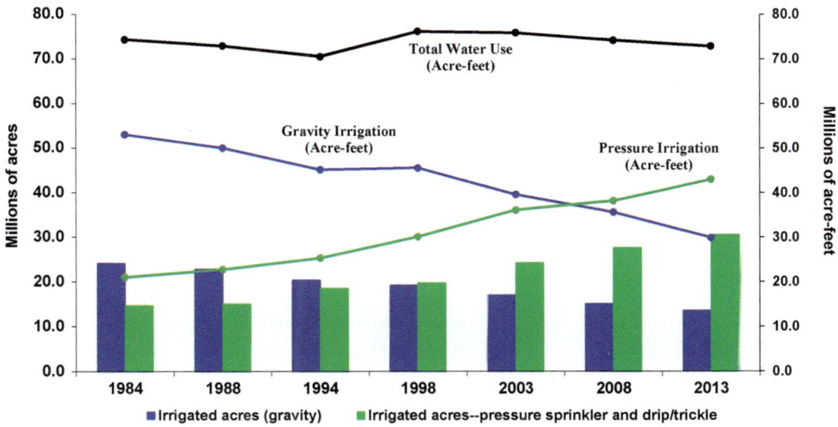


**FIGURE 6** Trends in irrigated acres and applied water use, 17 western states 1984–2013. Notes: water use information from USDA's FRIS reports on-farm water applied, not withdrawals. Also the area tracked includes all acres irrigated in the open. It excludes square feet and water use for horticulture under protection. *USDA, Economic Research Service calculations based on USDA, National Agricultural Statistics Service, 1984, 1988, 1994, 1998, 2003, 2008, and 2013, Farm and Ranch Irrigation Survey data.*

systems, low-pressure sprinkler, and low-energy precision application systems. Adoption of improved (more efficient) irrigation systems contributed to reducing agricultural water use, as fewer acre-feet were required to irrigate a greater number of acres using these systems. From 1984 to 2013, total acres irrigated (in the open) across the West increased by 1.7 million acres (from 38.1 to 39.8 million acres), while water applied for this agricultural production declined by nearly 1.4 maf (from 74.3 to 72.9 maf).

On-farm crop irrigation efficiency is measured as the fraction of applied water beneficially used by the crop, including the quantity of water required for crop ET (consumptive use) and water to leach salts from the crop-root zone (Howell, 2003; Burt et al., 1997).[22] Water applied to crops but not used for beneficial purposes is generally regarded as field loss, including water lost through excess evaporation and transpiration by noncropped biomass as well as surface runoff and percolation below the crop-root zone. Some portion of water loss to surface runoff and deep percolation may eventually return to the hydrologic system through surface return flow and/or aquifer recharge and may be available for other economic and environmental uses.

What happens to irrigation water that leaves the farm (ie, water not beneficially consumed through crop production) and its ultimate impact on local or regional water supplies depends on the many factors that influence the hydrologic water balance for the

22. This definition of crop irrigation efficiency is conceptually consistent with Howell's (2003) "seasonal irrigation efficiency" and the "irrigation efficiency" performance indicator presented by Burt et al. (1997). Depending upon the crop and region (and consistent with both references cited), crop beneficial use may also include water for cooling or frost protection of plants, seed bed preparation, enhancement of seed germination, and to meet ET requirements for plants beneficial to the crop, such as herbaceous windbreaks and cover crops.

watershed. Water balance accounts for where all the water within a watershed comes from and where it goes and is significantly influenced by soils, land cover, climate, water source, topography, and hydrologic characteristics both on and off the farm. Research demonstrates that while generally recognized as conserving water on the farm, improved on-farm irrigation efficiency *may or may not* contribute to water conservation at a basin scale (Clemmens et al., 2008; Evans and Sadler, 2008; Sadler et al., 2005; Fereres and Soriano, 2007; Geerts and Raes, 2009; CIT, 2011). This may require accompanying institutional measures that restrict agricultural consumptive use or reallocate efficiency savings within the basin. These studies do, however, reveal that improved on-farm irrigation can conserve water on and beyond the farm by:

1. reducing unnecessary evaporation and unwanted transpiration by weeds and other noncropped biomass within waterlogged parts of irrigated fields, along water supply ditches and canals, and within and along irrigation drainage pathways;
2. improving rainfall use with precipitation capture and moisture retention techniques (eg, land grading, snow fences, plant-row mulches, and furrow diking techniques);
3. reducing deep percolation water that is severely degraded in quality or uneconomic to recover;
4. reducing field runoff that is lost to the hydrologic system (ie, runoff water that is not accessible or reusable because of salinization or entry to a saline body);
5. reducing crop ET requirements for downstream irrigated agriculture (ie, by reducing saline return flows allows downstream irrigators to reduce their salt leaching requirements); and
6. reducing normal crop ET associated with crop stress under deficit irrigation (ie, the irrigator intentionally provides the crop with less than its full ET requirement, resulting in reduced yield but higher net economic returns).

These studies also indicate that, in many cases, conserved water to augment water supply in the river basin may not be the primary policy concern. Water conservation programs also focus on enhancing the viability and sustainability of the regional agricultural economy, improving the quality and availability of water supplies locally, improving the quality of return flows, and reducing environmental degradation of existing regional supplies. USGS National Water-Quality Assessment studies have identified irrigated agriculture as a key contributor to many of the nation's degraded surface-water bodies and groundwater aquifers because irrigation often makes heavier use of agricultural chemicals and because excess irrigation increases the hydrologic transport of agricultural chemicals, salts, and other soil-based chemicals potentially detrimental to water-based ecosystems (USGS, 2011b). Thus, even without adding to regional water supplies, water conservation programs encouraging improved on-farm irrigation efficiency can purposefully serve local and regional economic, water-quality, and environmental policy goals that contribute to farmer and societal welfare, improve fish and wildlife habitat, and reduce ecosystem and human health risks associated with environmental pollution. Such programs can also serve to help the USDA promote small farm, limited-resource, and socially disadvantaged farm policy goals.[23]

23. For these programs, see the website at: http://www.nrcs.usda.gov/wps/portal/nrcs/main/national/people/outreach/slbfr/.

The potential for continued improvement in on-farm irrigation efficiency to contribute to water conservation program goals relies a great deal on how efficient US irrigated agriculture is today. Because actual irrigation water use is rarely measured and actual consumptive use can vary significantly depending on agriclimatic conditions, the efficiency of irrigated agriculture (based on its traditional definition) cannot be readily measured. For an alternative measure, using farm-level data from USDA's Farm and Ranch Irrigation Survey (FRIS) from 1994 to 2008, Fig. 7 illustrates the relative efficiency of irrigated agriculture for the 17 western states based on the shares of irrigated acres where water is applied using more efficient irrigation systems (separately for gravity and pressure-sprinkler systems). Between 1994 and 1998, the share of western irrigated acres using improved gravity-flow systems increased from 21% to 25%. During this time period, the share of irrigated acres using improved pressure-sprinkler irrigation also increased and accounted for about 23% of total irrigated acres in 1998. Thus more efficient irrigation in 1998 (based on a system-based definition, unadjusted for on-farm water management) accounted for nearly 49% of irrigation in the West. From 1998 to 2008, however, the share of gravity-flow irrigated acres using improved gravity systems declined. At the same time, improved pressure-sprinkler irrigated acres continued to increase, although at a slower rate than in the earlier period. FRIS evidence reveals that while substantial technological innovation has already occurred in western irrigated agriculture, significant room for improvement in farm irrigation efficiency exists—as traditional gravity or less efficient pressure-sprinkler systems still account for over 50% of irrigated acres. Similarly, potential for improvement exists for irrigated agriculture in the eastern states where traditional, less efficient systems irrigate at least 48% of irrigated acres (USDA/NASS, 2014b). Historical transitions suggest that, while US irrigated agriculture is on a path toward greater sustainability, further progress will likely be needed as water demand and supply conditions evolve.

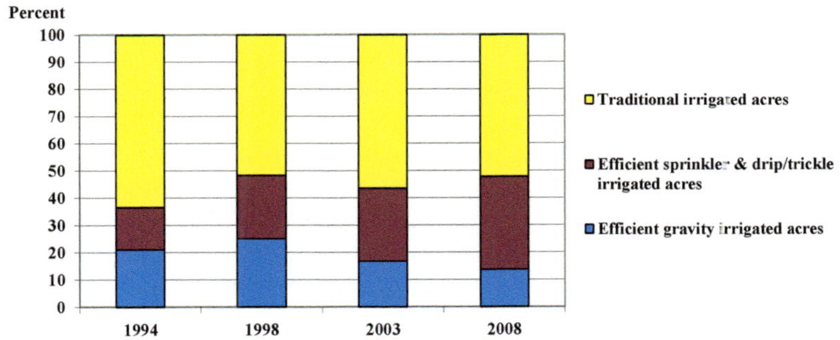


**FIGURE 7** Trends in the use of efficient irrigation systems, by system type, for the 17 western states, 1994–2008. *Efficient gravity irrigation* includes furrow irrigated acres using above- or below-ground pipe, or a lined open-ditch field water-delivery system, plus acres in flood irrigation (between borders or within basins) on farms using laser leveling and pipe or lined open-ditch field water-delivery systems. *Efficient pressure-sprinkler irrigation* includes acres using either drip/trickle and low-flow microsystems or lower pressure-sprinkler systems (pressure PSI < 30). *Traditional irrigation* included all remaining irrigated acres associated with traditional irrigation systems. *Reproduced from Schaible, G.D., Aillery, M.P., 2015. Irrigation and Water Use. USDA, Economic Research Service at: http://www.ers.usda.gov/topics/farm-practices-management/irrigation-water-use.aspx.*

However, adopting more efficient physical systems alone may not be enough in the face of increasing water scarcity, especially with new demands from climate change and an expanding energy sector. The sustainability of irrigated agriculture will depend increasingly on expanded adoption of more efficient "irrigation production systems" (Evans and Sadler, 2008; Sadler et al., 2005; Clemmens et al., 2008). A "production system" policy perspective encourages a continued shift from traditional, less efficient gravity/sprinkler irrigation to more efficient irrigation application systems, but with greater reliance on on-farm water management improvements that increase overall production efficiency beyond that normally attainable because of complementary investments in human capital, that is, helping farmers determine the optimal timing of irrigation and how much water to use by crop growth stage (Schaible et al., 2010).[24] Improved on-farm water management practices can help producers maximize the economic efficiency of their irrigation systems and the potential for real water savings through reduced system losses and managed reductions in crop consumptive use.

For irrigated agriculture in general, and for gravity irrigation in particular, FRIS survey data suggest that producers give more emphasis to such conventional practices as reducing irrigation set times, alternating furrow irrigation (for row crops), and using end-of-field dikes to restrict field runoff (USDA/NASS, 1996, 2004, 2010, 2014b). Use of tailwater pits for on-farm water reuse has declined across gravity irrigation, from 22% in 1994 to 8% by 2008, partly in response to irrigation application improvements that limit field runoff. In 2013, the combined application of these conventional water management practices was used on only 23% of gravity-irrigated acres (USDA/NASS, 2014b). Total gravity-irrigated acres that have been laser leveled or zero graded have declined from 27% of acreage in 1998 to 15% in 2013.

By 2013, less water-intensive gravity management practices, such as use of special furrowing techniques, shortened furrow lengths, PAM, and use of surge-flow or cablegation irrigation, were applied on a relatively small portion (ranging from 3% to 7%) of gravity-irrigated agriculture in the West. Less interest in these practices may reflect their expected economic impact at the farm level, either through increased costs for land preparation or for specialized furrow management equipment, particularly when expected profit margins are low.

Despite technological advances in crop and soil moisture sensing, irrigators across the United States continue to depend heavily on traditional decision-making methods in deciding when to irrigate a crop and by how much. In the West, most producers generally irrigate based on the visible "condition of the crop" or by "feeling the soil" for soil moisture content, or irrigation may be based on a calendar schedule or an "in-turn" (fixed rotation) delivery schedule for water supplied to the farm. For 2013, fewer than 12% of irrigators throughout the West used soil or plant moisture-sensing devices or commercial irrigation scheduling services (USDA/NASS, 2014b). Fewer than 2% of irrigators used computer-based simulation models to evaluate crop irrigation requirements based on consumptive use needs by crop growth stage and local weather conditions. Low adoption rates may be because these practices are much more human capital and management intensive than traditional water application decision tools. These more sophisticated tools

24. The sustainability of irrigated agriculture could also be enhanced through continued research and development of crop cultivars with improved tolerance to drought, heat, and salts, as well as shorter growing seasons. However, this particular issue is beyond the focus of this chapter.

may require more extensive technical training and support to increase their adoption. Similar relationships exist for the eastern states, except that irrigation decisions for this region generally are not based on water delivered within a fixed rotation to the farm because less than 5% of the region's irrigated acres use off-farm water sources.

Our findings suggesting significant potential for wider adoption of more efficient "irrigation production systems" are consistent with recommendations of the National Research Council report, *Toward Sustainable Agricultural Systems in the 21st Century* (NRC, 2010) and the USDA REE 2014 Action Plan (USDA REE, 2014). Both reports recommend the need for greater policy emphasis on integrating on-farm water conservation with watershed-level water management mechanisms that help facilitate optimal allocation of limited water supplies among competing demands, including use of conserved water rights, drought-year water banks, water option markets, contingent water markets, reservoir management, well drilling and/or groundwater pumping restrictions, and irrigated acreage retirement.[25]

Integrating agricultural water conservation programs with watershed-level water management tools allows for accounting for basin-level water balance by considering the fate of farm-level water savings/losses. Watershed-level water management tools help to create more efficient water allocations by encouraging basin stakeholders to recognize the opportunity value of water across competing uses and by facilitating water transfers through market-based trading and reallocation schemes.

## 5. IRRIGATION INVESTMENTS AND FUNDING SOURCES

While the need for continued improvements in water-conserving production systems in US irrigated agriculture is well established, water use efficiency gains depend primarily on irrigation investment decisions in the private farm sector. Approximately $2.6 billion was invested in irrigation systems in 2013 by irrigated farms across the United States (including both private expenditures and public funding assistance), compared with $2.1 billion in 2008 and $1.1 billion in 2003 (USDA-NASS, 2004, 2010, 2014b).[26] On-farm irrigation investments have tended to focus on more precise water application that satisfies crop requirements while minimizing field losses. Total irrigation investments across the western states for 2013 ($1.9 billion) accounted for 72% of irrigation investments across the United States. Upgrades in application equipment and machinery in 2013 (at $1.4 billion) accounted for 71% of total irrigation investments in the western states. New well construction or deepening of existing wells accounted for the next largest farm-level investment in the West ($340 million, representing 18% of total irrigation investment regionally). In terms of

25. "Sustainable agriculture" as a USDA policy goal was initiated with the Food, Agriculture, Conservation, and Trade Act of 1990, with the key objective to "protect and enhance America's water resources." In addition, USDA's Strategic Plan for FY 2014–18 also highlights the importance of using farm-level, watershed, and institutional measures as a strategic means to meet this goal (USDA, 2015).

26. Total irrigation investment expenditures reported by FRIS include those made by the farm and "the portion of the expenditures made by or shared with others (landlords or government agencies). Including programs such as Environmental Quality Incentive Program (EQIP)" (USDA-NASS, 2014b). While FRIS information does not allow for separation of private versus publicly financed investment expenditures, it does indicate the share of farms using such financial assistance (discussed in the next paragraph).

investment purpose, scheduled investments for equipment/machinery replacement or maintenance accounted for the largest share of investments on western irrigated farms ($851 million out of $1.9 billion). Nationally, upgrades in irrigation facilities and equipment specifically to improve water conservation accounted for $450.7 million in 2013 for the western states and $70.3 million for the eastern states, or roughly 24% and 10% of regional on-farm irrigation investment expenditures, respectively. The larger share (70%) of investments for land leveling or zero grading of cropland to improve the uniformity of applied water with gravity-flow systems across the West occurred on existing irrigated acres ($74 million out of $106 million); investments in land leveling to establish new irrigated acres accounted for less than 2% of total investment expenditures.

Most on-farm irrigation investment in the United States is financed privately. Of farms reporting irrigation improvements in 2013, only about 11% received public financial assistance. The Environmental Quality Incentives Program (EQIP), administered by USDA's Natural Resources Conservation Service (NRCS), is the nation's primary source of funding for agricultural conservation activities on working farms and ranches. In 2013, EQIP accounted for 28% and 2% of farms reporting public financial assistance for irrigation investments across the western and eastern states, respectively.[27] Other USDA financial assistance programs (eg, Conservation Stewardship Program, Wetlands Reserve Program, Conservation Reserve Program) accounted for 15% and 10% of farms reporting assistance within the western and eastern states, respectively, with the remaining funding provided by non-USDA programs (eg, Environmental Protection Agency, Bureau of Reclamation, as well as state and local water management and supply district programs).

## 6. WATER CONSERVATION POLICY: A WATERSHED PERSPECTIVE

USDA signaled a shift to a more watershed/institutional, stakeholder partnership focus for implementing its agricultural water conservation activities with the Agricultural Water Enhancement Program (AWEP) under the 2008 Food, Conservation, and Energy Act (2008 Farm Bill). AWEP, a voluntary conservation initiative, provided technical and financial assistance to producers to implement practices on agricultural land to conserve surface and groundwater and improve water quality. Producers applied for AWEP participation through USDA/NRCS watershed-level partnership agreements. From 2009 to 2013, USDA's NRCS entered into 100+ AWEP partnership agreements involving 6886 producer conservation contracts designed to enhance agricultural water conservation, with a total obligation commitment of $331.4 million (82.3% for financial assistance and 17.7% for technical assistance).[28]

27. Irrigated farms reporting EQIP funding assistance represented 3% of all US irrigated farms making irrigation investments in 2013. However, the statistics here represent irrigated farm participants in EQIP only for 2013 and do not reflect program participation over time.
28. AWEP, operated under USDA's EQIP program, involved numerous complex partnership agreements. These partnerships ranged from providing producers with assistance to convert from gravity irrigation to low-pressure sprinkler irrigation, to using irrigated acreage and water use restrictions and conserved water for instream flow uses, and to implementing managed drought-year water banks. For a description of AWEP and its partnership agreements, see http://www.nrcs.usda.gov/wps/portal/nrcs/detailfull/national/programs/?&cid=nrcs143_008334.

With passage of the Agricultural Act of 2014 (2014 Farm Bill), Congress established the broader USDA Regional Conservation Partnership Program (RCPP). RCPP is designed to help implement USDA resource conservation programs in a way that enhances farm land *and* water stewardship at the watershed/regional landscape scale. This program accomplishes this goal through USDA partnerships with farmers and other resource stakeholders within a watershed or multicounty/state region, leveraging federal, state, and local financial resources to assist producers with a broader set of land and water conservation activities designed to increase the restoration and sustainable use of soil, water, and wildlife and related natural resources across the landscape (USDA/NRCS, 2015). RCPP partnerships may include one or more of the following eligible partners: agricultural or silvicultural producer associations, farmer cooperatives or other groups of producers, state or local governments, American Indian tribes, municipal water treatment entities, water and irrigation districts, conservation-driven nongovernmental organizations, and institutions of higher education. Once established, eligible partnership participants, via actual conservation contracts or easement agreements, may include producers and landowners of agricultural land and nonindustrial private forestland.

For 2014–15, RCPP funded 114 approved projects[29] at a total obligation of $361.0 million (20 projects through the national competitive pool at $142.9 million; 24 projects through the CCA pool at $125.8 million; and 70 projects through the state competitive pool at $92.3 million). The number of partners on a given project range from 1 to 46, but average about 13. RCPP project costs range from $100,000 to $17.5 million, but average $7.1 million for national competitive projects, $5.2 million for CCA projects, and $1.3 million for state competitive projects. Improving water conservation and water quality associated with irrigated agriculture were significant objectives across 22 projects, funded at $66.4 million or 18.4% of RCPP project obligations for the 2014–15 period. Irrigation-oriented projects accounted for 16.8% of national project funding (five projects), 19.6% of CCA project funding (six projects), and 19.2% of the state competitive project funding (11 projects).[30] RCPP funding for 2016 projects is projected at $225 million with project proposal decisions expected by USDA's NRCS in early 2016. While the share of future RCPP funding involving irrigation water conservation/water quality objectives is yet to be determined, it is expected to play a continued prominent role in achieving USDA's landscape-based, resource conservation objectives.

Even with a watershed conservation focus, adoption of more efficient irrigation application systems will continue to be an important component of agricultural water conservation efforts. The sustainability of irrigated agriculture, however, could be further enhanced by more intensely integrating improved on-farm water management practices with high-efficiency irrigation application systems, that is, greater emphasis on promoting efficient

---

29. RCPP funding is allocated to partnership projects through three funding pools: (1) 25% for projects through a state competitive process administered by the USDA NRCS State Conservationist; (2) 35% to projects within one of up to eight Critical Conservation Areas (CCA's) designated by the Secretary of Agriculture; and (3) 40% percent for projects established via a national competitive process managed by USDA. Project partners are required to contribute to the cost of the project, conduct outreach and education to eligible producers, and for assessing project effects. For more discussion of RCPP, see the website: http://www.nrcs.usda.gov/wps/portal/nrcs/main/national/programs/farmbill/rcpp/.

30. For a more detailed discussion of RCPP-funded projects, see the USDA, NRCS website at: http://www.nrcs.usda.gov/wps/portal/nrcs/detail/national/programs/farmbill/rcpp/?cid=stelprdb1264664.

"irrigation production systems" rather than sole emphasis on efficient irrigation application systems. In addition, through collaborative federal, state, and local partnerships under USDA's RCPP, integration of on-farm conservation efforts with watershed-level water management tools will further encourage increased conservation of water resources for current, alternative, and future uses. Integrating irrigation efficiency improvements with other practices, such as deficit irrigation, acreage idling, and off-farm water transfers that compensate producers for water conservation gains, allows producers to balance yield declines with improvements in profitability through reduced costs of applying water and related inputs. Integrating on-farm conservation and federal/state institutional mechanisms (conserved water rights, drought-year water banks, contingent (option) water markets, reservoir management, as well as irrigated acreage and/or pumping restrictions) will likely also encourage producers and other stakeholders to interact jointly in determining market-based water reallocations.

Finally, designing agricultural water conservation policies that promote a more sustainable future for irrigated agriculture depends a great deal on improving the economic analysis of adaptation options within the irrigated farm sector. In an increasingly water scarce world, production system adaptation strategies are likely to involve complex production decisions on crop choice, water application rates, and adopting efficient irrigation technology and water management practices that adjust to changing water supply conditions over time. Economic analyses from a production system perspective could simultaneously consider all components of a producer's production decisions—crop choice, crop yield target, irrigation system type, and on-farm water management regime—combined with field-level physical/environmental characteristics and water supply conditions. As competing demands and climate change increasingly strain the water supply/demand environment for agriculture, economic analysis will be required to address the complexity of water conservation policy issues and their impact on agricultural production and regional resource use and quality. Such analyses, however, could also enhance the quality and reliability of information on irrigation choices, improving our understanding of irrigated agriculture's adaptability toward a more sustainable future.

## REFERENCES

Aillery, M., Shoemaker, R., Caswell, M., February 2001. Agriculture and ecosystem restoration in south Florida: assessing trade-offs from water-retention development in the everglades agricultural area. American Journal of Agricultural Economics 83 (1), 183–195.

Ayars, J.E., September 2010. Adapting Irrigated Agriculture to Drought in the San Joaquin Valley of California. Policy Note 03-0910. University of California-Riverside, Water Science and Policy Center, p. 4. at: http://wspc.ucr.edu/policy_notes/PN_03_0910_Drought.pdf.

Bartis, J.T., La Tourrete, T., Dixon, L., Peterson, D.J., Cecchine, G., 2005. Oil Shale Development in the United States: Prospects and Policy Issues. RAND Corporation, Arlington, VA, p. 68.

Bates, B.C., Kundzewicz, Z.W., Wu, S., Palutikof, J.P. (Eds.), 2008. Climate Change and Water, IPCC Technical Paper VI, Intergovernmental Panel on Climate Change. IPCC Secretariat, Geneva, p. 210.

Bernhardt, E.S., Palmer, M.A., Allan, J.D., Alexander, G., Barnas, K., Brooks, S., Carr, J., Clayton, S., Dahm, C., Follstad-Shah, J., Galat, D., Gloss, S., Goodwin, P., Hart, D., Hassett, B., Jenkinson, R., Katz, S., Kondolf, G.M., Lake, P.S., Lave, R., Meyer, J.L., O'Donnell, T.K., Pagano, L., Powell, B., Sudduth, E., April 2005. Synthesizing U.S. river restoration efforts. Science 308 (5722), 636–637.

Brekke, L.D., Kiang, J.E., Olsen, J.R., Pulwarty, R.S., Raff, D.A., Turnipseed, D.P., Webb, R.S., White, K.D., 2009. Climate Change and Water Resources Management—A Federal Perspective. U.S. Geological Survey, Circular 1331, p. 65.

Burt, C.M., Clemmens, A.J., Strelkoff, T.S., Solomon, K.H., Bliesner, R.D., Hardy, L.A., Howell, T.A., Eisenhauer, D.E., November 1997. Irrigation performance measures: efficiency and uniformity. Journal of Irrigation and Drainage Engineering 123 (6), 423−442.

CALFED Bay-Delta Program, 2010. CALFED Record of Decision at: http://calwater.ca.gov/calfed/objectives/ecosystem_restoration.html.

California Department of Conservation, 2011. 2006−2008 California Farmland Conversion Report, Farmland Division of Land Resource Protection. Mapping and Monitoring Program, Sacramento, CA, p. 97.

Center for Irrigation Technology (CIT), November 2011. Agricultural Water Use in California: A 2011 Update. Staff Report. California State University, p. 79.

Chesapeake Energy, January 2012. Water Use in Deep Shale Gas Exploration. Fact Sheet, p. 2. https://www.springsgov.com/units/boardscomm/OilGas/Water%20Use%20in%20Deep%20Shale%20Exploration%20Fact%20Sheet.pdf.

Chiu, Y.-W., Walseth, B., Suh, S., 2009. Water embodied in bio-ethanol in the United States. Environmental Science and Technology 43 (8), 2688−2692.

Christensen, N.S., Lettenmaier, D.P., 2007. A multimodel ensemble approach to assessment of climate change impacts on the hydrology and water resources of the Colorado river basin. Hydrology and Earth System Sciences 11, 1417−1434.

Claims Resolution Act of 2010, 2010. Titles III−VI, Public Law 111-291 (December 8), pp. 3073−3156.

Clemmens, A.J., Allen, R.G., Burt, C.M., 2008. Technical concepts related to conservation of irrigation and rainwater in agricultural systems. Water Resources Research 44, 16 at: http://onlinelibrary.wiley com/doi/10.1029/2007WR006095/epdf.

Congressional Budget Office (CBO), August 1997. Water Use Conflicts in the West: Implications of Reforming the Bureau of Reclamation's Water Supply Policies, United States Congress.

Dettinger, M., Earman, S., June 2007. Western groundwater and climate change−pivotal to supply sustainability or vulnerability in its own right? Ground Water: News and Views 4 (1), 4−5.

Evans, R.G., Sadler, E.J., 2008. Methods and technologies to improve efficiency of water use. Water Resources Research 44, 15 at: http://naldc.nal.usda.gov/download/21462/PDF.

Evett, S., Carman, D., Bucks, D., 2003. Expansion of irrigation in the Mid-South United States: water allocation and research issues. In: Proceedings, 2nd International Conference on Irrigation and Drainage, Water for a Sustainable World−Limited Supplies and Expanding Demand. U.S. Committee on Irrigation and Drainage, Phoenix, AZ, pp. 247−260. May 12−15.

Fereres, E., Soriano, M.A., 2007. Deficit irrigation for reducing agricultural water use. Journal of Experimental Botany, Special Issue 58 (2), 147−159.

Fischer Farm Services, 2011. Reasons to Invest in Mississippi Farmland. Aberdeen, MS. http://www.fischerfarmservices.com/reasons_farm.html.

Florida Department of Environmental Protection (DEP), March 2010. Sustaining Our Water Resources: Annual Report on Regional Water Supply Planning, p. 21.

Gallegos, T.J., Varela, B.A., Haines, S.S., Engle, M.A., June 16, 2015. Hydraulic fracturing water use variability in the United States and potential environmental implications. Water Resources Research 51, 1−7 at: http://onlinelibrary.wiley.com/doi/10.1002/2015WR017278/pdf.

Geerts, S., Raes, D., 2009. Deficit irrigation as an on-farm strategy to maximize crop water productivity in dry areas. Agricultural Water Management 96 (9), 1275−1284.

Green, T.R., Taniguchi, M., Kooi, H., August 2007. Potential impacts of climate change and human activity on subsurface water resources. Vadose Zone Journal 6 (3), 531−532.

Gregory, L., 2008. Technically open: the debate over Native American reserved groundwater rights. Journal of Land, Resources, and Environmental Law 28 (2), 361−373.

Ground Water Protection Council (GWPC) and All Consulting, April 2009. Modern Shale Gas Development in the United States: A Primer. U.S. Department of Energy, Office of Fossil Energy, National Energy Technology Laboratory, Contract DE-FG26-04NT15455, p. 96.

Hall, N.D., Stuntz, B.B., Abrams, R.H., Winter 2008. Climate change and freshwater resources. Natural Resources & Environment 22 (3), 30−35.

Hatfield, J.L., Boote, K.J., Fay, P.A., Hahn, G.L., Izaurralde, R.C., Kimball, B.A., Mader, T.L., Morgan, J.A., Ort, D.R., Polley, H.W., Thomson, A.M., Wolfe, D.W., May 2008. "Agriculture," the Effects of Climate Change on Agriculture, Land Resources, Water Resources, and Biodiversity in the United States. U.S. Climate Change Science Program and the Subcommittee on Global Change Research, Washington, DC, pp. 21–74.

Hoerling, M., Eischeid, J., 2007. Past peak water in the Southwest. Southwest Hydrology 6, 18.

Howell, T.A., 2003. Irrigation efficiency. Encyclopedia of Water Science 467–472.

Intergovernmental Panel on Climate Change (IPCC Report), 2007. Climate change 2007: impacts, adaptation and vulnerability. In: Parry, M.L., Canziani, O.F., Palutikof, J.P., van der Linden, J.P., Hanson, C.E. (Eds.), Fourth Assessment Report of the Intergovernmental Panel on Climate Change. Cambridge University Press, Cambridge, UK, p. 976.

Intergovernmental Panel on Climate Change (IPCC Report), 2014. Climate change 2014: impacts, adaptation, and vulnerability, part A: global and sectoral aspects. In: Field, C.B., Barros, V.R., Dokken, D.J., Mach, K.J., Mastrandrea, M.D., Bilir, T.E., Chatterjee, M., Ebi, K.L., Estrada, Y.O., Genova, R.C., Girma, B., Kissel, E.S., Levy, A.N., MacCracken, S., Mastrandrea, P.R., White, L.L. (Eds.), Contribution of Working Group II to the Fifth Assessment Report of the Intergovernmental Panel on Climate Change. Cambridge University Press, Cambridge, United Kingdom and New York, NY, USA, 1132 pp.

Izaurralde, R.C., Rosenberg, N.J., Brown, R.A., Thomson, A.M., 2003. Integrated assessment of Hadley Center (HadCM2) climate-change impacts on agricultural productivity and irrigation water supply in the conterminous United States, part II: regional agricultural production in 2030 and 2095. Agricultural and Forest Meteorology 117, 97–122.

Knowles, N., Dettinger, M.D., Cyan, D.R., September 2006. Trends in snowfall versus rainfall in the western United States. Journal of Climate 19 (18), 4545–4559.

Kuwayama, Y., January 7, 2016. Infographic: Exploring the Water-Energy Nexus: Water Use for Fossil Fuel Extraction and Processing. Resources for the Future, 1 pp. at: http://www.rff.org/research/publications/infographic-exploring-water-energy-nexus-water-use-fossil-fuel-extraction-and.

Landry, C., 1998. Market transfers of water for environmental protection in the western United States. Water Policy 1 (5), 457–469.

Lemke, P., Ren, J., Alley, R., Allison, I., Carrasco, J., Flato, G., Fujii, Y., Kaser, G., Mote, P., Thomas, R.H., Zhang, T., 2007. Observations: changes in snow, ice and frozen ground. In: Solomon, S., Quin, D., Manning, M., Chen, Z., Marquis, M., Averyt, K.B., Tignor, M., Miller, H.L. (Eds.), Climate Change 2007: The Physical Science Basis, a Contribution of Working Group I to the Fourth Assessment Report of the Intergovernmental Panel on Climate Change. Cambridge University Press, Cambridge and New York, pp. 337–384.

MacDonnell, L.J., 2009. Environmental flows in the rocky mountain west: a progress report. Wyoming Law Review 9 (2), 335–396.

Malcolm, S.A., Aillery, M., Weinberg, M., November 2009. Ethanol and a Changing Agricultural Landscape. ERR-86. U.S. Department of Agriculture, Economic Research Service, p. 58.

Marshall, E., Aillery, M.P., Malcolm, S., Williams, R., November 2015. Climate Change, Water Scarcity, and Adaptation in the U.S. Fieldcrop Sector. ERR-201. U.S. Department of Agriculture, Economic Research Service, p. 110.

Maupin, M.A., Kenny, J.F., Hutson, S.S., Lovelace, J.K., Barber, N.L., Linsey, K.S., 2014. Estimated Use of Water in the United States in 2010. U.S. Geological Survey, Circular 1405, at: http://pubs.usgs.gov/circ/1405/pdf/circ1405.pdf.

McCabe, G.J., Wolock, D.M., November 2007. Warming may create substantial water supply shortages in the Colorado river basin. Geophysical Research Letters 34 (L22708), 1–5.

Melillo, J.M., Richmond, Terese (T.C.), Yohe, G.W. (Eds.), 2014. Highlights of Climate Change Impacts in the United States: The Third National Climate Assessment. U.S. Global Change Research Program, 148 pp.

Midwest Irrigation, 2010. Midwest States Increase Irrigation. The Net Irrigate Blog. http://netirrigate.wordpress.com/tag/midwest-irrigation/.

Moore, M.R., Summer 1989. Native American water rights: efficiency and fairness. Natural Resources Journal 29, 763–791.

Mote, P.W., Hamlet, A.F., Clark, M.P., Lettenmaier, D.P., 2005. Declining mountain snowpack in western North America. Bulletin of the American Meteorological Society 86, 39–49.

Mote, P.W., Parson, E.A., Hamlet, A.F., Keeton, W.S., Lettenmaier, D., Mantau, N., Miles, E.L., Peterson, D.W., Peterson, D.L., Slaughter, R., Snover, A.K., 2003. Preparing for climate change: the water, salmon, and forests of the Pacific Northwest. Climate Change 61, 45–88.

National Energy Technology Laboratory (NETL), September 2008. Estimating Freshwater Needs to Meet Future Thermoelectric Generation Requirements, 2008 Update. DOE/NETL-400/2008/1339, p. 107. http://www.netl.doe.gov/File%20Library/Research/Energy%20Analysis/Publications/DOE-NETL-400-1339-2008WaterNeedsUpdate-Final-Sept08.pdf.

National Research Council (NRC), 1996. A New Era for Irrigation, Water Science and Technology Board. National Academies Press, Washington, DC.

National Research Council (NRC), 2010. Toward Sustainable Agricultural Systems in the 21st Century, Board on Agriculture and Natural Resources. National Academies Press, Washington, DC.

National Research Council (NRC), 2008. Water Implications of Biofuels Production in the United States, Water Science and Technology Board. National Academies Press, Washington, DC, p. 86.

National Water Assessment Group (NWAG Report), 2000. Water: The Potential Consequences of Climate Variability and Change, Peter Gleick (Project Lead). U.S. Global Change Research Program, Oakland, CA, p. 160.

Poff, N.L., Allan, J.D., Palmer, M.A., Hart, D.D., Richter, B.D., Arthington, A.H., Rogers, K.H., Meyer, J.L., Stanford, J.A., August 2003. River flows and water wars: emerging science for environmental decision making. Frontiers in Ecology and the Environment 1 (6), 298–306.

Rood, S.B., Samuelson, G.M., Weber, J.K., Wyrot, K.A., 2005. Twentieth-century decline in streamflows from the hydrographic apex of North America. Journal of Hydrology 306, 215–233.

Sadler, E.J., Evans, R.G., Stone, K.C., Camp, C.R., 2005. Opportunities for conservation with precision irrigation. Journal of Soil and Water Conservation 60 (6), 371–379.

Schaible, G.D., 2000. Economic and conservation tradeoffs of regulatory vs. incentive-based water policy in the Pacific Northwest. International Journal of Water Resources Development 16 (2), 221–238.

Schaible, G.D., 2013. Western Irrigated Agriculture. U.S. Department of Agriculture, Economic Research Service at: http://www.ers.usda.gov/data-products/western-irrigated-agriculture.aspx.

Schaible, G.D., Kim, C.S., Aillery, M.P., December 2010. Dynamic adjustment of irrigation technology/water management in western U.S. agriculture: toward a sustainable future. Canadian Journal of Agricultural Economics 58, 433–461.

Schaible, G.D., Aillery, M.P., 2006. Irrigation Water Management. In: Agricultural Resources and Environmental Indicators: 2006 Edition, pp. 134–143. EIB-16, U.S. Dept. of Agriculture, Economic Research Service. http://www.ers.usda.gov/media/872940/eib16.pdf.

Schaible, G.D., Aillery, M.P., 2015. Irrigation and Water Use. USDA, Economic Research Service at: http://www.ers.usda.gov/topics/farm-practices-management/irrigation-water-use.aspx.

Sophocleous, M., July-August 2007. The science and practice of environmental flows and the role of hydrogeologists. Ground Water 45 (4), 393–401.

U.S. Climate Change Science Program (U.S. CCSP) and the Subcommittee on Global Change Research. 2008. The Effects of Climate Change on Agriculture, Land Resources, Water Resources, and Biodiversity in the United States, Backlund, P., Janetos, A., Schimel, D., Hatfield, J., Boote, K., Fay, P., Hahn, L., Izaurralde, C., Kimball, B.A., Mader, T., Morgan, J., Ort, D., Polley, W., Thomson, A., Wolfe, D., Ryan, M.G., Archer, S.R., Birdsey, R., Dahm, C., Heath, L., Hicke, J., Hollinger, D. Huxman, T., Okin, G., Oren, R., Randerson, J., Schlesinger, W., Lettenmaier, D., Major, D., Poff, L., Running, S., Hansen, L., Inouye, D., Kelly, B.P., Meyerson, L., Peterson, B., Shaw, R, (Contrib.), U.S. Department of Agriculture, p. 362.

U.S. Department of Agriculture, National Agricultural Statistics Service (USDA/NASS), February 1996. Farm and Ranch Irrigation Survey (1994). In: Related Surveys, Part 1, (AC92-RS-1), vol. 3, p. 142 at: www.agcensus.usda.gov/Publications/1992/fris94/fris94.htm.

U.S. Department of Agriculture, National Agricultural Statistics Service (USDA/NASS), November 2004. Farm and Ranch Irrigation Survey (2003). In: Special Studies, Part 1, (AC-02-SS-1), vol. 3, p. 210 at: http://www.agcensus.usda.gov/Publications/2002/FRIS/.

U.S. Department of Agriculture, National Agricultural Statistics Service (USDA/NASS), February 2010. Farm and Ranch Irrigation Survey (2008). In: Special Studies of the 2007 Census of Agriculture, Part I (AC-07-SS-1), vol. 3, p. 268 at: http://www.agcensus.usda.gov/Publications/2007/Online_Highlights/Farm_and_Ranch_Irrigation_Survey/.

U.S. Department of Agriculture, National Agricultural Statistics Service (USDA/NASS), May 2014a. 2012 Census of Agriculture: U.S. Summary and State Data. In: Geographic Area Series, Part 51 (AC-12-A-51), vol. 1, p. 695, at: http://www.agcensus.usda.gov/Publications/2012/Full_Report/Volume_1,_Chapter_1_US/usv1.pdf.

U.S. Department of Agriculture, National Agricultural Statistics Service (USDA/NASS), November 2014b. 2013 Farm and Ranch Irrigation Survey. In: Special Studies, Part 1 (AC-12-SS-1), vol. 3, p. 259, at: http://www.agcensus.usda.gov/Publications/2012/Online_Resources/Farm_and_Ranch_Irrigation_Survey/.

U.S. Department of Agriculture, Natural Resources Conservation Service (USDA/NRCS), 2006. Conservation Practices that Save: Irrigation Water Management. In: Save Energy Save Money. http://www.nrcs.usda.gov/Internet/FSE_DOCUMENTS/nrcs143_023630.pdf.

U.S. Department of Agriculture, Natural Resources Conservation Service (USDA/NRCS), 2015. Regional Conservation Partnership Program (RCPP)—2014—15 Final Project Homepage. RCPP webpage at: http://www.nrcs.usda.gov/wps/portal/nrcs/detail/national/programs/farmbill/rcpp/?cid=stelprdb1264664.

U.S. Department of Agriculture, Research, Education, and Economics (USDA/REE), March 2014. Research, Education, and Economics Action Plan. Office of the Under Secretary for Research, Education, and Economics, p. 52.

U.S. Department of Agriculture (USDA), 2015. Strategic Plan FY 2014—2018. Office of the Secretary, p. 45. http://www.usda.gov/documents/usda-strategic-plan-fy-2014-2018.pdf.

U.S. Department of Energy (U.S. DOE), 2010. Concentrating Solar Power Commercial Application Study: Reducing Water Consumption of Concentrating Solar Power Electricity Generation. A Report to Congress, p. 35. https://www1.eere.energy.gov/solar/pdfs/csp_water_study.pdf.

U.S. Department of Interior, Bureau of Land Management (U.S. BLM), 2011. Oil Shale Fact Sheet. BLM Fact Sheet. http://www.blm.gov/pgdata/etc/medialib/blm/wo/Communications_Directorate/public_affairs/news_release_attachments.Par.96261.File.tmp/OilShaleFS_022011.pdf.

U.S. Department of Interior, Bureau of Reclamation (U.S. BoR), 2012a. Colorado River Basin Water Supply and Demand Study (Study Report) at: http://www.usbr.gov/lc/region/programs/crbstudy/finalreport/Study%20Report/CRBS_Study_Report_FINAL.pdf.

U.S. Department of Interior, Bureau of Reclamation (U.S. BoR), June 2012b. Interior's Indian Water Rights Settlement Program — Reclamation Support for Secretary's Program at: http://www.usbr.gov/native/waterrights/waterrights.html.

U.S. Department of Interior, Bureau of Reclamation (U.S. BoR), April 2011. Secure Water Act Section 9503(c)—Reclamation Climate Change and Water, Report to Congress. Patty Alexander, Brekke Levi, Gray Davis, Subhrendu Gangopadhyay, Katrina Grantz, Charles Hennig, Carly Jerla, Dagmar Llewellyn, Paul Miller, Tom Pruitt, David Raff, Tom Scott, Michael Tansey, and Toni Turner (Contrib.), Denver, CO, p. 205. at: http://www.usbr.gov/climate/SECURE/docs/SECUREWaterReport.pdf.

U.S. Department of Interior, Bureau of Reclamation (U.S. BoR), 2010. Interim Federal Action Plan Status Update for the California Bay Delta: 2011 and Beyond at: http://www.doi.gov/news/pressreleases/upload/Final-Status-Update-2010-12-15.pdf.

U.S. Department of Interior, Bureau of Reclamation (U.S. BoR), August 2009. Central Valley Project Improvement Act Program Activity Review Report. Final CPAR, at: http://www.usbr.gov/mp/cvpia/docs_reports/docs/2009_Final%20CPAR%20Report%208-25-09.pdf.

U.S. Department of Interior, U.S. Geological Survey (USGS), September 2009. Effects of Climate Variability and Change on Groundwater Resources of the United States. Fact Sheet 2009-3074, Gurdak, J.J., Hanson, R.T., Green, T.R., (Contrib.), p. 4.

U.S. Department of Interior, U.S. Geological Survey (USGS), 2011a. Groundwater Atlas of the United States, Alabama, Florida, Georgia, and South Carolina (HA 730-G) at: http://pubs.usgs.gov/ha/ha730/ch_g/G-text1.html; Iowa, Michigan, Minnesota, and Wisconsin (HA 730-J) at: http://pubs.usgs.gov/ha/ha730/ch_j/J-text1.html; and Arkansas, Louisiana, and Mississippi (HA 730-F) at: http://pubs.usgs.gov/ha/ha730/ch_f/F-text1.html.

U.S. Department of Interior, U.S. Geological Survey (USGS), 2011b. National Water-Quality Assessment (NAWQA) Studies at: http://water.usgs.gov/nawqa/bib.

U.S. Department of Interior, U.S. Geological Survey (USGS), September 2008. Water Use in Florida, 2005 and Trends 1950–2005. Fact Sheet 2008-3080. In: Marella, R.L. (Ed.), p. 7.

U.S. Government Accountability Office (U.S. GAO), October 2010. Energy-Water Nexus: A Better and Coordinated Understanding of Water Resources Could Help Mitigate the Impacts of Potential Oil Shale Development. GAO-11-35, p. 69.

U.S. Government Accountability Office (U.S. GAO), 2009. Energy-Water Nexus: Many Uncertainties Remain About National and Regional Effects of Increased Biofuel Production on Water Resources. GAO-10-116. U.S. House of Representatives, Committee on Science and Technology, p. 50.

U.S. Senate Committee on Indian Affairs, March 2012. Oversight Hearing on Indian Water Rights: Promoting the Negotiation and Implementation of Water Settlements in Indian Country. A webcast at: http://www.indian.senate.gov/hearing/oversight-hearing-indian-water-rights-promoting-negotiation-and-implementation-water.

Vories, E.D., Evett, S.R., December 5–8, 2010. Irrigation research needs in the USA mid-south and southeast, humid and subhumid regions. In: Dukes, M. (Ed.), Proceedings for the 5th National Decennial Irrigation Conference, ASABE # IRR10-8679. American Society of Agricultural and Biological Engineers (ASABE) and the Irrigation Association, Phoenix, AZ, pp. 1–12.

Wall, E., Smit, B., 2005. Climate change adaptation in light of sustainable agriculture. Journal of Sustainable Agriculture 27 (1), 113–123.

Wallander, S., Claassen, R., Nickerson, C., August 2011. The Ethanol Decade: An Expansion of U.S. Corn Production, 2000–09. EIB-79. U.S. Department of Agriculture, Economic Research Service, p. 16.

Western States Water Council, May 1984. Indian Water Rights in the West. Western Governors Association.

Zellmer, S., 2008. Legal tools for instream flow protection. In: Locke, A., Stalnaker, C., Zellmer, S., Williams, K., Beecher, H., Richards, T., Robertson, C., Wald, A., Paul, A., Annear, T., (Contrib.), Integrated Approaches to Riverine Resource Stewardship: Case Studies, Science, Law, People, and Policy, The Instream Flow Council, Cheyenne, WY.

Zellmer, S., 2009. United States: the emergence of environmental considerations. In: Dellapenna, J.W., Gupta, J. (Eds.), The Evolution of the Law and Politics of Water. Springer Science and Business Media B.V, pp. 205–223.

Chapter 2.1.2

# The Water–Energy Nexus and Irrigated Agriculture in the United States: Trends and Analyses

K. Schwabe[1], K. Knapp[1], I. Luviano[2]
[1]*University of California-Riverside, Riverside, CA, United States;* [2]*Michoacana San Nicolas de Hidalgo University, Morelia, Michoacán, Mexico*

## 1. INTRODUCTION

With rising populations coupled with concern over a changing climate, much attention has been directed at agricultural production and the current and future challenges it confronts in meeting societal demands for food and agricultural products (Whittlesey, 1987; Carruthers et al., 1997; Lotze-Campen et al., 2008; Foley et al., 2011). One strategy to meet such demand has been to increase the amount of cultivated area using irrigation, since irrigation has been shown to produce yields of up to three times greater per acre than rain-fed agriculture globally (Rosegrant et al., 2002).[1] Between 1950 and the early 1980s, acreage under irrigation nearly doubled in the United States, expanding from 26 to 49 million acres. With increased irrigated acreage, however, comes increased water withdrawals. As shown in a popular United States Geological Survey (USGS) figure (Fig. 1; Maupin et al., 2014), withdrawals for irrigation in the United States increased significantly ($\sim$69%) from 1950 to the early 1980s as well, and in lockstep with population growth, which grew by more than 50%.[2]

Continual expansion of irrigated agriculture as a singular strategy to meet rising food demand has challenges, particularly because of increased competition for water from the energy sector and for the environment. These two factors are likely largely behind the leveling of irrigation water withdrawals beginning in the early 1980s to 2010 even though the population rose by nearly 40% in the United States (Fig. 1). Certain types of energy production can be very water intensive as evidenced by the fact that thermo-electric energy production has surpassed irrigated agriculture since the mid-1960s as the

1. While the largest differences seem to appear in developing countries, smaller but significant differences still appear in developed countries.
2. Note that withdrawals are not the same as consumption and in many cases they can be significantly different (eg, in the case of hydropower).

**Competition for Water Resources. http://dx.doi.org/10.1016/B978-0-12-803237-4.00005-7**

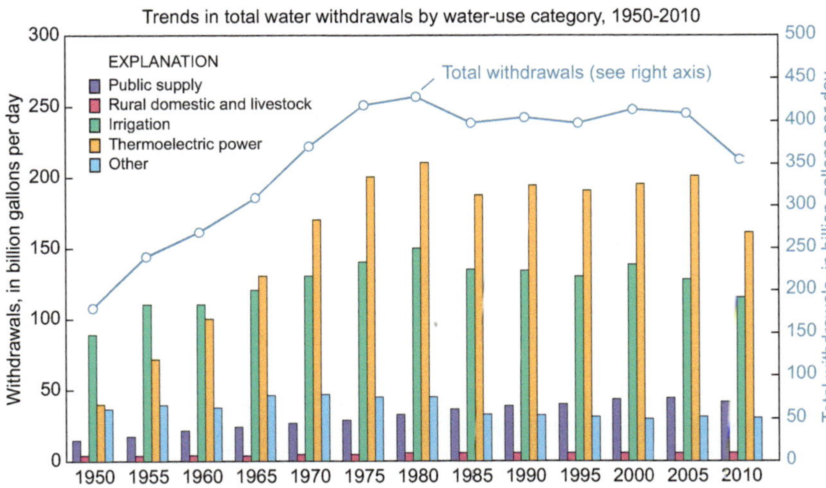


**FIGURE 1** USGS estimated trends in US water withdrawals by water use category, 1950–2010 (Maupin et al., 2014).

sector withdrawing the most water annually. Competition for surface water supplies for instream flows also increases water scarcity. Both sorts of competition likely contributed to reductions in aggregate irrigation water withdrawals, particularly surface water withdrawals with some corresponding increase in groundwater usage.[3] Indeed, over the period from 1950 to 1975 in which water withdrawals from thermoelectric power increased significantly and surpassed agriculture, groundwater withdrawals in the western United States grew from 18.2 million acre-feet (MAF) to 56 MAF (Lacewell and Collins, 1987), a move that likely increased on-farm water costs for growers that have access to both surface and groundwater supplies.[4] Consequently, and as emphasized in Waskom et al. (2014), both competition and rising costs over the past few decades have led producers to increasingly invest in intensification rather than expansion to meet demand and adapt.

While energy generation requires significant amounts of water, irrigated agriculture requires significant amounts of energy. For instance, in California, over 10,000 GWh of energy—nearly 20% of California's electricity production—is used for moving and pumping water for agriculture (CEC, 2005).[5] Overlooked in these estimates is the indirect energy embedded in fertilizer and pesticide use, which comprises nearly 30% of all energy used in

3. Thermoelectric power primarily relies on surface water withdrawals. As Maupin et al. (2014) note, over 99% of thermoelectric power water withdrawals is from surface water supplies (with 73% of those supplies emanating from freshwater sources).
4. For instance, Harman (1987) writes that nearly two-thirds of the water pumped from groundwater sources was about 80% more expensive than surface water supplies.
5. The California State Water Project, which delivers 30% of its water supplies to nearly 750,000 acres of irrigated farmland, is the largest single energy user in the state, consuming about 2.5% of all electricity produced.

agriculture in the United States (Miranowski, 2005). Accordingly, changes in energy prices can impact irrigated agriculture both directly (ie, through on-farm energy purchases) and indirectly (ie, through inputs that require significant energy to produce). These relationships can be significant since energy generation, similar to irrigated agriculture, has also had its share of challenges, including increased competition for water and environment concerns. Again referring to Fig. 1, water withdrawals for thermal energy production have also leveled off, if not declined slightly, since the 1980s. As Maupin et al. (2014) explain, this reduction in thermoelectric water withdrawals was caused by a combination of factors, including more power plants that use recirculating or dry cooling systems and concerns over the environment, including both aquatic life and greenhouse gas emissions from thermoelectric power plants burning fossil fuels.

With these considerations and concerns in mind, the objectives of this chapter are twofold. First, we will highlight recent trends in irrigated agricultural production and related factors in an attempt to shed light on the directions and developments among irrigated agricultural relationships across different regions within the United States over the past 35 years. By shedding light on these relationships, we can identify similarities and differences across these regions since the late 1970s that may or may not conform to our expectations and thus provide motivation for more in-depth and targeted research. We limit our analysis from the late 1970s to the present given the significant changes in energy policy and environmental regulations since the mid-1970s.[6] Second, we will investigate, using a dynamic economic–hydrologic model of regional irrigated agricultural production with access to both surface and groundwater supplies, the impacts of changes in (1) surface water and groundwater costs from energy price increases and (2) surface water availability on agricultural profits and production, water use efficiency and management, and groundwater levels. The relationship between water and energy in irrigated agricultural production is complex and influenced by relative rates of economic and technical substitution, among many other factors. As such, the programming model we develop will help us account for this complexity and provide insight into a wide array of impacts from changes in water supply costs and availabilities.

The next section highlights recent trends from the late 1970s to the present, while Section 3 provides a discussion of the dynamic economic–hydrologic model of irrigated agricultural production. Section 4 presents results from an analysis of changes in water prices and availabilities on irrigated agricultural management, profits, and groundwater levels. In Section 5, a summary of the findings and chapter is presented.

## 2. TRENDS IN US IRRIGATED AGRICULTURE

The graphs, tables, and discussions in this section are based on data gathered from the Farm and Ranch Irrigation Survey (FRIS) and published by the United States Department of Agriculture (USDA) Economic Research Service. We limit our focus to the years from 1979—after the 1970s energy crises—to 2013. Over this period, the USDA published eight FRIS—1979, 1984, 1988, 1994, 1998, 2003, 2008, and 2013. Considering trends over the

6. Two excellent resources that investigate the water–energy nexus in irrigated agriculture from the 1950s through the energy crises of the 1970s and soon thereafter include Lockeretz (1977) and Whittlesey (1987). Waskom et al. (2014) provide breadth and depth into the food–energy–water nexus discussion.

past 35 years seems reasonable given advances in the technologies and changes in the policies surrounding both water and energy since the mid-1970s. In some instances, aggregate US statistics will be presented, but the bulk of the analysis is presented on a regional basis. Within FRIS, data are tabulated at the national and state levels, but also regionally in what they term Water Resource Region (WRR). A WRR is defined by topographic drainage characteristics. Within the continental United States there are 18 individual WRRs.

For the purpose of this chapter, we focus on four WRRs that differ significantly in terms of prominent energy source, climate, water supplies, and regulations/governance.[7] The four WRRs include: the California region (CAL), which encompasses the drainage basins in the United States that discharge into the Pacific Ocean and whose point of discharge is within California; the Pacific-Northwest WRR (PNW), which includes drainage basins that discharge into the Straights of Georgia and Juan de Fuca and the Pacific Ocean, and includes regions in Washington and Oregon; the Arkansas-White-Red WRR (AWR), which includes drainage basins that include parts of the Arkansas River, Red River, and White River; and the South Atlantic-Gulf WRR (SAG), which accounts for drainage basins that discharge into the Atlantic Ocean and the Gulf of Mexico and whose points of discharge are located within and between North Carolina, Mississippi, and Florida. Combined, these regions cover a significant part of the irrigated acreage in the United States. For instance, in 2013 these four WRRs comprised nearly 50% of the 55 million acres of irrigated farmland in the United States (FRIS, 2013). For data that include costs or prices, such data are transformed into 2013 dollars using the historic producer price index series from the US Bureau of Labor Statistics (USBLS, 2015).[8]

## 2.1 Irrigated Acreage, Water Rates, and Energy Costs per Acre for the United States

Fig. 2 presents the trends in irrigated acreage and applied water rates in the United States since 1979, including for farms with access to off-farm water sources.[9] Also presented in Fig. 2A (with units presented on the right vertical axis) is the FRIS estimate of the energy costs per acre as a percentage change from 1979 values (which was $50.62 in 2013 dollars).[10] As shown in Fig. 2A, overall irrigated acreage has increased slightly to over 11% since the late 1970s, while acreage irrigated with some amount of off-farm water supply has experienced a near 18% decline (from 17.5 million down to

7. Whittlesey (1987) highlights the variability in energy source usage across the United States and, consequently, energy costs.

8. Additionally, each FRIS report presents sample characteristics and measures of precision for each of the variables.

9. As noted in the FRIS reports in defining off-farm water supply (eg, USDA, 2013; p. B-6), "Off-farm water supply is water from off-farm water suppliers, such as the U.S. Bureau of Reclamation; irrigation districts; mutual, private, cooperative, or neighborhood ditches; commercial companies; or community water systems. It includes reclaimed water from off-farm livestock facilities, municipal, industrial, and other reclaimed water sources."

10. On-farm energy costs are presented as per acre costs and include on-farm energy expenses for pumping irrigation water from both wells and surface water yet exclude energy expenses associated with solar-powered pumps.

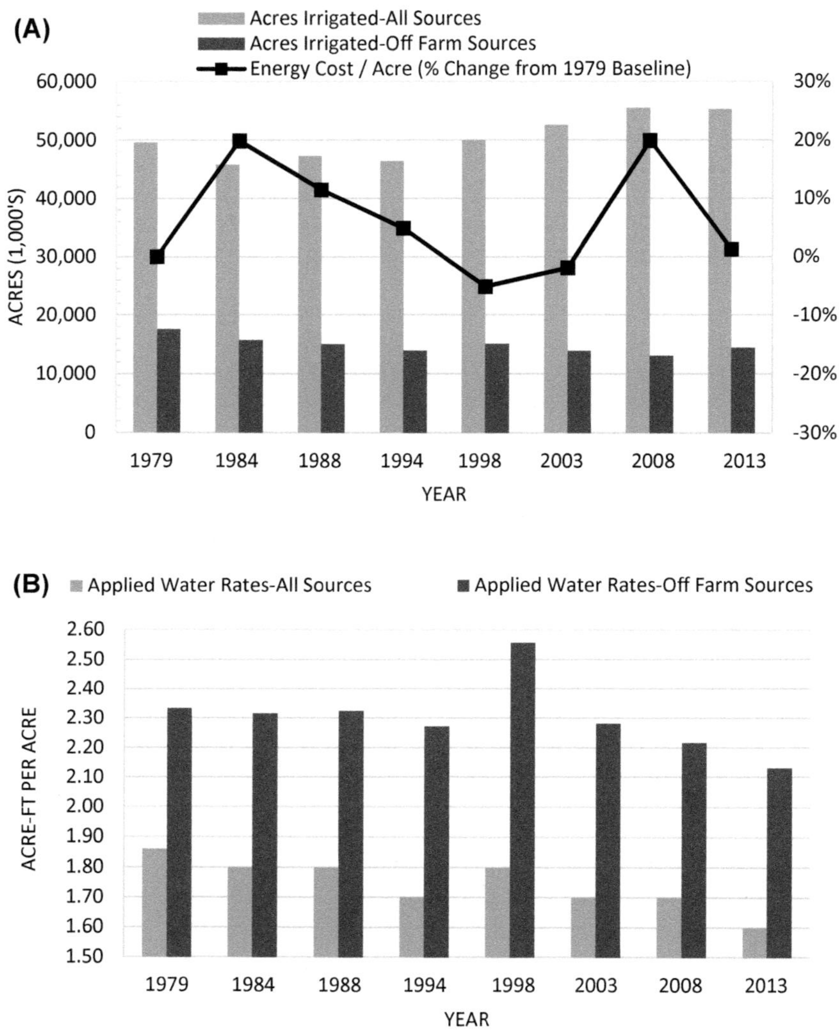


FIGURE 2 Irrigated agricultural trends in the United States at the national level: 1979–2013. (A) Irrigated acreage and energy costs. (B) Applied water rates. Calculated by authors using USDA FRIS data (various years). All prices were transformed into 2013 dollars using the Producer Price Index (USBLS, 2015). On-farm energy costs per acre are measured as a percentage change from 1979 levels.

14.5 million). Over this period, as shown in Fig. 2B, applied water rates for land without access to off-farm water supplies declined by less than 9% while rates for farms with access to off-farm sources declined by nearly 14%. Noticeable in Fig. 2B is the significantly higher application rates for farms with access to off-farm water supplies. As shown next, most of the acreage with access to off-farm water supplies is in the west and thus more reliant on irrigation as a means to meet full crop requirements than elsewhere in the country.

Energy costs per acre (in real, not nominal, terms) are quite variable over the period examined. The 1998 FRIS shows a spike in off-farm applied water rates per acre, coinciding with the year with the lowest per acre energy costs (over 5% less relative to 1979 values). Beyond this, there does not appear to be any clear trend based on these data alone with the value in 2013 estimated as only 1% higher in real terms than the value in 1979. This is not surprising given there are significant differences in both water supply (eg, groundwater, surface water) and energy (eg, diesel, wind, thermoelectric, hydropower, natural gas) prices across the country.

To better appreciate how energy costs per acre vary across regions and time for different energy sources, Table 1 presents USDA FRIS energy costs per acre for four water resource regions—CAL, PNW, AWR, and SAG—and four energy source categories—All,

**TABLE 1** Energy Cost per Acre by US Water Resources Region and Energy Type (1979–2013)[a]

| | | Region | | | |
|---|---|---|---|---|---|
| | Year | CAL | PNW | AWR | SAG |
| **(a) All Sources** | | | | | |
| **Baseline** | **1979** | **$65.77** | **$42.99** | **$48.09** | **$36.69** |
| % Change from baseline | 1984 | 35% | 15% | 18% | 14% |
| | 1988 | 49% | 35% | −5% | −11% |
| | 1994 | 74% | 35% | −4% | −20% |
| | 1998 | 40% | 19% | 0% | −15% |
| | 2003 | 22% | 31% | 35% | −34% |
| | 2008 | 37% | 15% | 21% | 50% |
| | 2013 | 29% | 33% | 19% | 8% |
| **(b) Electricity** | | | | | |
| **Baseline** | **1979** | **$68.18** | **$43.68** | **$48.24** | **$30.45** |
| % Change from baseline | 1984 | 33% | 13% | 26% | 56% |
| | 1988 | 52% | 38% | 2% | 32% |
| | 1994 | 78% | 34% | −4% | 20% |
| | 1998 | 52% | 18% | 13% | 40% |
| | 2003 | 27% | 30% | 0% | 7% |
| | 2008 | 28% | 13% | 55% | 36% |
| | 2013 | 33% | 32% | 25% | 23% |

*Continued*

**TABLE 1** Energy Cost per Acre by US Water Resources Region and Energy Type (1979–2013)[a]—cont'd

| | | Region | | | |
|---|---|---|---|---|---|
| | Year | CAL | PNW | AWR | SAG |
| **(c) Natural Gas** | | | | | |
| **Baseline** | **1979** | **$103.14** | **$37.62** | **$47.76** | **$60.37** |
| % Change from baseline | 1984 | 38% | 11% | 19% | 48% |
| | 1988 | −6% | 50% | −5% | 69% |
| | 1994 | 19% | −10% | 1% | 70% |
| | 1998 | −41% | −32% | 6% | −64% |
| | 2003 | 15% | −100% | 85% | −29% |
| | 2008 | −16% | 81% | 159% | 120% |
| | 2013 | −43% | −31% | −3% | −60% |
| **(d) Diesel** | | | | | |
| **Baseline** | **1979** | **$33.19** | **$22.73** | **$51.43** | **$35.98** |
| % Change from baseline | 1984 | 54% | 67% | 7% | 0% |
| | 1988 | 48% | 4% | −36% | −19% |
| | 1994 | 150% | 52% | −22% | −32% |
| | 1998 | 22% | 4% | −44% | −28% |
| | 2003 | 60% | 41% | −20% | −34% |
| | 2008 | 214% | 130% | 15% | 80% |
| | 2013 | 84% | 83% | −28% | 15% |

*AWR*, Arkansas-White-Red; *CAL*, California; *PNW*, Pacific Northwest; *SAG*, South Atlantic-Gulf.
[a]*Calculated by authors using USDA FRIS data (various years). Baseline estimates all are in 2013 dollars using Producer Price Index (US BLS, 2015).*

Electricity, Natural Gas, and Diesel. The 1979 baseline estimates are presented, followed by the percentage change from the baseline for each of the subsequent FRIS surveys. Two general conclusions can be drawn from Table 1. First, while all regions incurred higher per acre energy costs on average in 2013 relative to 1979, there is significant variability across the region, source, and time. For instance, for all four regions per acre energy costs have decreased for acreage reliant on natural gas, but increased for acreage reliant on electricity. Per acre costs on acre reliant on diesel has increased by the largest percentages—approximately 84% in CAL and PNW. Second, the CAL region incurs consistently higher energy costs per acre than the other regions. Indeed, the largest spikes occur in the CAL region on acreage reliant on diesel for pumping.

## 2.2 Regional Irrigated Acreage, Water Use, and Energy Costs

To better understand how irrigation trends vary across regions, Fig. 3 presents a comparison of trends in irrigated acreage and applied water rates by region. Included on the acreage graphs for each region is a graph of the percent change in energy costs per acre (measured using the right-hand side vertical axis) relative to the 1979 baseline; included on the applied water rates graphs are each region's per acre expenses for irrigation water from off-farm suppliers ($/acre). As shown, while overall irrigated acreage fell in the CAL and PNW regions by 9.3% and 5.5%, respectively, between 1979 and 2013 (Fig. 3A and C), it rose slightly in the AWR region (2.9%) and significantly (by nearly 15%) in the SAG region (Fig. 3E and G). Yet, acreage irrigated with water from off-farm supplies decreased in all regions, falling by 15% in the CAL region and by 43% in AWR region relative to 1979.[11]

Applied water rates varied significantly across the regions. The CAL region has the highest average water application rates per acre, ranging from 2.78 to 3.10 acre-feet per acre, with significantly lower levels in PNW (1.90–2.02), AWR (1.20–1.54), and SAG (0.80–1.60). CAL is the only region with an increase in applied water rates over the 35-year period, with an overall increase of 11.51%, while the SAG region experienced the greatest decrease (~40%). For application rates associated with off-farm water supplies, each region experienced a decline in rates, ranging from around 5% in the CAL region to 23% in the SAG region.

Two potential drivers of irrigation decisions and application rates are the costs of pumping water and the price of water. The black lines in Fig. 3 provide estimates of the trends in (1) the on-farm energy costs per acre for irrigation pumping (top graphs) and (2) the off-farm water costs per acre (bottom graphs) over time. As shown, on-farm energy costs per (irrigated) acre generally increased, an outcome that would be associated with increases in pumping pressure (irrigation system efficiency) or groundwater pumping, either of which are often associated with lower application rates, ceteris paribus. In percentage terms, energy costs per acre increased the most in western regions (~30% in the CAL and PNW regions) and least in the SAG region (~5%). Yet, given the significant variability in real energy costs per acre from one FRIS survey to another, the trends are poor predicators of what to expect from one year to another.

Off-farm water costs per acre by region are presented on the bottom graphs in Fig. 3. As shown, off-farm water costs increased by 57% and 126% in the CAL and SAG regions, respectively, but decreased in the PNW and AWR regions by 20% and 52%, respectively. As shown, however, and similar to the energy costs, there is significant variability across surveys for these estimates. Noticeable in Fig. 3 is the significant drop in off-farm water costs for the CAL, PNW, and AWR regions from 2008 to 2013. Whether such a precipitous drop is because of the recent drought, which leads to more groundwater pumping and less reliance on off-farm sources, the economic downturn in the United States between those periods, or some other factors remains unclear and is certainly worthy of additional attention.

Increases in irrigation efficiency often result in increased energy costs per acre and lower application rates. Fig. 4 presents estimates of the trends in irrigated acreage devoted

11. It should be emphasized that while many of the figures are summarized by presenting a percentage that represents the change in the factor from 1979 to 2013, in many cases there was significant variability across years (as can be seen from the figures). Consequently, the percentages highlighted in the text should not be considered to represent long-term trends.

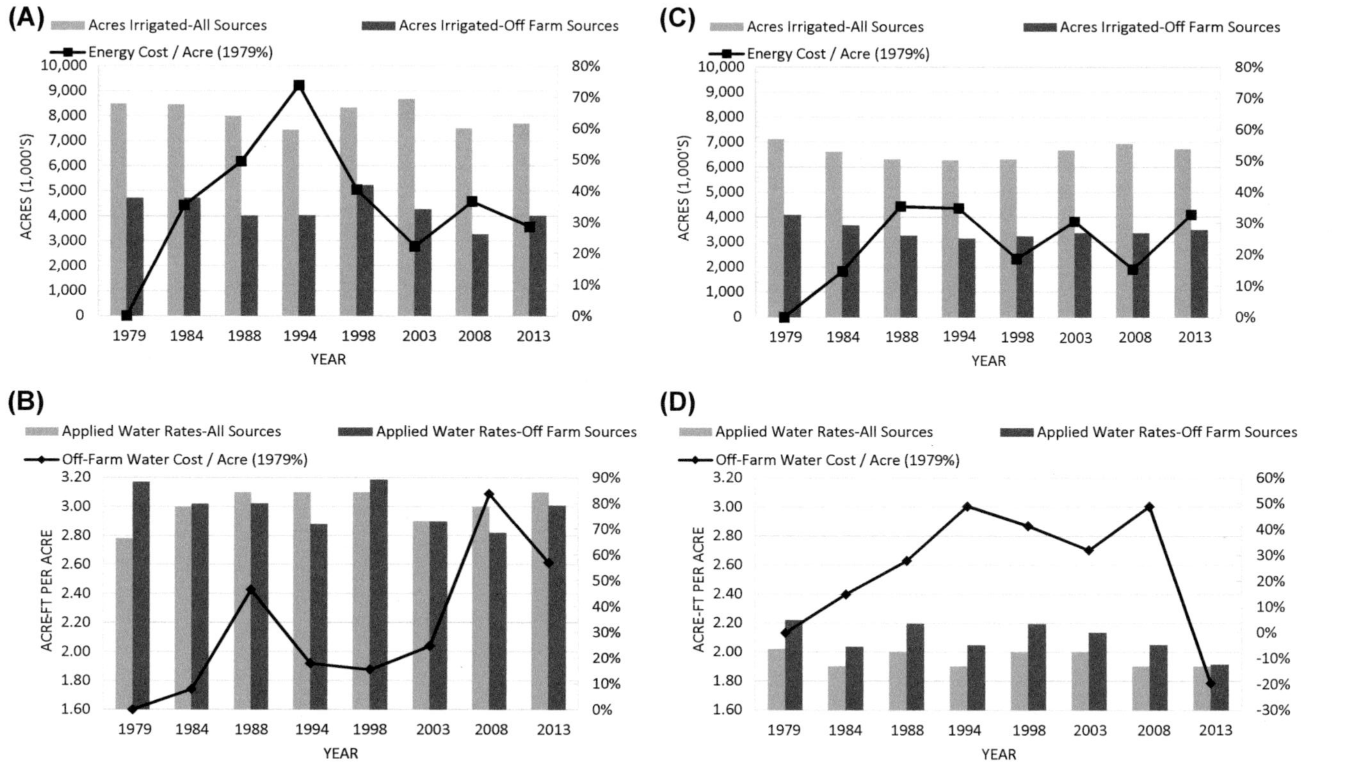


FIGURE 3 Irrigated acreage, water use, and energy costs for selected US regions. (A) Irrigated acreage and energy costs (CAL). (B) Applied water rates and off-farm water costs (CAL). (C) Irrigated acreage and energy costs (PNW). (D) Applied water rates and off-farm water costs (PNW). (E) Irrigated acreage and energy costs (AWR). (F) Applied water rates and off-farm water costs (AWR). (G) Irrigated acreage and energy costs (SAG). (H) Applied water rates and off-farm water costs (SAG). Calculated by authors using USDA FRIS data (various years). *AWR*, Arkansas-White-Red Water Resource Region (WRR); *CAL*, California WRR; *PNW*, Pacific Northwest WRR; *SAG*, South Atlantic-Gulf WR. All prices converted into 2013 dollars using Producer Price Index (USBLS, 2015). Water and on-farm energy costs associated with pumping on a per acre basis with units measured on the vertical left-hand side axis.

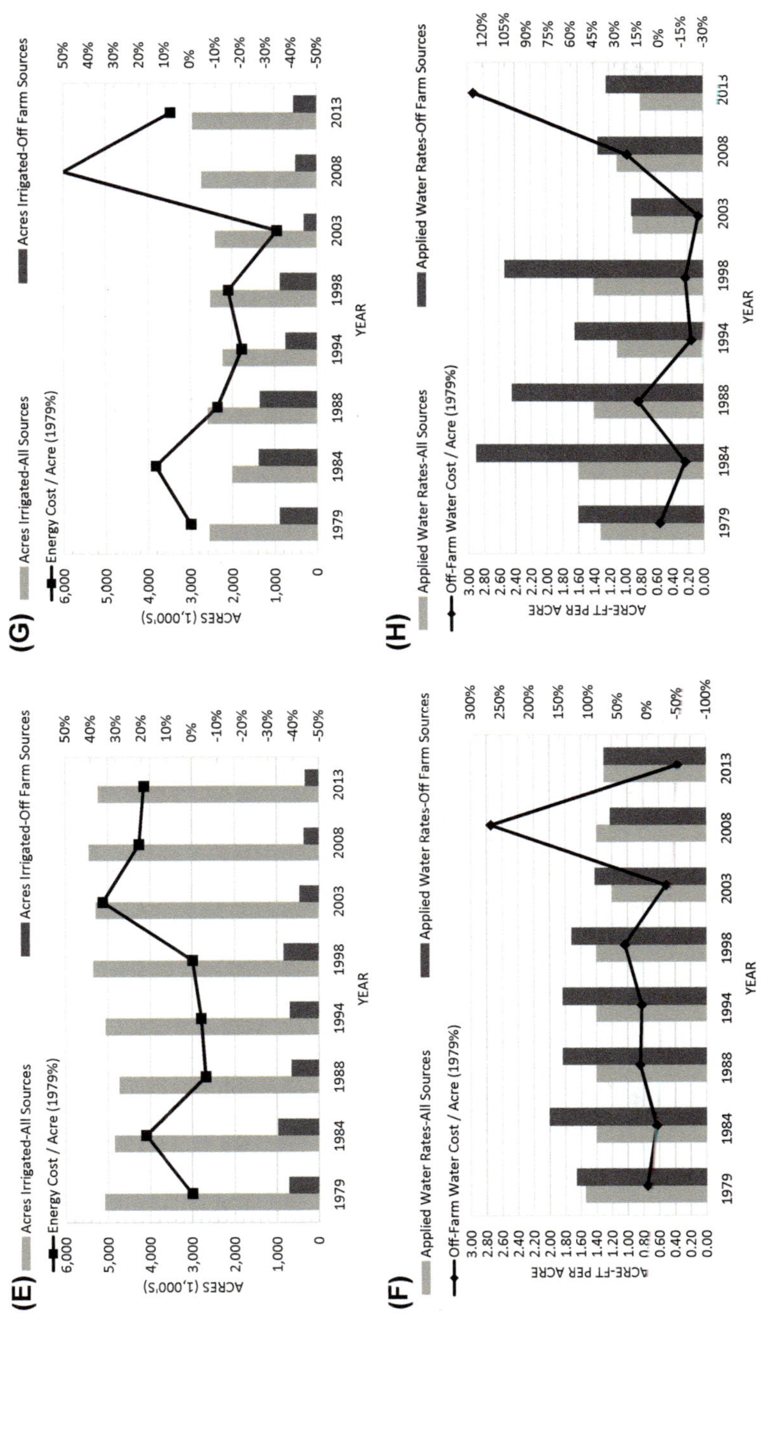


**FIGURE 3** cont'd

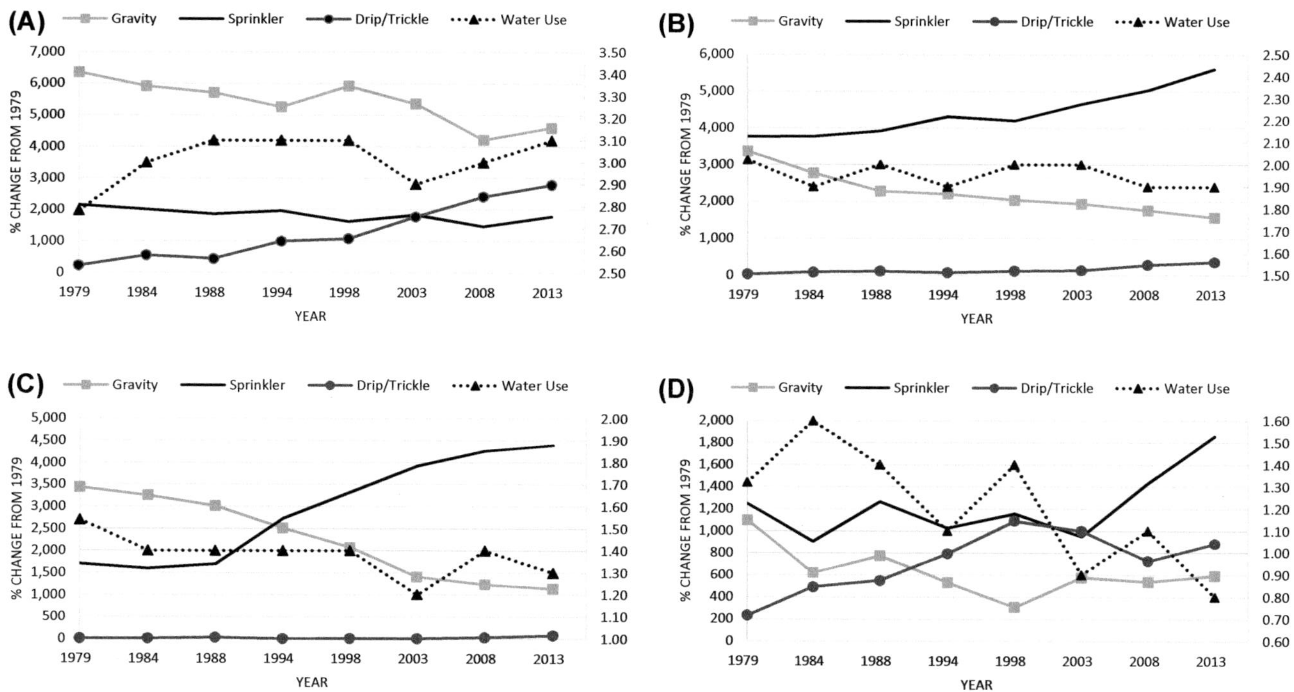


FIGURE 4 Irrigated acreage by irrigation system for selected US regions. (A) CAL WRR. (B) PNW WRR. (C) AWR WRR. (D) SAG WRR. Calculated by authors using USDA FRIS data (various years). *AWR*, Arkansas-White-Red; *CAL*, California; *PNW*, Pacific Northwest; *SAG*, South Atlantic-Gulf; *WRR*, water resources region. Water use is in acre-feet per acre (vertical axis on left-hand side).

to three different irrigation categories: gravity systems, sprinkler systems, and drip or trickle systems.[12] In comparing systems across each of the WRRs, note that in each case the amount of acreage devoted to gravity irrigation declined, from around 28% in the CAL region to nearly 67% in the AWR region. These two regions alone contain around 4 million fewer acres of gravity irrigation as of 2013 relative to the late 1970s. Much of the acreage losses in gravity systems in the AWR region were replaced by acreage under sprinkler systems, which experienced a near 160% increase. Indeed, other than the CAL region, which experienced about a 17% decline in sprinkler acreage, each of the other regions saw acreage under sprinklers increase. In terms of acreage under drip, trickle, or low-flow microsprinkler, each region experienced an increase, with the CAL region adding over 2.5 million acres and the SAG region adding about 650,000 acres.

Each graph in Fig. 4 also includes estimates of water use per acre over the eight FRIS periods. Water use per acre is often a measure of efficiency and/or intensification. Based on that definition alone, the PNW, SAG, and AWR regions have experienced water use efficiency improvements ranging from around 6% (PNW) to 40% (SAG) as measured by declining applied water rates since the late 1970s. Surprisingly, the CAL region, which underwent an overall decrease in irrigated acreage by over 9% yet invested significantly in high-efficiency irrigation systems (eg, drip, trickle), experienced an increase in applied water rates by approximately 12%. While the figures are not shown, the applied water rates for the drip and trickle systems in California decreased by over 10%, but rates for sprinkler and gravity systems have increased by 8% and 5.6%, respectively, since the late 1970s. Furthermore, as illustrated in Fig. 3, overall applied water rates for farms with access to off-farm water supplies decreased by nearly 9%—for a total reduction of over 10 MAF per year—in the CAL region. Obviously, changes in crop type might be driving this result as California has invested heavily in almond and other perennial crop acreage in recent years. For instance, between 1980 and 2003, acreage devoted to almonds and wine grapes increased by 69% and 116%, respectively (Lobell et al., 2006).[13]

## 3. IRRIGATED AGRICULTURAL PRODUCTION MODEL: AN EMPIRICAL INVESTIGATION

For an evaluation of the interactions between water and energy within an irrigated agricultural context, a hydroeconomic model of irrigated agricultural production is developed. Conceptually, the model consists of an irrigated agricultural region overlying an aquifer. Growers have access to two sources of water—surface water and groundwater. A portion of applied irrigation water goes to crop evapotranspiration (ET). Irrigation system nonuniformities and salt leaching requirements, however, generate deep percolation flows that add to the aquifer. In addition to deep percolation flows from both surface water and

12. In the 2013 FRIS the drip and trickle category also included low-flow microsprinkler acreage. Gravity systems refer to traditional irrigation methods such as furrow or flood systems in which water is applied at the top of the field and then is distributed by gravity over the remainder of the field. Sprinkler systems might include solid-set or linear move systems in which the system moves across the field with nozzles emitting over a limited range. Drip systems involve plastic tubing to deliver the water to very precise points. These systems have roughly increasing capital requirements in the order given.
13. Later we consider a coupled production/aquifer model with variable crop areas, irrigation systems, and applied water rates to investigate underlying determinants of these trends.

groundwater applications, additions to the aquifer include natural recharge and conveyance losses from surface water imports. Conversely, in addition to natural leakage, the aquifer water table height declines with groundwater pumping (extractions) for irrigation. Major energy costs associated with irrigated agriculture in the region are associated with conveying surface water and pumping groundwater.

Empirically, our application focuses on Kern County, California, with approximately 1 million acres of irrigable farmland. The aquifer in Kern County can be considered a poorly regulated common property resource given there are over 2000 individual irrigators pumping from the aquifer and for which there is little monitoring and organized management.[14] Because of environmental concerns and recurring drought over the past 30 years, surface water deliveries for irrigation within the region have been significantly reduced. Groundwater levels have been reduced substantially over the past three decades as well. The challenges confronting Kern County, then, are not uncommon elsewhere as most irrigated agricultural regions worldwide are threatened by lower surface water supplies and overdrafted aquifers. These aquifers are often poorly managed and monitored, essentially being treated as a common property resource. Consequently, exploring how changes in energy costs or surface water availability affect irrigated agricultural production and water management can be useful in understanding issues surrounding groundwater use and irrigated agriculture both in California and elsewhere.

The general irrigated agricultural production model is developed next. Following this, the dynamic economic–hydrologic model of irrigated agricultural production is used to investigate the implications of changes in energy costs associated with both surface and groundwater supplies, reductions in surface water availability, and changes in the biophysical characteristics of the system on groundwater use, water table height, and annual net benefits. The last section concludes with a summary of the qualitative and quantitative insights.

## 3.1 Model

The regional agricultural production model developed next follows Kan et al. (2002) for the crop–water production functions, and Knapp et al. (2003), Schwabe et al. (2006), and Knapp and Schwabe (2015) for the regional programming model. Annual net benefits from crop production in year $t$ are defined in Eq. (1).

$$\pi_t = \sum_j \sum_k \pi_{jkt} x_{jkt} - c_{gw}\left[h_t, q_{gt}\right] - c_{sw}[q_{st}] \tag{1}$$

Indices $j$ denotes crop type, $k$ denotes irrigation system type, $s$ denotes surface water, and $g$ denotes groundwater, and per-acre net returns are $\pi_{jkt} = (p_{cj}y_{jkt} - \gamma_{jk} - \gamma_{wjk}w_{jkt})$. Variables are $y_{jkt}$, crop yield; $w_{jkt}$, applied water depth (feet); and $x_{jkt}$, cropped area (acres). Parameters are $p_{cj}$, crop price (\$/ton); $\gamma_{jk}$, nonwater production cost (\$/acre); and $\gamma_{wjk}$, pressurization cost (\$/acre-foot). Surface and groundwater costs, both of which include energy costs, are represented by $c_{gw}$ and $c_{sw}$, respectively, with the former a function of water table height, $h$.

14. While there is progress in California legislation to begin monitoring groundwater use and developing plans for sustainable groundwater use, for most aquifers in this basin, little collective management and monitoring occur presently.

Crop–water production functions were developed in Kan et al. (2002) and used more recently in Knapp and Schwabe (2014) and Schwabe and Knapp (2015). These functions give crop yield and deep percolation flows as a function of applied water depth and salt concentration, and vary by crop, irrigation system, and time. Deep percolation flows, $d_{jkt}$, adhere to mass balance conditions and are equal to the difference between applied water and ET. An underlying production function model is specified utilizing concepts from the soil science, agronomy, and irrigation engineering literature. At the plant level, crop ET as a function of irrigation quantity and quality is derived utilizing a steady-state soil salinity model and crop salinity response functions. Crop yield then depends on ET. Field-scale water infiltration follows a lognormal distribution with standard deviation calibrated to reported Christiansen uniformity coefficients associated with the different irrigation systems. Data for the model come from a wide range of experimental trials. This system is then simulated over a wide range of irrigation water quantities and salinities, and then fit to nonlinear estimating equations. Kan et al. (2002) provide the details. A salinity concentration in the surface and groundwater source is assumed constant and equal to 0.7 dS/m. Land constraints are defined so the sum of crop-irrigation system areas in any period $t$ cannot exceed total land available for irrigated production $\overline{x}$ (million acres):

$$\sum_j \sum_k x_{jkt} \leq \overline{x} \tag{2}$$

Rotational constraints are also imposed on individual crops as:

$$\underline{x}_j \leq \sum_k x_{jkt} \leq \overline{x}_j \tag{3}$$

where $\underline{x}$ and $\overline{x}$ are lower and upper bounds, respectively. Historical ranges over the past 20 years are used to develop these values.

The regional water constraint in this model is defined in Eq. (4):

$$\sum_{j=1}^{nc} \sum_{k=1}^{K} w_{jkt} x_{jkt} \leq q_{st} + q_{gt}. \tag{4}$$

Eq. (4) implies that water use for irrigation is less than or equal to total supply, where total supply is combined surface and groundwater. Surface water deliveries are subject to the constraint that $q_{st} \leq (1 - \beta_s)\overline{q}_{st}$; $\overline{q}_{st}$ is the maximum amount of surface water available to the region.

The equation of motion describing water table elevation, $h_t$, response to extractions and deep percolation flows is defined as:

$$h_{t+1} = h_t + \frac{1}{As^y}\left(\omega + \beta_s q_{st} + \sum_j \sum_k d_{jkt} x_{jkt} - q_{gt}\right) \tag{5}$$

where $A$ and $s^y$ are defined previously, $\omega$ is natural recharge, and $\beta_s$ is the surface water infiltration coefficient. Here the water table rises with surface water imports (canal losses) and deep percolation, and falls with extractions for irrigation. The water table elevation is constrained by $\underline{h} \leq h_t \leq \overline{h}$, where $\underline{h}$ is determined by the lower confining layer and $\overline{h}$ is determined by the rootzone depth. The lower bound limits groundwater extractions to the available supply, while an upper bound limits net deep percolation flows to the maximum storage capacity consistent with maintaining a sufficient rootzone depth for crop production.

## 3.2 Data

The six crops considered are cotton, tomatoes, wheat, lettuce, alfalfa, and Bermuda grass. Cost, price, and production data come from a variety of sources, including Kan et al. (2002), Schwabe et al. (2006), Knapp and Baerenklau (2006), and Knapp and Schwabe (2015). Market prices for each cropping system are derived from county agricultural commissioner crop reports. Nonwater production costs account for planting, land preparation, weed cultivation, fertilizer, and tile and drainage systems. Harvest costs include both a yield-related variable component and a fixed per acre component. Irrigation system data are generally from Posnikoff and Knapp (1996), with adjustments for inflation. The six irrigation systems considered include furrow with ½ mile and ¼ mile runs, low-energy precision application, linear move, sprinkler, and drip. In addition to the crops just mentioned, a minimum of 12% of the regional agricultural land is designated as fallowed and 5% set aside as permanent tree crops.

Water costs, which are represented by the last two terms in Eq. (1), are derived from two sources. Following Knapp et al. (2003), groundwater pumping costs, $c_{gw}(.)$, are defined as:

$$c_{gw}(h, w^g) = (k + e \times \Delta h_{cd})w^g + e\left(\overline{h} - h\right)w^g + e\frac{\left(w^g\right)^2}{2As^y} \tag{6}$$

where $k$ is the average cost per acre-foot of groundwater extraction related to equipment use and is equal to \$15.04, $e$ denotes pumping costs per unit of lift per unit of water and is assigned a value of 0.14, $\Delta h_{cd}$ is additional drawndown because of the cone of depression associated with running the pumps and is equal to 60 ft, and $\overline{h}$ is height of the land surface. $w^g$ is total amount of groundwater pumped, $A$ is aquifer area, and $s^y$ is aquifer-specific yield. The first term on the right side of Eq. (6) captures the O&M costs associated with the well and pump along with the energy costs associated with water table drawdown below the water table surface during pumping. The second and third terms capture the nonlinear energy costs with lifting the water to the land surface from the water table surface. Here the pumping cost parameter $e$ is calculated as an energy requirement to lift a unit of water a unit distance, divided by typical efficiencies associated with physical pumps and then multiplied by the cost of energy per unit of energy. To illustrate the implications of changes in energy-related costs associated with groundwater pumping, we vary the pumping costs, $e$.

For surface water costs, growers in Kern County receive water from a variety of sources, with the costs to growers of surface water varying spatially and temporally within Kern County. For example, Dale et al. (2008) estimate the embedded energy costs and the wholesale water price for agricultural water in the Central Valley, California, based on the conveyance distance of the surface water. For local surface water supplies, the energy cost and water price is \$3/acre-foot and \$14/acre-foot, respectively. For water that is conveyed medium distances, the energy cost and water price is \$29/acre-foot and \$63/acre-foot, respectively, while for surface water that is conveyed from distance sources, the energy costs and water price are $61/acre-foot and $117/acre-foot, respectively. On average across these three sources of water, energy costs comprise approximately 40% of the overall surface water price. The average price over these three sources is approximately $65/acre-foot. A baseline surface water price of $65/acre-foot is assumed. To analyze the implications of changing energy prices as they relate to surface water costs, sensitivity analysis over the fraction of the overall surface water price that is related to energy costs is performed.[15]

Aquifer characteristics for the Kern County empirical application come from Knapp and Olson (1995). The aquifer specific yield ($s^y$) is 0.13, with aquifer lower and upper bounds at −233 ft below mean sea level (msl) and 375 ft above msl, respectively. The surface water infiltration coefficient is 0.3, while natural recharge in the region is 0.052 acre-foot per year. Aquifer area extends beyond the regional irrigated area at 1.29 million acres.

## 3.3 Decision Variables and Management Regime

Given the setup in Section 3.1, the decision variables in the model for each time period $t$ include crop area, irrigation technology, applied water depth of groundwater and surface water quantities for each crop-irrigation system combination, and aggregate surface and groundwater quantities. As mentioned earlier, the aquifer is treated as a common property resource. To model common property usage, a period-by-period optimization framework is chosen in which irrigators select their decision variables to maximized annual net benefits [Eq. (1)] over a unit area based on the water table height at the beginning of the period and the previously mentioned land and water constraints. The analysis is taken over a 100-year period, a time period intended to illustrate in general terms the relationship between energy costs, groundwater use, water table height, and agricultural production.

# 4. REGIONAL GROUNDWATER USAGE: WATER AVAILABILITY AND ENERGY COSTS

Figs. 5 and 6 present the results from an analysis of changes in energy prices and surface water availability on irrigated agricultural profits and management, and water table height. Fig. 5 presents results given an initial aquifer level at 300 ft above msl. Given the aquifer depth ranges from 385 ft above msl to 233 ft below msl, the series of results presented in Fig. 5 is for a relatively full aquifer. Fig. 6, alternatively, presents the results for an initial water table height significantly lower, at msl. Given the concern over depleting aquifers worldwide, being able to compare the results from such an analysis under two different aquifer conditions may be useful.

Within each set of analyses, four scenarios are presented. The baseline scenario (*Baseline*) is the result from the current model as described earlier. The second scenario (*40%EnergyPrice*) consists of increasing the electricity price by 40%. This price increase impacts both surface water costs and groundwater costs (via increased pumping costs). The third scenario (*40%SurfaceWater*) consists of a reduction in the available supply of surface

15. Assuming a $65/acre-foot price for water is within the range identified by Wichelns (2010) as well.

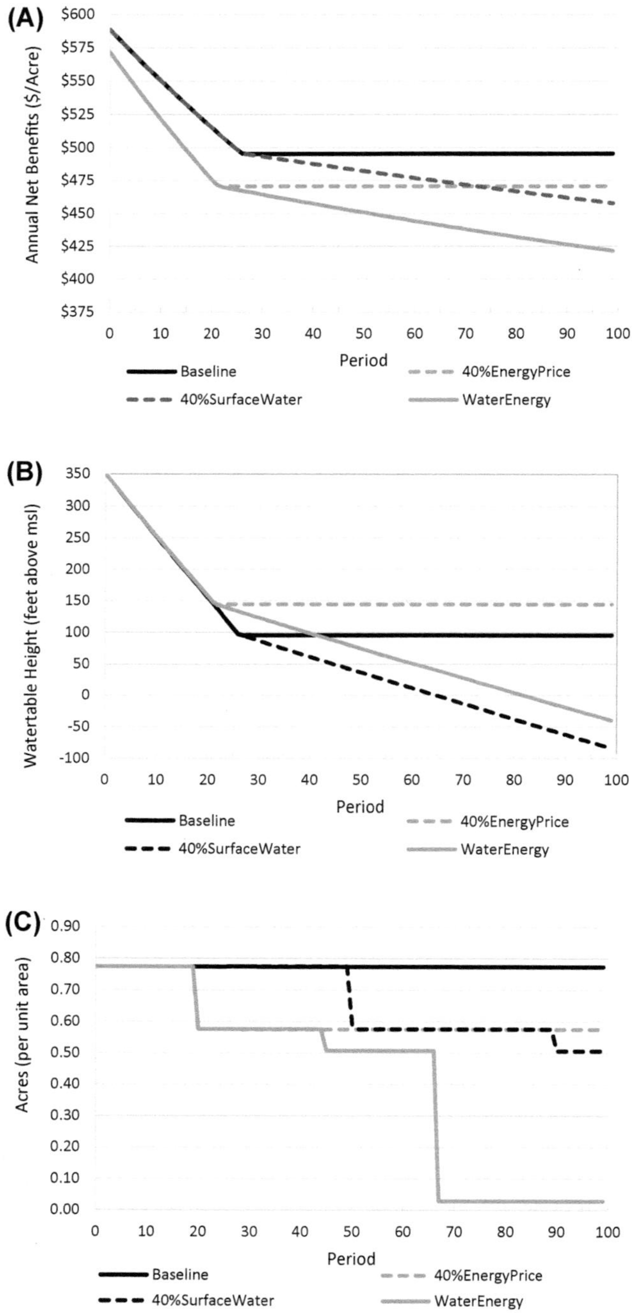


FIGURE 5 Time profiles for an initial high water table level using the Kern County Groundwater Model. (A) Annual net benefits. (B) Water table height. (C) Acreage irrigated using furrow $^1/_2$ mile. (D) Acreage irrigated with surface water. (E) Acreage irrigated with groundwater. (F) Applied water rates with surface water (acre-foot/acre). (G) Applied water rates with groundwater (acre-foot/acre).

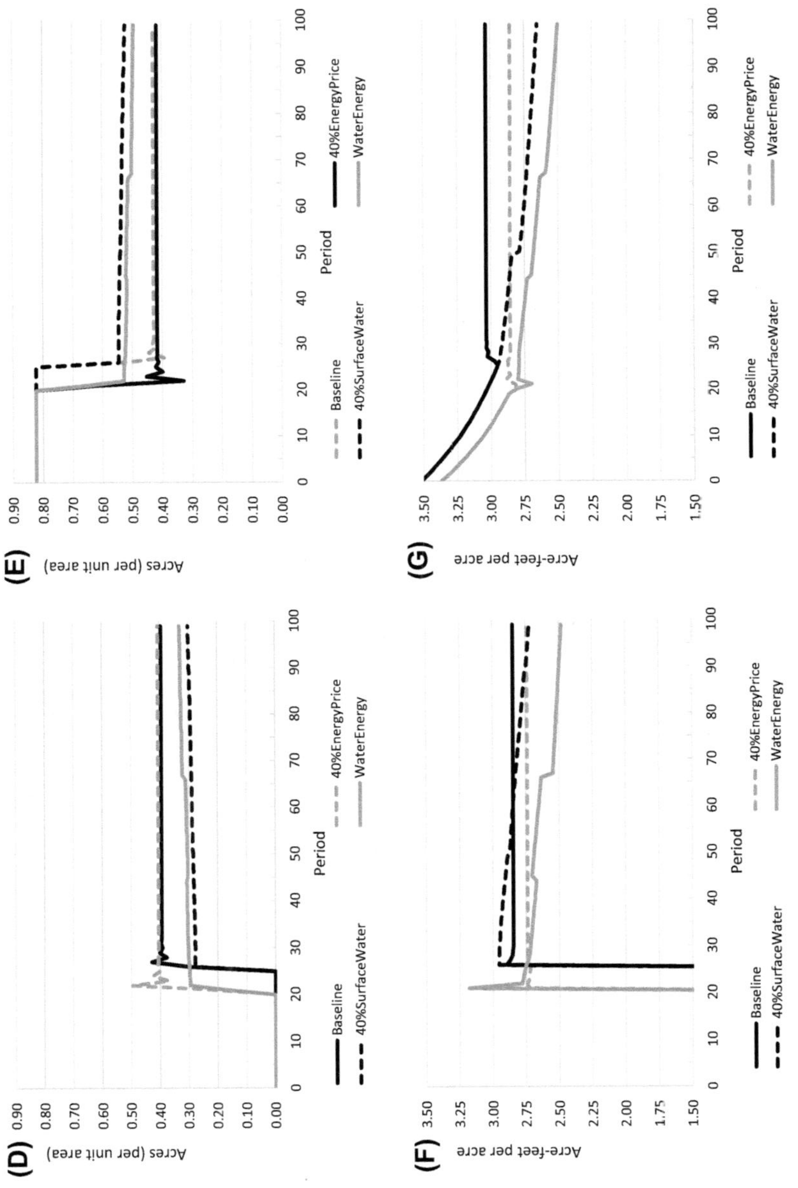


FIGURE 5 cont'd

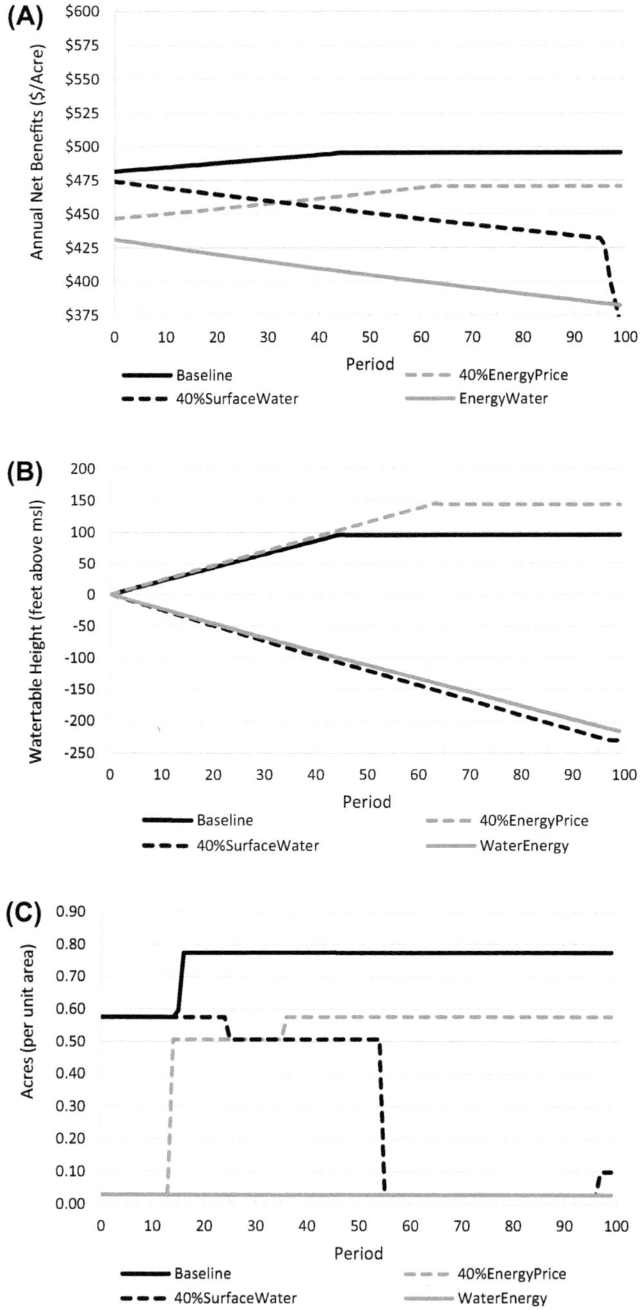


FIGURE 6 Time profiles for an initial low water table level using Kern County Groundwater Model. (A) Annual net benefits. (B) Water table height. (C) Acreage irrigated using furrow $^1/_2$ mile. (D) Acreage irrigated with surface water. (E) Acreage irrigated with groundwater. (F) Applied water rates with surface water (acre-foot/acre). (G) Applied water rates with groundwater (acre-foot/acre).

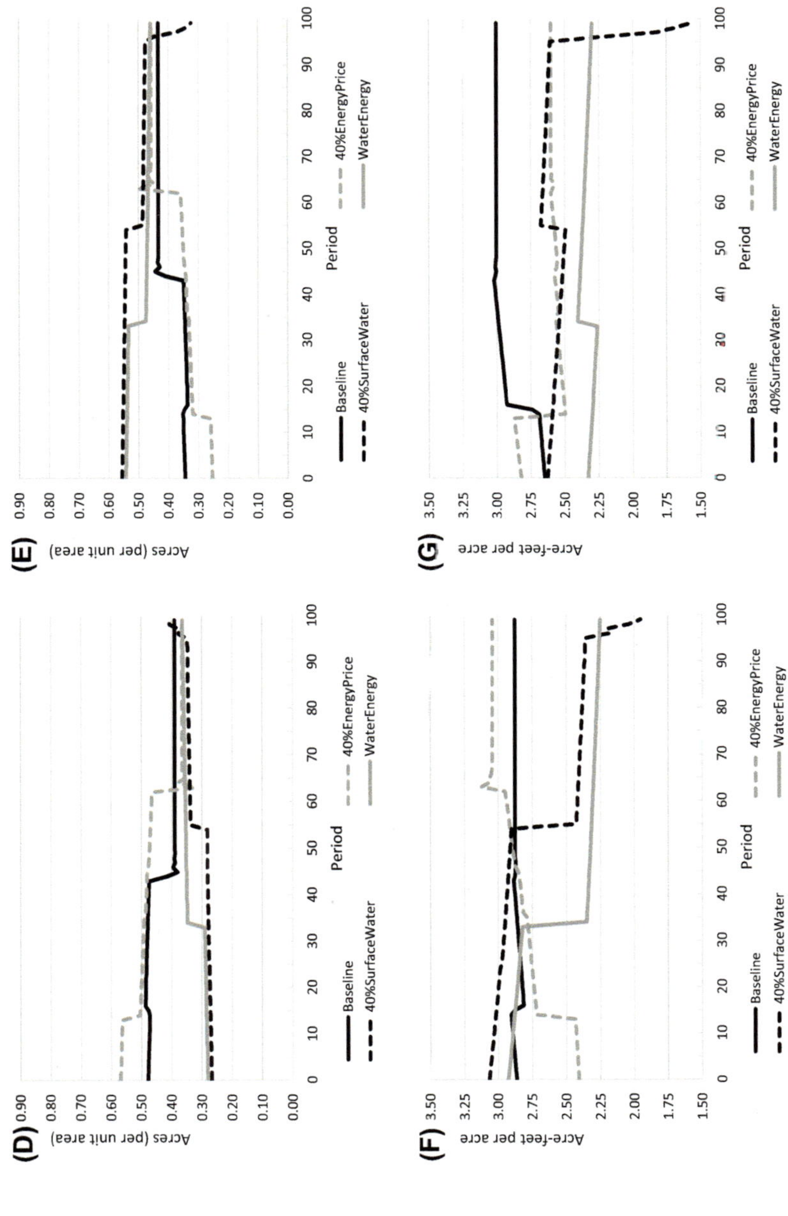


FIGURE 6 cont'd

water by 40% relative to the baseline. The final scenario—(*WaterEnergy*) consists of combining both the second and third scenarios, that is, energy prices increase by 40% and surface water supply reductions by 40%. These scenarios are meant to represent trends that have and will continue to impact irrigated agriculture from an energy and water availability perspective.

## 4.1 Abundant Groundwater Supplies

As shown in Fig. 5A and B, and with an initially high water table height, annual net benefits and water table levels initially decrease under each of the scenarios, with those scenarios that confront lower energy costs producing higher annual net benefits initially. A steady state is reached for both the baseline and the *40%EnergyPrice* scenario at around 20–30 periods. For those scenarios in which surface water quantity is constrained, annual net benefits and water table levels continue to decline, even after 100 years. Obviously, the 40% reduction in surface water supplies poses a greater challenge to irrigated agriculture based on our model than a 40% increase in energy prices.

Fig. 5C illustrates that as the water table is being drawn down, and groundwater pumping costs increase—by over 70% under all scenarios—less acreage is devoted to the least water-efficient irrigation system—furrow ½ mile. So, while initially all acreage is irrigated with furrow ½ mile, energy price increases or surface water supply reductions lead to adoption of more water-efficient irrigation systems. The most significant change, not surprisingly, is when both energy prices increase and surface water supplies are reduced. Under this *WaterEnergy* scenario, less than 5% of the acreage remains in furrow ½ mile.

Fig. 5D–G illustrate how acreage irrigated with surface water and groundwater, and their respective application rates, change over time under each scenario. With a nearly full aquifer, initially all cultivated acreage is irrigated with groundwater given that the pumping costs associated with such a small lift are low.[16] Yet, after about 20 periods, the costs of groundwater reach such a level that the model begins to allocate more acreage to surface water irrigation, first under those scenarios in which surface water supplies are reduced. While surface water irrigation is adopted earlier under the quantity restriction scenarios than the baseline or *40%EnergyPrice* scenario, less surface water acreage is adopted as well. Indeed, for those situations when surface water supplies are more binding, the majority of acreage continues to be irrigated with groundwater.

Fig. 5F and G present the applied water rates for acreage irrigation with surface water and groundwater, respectively. Initially, groundwater application rates are quite high but level out and reach a steady state for the baseline and price-only scenarios at around period 20. But for the scenarios in which surface water availability is reduced, application rates for both groundwater and surface water continue to decline. This is not surprising given that, as shown in Fig. 5C, there is a continual reduction in the amount of acreage irrigated with furrow ½ mile. It is also noteworthy that for the first 20 years, over which there is no change in irrigation system efficiency, applied groundwater rates decline. Clearly, the higher pumping costs create the incentive to reduce groundwater application rates through deficit irrigation.

16. Recall that at least 5% of the land must be fallowed and 12% is in permanent tree crops.

## 4.2 Scarce Groundwater Supplies

Fig. 6 presents a similar analysis of the four scenarios but with a much lower initial water table level. As shown, the results are significantly different, particularly for those scenarios in which surface water availability is not reduced. For those scenarios in which surface water is not further constrained, both annual net benefits and the water table rise. In these scenarios, deep percolation flows add more to the water table than groundwater use extracts. Annual net benefits increase as the pumping costs decrease because of a rising water table. Similar to Fig. 5A and B, the baseline and *40%EnergyPrice* scenario reach a steady state between 30 and 40 periods. Under the two scenarios with surface water reduced by 40%, annual net benefits decline as the declining water table requires increased pumping costs.

Fig. 6C illustrates that with a lower initial water table relative to that in Fig. 5, less acreage is devoted to furrow ½ mile initially. Indeed, for the *WaterEnergy* scenario less than 5% of the acreage is devoted to furrow ½ mile for the entire 100 periods. For the baseline scenario, there is an increase in acreage over time to the less water-conserving irrigation technology, a response likely caused by the rising water table and consequent lower pumping costs. For the *40%EnergyPrice* scenario, furrow ½ mile is also increasingly adopted over time. Yet, for the *40%SurfaceWater* scenario, irrigation system efficiency continues to improve, likely as a result of a declining water table and rising groundwater pumping costs.

Fig. 6D–G illustrate how groundwater and surface water acreage and application rates change over time across the scenarios. When surface water availability is reduced, less acreage is devoted to irrigation with surface water relative to acreage with groundwater, and vice versa. Yet, for those scenarios for which the groundwater table rises over time and, consequently, groundwater becomes relatively less expensive, more acreage is irrigated with groundwater and less with surface water over time. Similarly, as the water table decreases and costs of pumping groundwater increase, less acreage is irrigated with groundwater and more with surface water over time. In terms of water application rates, a similar trend occurs. As the water table declines, both surface water and groundwater application rates decrease; conversely, as the groundwater table rises, application rates for both water sources rise. As expected, as the price of energy increases—under both the *40% SurfaceWater* and *WaterEnergy* scenarios—water application rates are lower. Finally, comparing irrigation system adoption with water application rates illustrates a one-to-one qualitative correspondence.

# 5. CONCLUSIONS

The relationship between water and energy presents an interesting challenge for irrigated agriculture. As the demand for food and agricultural products rises, and given the yield benefits from irrigated agriculture relative to rain-fed agriculture, it is likely that society will demand more from irrigated agricultural production. Ceteris paribus, more food production requires more water and more energy. Meeting societal agricultural demands through increases in the "scale" of input use alone, however, is unlikely given current and future levels of water scarcity relative to past periods. Alternatively, increases in food production will need to be met through changes in the efficiency in which water is used (ie, the intensity effect) and perhaps in changing the basket of goods that are consumed/produced (ie, the composition effect). Yet, intensification often requires more energy. With energy often a competitor with agriculture for water, coupled with concerns over fossil fuel use, changes in the energy sector and future energy costs will present unique challenges to this strategy as well.

What makes this problem even more complicated, if not interesting, is how water and energy are intricately related to one another. That is, significant amounts of water are required to produce energy, and significant amounts of energy are required for the conveyance and movement of water. While there is significant insight into how agricultural production and sustainability can or will persist with changes in either water or energy inputs, they really are intricately intertwined in agricultural production. Consequently, agricultural as well as energy, water, and environmental policymakers will increasingly need to be cognizant of these intricacies and complexities if they intend to develop effective policies.

In an effort to provide some additional clarity and understanding of these relationships as they pertain to irrigated agricultural production in the United States, this chapter has highlighted trends in irrigated agricultural production—water use, acreage, irrigation system adoption, and both energy and water costs—over the past 35 years. Overall irrigated acreage has increased slightly since the late 1970s, and there has been a significant increase in the adoption of more efficient irrigation systems. More efficient irrigation systems are likely one reason for the overall decline in water application rates. A decrease in the acreage irrigated with off-farm water sources may be another given that applied water rates are shown to be higher on average for farms that have access to off-farm water sources than farms that do not have such access. Environmental restrictions on the import from such sources and rising energy costs associated with electricity could be factors motivating such changes. While overall energy costs per acre in the United States are nearly what they were 35 years ago (in real terms), this result overlooks two important factors. There is significant variability in energy costs over time and there is significant variability across regions and energy source. For instance, in CAL and the PNW regions, energy costs per acre for irrigation reliant on diesel fuel have increased by around 84% relative to 1979 values, but for acreage reliant on natural gas, such costs have decreased by over 30%. One consistent result across the four regions analyzed is that the per acre costs for acreage irrigated using electricity have risen.

This chapter also provided a dynamic analysis—using an economic hydrologic regional programming model of irrigated agricultural production—of how changes in energy prices and surface water restrictions influence net benefits, groundwater levels, and irrigation management over time under two different groundwater initial conditions. Reductions in the availability of surface water supplies was shown to have significant impact on water management and groundwater levels. Relative to a change in energy prices alone, groundwater tables declined more rapidly, energy costs rose more precipitously (because of increased pumping), and the adoption of more efficient irrigation systems took place sooner. Given that there were significant opportunities to change water application rates, switch irrigation efficiencies, and change crop mix, increases in energy prices alone impacted annual net benefits, but steady-state solutions were reached relatively quickly (as compared to the scenarios that included surface water supply reductions). Finally, as groundwater levels become more depleted, we can expect energy costs to rise, quicker and wider adoption of more efficient irrigation systems, and greater incentives to utilize surface water supplies. Obviously, if the surface water supplies are limited, or are responsible for the greater reliance on groundwater supplies, the impacts on irrigated agriculture will increase. Overall, then, while irrigated agricultural acreage continues to expand in the United States and responds with increased on-farm efficiencies, greater competition for increasingly scarce water resources and concerns over energy production and greenhouse gas emissions portend continual challenges.

## REFERENCES

California Energy Commission (CEC), 2005. California's Water-Energy Relationship. Final Staff Report (CEC-700-2005-011-SF). Sacramento, CA.

Carruthers, I., Rosegrant, M., Seckler, D., 1997. Irrigation and food security in the twenty-first century. Irrigation and Drainage Systems 11, 83–101.

Dale, L., Fujita, S., Hagan, J., Hanemann, W.M., 2008. The interaction of water and energy in California: climate change and price impacts. In: Energy Analysis and Environmental Impacts Division. Berkeley Lab. University of California, 37 pp.

Foley, J.A., et al., 2011. Solutions for a cultivated planet. Nature 478 (7369), 337–342.

Harman, W., 1987. Management systems for improving on-farm energy and water use efficiency. In: Whittlesey (Ed.), Energy and Water Management in Western Irrigated Agriculture. Westview Press, London, pp. 129–154.

Kan, I., Schwabe, K.A., Knapp, K.C., 2002. Microeconomics of irrigation with saline water. Journal of Agricultural and Resource Economics 27, 16–39.

Knapp, K.C., Baerenklau, K.A., 2006. Groundwater quantity and quality management: agricultural production and aquifer salinization over long time scales. Journal of Agricultural and Resource Economics 31 (3), 616–641.

Knapp, K.C., Olson, L., 1995. The economics of conjunctive groundwater management with stochastic surface supplies. Journal of Environmental Economics and Management 28, 340–356.

Knapp, K.C., Weinberg, M., Howitt, R., Posnikoff, J., 2003. Water transfers, agriculture, and groundwater management: a dynamic economic analysis. Journal of Environmental Management 67, 291–301.

Knapp, K.C., Schwabe, K., 2015. Groundwater use and irrigated agriculture in California: dynamics, uncertainty, and conjunctive use. In: Burnett, K., Howitt, R., Rousmasset, J., Wada, C. (Eds.), The Handbook of Water Economics and Institutions. Routledge Press, New York, pp. 134–162.

Lacewell, R., Collins, G., 1987. Energy inputs on western groundwater irrigated areas. In: Whittlesey (Ed.), Energy and Water Management in Western Irrigated Agriculture. Westview Press, London, pp. 155–175.

Lobell, D.B., Bala, G., Duffy, P., 2006. Biogeophysical impacts of cropland management changes on climate. Geophysical Research Letters 33, L06708. http://dx.doi.org/10.1029/2005GL025492.

Lockeretz, W., 1977. Agriculture and Energy. Academic Press, New York.

Lotze-Campen, H., Muller, C., Bondeau, A., Rost, S., Popp, A., Lucht, W., 2008. Global food demand, productivity growth, and the scarcity of land and water resources: a spatially explicit mathematical programming approach. Agricultural Economics 39, 325–338.

Maupin, M., Kenny, J., Hutson, S., Lovelace, J., Barber, N., Linsey, K., 2014. Estimated Use of Water in the United States in 2010. U.S. Geological Survey Circular 1405, 56 pp. http//dx.doi.org/10.3133/cir1405.

Miranowski, J., 2005. Energy consumption in US agriculture. In: Agriculture as a Producer and Consumer of Energy. CABI Publishing, Cambridge, MA, pp. 68–111.

Posnikoff, J., Knapp, K.C., 1996. Regional drainwater management: source control, agroforestry, and evaporation ponds. Journal of Agricultural and Resource Economics 21 (2), 277–293.

Rosegrant, M., Cai, X., Cline, S., Nakagawa, N., 2002. Role of rainfed agriculture in the future of global food production. In: EPTD Discussion Paper No. 90. International Food Policy Research Institute, Washington DC.

Schwabe, K., Kan, I., Knapp, K.C., 2006. Integrated drainwater management in irrigated agriculture. American Journal of Agricultural Economics 88 (1), 133–149.

Schwabe, K., Knapp, K.C., 2015. Salinity and groundwater management: a hydro-economic analysis. In: Dinar, A., Schwabe, K. (Eds.), The Handbook of Water Economics. Edward Elgar.

United States Bureau of Labor Statistics (USBLS), 2015. Producer Price Indexes. http://www.bls.gov/ppi/tables.htm.

United States Department of Agriculture (USDA), 1979. Farm and Ranch Irrigation Survey (FRIS). Economic Research Service, Washington, DC.

United States Department of Agriculture (USDA), 1984. Farm and Ranch Irrigation Survey (FRIS). Economic Research Service, Washington, DC.

United States Department of Agriculture (USDA), 1988. Farm and Ranch Irrigation Survey (FRIS). Economic Research Service, Washington, DC.

United States Department of Agriculture (USDA), 1994. Farm and Ranch Irrigation Survey (FRIS). Economic Research Service, Washington, DC.

United States Department of Agriculture (USDA), 1998. Farm and Ranch Irrigation Survey (FRIS). Economic Research Service, Washington, DC.

United States Department of Agriculture (USDA), 2003. Farm and Ranch Irrigation Survey (FRIS). Economic Research Service, Washington, DC.

United States Department of Agriculture (USDA), 2008. Farm and Ranch Irrigation Survey (FRIS). Economic Research Service, Washington, DC.

United States Department of Agriculture (USDA), 2013. Farm and Ranch Irrigation Survey (FRIS). Economic Research Service, Washington, DC.

Waskom, R., Akhbari, M., Grigg, N., 2014. U.S. Perspective on Water-Energy-Food Nexus. Colorado Water Institute Information Series No. 116. Colorado State University.

Whittlesey, N., 1987. Energy and Water Management in Western Irrigated Agriculture. Westview Press, London.

Wichelns, D., 2010. Agricultural Water Pricing: United States. Hanover College, United States. Available at: www.oecd.org/water, 27 pp.

Chapter 2.1.3

# The Water—Energy Nexus in Europe and Spain: An Institutional Analysis From the Perspective of the Spanish Irrigation Sector

**S. Villamayor-Tomas**
*Thaer Institute, IRI THESys; Humboldt University of Berlin, Berlin, Germany*

## 1. INTRODUCTION

The water—energy nexus approach is becoming increasingly popular among practitioners and scholars, as illustrated by the number of publications devoted to the topic in recent years (Hellegers et al., 2008; Hussey and Pittock, 2012; Waughray, 2011). Part of the popularity of the approach responds to the production and consumption tradeoffs that have emerged with the increase of scarcity and competition over the last few decades. The nexus highlights the need to study the use and management of both resources together. Although sectors such as water and energy can be analyzed independently, doing so would overlook the multiplicity of feedbacks and interdependencies that jointly affect their sustainability.

There are some excellent reviews of the water—energy nexus in European countries. In Spain the works of Hardy et al. (2010, 2012); Hardy and Garrido (2012) and Corominas (2010) have set an important precedent. Both sets of studies distinguish the "energy for water" perspective from the "water for energy" perspective. As they report, approximately 5.8% of total electricity demand in Spain is because of the water sector. Extraction and water treatment are the most electricity demanding stages of the water cycle, accounting for 64% of the total demand. Irrigated agriculture is one of the Spanish water sectors that show the largest growth in energy requirements, accounting for 40% of total water-related electricity demand. On the other hand, the energy sector accounts for only 3.2% of the total water usage, but in terms of extracted volumes it reaches 25% of the total water withdrawn. Excluding hydropower, which is by far the main source of water withdrawals (24,400 MCM in 2008), nuclear power accounts for 50% of water withdrawals by the energy sector, followed by thermal (30%) and gas (15%).

**Competition for Water Resources. http://dx.doi.org/10.1016/B978-0-12-803237-4.00006-9**

This chapter aims to complement existing reviews of the water–energy nexus in Spain in three ways. First, the text offers a broad characterization of energy and water institutions at multiple levels, from the European down to the local level. Second, the chapter provides a historical perspective to governance in the two sectors, including important reforms that took place from the late 1990s to the late 2000s. Last but not least, the chapter sheds light on the cross-scale interaction of water and energy institutions over time, through a focus on the use of energy for water extraction and allocation in the Spanish irrigation sector.

The Spanish water governance system is internationally renowned for the long-standing governmental promotion of water infrastructure, the organization of water management into river basin organizations (RBOs), and the high autonomy of irrigation communities (ie, associations). The roots of the system can be traced back to the early 20th century, when, motivated by the economic crisis, the government took over the promotion of irrigation and hydropower generation via investments in infrastructure. So-called hydroagricultural plans (Perez Picazo and Lemeunier, 2000) and a series of RBOs responsible for their implementation became the centerpieces of the policy. This water supply paradigm started to be challenged in the 1990s, with the push for so-called demand management measures seeking water efficiency and water savings. As further explained in this chapter, the European Water Framework Directive (EWFD) consolidated that trend and pushed it forward.

The Spanish electricity sector has also gone through changes over the last decade, including a market transition and a renewable energy transition. Until 1997, the electricity system was regulated. The government fixed the electricity price based on generation, transport, and distribution costs from a set of private electricity firms. The market reform was launched at the European and national levels by the end of the 1990s. It suppressed part of the regulatory role of the government and introduced competition and a market-based energy pricing system in the generation and commercialization stages of the chain. The National Market and Competition Commission became the body responsible for competition and transparency of the pricing system. Electricity prices were supposed to decrease over the long run; however, they have tended to increase (Robinson, 2014). The renewable transition unfolded in parallel to the market transition. The enormous financial effort made by the government put Spain at the forefront of renewable electricity generation worldwide; however, this effort has also created finance issues in the sector.

The Spanish irrigation sector represents almost a third of the total irrigated area in the European Union. This is a result of both the climatological conditions of the country that make water a naturally scarce resource, as well as the previously mentioned public investments in the promotion of irrigation infrastructure (López-Gunn et al., 2012). Irrigation in Spain accounts for 60% of total agricultural production and 80% of total farm exports, but also consumes 70% of total water resources in a typical year. The crisis of the water supply paradigm translated in the irrigation sector to new infrastructure investments, including improvements in existing infrastructure as well as the introduction of sprinkler and drip irrigation technologies.

As mentioned, the irrigation sector is also an important consumer of energy. By 2013, energy consumption in the sector was slightly below 6670 GWh (Berbel et al., 2014). Approximately 70% of the energy consumed is electricity and the rest is diesel. Energy in irrigation systems is for the most part used to pump groundwater and distribute it through sprinkler and drip irrigation systems. In 2008, the government suppressed the preferential tariff that irrigation communities had enjoyed in the past and pushed them to acquire electricity at market prices, which were by then considerably higher than the tariff. The 2008 shock created a still unsolved financial crisis in the sector. This chapter focuses on the crisis in the irrigation sector to shed light on the water–energy nexus.

Many nexus studies begin by identifying two or more resources that are often broadly defined in terms of water or energy within a particular geographic or sociopolitical boundary. They then continue by quantifying resource use and the ways in which the use of one resource affects the other to affect the sustainability of the interconnected sectors. Few studies consider the role of institutions as the focus of their analysis. This is not a trivial omission, as one of the central questions for nexus scholars is to assess the social and ecological effects of alternative policies (Scott et al., 2011).

The institutional analysis proposed in this chapter aims to contribute to fill that gap, in an attempt "to move the water–nexus construct beyond an input–output relationship into the realm of resource governance" (Scott et al., 2011). From an institutional perspective, policies are understood as amalgams of institutions that shape either directly or indirectly the use and production of water and energy resources and thus potentially mediate the emergence of tradeoffs and synergies across different production chains (Villamayor-Tomas et al., 2015). Institutions are viewed as particularly important as they structure the incentives that actors face when they make choices from among a set of alternatives. These institutions are themselves the result of social processes and "consist of both informal constraints (sanctions, taboos, customs, traditions, codes of conduct), and formal rules (constitutions, laws, property rights)" (North, 1990).

The study presented here follows a sort of before and after design. Water and electricity policies are reviewed at the European, national, and local levels, before and after the 2008 price shock in the irrigation sector. Data come from two sources. On the one hand, the chapter is a synthesis of a number of studies and reviews on the water–energy nexus in Spain and in the irrigation sector. On the other hand, the chapter uses new data, mostly coming from governmental and stakeholder documents, press, and informal interviews, on institutional developments that followed the 2008 shock.

## 2. WATER AND ENERGY REFORMS IN THE EUROPEAN UNION AND SPAIN: A LOOK FROM THE TOP DOWN

Current affairs in the Spanish irrigation sector can be understood by looking at water and energy policy and the reforms launched at the European and national levels.

### 2.1 Water Reform in Europe and Spain

The EWFD, approved in 2000, embodies the most recent and overarching water management reform across European countries. One of the central tenets of the EWFD is the principle of cost recovery of water services, including environmental and resource costs (Gómez-Limón et al., 2002). The inclusion of environmental costs in the cost recovery principle is not trivial. According to the Directive, the costs of investments in maintaining the good ecological status (GES) of water bodies should not severely outweigh the benefits. This has faced a number of implementation challenges because of lack of clear definitions of GES and appropriate methods to calculate environmental benefits and costs (Martin-Ortega, 2012).

The EWFD was transposed to the Spanish system in 2003. The transposition was not particularly smooth. The principle of cost recovery generated certain resistances given the traditionally strong role played by government in the promotion of water infrastructure (La Roca, 2011). According to the Spanish rule, hydrologic plans would need to include estimates of water service costs, environmental costs, water service revenues, planned investments, and a balance of cost recovery. Although this has been accomplished, there are issues related to a number of exceptions to the cost-recovery prescription and the methodology used by RBOs to assess costs and revenues (La Roca, 2011).

## 2.2 Energy Liberalization and Diversification Reforms in the European Union and Spain

The EU regulations to create a single energy market have exerted notable influence over the energy sector of Member States. By the early 1980s, neoliberal arguments about the benefits of competition and privatization and the development of more efficient generation technologies opened a window of opportunity for European reformists to promote a liberalization reform across Member States (Serrallés, 2006). In 1996, a Franco-German deal paved the way for the approval of an Electricity Directive (EC, 1996). The Directive called for an increase in efficiency via the unbundling and open access to the transmission and distribution networks "while reinforcing security of supply and the competitiveness of the European economy and respecting environmental protection" (Preamble 4).

In 2003, a new Directive (EC, 2003) further expanded the reform, requiring legal unbundling of network activities from generation and supply, establishing a regulator in all Member States, imposing transparent network tariffs, and establishing deadlines for opening the electricity market to all nonhousehold consumers in 2004, and to all consumers in 2007 (Jordana et al., 2006)

In Spain, Law 40/1994 mandated the legal unbundling of the transmission network and created an independent joint public–private operator to regulate its use called Red Electrica Española. Regarding generation, the law did not create maximum market share for generating companies, which encouraged firms to start a process of consolidation. From over 35 independent regional generation companies in operation in 1990, only five were left by 2002. The three largest companies in 2005 (ENDESA, Iberdrola, and Union Fenosa) accounted for 83% of the installed generation capacity (Serrallés, 2006). A new electricity law in 1997 and an agreement between the Ministry of Industry and Energy and the electricity sector in 1998 accelerated the liberalization process even beyond the EU minimum requirements (BOE, 1997). Since 2003, all consumers have been eligible to freely choose their supplier.

European environmental policy has also exerted notable influence over energy sector reforms of Member States. In a context of increasing concern about climate change, the development of renewable energy has become one of the central objectives of the European Commission (Míguez et al., 2006). In 2001 and 2003, two directives set the path for the transition to renewables, suggesting that Member States generate at least 22% of total electricity consumed and around 6% of all fossil fuel-primary energy consumed from renewable sources. In 2007 the European Council approved the so-called decision 20-20-20, which prescribed a 20% share of renewables in electricity consumption, a 20% reduction of greenhouse gases, and an increase in 20% in energy efficiency. This commitment was integrated in a new Renewable Energy Directive in 2009 (EC, 2009), which also prescribed the approval of Action Plans by Member States to accomplish those goals.

In Spain the promotion of renewables can be traced back to the 1986 Renewable Energy Plan and two National Energy Plans (1991–2000 and 2000–2003 editions). The 1997 electricity law and the Program for the Encouragement of Renewable Energy in Spain (2000–2010) set a target of 12% of primary energy consumption from renewables by 2010 (BOE, 1997). In 2005 a new National Plan of Renewable Energy integrated updates from EU regulations, increasing the renewable share of electricity consumption to 30% and setting a 6% biofuel share in the transportation sector. The public investment from 2005 to 2010 was notable, including subsidies to investments (€198M), fiscal incentives (€1116M), and price premiums (€5572M) for a total of €6886M. Private investments for the same reached €36462M (AEVAL, 2011). By 2010 the share of renewables was 11.3% for primary energy and 30.6% for electricity consumption. Most notably, the share of renewables in the electricity market surpassed that of nuclear electricity, and the share of wind production was twice that of coal with a 16% share. Spain ranked third in the world in wind generation capacity and among the top five countries in amount of investments (AEVAL, 2011).

Although the European Union anticipated that competition will result in a reduction in prices both for industrial and residential consumers, prices have tended to increase across the board (Robinson, 2014). The reasons for the increases vary by country across the European Union. In the Spanish case, much of the increase is because of the so-called "governmental levy" (Robinson, 2014). The electricity bill in Spain consists of two components, the power component and the energy component. Both components are partially regulated by the government through an "access" fee, which recovers mostly transport and distribution costs, price premiums to renewables (ie, not uncovered via public funding), and compensates electricity companies for the so-called "tariff deficit" (owed by the government to electricity generation companies for maintaining the access fee below the level required to recover the theoretical cost of production and distribution). In 2013, all costs amounted to €19,658M, 45% and 27% of which corresponded to the renewables premium and distribution costs, respectively (Energia y Sociedad, 2014b). From 2008 to 2012, the weight of the government fee on the final price increased by 68% (from €19/MWh to €32/MWh) for mid-sized industrial users, and by 130% (from €49/MWh to €113/MWh) for domestic users (Robinson, 2014).

## 3. THE WATER–ENERGY NEXUS AT THE LOCAL LEVEL IN SPAIN: THE CASE OF IRRIGATION

The water reform triggered important modernization efforts in the Spanish irrigation sector. This made the sector more robust to water scarcity but also more vulnerable to the new market dynamics of the energy sector.

### 3.1 "Modernization" Reform in the Spanish Irrigation Sector

Water policy reforms at the European level paved the way to reform the irrigation sector (Corominas, 2010; Dumont et al., 2013; Hardy et al., 2010). The growth of the service sector and cities as well as a series of severe droughts in the previous decade had already triggered debates about the allocation of water across sectors. The irrigation sector was increasingly seen as an old fashioned, wasteful, and inefficient user of water (López-Gunn et al., 2012). The European reforms added concerns about environmental impact and competitiveness (Hardy and Garrido, 2012).

In recognition of this, the government initiated a series of programs for the modernization of the sector. After a failing *National Irrigation Plan—Horizon 2005* and a number of subsequent studies, the government approved in 2002 the *National Irrigation Plan—Horizon 2008* (MAPA, 2001). In clear resonance with European regulations, the Plan aimed to secure the agrofood value chain, improve living conditions of farmers, increase water efficiency and conservation, and include environmental criteria in the management of irrigation systems. For this purpose, the Plan included farmer training programs and infrastructure improvement investments ranging from the lining of old canals and improvement of storage facilities to the promotion of sprinkler and drip irrigation technologies (MAPA, 2001). The Plan was for the most part implemented in the form of public subsidies for infrastructure improvement projects that would be carried out by irrigation associations on a voluntary basis. Depending on the project, the subsidies reached up to 60% of the total projects costs (MAPA, 2001).

In 2005 the country witnessed a severe drought that triggered a new step in the modernization process. The Royal Decree 10/2005 prescribed drought mitigation measures, some of which included infrastructure improvements. In 2006 the government approved the *Shock Plan for Irrigation Modernization*, which aimed for additional water savings of 1420 $hm^3$/year on top of the planned water savings foreseen by the 2002 Plan. The Modernization Plan counted on financial support from the European Rural Development Fund, with the understanding that such modernization contributed to the implementation of the EWFD (MAGRAMA, 2012b).

The Plans found much support in the old hydroagricultural policy community. Farmer organizations and the Federation of Irrigation Associations (FENACORE) understood the process as an opportunity to increase water productivity and improve the image of the irrigation sector vis-à-vis future investments in the sector. The Plans were generally executed by state companies, mostly staffed by the corps of engineers. The companies designed and implemented projects as contractors for irrigation communities while also acting as catalysts for state investment (Eimil, 2003).

According to the Spanish Department of Agriculture, more than €3.800M has been invested in the "modernization" of water storage and conveyance infrastructure since 2000. The EU Rural Development Fund has covered €925M, the central and regional governments €1.718M, and farmers (irrigation associations) the remaining amount. Additionally, farmers have invested €5000M in in-plot infrastructure improvements, mostly associated with the introduction of sprinkler and drip irrigation technologies (Union, 2014).

As a result, from 2004 to 2013 flood irrigation decreased from more than 1.23M ha to around 1M ha (−20%), sprinkler remained constant at around 530,000 ha, and dripping irrigation increased from 1.19 to 1.7M ha (+42%) (Hardy et al., 2010; MAGRAMA, 2015).

Outcomes of the irrigation reform have been only mixed. At the national level, water savings seem to have increased. Water use has decreased from 17,681 $hm^3$ in 1999 to 15,833 $hm^3$ in 2012, and irrigated land has increased from 3.3M ha in 2002 to 3.6M ha in 2014 (Hernandez Garcia, 2014). There has also been a diversification in the water sources. Notably, groundwater has reached a 30% share of total water used and the use of transferred, desalinized, and treated water collectively has reached 4% (Corominas, 2010).

At the basin level, results are also mixed. In some basins, for example, basins in Andalucía, modernization has led to water savings of more than 1200 $hm^3/ha$ (Corominas, 2010). In other basins, water productivity has not offset increases in irrigated land or intensification of irrigation, resulting in net water use increases and/or decreases in irrigation returns (Cots, 2011; Dumont et al., 2013; Lecina et al., 2010). This rebound effect has generally been associated with property rights and economic issues. First, water efficiency gains have been used to satisfy otherwise unfulfilled water use rights. Second, studies have shown that modernized systems may not be able to increase profitability with the same level of consumptive use, thus forcing farmers to increase water use to pay their investment back (Dumont et al., 2013).

## 3.2 The Electricity Tariff in the Agricultural Sector

Traditionally, the irrigation sector had enjoyed a preferential tariff (the *R tariff*), regulated by the Department of Industry and published every year in the National Legislative Bulletin (BOE). Since the approval of the energy reform in the late 1990s, the government issued a number of moratoria for electricity users to join the new system. The irrigation sector benefited from the moratoria until 2008, when the *R tariff* was eliminated (BOE, 2007, 2008). From 2008 to 2009 the average power component price paid in the irrigation sector increased by 350% and continued to do so by around 25% every 6 months at least until 2010 (Ederra and Murugarren, 2010). The energy component price increased more progressively (Ederra and Murugarren, 2010), but also faced volatility issues associated with supply and demand dynamics (Robinson, 2014).

## 3.3 A Crisis of Sustainability in the Irrigation Sector

The year 2008 marked the beginning of a financial crisis in the Spanish irrigation sector. The crisis disproportionately affected irrigation systems and communities that had invested the most in modernization. As reported by FENACORE, from 2005 to 2014 fixed costs associated with the electric tariff (ie, the power term) increased by more than 1000% (from €390M to more than €700M). Accordingly, irrigation costs increased from 7% to 40% of total production costs (Berbel et al., 2014; García de Durango, 2014; RRAA, 2009b). In a 1-year period (from 2008 to 2009), the energy bill in many irrigation communities rose to reach three times the cost of water (200–300 €/ha vs 80–100 €/ha, respectively) (López-Gunn et al., 2012). This, together with the high loans farmers and communities had to acquire to face the cost of modernization, has led many communities to quite difficult economic situations (ASAJA, 2013; García de Durango, 2014).

The EWFD full cost recovery principle limits governmental support so farmers are to a great extent on their own to pay back their debt (Hardy et al., 2010). Moreover, the increase in the energy bill has not come with equivalent increases in crop market prices, rather the contrary (Ederra and Murugarren, 2010). Overall, the situation has caused energy to become a limiting factor for agricultural production (Mayor et al., 2015), putting a number of irrigation systems under risk of abandonment (Corominas, 2010). As of today, the discontent among farmers keeps growing across the country, as shown in the number of public protests and petitions addressed to the government (García de Durango, 2014).

One side of the crisis is rather technical and has to do with the inverse relationship between water and energy efficiency in the Spanish irrigation sector (Corominas, 2010).

Drip irrigation systems in Spain are more water efficient than sprinkler and flood irrigation, but also consume more energy per unit of water. Water use in gravity and sprinkler irrigation systems reaches 7500 and 6500 $m^3$/ha, versus 5000 $m^3$/ha in drip systems. Alternatively, energy use for flood and sprinkler ranges from 0.15 to 0.49 kWh/$m^3$ depending on the source of water, versus 0.28 to 0.68 kWh/$m^3$ for drip systems (Corominas, 2010). Thus from 2000 to 2007 the 40% increase in drip systems (Hardy et al., 2012) translated to a 7% decrease in water use per hectare (from 7000 to 6500 $m^3$/ha) (Corominas, 2010); however, that also meant an increase in electricity needs (by 20%, from 4893 to 5866 GWh) and an increase in energy use per hectare (by 9%, from 1435 to 1560 kWh/ha) (Corominas, 2010).

The other side of the crisis is rather institutional and has to do with the dramatic increase of electricity prices in the irrigation sector and the difficulties faced by irrigation communities in dealing with the energy market and the new tariff options. From 2008 to 2009 the communities had to confront both the transition from the *R tariff* to the market system, as well as a dramatic increase in government fees (ie, the "access" fee). The decision of the government to increase the access fee in 2008 has its own story. Since the energy reform, and particularly after the year 2000, the government had kept its fee systematically lower than the costs of maintaining the system. Although this contributed to controlling inflation and protecting the competitiveness of the industry, it also resulted in a debt with electricity companies (ie, the "tariff deficit"). From 2004 to 2005 the debt increased by €4000M, for a cumulative deficit of €5368M since 2000. The lack of legal tools to force electricity companies to reveal their cost structure prevented a revision of the debt. After a series of unsuccessful attempts to solve the problem, the government opted for increasing its fee in 2008 and successive years with the goal of cancelling the deficit in 5 years (Energia y Sociedad, 2014b).

Irrigation communities found themselves rather unprepared to deal with the energy market and energy prices from one year to another (Ederra and Murugarren, 2010). Neither the government nor the communities planned for a transition phase during the almost 10 years that elapsed from the approval of the energy reform until 2008. Indeed, few communities voluntarily joined the market before 2008. By 2008, many remaining communities lacked the necessary expertise to properly evaluate offers from energy retailers, and this diminished their bargaining power and capacity to sign efficient contracts. By the same token, lack of common understanding among farmers in the communities caused delays in decision making and the formalization of contracts with the energy retailers, which in turn translated into important penalties by the retailers (Ederra and Murugarren, 2010).[1] Also communities found it difficult to adjust their irrigation (ie, water pumping) schedules to the periods when energy was cheaper because those periods did not match the periods when farmers were used to irrigating (Ederra and Murugarren, 2010; RRAA, 2008). Similarly, communities had to contract power capacity with retailers without much information about their collective needs. At best, many communities overcontracted capacity and failed to effectively use up all the contracted

1. Following regulations, communities that would not abandon the *R tariff* voluntarily before July 1, 2008 were automatically transferred to the most similar tariff under the market system, with a monthly increase of 5% in the tariff, until they formalized a new contract with the retailer (BOE, 2008).

capacity. At worst, the communities undercontracted capacity and had to face penalties by retailers for exceeding power use (Ederra and Murugarren, 2010).[2]

## 4. NEW GOVERNANCE ARRANGEMENTS IN THE IRRIGATION–ENERGY NEXUS: A LOOK FROM THE BOTTOM UP

The crisis since 2008 has generated a series of institutional responses at the local, regional, and national levels.

### 4.1 Local Responses

At the local level, irrigation communities have taken the lead via a number of collective action responses, all of them oriented toward mitigating the impact of energy price increases and adapting to the new institutional scenario. One of the first responses of many communities to the crisis was the redesign of irrigation schedules to reduce the energy bill. They have done so by organizing different irrigation schedules for different groups of farmers depending on their electricity needs (eg, sprinkler and higher elevation plots vs others), as well as by concentrating water use during the lowest price periods (Abadia et al., 2010; Rocamora et al., 2008; RRAA, 2009b). The strategic use of in-system water storage pools and the individual metering and billing of energy use has also contributed to the adaptation (Rocamora et al., 2008).

Energy audits are also spreading quickly among communities, both to assess the feasibility and design of modernization projects, as well as to optimize those that have been implemented (Abadia et al., 2010; Carrillo-Cobo et al., 2010; Jiménez et al., 2014; Mayor et al., 2015; Moreno et al., 2010; Rocamora et al., 2008; Rodríguez Díaz et al., 2011; RRAA, 2009a). According to experts, the energy saving potential of audits in Spanish irrigation communities can reach up to 20% of total energy use and up to 25% of the energy bill (Hernandez Garcia, 2014); however, for audits to be effective, communities also need to have the capacity to implement necessary changes in the system, and this can be both costly and difficult from an organizational point of view (Moval-Agroingenieria, 2014).

Another successful institutional response has been the collective bargaining and contracting of energy. Communities have self-organized at both the local and national levels (FENACORE, 2013; RRAA, 2012). The national level experience is remarkable. It was initially born as a collaborative venture in 2013, coordinated by FENACORE between a retailer firm and a handful of communities from Andalucía (Parias Fernandez de Heredia, 2014). In 2014, the initiative coordinated more than 20 communities representing 100,000 ha, and generated savings of 9% in the electricity bill on average per community and up to 30% for some communities. The strategies followed to increased savings have included the counseling of communities, collective bargaining with wholesalers, strategic purchase of energy in the futures market, and bilateral agreements with energy generators (Europa-Press, 2015; Parias Fernandez de Heredia, 2014). According to estimations by

2. According to regulations, users that would demand more power than contracted with the retailers would face a penalty equal to the demand excess every 15 min (BOE, 2001).

FENACORE, collective bargaining could save the ensemble of Spanish irrigation communities around €56M (FENACORE, 2013). By 2016, FENACORE has indeed estimated that more than 200 communities representing close to 1M ha will join the enterprise (iagua, 2014).

Finally, a possibility proposed by FENACORE is to turn water users' communities into renewable electricity producers so they can consume their own electricity (Retema, 2015). The irrigation period in Spain usually begins in Mar. and ends in Oct. The number of sunshine hours and water flow (ie, through the canals) during that period is high; thus the electricity that communities would produce should be enough to supply their own needs (Hardy and Garrido, 2012). Auto-production is one type of distributed energy system, characterized by small generators that are near the points of energy consumption. Distributed systems and auto-production make sense in the context of "net balance" systems, which allow auto-producers to deliver energy surpluses to the grid in exchange for discounts in the electricity bill or for energy during periods of production deficit (Energia y Sociedad, 2014a). The price of photovoltaic panels, small windmills, and batteries has notably decreased over time, making auto-production a real possibility. In the irrigation sector, this adds to the existing capacity to use the canals to generate hydropower.

## 4.2 Regional and National Responses

At the regional level, there have been initiatives in both the water and energy sectors. Much of the energy consumed in sprinkler and drip systems has to do with the pumping of underground water. Unfortunately, underground water management in Spain shows serious issues of governability and overextraction (Garrido et al., 2006; Ross and Martinez-Santos, 2010). To cope with this issue, regional and basin authorities across the country have promoted the creation of groundwater user associations (ie, communities) that take responsibility over water management (Rica, López-Gunn and Llamas, 2012). Additionally, water authorities in some Spanish basins have provided irrigation communities with technical and administrative support to reduce pumping; assisted the communities in finding low electricity-demand water sources (Mayor et al., 2015); sold electricity at reduced prices (ie, electricity coming from hydropower plants located in basins) (CHE, 2015); and facilitated the promotion of large-scale, renewable energy projects in the sector (Extremadura, 2015; N.A., 2009).

At the national level, efforts have also been oriented to reduce the energy costs borne by irrigation communities. Measures include, for example, the sponsoring of conferences on the topic (IDAE, 2011) and electricity tax deductions (MAGRAMA, 2014). Most important are, however, the efforts made to better integrate irrigation and energy policies. An example is the *Strategy for Sustainable Modernization Horizon 2015* (MARM, 2009). According to the Strategy, every modernization project should include studies that guarantee the provision of necessary infrastructure for the distribution of electricity as well as the use of the most up-to-date technological devices to optimize energy use (Hardy et al., 2010). Another example is the water-related indicators proposed in the National Energy Plan to assess the sustainability of the energy sector (MINETUR, 2011). A final example is the inclusion in the new National Hydrologic Plan of a chapter looking at the interactions between water and other policies, including energy policy (Mayor et al., 2015).

## 5. INSTITUTIONAL CHALLENGES

Broadly speaking, the water (ie, irrigation) and energy policy sectors in Spain need to be better integrated (Mayor et al., 2015). The previously reviewed responses reveal the existence of a number of institutional challenges that would need to be overcome to further move forward in that direction.

### 5.1 Water Property Right Reforms to Reduce Water Use

The increased return to water use resulting from the introduction of drip and sprinkler technologies has encouraged the intensification of irrigation and extension of irrigated land and, in turn, the use of more energy intensive water sources such as groundwater (Dumont et al., 2013). As already mentioned, this has partially to do with an issue of water property rights in the irrigation sector. In a nutshell, the modernization process has not been followed by a proper revision of water use rights in the sector. The "Alberca" program was launched by the Department of Agriculture, Food and Environment to update water use rights at the national level across all types of uses (MAGRAMA, 2012a). The program has withdrawn a notable number of unused rights from the system; however, most of these correspond to abandoned agricultural land or water uses such as old mills. Although this is a necessary step in the advent of new modernization investments, it is unlikely that landowners who do use their agricultural land and are interested in modernizing are going to give away their rights. This is even more the case when irrigation communities have not been able to fulfill those rights in some basins because of structural problems of water scarcity (Tàbara et al., 2004). A similar situation indeed occurs in the context of water rights markets in Spain and other countries, where farmers and communities are rather reluctant to lease their rights for fear of becoming vulnerable to water use rights downgrades later on (Palomo-Hierro et al., 2015).

### 5.2 Socioecological Diversity

A second important challenge is that of socioecological diversity. There are a variety of factors that affect water and energy consumption at the local level, including the crops planted, the water source, the technology used, the topography of the terrain, and management aspects (Rodríguez-Díaz et al., 2008). As shown by some scholars, applying the same governance arrangements (eg, energy contracts, water and energy management rules) to the combined used of water and energy can lead to unsatisfactory results (Abadia et al., 2010). This means that, to a great extent, solutions have to be tailored to the characteristics of each local context. This implies a strong involvement of water and energy users in the conjunctive management of both resources. The promotion of auto-production and participatory policy making are two potential ways to move in that direction, but, as explained next, there are also barriers to overcome.

### 5.3 Participatory Energy Policy Making

Irrigation communities have long enjoyed a privileged access to the decision-making process. Water planning in the early 20th century was centered on the RBOs, and consisted mostly of promoting hydraulic and irrigation infrastructure. Accordingly, much of the decision making was exclusive to a close policy community composed of hydropower firms, irrigation communities, and the corps of engineers (Perez Picazo and Lemeunier, 2000). With the transition to democracy and the approval of the Water Act of 1986, the planning process opened up, including the implementation of public information procedures and stakeholder-based water councils at the basin and national levels (BOE, 1985). Additionally, irrigation communities have enjoyed a great deal of autonomy from early on. They are depositaries of collective use rights granted by the RBOs and as such have authority to manage water within their jurisdictions. Also they can form federations to coordinate water allocation and infrastructure maintenance within subbasins and basins in collaboration with the RBOs.

The situation in the energy sector is different. Although irrigation communities can produce energy (ie, via small hydropower dams in their canals) and use it for self-consumption they are not formally recognized as energy producers by the government. They are recognized as a special energy user group and have accordingly enjoyed the favor of government in different ways (eg, the *R tariff*, tax deductions); however, they do not participate in the energy planning decision-making process. Indeed, neither consumer groups nor regional authorities at large have a formal seat in the decision-making process other than via claims during public information periods (Ruiz de Apodaca Espinosa, 2010). Energy plans are elaborated by the central government and then approved by the executive board and the Congress (Ruiz de Apodaca Espinosa, 2010). There is a National Energy Commission, but it has a purely technical advisory role, that is, it does not include political or stakeholder representation and focuses on analyzing prices and technical norms. Including the participation of stakeholders in the planning process would not only allow policymakers to detect issues and conflicts derived from nexus tradeoffs (Mayor et al., 2015), but maybe also contribute to the transparency of a process questioned for its opacity and the disproportionate influence of big electricity companies (Hernandez Aguado, 2014).

## 5.4 Self-Consumption of Energy/Distributed Energy in the Irrigation Sector

Auto-production also faces some institutional barriers to expansion. Although technologically possible, the option is not fully cost effective for irrigation communities. This is because of current regulations that de facto penalize the option. According to the Decree signed in 2015, auto-producers have to pay a variety of fees associated with the public maintenance of the electricity system that add to the existing "access" fee (BOE, 2015). Those added fees include a "support fee" for the "backup function" fulfilled by the government in case of unexpected domestic shortages, and penalties for the use of storage batteries that are connected to the grid. Additionally, the Decree prohibits that auto-producers make agreements to pool production and demand; or that they use their energy surpluses to supply the grid in exchange for reductions in the energy bill (ie, net balance). Finally, auto-producers have to commission feasibility studies, which can be particularly costly (Miguel, 2015). The situation in Spain is in stark contrast with that of many other European countries, where the system of net balance is not only allowed but also promoted (EC, 2015; Ropenus and Skytte, 2005).

The government fears that an expansion of auto-production jeopardizes its ability to finance the system and keep the "tariff debt" under control. This, however, is only partially justifiable because the "tariff debt" is caused by preexisting structural problems such as the excess production capacity of the system. The promotion of renewables and combined cycle plants in Spain has resulted in a considerable growth of production capacity. By 2013, the country reached a peak production capacity of 108,000 MW, almost twice that in 2000; however, the peak demand has never gone beyond 46,000 MW. This means that a good portion of the production capacity in Spain (mostly coal and combined cycle plants) is underutilized; however, the infrastructure still needs to be maintained and this is particularly costly (Martil, 2014).

A European energy network could make Spain a net exporter of energy and help balance production and maintenance costs (today net electric transactions amount to just 2.4% of total generation) (Jiménez et al., 2014); however, the long-promised transboundary grid has its own institutional barriers (Serrallés, 2006). European governments are keeping their grip on market competition for reasons of security of supply and protection of their electricity companies (Karan and Kazdağli, 2011). Some governments have favored the emergence of national champions that would be much better positioned in a future European electricity market than those from other countries where the sector is less oligopolistic. Additionally, European countries are diverse in terms of their mix of electricity generation. France, for example, strongly relies on nuclear power, which positions it well to cope with supply issues on its own. Alternatively, Spain strongly relies on quite volatile sources such as solar, hydropower, and wind and would therefore benefit from a cross-boundary electricity market. Not without reason, Spain has set up a common Iberian electricity market with Portugal and has strong ambitions in developing it further.

## 6. CONCLUSIONS

This chapter provided a bottom-up and multilevel institutional analysis of the water–energy nexus in Spain, with a focus on the use of energy for water extraction and distribution in the irrigation sector. The analysis revealed important institutional interplays at different levels of governance, from the European to the local level. The concurrent reforms of the energy and water policy at the European level converge in their aim to promote efficient allocation of said resources and competitiveness; at the national and local levels, however, the reforms have led many farmers to a crisis of financial sustainability. On the one hand, the active promotion of new "modernized" irrigation technologies in the irrigation sector has contributed to an increase in water efficiency but also in energy dependence and use by farmers. On the other hand, the enthusiastic promotion of energy liberalization and renewables has resulted in a structural problem of high electricity prices. The crisis in the irrigation sector has also led to institutional responses at different governance levels that aim to better integrate energy and water use. Cooperative responses from local irrigation communities are most salient, and include the adaptation of irrigation schedules to energy price schedules, the commissioning of energy audits for irrigation systems, the collective purchase of energy, and the self-organization for auto-production. The role of basin and regional authorities is also undeniable and has been oriented mostly to support the communities' initiatives. At the national level, there have been some efforts to better integrate the irrigation and energy sectors, such as the prescription of energy audits in modernization projects and the promotion of water-based sustainability indicators in energy planning. Finally, the analysis revealed a number of institutional challenges for future discussion. Those challenges include a water rights reform that capitalizes on modernization efforts to conserve water, the participation of local stakeholders (eg, irrigation communities) in energy planning, and the effective promotion of a distributed energy network within Spain and at the European level.

The interplay between modernization and energy intensification is not unique to Spain. A number of Southern European and Mediterranean countries are facing similar tradeoffs (Daccache et al., 2014). The Plan Bleu, issued by the regional center of the United Nations Environmental Program, has shown that demand for water (domestic, industrial, and irrigation only) in the Southern and Eastern Mediterranean countries will rise from 150 $km^3$/year in 2005 to 200 $km^3$/year in 2025, while demand for electricity for the same water uses and the same geographical area will rise from 20 TWh/year in 2005 to 200 TWh/year in 2025 (Bleu, 2009). This makes the irrigation sector in the region particularly vulnerable to energy price increases. Not without reason, leaders from the sector in a number of Mediterranean countries associated with the Euro-Mediterranean Irrigators Community are considering joining Spanish irrigation communities in their efforts to acquire energy collectively (iAgua, 2014).

The analysis of this chapter is limited to the irrigation sector and the understanding of energy as an input for water extraction and allocation. This narrow focus facilitates the tracing of the processes linking institutions, behavior, and outcomes. The nexus, however, can also be studied from a "water for energy" approach (Hardy et al., 2012) and from the perspective of value chains that involve multiple stages of water and energy production and use (Villamayor-Tomas et al., 2015). Also irrigation is only one entry point to the nexus. Other relevant entry points in Spain include issues associated with reservoir management in the context of increased competition for water resources; discussions about the energy implications of alternative water production systems (eg, desalination, urban waste water treatment); the water implications of promoting biofuels; or the impact on water and energy availability from climate change and extreme weather events (Hardy and Garrido, 2012; Hardy et al., 2010). Further institutional analyses of the nexus from each of these entry points shall contribute to a better understanding of challenges and opportunities of integrated management.

## REFERENCES

Abadia, R., Rocamora, C., Corcoles, J.I., Ruiz-Canales, A., Martinez-Romero, A., Moreno, M.A., 2010. Comparative analysis of energy efficiency in water users associations. Spanish Journal of Agricultural Research 8. http://dx.doi.org/10.5424/sjar/201008S2-1356134-142.

AEVAL, 2011. Evaluacion del plan espanol de energias renovables 2005–2010. Agencia Estatal de Evaluación de las Políticas Públicas y la Calidad de Los Servicios, Madrid, p. 252.

ASAJA, 2013. El incremento en las tarifas eléctricas asfixia a los agricultores de regadío y amenaza la continuidad de 50.000 explotaciones que generan 128.000 empleos en Andalucía. Retrieved 3 December, 2015, from. http://www.asaja.com/publicaciones/el_incremento_en_las_tarifas_electricas_asfixia_a_los_agricultores_de_regadio_y_amenaza_la_continuidad_de_50_000_explotaciones_que_generan_128_000_empleos_en_andalucia_1921.

Berbel, J., Gutierrez-Martín, C., Camacho, C., Montesinos, P., Rodríguez-Díaz, J.A., 2014. Efectos de la modernizacion de regadios en consumo de agua, enegia y coste. In: Paper presented at the XIII Congreso de Comunidades de Regantes, Huelva.

Bleu, P., 2009. State of the Environment and Development in the Mediterranean. UNEP/MAP-Plan Bleu, Athens.

BOE, 2015. Real Decreto 900/2015, de 9 de octubre, por el que se regulan las condiciones administrativas, técnicas y económicas de las modalidades de suministro de energía eléctrica con autoconsumo y de producción con autoconsumo. Boletin Oficial del Estado, Madrid.

Carrillo-Cobo, M.T., Rodriguez-Diaz, J.A., Camacho-Poyato, E., 2010. The role of energy audits in irrigated areas. The case of 'Fuente Palmera' irrigation district (Spain). Spanish Journal of Agricultural Research 8, 10. http://dx.doi.org/10.5424/sjar/201008S2-1358.

CHE, 2015. Confederación del Ebro firma con Ayuntamientos y Riegos del Alto Aragón el primer convenio para la cesión de la energía reservada al Estado. Retrieved 22 September, 2015, from. http://www.chebro.es/contenido.visualizar.do?idContenido=44203.

Corominas, J., 2010. Agua y energía en el riego, en la época de la sostenibilidad. Ingeniería del agua 17 (3).

Cots, L., 2011. Desarrollo y calibración de un modelo de simulación de recursos hídricos aplicado a la cuenca del río Corb dentro de la zona regable de los canales de Urgell (Lleida) (PhD). University of Lleida, Lleida.

Daccache, A., Ciurana, J.S., Diaz, J.A., Rodriguez, Knox, J.W., 2014. Water and energy footprint of irrigated agriculture in the Mediterranean region. Environmental Research Letters 9 (12), 124014.

Directive 2003/54/EC of the European Parliament and of the Council of 26 June 2003 Concerning Common Rules for the Internal Market in Electricity and Repealing Directive 96/92/EC - Statements Made with Regard to Decommissioning and Waste Management Activities L 176/37 C.F.R, 2003.

Directive 2009/28/EC of the European Parliament and of the Council of 23 April 2009 on the Promotion of the Use of Energy from Renewable Sources and Amending and Subsequently Repealing Directives 2001/77/EC and 2003/30/EC, 2009.

Directive 96/92/EC of the European Parliament and of the Council of 19 December 1996 Concerning Common Rules for the Internal Market in Electricity, L 027 C.F.R, 1996.

Dumont, A., Mayor, B., López-Gunn, E., 2013. Is the rebound effect or Jevons paradox a useful concept for better management of water resources? Insights from the irrigation modernisation process in Spain. Aquatic Procedia 1 (0), 64–76. http://dx.doi.org/10.1016/j.aqpro.2013.07.006.

EC, 2015. Best Practices on Renewable Energy Self Consumption. European Commission, Brussels.

Ederra, I., Murugarren, N., 2010. La nueva tarifa eléctrica: la escalada de precios del agua. Servicio de Asesoramiento al Regante. Riegos de Navarra, SA.

Eimil, C.D., 2003. La modernización del regadío a través de las SEIASAS. Agricultura: Revista agropecuaria 850, 244–246.

Energia y Sociedad, 2014a. 2. NormativaManual de energia: Energia y Sociedad. Retrieved from. http://www.energiaysociedad.es/ficha/2-2-el-marco-normativo-espanol.

Energia y Sociedad, 2014b. 7. Peajes de acceso, cargos y déficit tarifarioManual de la electricidad: Energia y Sociedad. Retrieved from. http://www.energiaysociedad.es/ficha/peajes-de-acceso-y-deficit-tarifario.

Europa-Press, 2015. La central de compras de Fenacore ahorrará a sus asociados cerca de 120.000 euros en la actual campaña de riego. Europa Press. From. http://www.europapress.es/andalucia/sevilla-00357/noticia-central-compras-fenacore-ahorrara-asociados-cerca-120000-euros-actual-campana-riego-20150605161639.html.

Extremadura, El correo de, 2015. El Gobierno extremeño prepara un proyecto piloto para usas energía renovable en el riego y abaratar costes. El correo de extremadura. Retrieved from. http://www.elcorreoextremadura.com/noticias_region/2015-04-15/1/7825/el-gobierno-extremeno-prepara-un-proyecto-piloto-para-usasenergia-renovable-en-el-riego-yabaratar-costes.html.

FENACORE, 2013. FENACORE crea una central de compras para ahorrar energía. Retrieved 7 December, 2015, from. http://www.fenacore.org/escaparate/noticias.cgi?idnoticias=192657.

García de Durango, Á., 2014. La subida de las tarifas eléctricas ahoga a los regantes españoles. *iagua*. Retrieved 10 December, 2015, from. http://www.iagua.es/noticias/tarifas/14/02/11/la-factura-electrica-de-los-regantes-se-ha-incrementado-en-mas-de-un-80-desde-2008-45084.

Garrido, A., Martínez-Santos, P., Llamas, M.R., 2006. Groundwater irrigation and its implications for water policy in semiarid countries: the Spanish experience. Hydrogeology Journal 14 (3), 340–349. http://dx.doi.org/10.1007/s10040-005-0006-z.

Gómez-Limón, J.A., Arriaza, M., Berbel, J., 2002. Conflicting implementation of agricultural and water policies in irrigated areas in the EU. Journal of Agricultural Economics-Reading 53 (2), 259–282.

Hardy, L., Garrido, A., 2012. Challenges and opportunities related to the Spanish water-energy nexus. In: De Stefano, L., Llamas, M.R. (Eds.), Water, Agriculture and the Environment in Spain: Can We Square the Circle? CRC Press.

Hardy, L., Garrido, A., Sirgado, L.J., 2010. Análisis y evaluación de las relaciones entre el agua y la energía en España. Fundación Marcelino Botín Santander, Spain.

Hardy, L., Garrido, A., Juana, L., 2012. Evaluation of Spain's water-energy nexus. International Journal of Water Resources Development 28 (1), 151–170.

Hellegers, P., Zilberman, D., Steduto, P., McCornick, P., 2008. Interactions between water, energy, food and environment: evolving perspectives and policy issues. Water Policy 10 (Suppl. 1).

Hernandez Aguado, I., 2014. Una evaluación del lobby en España: análisis y propuestas: Transparency International España.

Hernandez Garcia, A., 2014. Modernizacion de regadios, situacion actual y retos de futuro, *iagua*. Retrieved from. http://www.iagua.es/blogs/alberto-hernandez-garcia/modernizacion-regadios-situacion-actual-y-retos-futuro.

Hussey, K., Pittock, J., 2012. The energy–water nexus: managing the links between energy and water for a sustainable future. Ecology and Society 17 (1), 31.

iAgua, 2014. 200 comunidades de regantes comprarán energía de forma conjunta en 2016 para ahorrar 5 millones de euros anuales. Retrieved 30 September, 2015, from. http://www.iagua.es/noticias/espana/fenacore/14/12/17/200-comunidades-regantes-compraran-energia-forma-conjunta-ahorrar.

IDAE, 2011. I Congreso Nacional de Regadíos y Energías Renovables. Retrieved 4 January, 2016, from. http://www.idae.es/index.Php/id.175/mod.eventos/mem.detalle.

Jiménez, S., Langarita, R., Sarasa, C., 2014. Evaluación del papel de los costes energéticos en las comunidades agrarias: el caso de la Comunidad General de Riegos del Alto Aragón. In: Paper Presented at the International Conference on Regional Science, Zaragoza. http://www.reunionesdeestudiosregionales.org/Zaragoza2014/htdocs/pdf/p1049.pdf.

Jordana, J., Levi-Faur, D., Puig, I., 2006. The limits of Europeanization: regulatory reforms in the Spanish and Portuguese telecommunications and electricity sectors. Governance 19 (3), 437–464. http://dx.doi.org/10.1111/j.1468-0491.2006.00325.x.

Karan, M.B., Kazdağli, H., 2011. The development of energy markets in Europe. In: Financial Aspects in Energy. Springer, pp. 11–32.

La Roca, F., 2011. Balance de la política de recuperación de los costes de los servicios del agua en el Estado Español, 2000–2010.

Lecina, S., Isidoro, D., Playán, E., Aragüés, R., 2010. Irrigation modernization and water conservation in Spain: the case of Riegos del Alto Aragón. Agricultural Water Management 97 (10), 1663–1675.

Ley 29/1985, de 2 de agosto, de Aguas., 29/1985 C.F.R, 1985.

Ley 54/1997, de 27 de noviembre, del Sector Eléctrico., 285 C.F.R, 1997.

López-Gunn, E., Mayor, B., Dumont, A., 2012. Implications of the modernization of irrigation systems. In: De Stefano, L., Llamas, M.R. (Eds.), Water, Agriculture and the Environment in Spain: Can We Square the Circle? CRC Press, pp. 241–258.

MAGRAMA, 2012a. Actualización Registros de Aguas: Programa ALBERCA. Retrieved 7 November, 2015, from. http://www.magrama.gob.es/es/agua/temas/concesiones-y-autorizaciones/uso-privativo-del-agua-registro-del-aguas/alberca/.

MAGRAMA, 2012b. Marco Nacional de Desarrollo Rural 2007–2013. Ministerio de Agricultura y Medio Ambiente, Madrid.

MAGRAMA, 2014. García Tejerina anuncia una exención del 85% del impuesto especial de la electricidad para los regantes y un ajuste en el régimen de módulos del IRPF. Retrieved 1 November, 2015, from. http://www.magrama.gob.es/es/prensa/noticias/-garc%C3%ADa-tejerina-anuncia-una-exenci%C3%B3n-del-85-del-impuesto-especial-de-la-electricidad-para-los-regantes-y-un-ajuste-en-el-r%C3%A9gimen-de-m%C3%B3dulos-del/tcm7-346527-16.

MAGRAMA, 2015. Encuesta sobre superficies y rendimientos: Informe sobre regadios en Espana. Ministerio de Agricultura., Alimentacion y Medio Ambiente, Madrid, p. 40.

MARM, 2009. Estrategia nacional para la modernización sostenible de los regadíos H2015, Documento de inicio para consulta sobre evaluación ambiental. Dirección generalcdel agua, Ministerio de Medio Ambiente, Medio Rural y Marino, Madrid.

Martil, I., 2014. Cómo es el sistema de producción de energía eléctrica en España? Retrieved from. http://blogs.publico.es/econonuestra/2014/10/01/como-es-el-sistema-de-produccion-de-energia-electrica-en-espana/.

Martin-Ortega, J., 2012. Economic prescriptions and policy applications in the implementation of the European water framework directive. Environmental Science & Policy 24, 83–91. http://dx.doi.org/10.1016/j.envsci.2012.06.002.

Mayor, B., López-Gunn, E., Villarroya, F.I., Montero, E., 2015. Application of a water–energy–food nexus framework for the Duero river basin in Spain. Water International 1–18. http://dx.doi.org/10.1080/02508060.2015.1071512.

Miguel, M., 2015, 14 of June 2015. El Gobierno estrecha el cerco al autoconsumo eléctrico: las claves del nuevo decreto. 20 Minutos. Retrieved from. http://www.20minutos.es/noticia/2488992/0/autoconsumo/electricidad/gobierno/.

Míguez, J.L., López-González, L.M., Sala, J.M., Porteiro, J., Granada, E., Morán, J.C., Juárez, M.C., 2006. Review of compliance with EU-2010 targets on renewable energy in Galicia (Spain). Renewable and Sustainable Energy Reviews 10 (3), 225–247. http://dx.doi.org/10.1016/j.rser.2004.09.009.

MINETUR, 2011. Planificación de los sectores de electricidad y gas 2012–2020 (PSEG). Ministerio de Industria, Energía y Turismo, Madrid.

Moreno, M.A., Ortega, J.F., Córcoles, J.I., Martínez, A., Tarjuelo, J.M., 2010. Energy analysis of irrigation delivery systems: monitoring and evaluation of proposed measures for improving energy efficiency. Irrigation science 28 (5), 445–460.

Moval-Agroingenieria, 2014. Estrategia de Ahorro y Eficiencia Energética en regadío. Retrieved 5 December, 2015, from. http://moval.es/2014/01/estrategia-de-ahorro-y-eficiencia-energetica-en-regadio/.

N.A., 2009. Aragon-El Gobierno quiere a los regantes en los proyectos de parques eolicos. El diario de Teruel. Retrieved from. http://www.aeeolica.org/es/new/aragn-el-gobierno-quiere-a-los-regantes-en-los-proyectos-de-parques-elicos/.

North, D., 1990. Institutions, Institutional Change and Economic Performance. Cambridge University Press, Cambridge.

ORDEN ITC/1857/2008, 2008. de 26 de junio, por la que se revisan las tarifas eléctricas a partir del 1 de julio de 2008., BOE-A-2008–10968 C.F.R.

Palomo-Hierro, S., Gómez-Limón, J.A., Riesgo, L., 2015. Water markets in Spain: performance and challenges. Water 7 (2), 652–678.

Parias Fernandez de Heredia, P., 2014. La Importancia de la Energía en el Regadío. In: Paper Presented at the Congreso Nacional de Medio Ambiente, Madrid.

Perez Picazo, M.T., Lemeunier, G., 2000. Formation et mise en cause du modèle de gestion hydraulique espagnol de 1780 a 2000. Economies et Sociétés, Hors-Série 37.

Plan Nacional de Regadíos Horizonte 2008, 2001.

Real Decreto, 2001. 1164/2001, de 26 de octubre, por el que se establecen tarifas de acceso a las redes de transporte y distribución de energía eléctrica., 268 C.F.R.

Real Decreto, 2007. 871/2007, de 29 de junio, por el que se ajustan las tarifas eléctricas a partir del 1 de julio de 2007., BOE-A-2007–12759 C.F.R.

Retema, 2015. ITE y FENACORE firman un convenio para la integración de energías renovables en las comunidades de regantes. Revista Tecnica de Medio Ambiente.

Robinson, D., 2014. Analisis Comparativo de los Precios de la Electricidad en la Union Europea: Una Perspectiva Española. Eurocofin.

Rica, M., López-Gunn, E., Llamas, R., 2012. Analysis of the emergence and evolution of collective action: An empirical case of Spanish groundwater user associations. Irrigation and Drainage 61 (S1), 115–125.

Rocamora, M.C., Abadia, R., Ruiz-Canales, A., 2008. Ahorro y Eficiencia Energética en las Comunidades de Regantes *Ahorro y Eficiencia Energética en la Agricultura*: Instituto para la Diversificación y Ahorro de la Energía. Ministerio de Industria, Turismo y Comercio.

Rodríguez Díaz, J.A., Camacho Poyato, E., Blanco Pérez, M., 2011. Evaluation of water and energy use in pressurized irrigation networks in Southern Spain. Journal of Irrigation and Drainage Engineering 137 (10), 644–650.

Rodríguez-Díaz, J.A., Camacho-Poyato, E., López-Luque, R., Pérez-Urrestarazu, L., 2008. Benchmarking and multivariate data analysis techniques for improving the efficiency of irrigation districts: an application in spain. Agricultural Systems 96 (1–3), 250–259. http://dx.doi.org/10.1016/j.agsy.2007.07.010.

Ropenus, S., Skytte, K., 2005. Regulatory review and barriers for the electricity supply system for distributed generation in EU-15. In: Paper Presented at the Future Power Systems, 2005 International Conference on.

Ross, A., Martinez-Santos, P., 2010. The challenge of groundwater governance: case studies from Spain and Australia. Regional Environmental Change 10 (4), 299–310.

RRAA, 2008. Las tarifas de riego desaparecen. Riegos del Alto Aragon 1.

RRAA, 2009a. Asesoría para conseguir una alta eficiencia en las instalaciones de riego a presión. Riegos del Alto Aragon 23.

RRAA, 2009b. Memoria 2009, 122.

RRAA, 2012. Memoria 2012. Huesca: Comunidad general de regantes de Riegos del Alto Aragon.

Ruiz de Apodaca Espinosa, Á., 2010. La planificación energética y su sometimiento a evaluación ambiental estratégica. Revista Catalana de Dret Ambiental 1 (1).

Scott, C.A., Pierce, S.A., Pasqualetti, M.J., Jones, A.L., Montz, B.E., Hoover, J.H., 2011. Policy and institutional dimensions of the water–energy nexus. Energy Policy 39 (10), 6622–6630.

Serrallés, R.J., 2006. Electric energy restructuring in the European Union: integration, subsidiarity and the challenge of harmonization. Energy Policy 34 (16), 2542–2551. http://dx.doi.org/10.1016/j.enpol.2004.08.041.

Tàbara, J.D., van Woerden, F., Costejà i Florensa, M., 2004. Las culturas del agua en la prensa española. Los marcos culturales en la comunicación sobre el Plan Hidrológico Nacional. In: Paper presented at the Papers: revista de sociologia.

Union, J.A., 2014. El regadío y su encrucijada El País. Retrieved from. http://economia.elpais.com/economia/2014/08/07/actualidad/1407421568_757062.html.

Villamayor-Tomas, S., Grundmann, P., Epstein, G., Evans, T., Kimmich, C., 2015. The water-energy-food security nexus through the lenses of the value chain and IAD frameworks. Water Alternatives 8 (1), 735–755.

Waughray, D., 2011. Water Security: The Water-food-energy-climate Nexus. Island Press, Washington DC.

SUBPART 2.2

# Water for Energy Production

Chapter 2.2.1

# Water Scarcity and Conservation Along the Biofuel Supply Chain in the United States: From Farm to Refinery

J.S. Bergtold[1], A.C. Sant'Anna[1], N. Miller[1], S. Ramsey[1], J.E. Fewell[2]
[1]*Kansas State University, Manhattan, KS, United States;* [2]*Lake Region State College, Devils Lake, ND, United States*

## 1. INTRODUCTION

Biofuel production has increased in the past couple of decades with the aim of protecting the environment and ensuring energy independence. In the United States, the Renewable Fuels Standard Program was implemented to promote biofuel production through the setting of blending targets and subsidies for biofuel production. The Energy Independence and Security Act set the goal of producing 36 billion gallons per year of renewable fuel by 2022 (US Congress, House of Representatives, 2007). These policies have provided a significant policy-driven demand for biofuel production (especially ethanol). In addition, these policies have focused on the use of alternative biofuel feedstocks (eg, oilseeds, cellulosic, etc.) to meet this higher biofuel demand, placing a demand on the agricultural sector to help meet our energy needs.

The predominant biofuels used in the United States are ethanol and biodiesel. Ethanol is mostly blended with gasoline, while biodiesel is blended with diesel or consumed directly. In the United States, the main feedstocks are corn and soybeans for ethanol and biodiesel production, respectively (Dufey and Grieg-Gran, 2006). Nevertheless, there has been a recent a surge in ethanol produced from lignocellulosic feedstocks and other bioenergy crops, such as corn stover, sorghum, and switchgrass (second-generation biofuels).

Significant growth in the biofuels industry can have both positive and negative environmental effects. Biofuels produce lower carbon dioxide emissions than fossil fuels, but expansion may put stress on land and water resources. Water is consumed at all stages of biofuel production: at the agricultural level water is demanded for irrigation; at the industrial level it is used in the cooling and drying processes; and water is present in the biofuel that reaches the consumer. Feedstock production, at the agricultural level, impacts water supply and scarcity because of irrigation demands, and water quality because of runoff and sedimentation (Gramig et al., 2013; Service, 2009). At the industry level, water supply is

**Competition for Water Resources. http://dx.doi.org/10.1016/B978-0-12-803237-4.00007-0**

impacted because of its use in cooling and drying processes, while incorrect waste water disposal impacts water quality.

There are measures that can offset the impact of biofuel production on water scarcity, supply, and quality. In agriculture, best management practices and the adoption of water efficient irrigation technologies can reduce water usage, while residue management practices can reduce runoff and sedimentation (Gramig et al., 2013; Stone et al., 2010). Water demand can be reduced by motivating biofuel production with feedstocks that require less water, fertilizer, herbicides, and pesticides, such as switchgrass or sorghum varieties. Water requirements can also be reduced by promoting feedstock cultivation in areas with higher precipitation. Refineries can reduce water usage by using treated municipal water or by recycling their own used water (Varghese, 2007).

This chapter provides a summary analysis of the various impacts of biofuel production on water usage and scarcity in the United States along the biofuel supply chain. Consideration is given to practices and policies that promote water conservation along the supply chain. The goal of the chapter is to make readers aware of how biofuels' expansion could be promoted while being conscientious of its impacts on water.

## 2. BIOFUELS PRODUCTION AND AGRICULTURE

Since 2007, total US production of ethanol has doubled, producing more than 14 billion gallons of ethanol a year (USDA, FAS, 2015). The United States is a large producer of biodiesel as well. Growth in the biofuels industry has been driven by government policies and has had a significant impact on the agricultural economy. This section of the chapter examines the market for biofuel production and its linkages to the agricultural sector.

### 2.1 Market Demand and Government Policy

The Renewable Fuel Standard (RFS) program, implemented by the US Environmental Protection Agency (EPA), was created under the Energy Policy Act of 2005 and expanded in the Energy Independence and Security Act of 2007. The RFS program obligates refiners and importers of gasoline and diesel to blend renewable fuels into petroleum-based fuels, heating oil, or jet fuel. Since 2007, the program sets yearly volume requirements for the production of renewable fuels. The goal is to reach a yearly production of 36 billion gallons of renewable fuel in 2022, more than doubling the 2012 level of production of approximately 13 billion gallons (EPA, 2016a). The increased level of production is mandated to be produced from cellulosic feedstocks, dedicated bioenergy crops, and other "advanced" feedstocks as designated by the EPA (EIA, 2013; EPA, 2016a). At the time the RFS was implemented, standard vehicles were thought to be unable to handle more than a 10% ethanol blend (E10) because of technological obstacles. Warranties issued by vehicle and engine companies decline to cover damage to cars using higher levels of ethanol blends (Schnepf, 2011). These technological obstacles served as a "blend wall," limiting demand for ethanol. Despite these barriers, further testing by the US Department of Energy, indicates a 15% ethanol blend (E15) does not produce adverse effects. The EPA has issued partial waivers for gasoline containing up to E15 (Schnepf and Yacobucci, 2013). This raises the possibility of the EPA revising ethanol mandates upward to include E15. Such a policy change would increase demand for E15 and in turn ethanol (but decrease demand for E85) (Zhang et al., 2010).

A consequence of the RFS program is that ethanol acts as a complement to gasoline. Luchansky and Monks (2009) found a positive relationship between ethanol and gasoline prices, but this relationship may not hold for higher ethanol blends. Additionally, increased E85 demand led to further investment in infrastructure for supplying E85 fueling stations and increased use of flex fuel vehicles (Liu and Greene, 2014). Though ethanol demand is constrained domestically, the export market offers a potential area of growth. Canada and Brazil comprise the largest markets for American exports (Canada held 29.5% and Brazil 17.6% of US ethanol exports for the first 9 months of 2015) (Meyer and Paulson, 2014). Outside of ethanol, biodiesel and aviation biofuel present substantial areas of additional biofuel demand. Consumption of biodiesel increased from 10 million gallons in 2001 to 1.4 billion gallons in 2014 (EIA, 2016).

Apart from the RFS, the US government also seeks to promote biofuel research, production, blending, and/or commercialization through market incentives. Production of biofuels from feedstock other than corn is the focus of current policies, which include:

- biodiesel blenders receiving a tax incentive of $1 per gallon of biodiesel blended with petroleum diesel (Biodiesel Mixture Excise Tax Credit);
- second-generation biofuel producers being eligible to receive up to $1.01 per gallon of second-generation biofuel (Second Generation Biofuel Producer Tax Credit);
- advanced biofuels producers receiving payments to support their production (Advanced Biofuel Production Payments);
- loan guarantees for biorefineries that produce renewable fuels from a biomass other than corn kernel starch (Advanced Biofuel Production Grants and Loan Guarantees); and
- incentives for farmers to produce and deliver biomass feedstock to biofuel production facilities (Biomass Crop Assistance Program) (DOE, 2016).

This focus on other types of feedstocks could reduce pressure on water if the policy implemented aims at feedstocks with lower water requirements.

## 2.2 Market Supply and Biofuel Production

Corn-based ethanol production increased by 3 billion gallons between 1993 and 2003. With the implementation of the RFS in 2005, ethanol production rapidly increased by over 10 billion gallons between 2005 and 2011. Consumption followed the same pattern. In comparison, biodiesel production increased from 900 million gallons in 2005 to just over 1 billion gallons in 2013. Table 1 provides an overview of production of ethanol and biodiesel by feedstock in the United States. Again the table shows that corn and soybean are the dominant feedstocks for producing biofuels.

The geographic distribution of ethanol plants, biodiesel plants, and integrated biorefineries is varied. Most ethanol plants are located in the "Corn Belt" states of Iowa, Nebraska, Minnesota, and Illinois. Biodiesel plants, on the other hand, are less clustered—with North Carolina, Texas, Iowa, and California having a large number of plants (Fig. 1). Integrated biorefineries (mainly pilot facilities able to convert a large range of bioresources) are lesser in number and more spread out. Among ethanol plants, there is also diversity in the use of noncorn feedstocks.

In Kansas, four out of the 12 plants can produce ethanol from grain sorghum, while in Colorado, one out of the four plants uses waste beer as a raw material for ethanol

**TABLE 1** Total Biofuel Production in the United States By Year and by Type of Feedstock

| Fuel | Feedstock | US Production (Million Gallons) | Year | Sources |
|---|---|---|---|---|
| Ethanol | Corn | 14,659 | 2015 | USDA (2016a) |
| | Sorghum, wheat, barley, and brewery waste | >450 | 2009 | Schnepf (2011) |
| | Cane sugar | 1.5 | 2009 | Schnepf (2011) |
| | Second generation | 110 | 2015 | USDA (2016a,b) |
| Biodiesel | Soybean oil | 1240 | 2015 | USDA (2016b) |
| | Other vegetable oils | <10 | 2011 | Schnepf (2011) |
| | Recycled grease | <10 | 2011 | Schnepf (2011) |

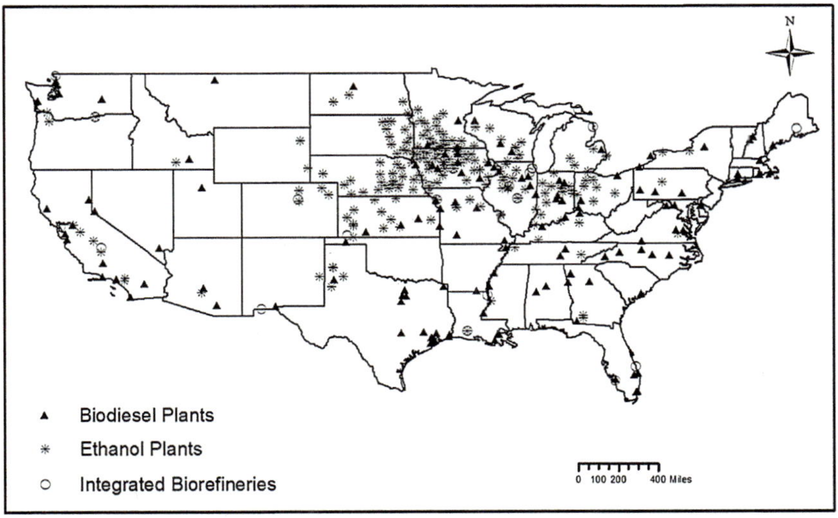


**FIGURE 1** Location of biodiesel plants, ethanol plants, and integrated biorefineries in the continental United States. *Adapted from National Renewable Energy Laboratory (NREL), 2014. The Biofuel Atlas. https://maps.nrel.gov/bioenergyatlas/. Map created using ArcGIS software by Esri. ArcGIS and ArcMap are the intellectual property of Esri and are used herein under license. Copyright © Esri. All rights reserved. For more information about Esri software, please visit www.esri.com.*

production. Other raw materials used by ethanol plants shown in Fig. 1 are cheese whey, beverage waste, sugarcane, barley, wood waste, and waste sugar (NREL, 2014). Larger ethanol plants have, on average, a 120 million gallon capacity per year, while smaller plants have, on average, a 5 million gallon capacity per year. Biodiesel plants produce from 30,000 to 100 million gallons per year. Integrated plants, which use agricultural waste, algae, and energy crops as feedstock, have a capacity of producing 1–25 million gallons per year (NREL, 2014).

## 2.3 Agriculture and Feedstock Production

There are a number of biofuel feedstock categories (Table 2). Traditional feedstocks for producing ethanol and biodiesel include cereal grains, such as corn and grain sorghum, and oilseeds, such as soybean, sunflower, canola, and rapeseed. Advanced biofuel feedstocks have primarily included cellulosic sources for producing ethanol, but these include advanced feedstocks for producing biodiesel and related products as well. These feedstocks comprise other oilseeds, agricultural residues, annual bioenergy crops, perennial bioenergy crops, and other sources such as sugarcane and algae.

In the United States, farmers reacted to the increased demand for corn for ethanol by increasing the total number of acres planted (Wallander et al., 2011). In 2001, 75 million corn acres were planted, increasing to a peak of 93.5 million acres in 2007. From 2008 to 2011, acres planted trailed off until rising to a new high of 95 million acres in 2013. Since 2010, about 5 billion bushels of harvested corn are used for producing ethanol per year, representing approximately 40% of total US corn production. Soybean production rose from 74 million acres in 2001 to 83 million acres in 2014. The amount of soybean oil that goes toward biodiesel production has doubled since 2010, from 2.8 billion pounds to 4.7 billion pounds, 25% of total US production in 2016 (USDA, 2016b).

Farmers have also significantly altered their production practices from 2000 to 2009, switching from more diverse crop production to corn and soybean monocropping. Also, during this period, double cropping (ie, planting two crops in 1 year) became increasingly more common (Wallander et al., 2011). This specialization in production can place additional pressure on traditional and marginal cropland and associated natural resources (Wright and Wimberly, 2013).

# 3. WATER USAGE IN BIOFUEL AND FEEDSTOCK PRODUCTION

Water is a scarce resource in the areas of the United States where biofuel feedstock and biofuel production predominantly take place. Biofuel crops are irrigated with 2% of all the water withdrawn for irrigation worldwide (De Fraiture et al., 2008). This section of the chapter examines the linkages between water usage and biofuel feedstock and biofuel production along the biofuel supply chain in the United States.

## 3.1 Brief Overview of Water Use for Agriculture in the United States

In 2010, the top three water uses in the United States were for thermoelectric production (49%), irrigation (31%), and industrial manufacturing purposes (15%). The remaining 18% of water usage was split between public supplies, domestic use, aquaculture, mining, and livestock (Maupin et al., 2014). Irrigation water use includes freshwater irrigation in agriculture and horticulture, as well as water used in the irrigation of parks, cemeteries, nurseries, golf courses, and other areas. Water used for irrigation in agriculture includes water used

**TABLE 2 Biofuel Feedstock and Water Usage**

| Feedstock Type | Feedstock Sources | Primary Biofuel Product | Water Usage for Irrigated Production |
|---|---|---|---|
| Cereal grains | Corn<br>Grain sorghum | Ethanol | Corn:<br>14 in./acre[a]<br>Grain sorghum:<br>13 in./acre[b] |
| Soybeans | – | Biodiesel | 10 in./acre[c] |
| Other oilseeds | Rapeseed and canola, safflower and sunflower | Biodiesel<br>Heat oil<br>Aviation biofuel | Canola: 10–12 in./acre[d]<br>Sunflower: 12 in./acre[d] |
| Agricultural residues | Corn stover, wheat straw, sorghum stover | Cellulosic ethanol | – |
| Annual bioenergy crops | Rotational crops:<br>Corn silage<br>Forage sorghum, energy sorghum | Cellulosic ethanol | Corn silage:<br>26 in./acre[e]<br>Forage sorghum:<br>15 in./acre[f] |
| Perennial bioenergy crops | Switchgrass<br>Miscanthus | Cellulosic ethanol | Switchgrass: 20 in./acre[g]<br>Miscanthus:<br>30 in./acre[h] |
| Sugarcane | – | Ethanol | Sugarcane:<br>43 in./acre[i] |

[a]*Kansas State University Agricultural Experiment Station and Cooperative Extension Service, 2007. Corn Production Handbook. C-560, Manhattan, KS, September.*
[b]*Kansas State University Agricultural Experiment Station and Cooperative Extension Service, 1998. Grain Sorghum Production Handbook. C-687, Manhattan, KS, May.*
[c]*Anderson, M., 1994. Agricultural Resources and Environmental Indicators. Washington, DC. US Department of Agriculture, Agricultural Handbook AH-705, December.*
[d]*Aiken, R.M., F.R. Lamm, Aboukheira, A.A., 2011. Water use of oilseed crops. In: Proceedings of the 23rd Annual Central Plains Irrigation Conference. http://water.columbia.edu/files/2011/11/Aboukheira2011Water%20Use.pdf.*
[e]*Bean, B., Marsalis, M., 2012. Corn and sorghum silage production considerations. In: High Plains Dairy Conference. http://amarillo.tamu.edu/files/2010/11/CornandSorghumSilageProductionConsiderations.pdf.*
[f]*Newman, Y.J. Erickson, W. Vermerris, Wright, D., 2010. Forage Sorghum (Sorghum bicolor): Overview and Management. University of Florida: IFAS Extension. https://edis.ifas.ufl.edu/pdffiles/AG/AG34300.pdf.*
[g]*Kansas State University Agricultural Experiment Station and Cooperative Extension Service, 2011. Kansas Switchgrass Production Handbook. Manhattan KS, November.*
[h]*US Department of Agriculture, 2011. Planting and Managing Miscanthus as a Biomass Energy Crop. National Resources Conservation Service. Technical Note No. 4. http://www.nrcs.usda.gov/Internet/FSE_DOCUMENTS/stelprdb1044768.pdf.*
[i]*Carr, M.K.V., Knox, J.W., 2011. The Water Relations and Irrigation Requirements of Sugar Cane (Saccharum officinarum): A Review. Experimental Agriculture 47: 1–25.*

for preirrigation, crop cooling, field preparation, frost protection, chemical application, and harvesting. In 2010, more acres were irrigated than in 2005: 62.4 million acres compared to 61.5 million acres, respectively (Maupin et al., 2014). Arid western states displayed higher irrigated water application rates.

## 3.2 Water Usage Across the Biofuel Production Supply Chain

Water is consumed in all stages of the biofuel production supply chain, illustrated in Fig. 2. The supply chain can be subdivided into water used by (1) the agricultural sector, (2) the biofuel industry, and (3) consumers.

At the agricultural level, the production of feedstock requires water as an input, which comes from irrigation or from rainfall. Irrigated water is withdrawn from either groundwater or surface water sources. Whether from rainfall or irrigation, only some of the water is absorbed by the feedstock, the rest being lost through evapotranspiration; as runoff into water bodies; or through the soil. Runoff can carry excess fertilizer and pesticides into surface or underground water sources, possibly compromising water quality. This in turn can cause the growth of excess algae, their decomposition, and consequential consumption of oxygen in the water, creating "dead zones" where fish cannot live (NRC, 2008). Runoff can be reduced with practices that improve fertilizer application efficiency by considering each crop's requirements and with residue management practices. The water that is absorbed by the plant or lost through evapotranspiration is called "consumptive use," having no reuse during one hydrologic cycle (Fingerman et al., 2010). Evapotranspiration rates vary across biofuel feedstocks.

Biofuel refineries use water for heating, cooling, and drying. This water is usually withdrawn from wells or surface water sources. During ethanol production some water is lost through steam and evaporation (NRC, 2008), but part of the water used by ethanol refineries can be recycled and reused. The two main waste streams from ethanol production

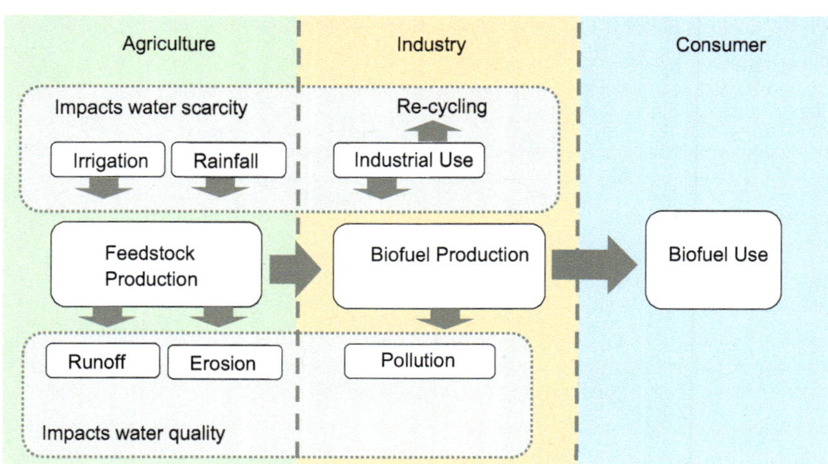


FIGURE 2 Water usage across the biofuel supply chain. *Adapted from Fingerman, K.R., Torn, M.S., O'Hare, M.H., Kammen, D.M., 2010. Accounting for the water impacts of ethanol production. Environmental Research Letters 5 (1), 1–7.*

are the discharge of salts from cooling towers and boilers and brine effluent from processing of pure water (NRC, 2008). In contrast to agriculture, the amount of water a refinery needs for ethanol production is independent of its location.

King and Webber (2008) provide estimates for water usage per mile driven for light-duty vehicles. Ethanol and biodiesel from irrigated feedstocks have an average water consumption and withdrawal of 28–36 gallons/mile for corn and 8–10 gallons/mile for soybeans (King and Webber, 2008). Another impact of water use in biofuels is with the blending with other fuels consumed. For example, a higher presence of ethanol in gasoline sold enhances the dissolution of the fossil fuel, such that in the event of fuel spills, toxic compounds from the fuel may reach further distances, impacting water quality in a larger area (NRC, 2008).

## 3.3 Water Usage in Feedstock Production

Water is a valuable input in crop production. In 2007, 40% of the value of US agricultural production came from irrigated farms (Schaible and Aillery, 2012). Irrigation demand is dependent on location, atmospheric conditions, and crop growth stage (Wu et al., 2009). The majority of irrigation occurs in the states where the average yearly precipitation is below 20 in. (Maupin et al., 2014). Out of the major biofuel feedstocks, corn is the most likely to come from irrigated land.

The impact of biofuel feedstock production is not limited to water use. Feedstock production requires fertilizers, herbicides, and pesticides. Given that the amount applied of these inputs is not totally absorbed by the crop, the remainder can be carried away by runoff into surface and groundwater sources. As mentioned earlier, elevated nitrogen runoff can result in "dead zones," which have occurred where the Mississippi River empties into the Gulf of Mexico and the Chesapeake Bay in Virginia (NRC, 2008). Increased biofuel feedstock production with higher rates of input usage may pose further water quality risks in the future.

Water requirements for and impacts on water from biofuel feedstock production are feedstock specific. For example, runoff containing excess herbicides and pesticides is less likely from fields planted to soybean than corn, since corn production requires higher amounts of fertilizer and pesticide (NRC, 2008). Even lower application rates are required for native grasses, which can serve as a cellulosic biofuel feedstock.

### *3.3.1 Corn Grain*

Corn ethanol production impacts water availability when produced in states that require supplemental irrigation (eg, Nebraska, Kansas, and other states in the Midwest and High Plains) (Stone et al., 2010). The impact on water scarcity has become more significant with the expansion of corn production into areas that were traditionally used for dryland farming (Barton and Clark, 2014). In the High Plains, the main supply of water for agricultural uses is the Ogallala Aquifer, located beneath 45 million hectares spanning eight states. Since the Ogallala recharges slower than the rate of water withdrawals, there has been a decline in water tables across the aquifer over time (Stone et al., 2010). Around 70% of the corn grown in the United States requires 10–17 L of total water (rainfall plus irrigation) to produce 1 L of ethanol from corn (Wu et al., 2009). Growing 160 bushels of corn per acre in Nebraska requires approximately 25 in. of water. Thus each cubic meter of corn used for ethanol needs 1553 $m^3$ of water (Stone et al., 2010). Given the low amounts of rainfall in the western United States and on the High Plains, when corn prices are high, there is a strong economic incentive to irrigate corn to increase yields and crop revenue.

The majority of irrigation in agriculture goes to corn production: 15.4 million acre-feet annually (Barton and Clark, 2014). Demand for irrigation water for corn has increased, primarily because of an increase in land used to grow corn. Additionally, 87% of irrigated corn is grown in areas under water stress: Nebraska, Kansas, California, Colorado, and Texas (Barton and Clark, 2014). The majority of ethanol plants are concentrated in the US Corn Belt, a region with mainly rain-fed corn. Corn production in rain-fed areas (eg, states of Iowa, Indiana, Illinois, Ohio, Missouri, Minnesota, Wisconsin, and Michigan) requires less water per kilogram of corn produced, 3–6 L (Wu et al., 2009). Taking 2014 as an example, 5.2 billion bushels of corn went into the production of ethanol, equating to water requirements of approximately 814 million acre-inches of water (using the water requirements from Nebraska).

While the states in the Corn Belt may not be situated in areas with very high water stress (with the exception of maybe Nebraska and parts of Kansas) they are located in an area with high nitrogen pollution from agriculture (Barton and Clark, 2014). Expansions of corn production in these states could exacerbate existing problems, given the high application rates of fertilizer and pesticides for corn (NRC, 2008).

### 3.3.2 Grain Sorghum

Grain sorghum or milo is another crop that has been used for ethanol production, primarily in Kansas. Over 6.4 million acres of grain sorghum were harvested in 2014, amounting to 433 million bushels of production (USDA, 2015). Grain sorghum is a desirable crop, highly resistant to drought and can withstand water logging better than other cereal crops. In addition, sorghum varieties have been known to thrive better on marginal lands than other cereal crops (Saballos, 2008). Reduced water and input use by sorghum has made the EPA consider sorghum an approved advanced biofuel feedstock when combined with other greenhouse gas-reducing technologies (Schill, 2012).

### 3.3.3 Soybean and Other Oilseeds

The United States produces around a third of the world's soybean. Among the oilseed crops, soybean contains a small amount of oil (18%), in comparison to canola or rapeseed (40%), but still remains the largest feedstock for biodiesel production in the United States (Hay, n.d.). Production of biodiesel from oilseeds is under heavy competition as demand for vegetable oils has increased in recent years, limiting biodiesel consumption to about 5% of petroleum diesel fuel consumption (Van Gerpen, 2007). A potential alternative oilseed that shows promise is rapeseed. Stephenson et al. (2008) show, using life cycle analysis, that production of biodiesel from rapeseed may provide greater reductions of greenhouse gas emissions compared to low sulfur diesel alternatives. Pimentel and Patzek (2005) explain that soybean can yield (without nitrogen) up to 2668 kg/ha and it takes 5556 kg of soybean to produce 1000 kg of oil. This amounts to about 484 kg/ha of oil from soybean. In contrast, canola or rapeseed can yield upward of 3000–4000 kg/ha, producing 1345–1794 kg of oil per hectare (Herkes, 2014). Thus more hectares of soybean are needed to produce 1000 kg of oil than that of canola or rapeseed.

Soybean is often rotated with corn, although rotations with wheat, sorghum, and other crops are common. In 2012, 14% of all irrigated acres were planted to soybean, representing the second largest irrigated cash row crop in western states where water is scarcer (Schaible and Aillery, 2015). The amount of water required for soybean production, in comparison to corn production, depends on soil and climate. For instance, corn grown in the Pacific and Mountain regions needs less irrigation than soybean grown in the same areas. A reversed situation exists in the Northern and Southern Plains (Wu et al., 2009).

### 3.3.4 Agricultural Residues

Wood and agricultural residuals can be used to produce cellulosic ethanol, including corn stover and cobs, sorghum stover, wheat straw, among others. Much of the available agricultural and crop residues are located in the Great Plains, Midwestern United States, and along the Mississippi River. The removal of crop biomass during or after harvest may provide a farmer with an additional revenue source, but it may also have negative side effects, such as a decrease in soil organic matter (SOM) and an increase in runoff rates (USDA, NRCS, 2014a). Decreased soil organic matter or increased runoff rates can lead to immediate and long-term costs to farmers.

When crop residues are removed for sale, producers are reducing the amount of SOM that will be present in the future. One consequence is reduced water-holding capacity or water infiltration rates (USDA, NRCS, 2014a). Water-holding capacity refers to the amount of water stored in soils that is available for crop use (Currel, 2011) and water infiltration refers to the portion of water that makes its way into the subsurface soil and rock (USGS, 2016). Simultaneous reductions in both water-holding capacity and infiltration are likely to compound the overall impact on crops. A reduction in SOM reduces access to nutrients for plants in the soil as well (USDA, NRCS, 2014a). If organic matter is diminished, farmers may have to increase applied fertilization rates to compensate.

Runoff is another potential consequence of residue removal. Runoff from rain or snow melt can cause erosion resulting in several negative impacts, including further reductions in SOM (Pimentel et al., 1995). Pimentel et al. (1995) suggest that eroded soil (sediment carried away by runoff) typically contains about three times more nutrients than the soil left behind. Additionally, as water runoff increases, less water enters the soil, leaving less for future crop use. Studies indicate that runoff rates of 20–30% can create significant water shortages (Pimentel et al., 1995; Elwell, 1985). In addition, runoff can carry sediment, nutrients, and pesticides into water bodies impacting local biodiversity and water quality.

### 3.3.5 Dedicated Annual Bioenergy Crops

Dedicated annual bioenergy crops provide a potentially flexible alternative cash crop for farm managers that can be grown in traditional crop rotations. For example, sweet, energy, or forage sorghum varieties may serve as annual bioenergy crops. There are several advantages to these crops: production of high amounts of biomass, drought tolerance, and the ability to be included into existing crop rotations (Calvino and Messing, 2012). For example, sweet sorghum may be an ideal feedstock for biofuel. It is more drought tolerant, requires 36% of the fertilizer used in corn, grows rapidly, and matures early (Stone et al., 2010). For biofuel production, it has been reported in some sorghum varieties that sugar and starch content may be higher in stems that are under drought stress than in stems under well-watered conditions, resulting in equal sugar yield despite reductions in plant growth caused by drought conditions. Sorghum crops have large root systems that can obtain nutrients and water even in poorer soils, making them more robust in marginal conditions (Saballos, 2008).

### 3.3.6 Switchgrass and Other Dedicated Perennial Bioenergy Crops

Dedicated perennial bioenergy crops (eg, switchgrass, miscanthus, and perennial grasses) have distinct advantages and disadvantages. These crops may require a longer-term commitment by the farmer and require up to 2–3 years to become fully established.

Switchgrass, for instance, needs relatively low amounts of water and nutrient inputs to be productive, making it a good choice to plant on marginal land (Sanderson and Adler, 2008). It yields from 4.5 to 8 tons per acre without irrigation (Wu et al., 2009) and a ton of switchgrass can produce from 65 to 90 gallons of ethanol (DOE, 2001). The plant can withstand drought periods at the cost of yield reductions (Holman et al., 2011). Net water consumed in the production of switchgrass ethanol ranges from 1.9 to 9.8 gallons of water to 1 gallon of ethanol. Switchgrass has the advantage that it is deep rooted and uses water and nutrients more efficiently than other biofuel feedstocks (Wu et al., 2009). Furthermore, switchgrass can have positive effects on soil erosion and provide numerous environmental benefits.

## 3.4 Water Usage in Biofuel Production

Biofuel plants use less water than the agricultural sector (Varghese, 2007). Although water consumption at the industrial level is lower, it can still have a local impact as biofuel plants are spatially more concentrated than feedstock production (Fingerman, 2012). In the production of corn ethanol, water is distributed among heating (3%), cooling (53%), and drying (42%) processes and production of the coproduct, distillers dried grains with solubles (2%) (Wu et al., 2009)

Corn ethanol in the United States can be produced from wet mill or dry grind processes. The production of second-generation ethanol is similar to that of corn ethanol, though more complex. Water consumption occurs through evaporation, incorporation into the final product, water discharge, and blow-down from the boiler. Corn ethanol consumes up to 324 L of water to produce 1 L of ethanol, while ethanol produced from switchgrass requires 1.9–9.8 L of water to produce a liter of ethanol (Wu et al., 2009). The biochemical conversion of switchgrass technique needs 9.8 L of water while the thermochemical conversion using gasification requires 1.9 L of water (Wu et al., 2009). Thus refineries require pure water for the production of biofuels, which can be withdrawn from wells (groundwater) or surface water. Generally, a water permit is required (NRC, 2008).

# 4. WATER CONSERVATION STRATEGIES AND OPTIONS ACROSS THE BIOFUEL SUPPLY CHAIN

How biofuel production will impact water resources depends greatly on the type of feedstock produced and the location of its production. Table 3 provides a measure of the water requirements for the production of different feedstocks and its conversion to biofuel. The impact of biofuel production on water use can be reduced by implementing various techniques at the agricultural and industrial levels.

The impact on water from biofuel feedstock production can be lessened in several ways at the agricultural level. Among these are methods that conserve irrigation water, prevent erosion, increase fertilization efficiency, and utilize precision agriculture tools. By incorporating such strategies, producers can harvest feedstocks with less concern over exposure to water-based risks. Multiple practices can help increase water use efficiency. Subsurface

**TABLE 3** Water Requirements for Feedstock Production and for Biofuel Production

| Crop | Water Requirements ($m^3$ Water/Mg Crop) | Biofuel Conversion (L Fuel/Mg Crop) | Crop Water Requirement for Biofuel ($m^3$ Water/ Mg Fuel) |
|---|---|---|---|
| World corn (grain) | 833 | 409 | 2580 |
| World sugarcane | 154 | 334 | 580 |
| Nebraska corn (grain) | 634 | 409 | 1968 |
| Corn stover | 634 | 326 | 2465 |
| Switchgrass | 525 | 336 | 1980 |
| Grain sorghum | 2672 | 358 | 9460 |
| Sweet sorghum | 175 | 238 | 931 |

Adapted from Stone, K.C., P.G. Hunt, K.B. Cantrell, Ro, K.S., 2010. The potential impacts of biomass feedstock production on water resource availability. Bioresource Technology 101, 2014–2025.

drip irrigation, rainfall harvesting, use of reclaimed water, and microwave remote sensing to monitor soil moisture and weather are just a few options available (NRC, 2008). Techniques that require fewer underground water withdrawals leave more water available long term compared to techniques that lower surface water withdrawals (NRC, 2008). The importance of such tradeoffs is likely to vary, perhaps both geographically and between individual farmers.

There are many approaches to combat erosion impacts on water. Conservation buffers or the use of conservation tillage such as no-till or strip-till, for example, are all effective in reducing erosion and sediment runoff. In the case of conservation tillage practices, policies such as the use of incentive payments have been available since 1985 and are credited with decreasing annual cropland erosion from 2.98 billion tons annually in 1982 to 1.67 billion tons in 2012 (USDA, 2015). Other options to combat erosion include focusing production on feedstocks that require less irrigation, fertilizer, and pesticides and provide better protection against erosion or produce higher amounts of biofuel per planted acre.

The primary focus of this section will be on the agricultural sector, given its predominance in the quantity of water used along the biofuel supply chain, but strategies and policies will be discussed at the industrial level as well.

## 4.1 Irrigation and Water Conservation

Chiu et al. (2009) estimate that a liter of ethanol requires from 5 to 2138 L of water. This estimation is from the farm to the pump and varies depending on the irrigation practices used on the farm. The authors find that water conservation measures in biofuel production should focus on reducing irrigation amounts instead of on the water usage at the biorefinery level.

Precision agriculture technologies, particularly with respect to field operations and input use, have made vast improvements in farmers' efficiency. These technologies involve applying the right product in the right place in the right amount at the right time (Ess and Morgan, 2010), enabling producers to maximize production and minimize costs (McBratney et al., 2005). Applying the correct amount of water in the right parts of a field saves water, prevents runoff, and minimizes pumping costs (Howell, 2001). However, farmers have been slow to adopt precision technologies because of variations in farm type, size, technology cost, and uncertainty in terms of returns to the technology (Evans et al., 2013; Schoengold and Sunding, 2014; Lichtenberg et al., 2015).

In general, site-specific water application has become more widely available, but has not yet been widely adopted because of high marginal costs (Evans et al., 2013). Thus it has been used primarily to reduce water on noncropped areas. Water conservation from precision technologies will come from not watering noncropped areas. To do this, irrigation systems will need to be programmed to water only cropped areas and to apply only the amount needed by the crop (Sadler et al., 2005). In addition, applying proper water amounts based on soil types will reduce the potential for runoff, which is especially important when chemicals or fertilizers are applied through the irrigation system (Basso et al., 2013). Adoption should increase as restrictions on "wasteful" water use continue to mount, but a positive return will need to be seen before widespread adoption occurs. Farmers using best management practices are more likely to adopt precision irrigation technology since they are more likely to recognize its environmental and economic benefits (Pannell et al., 2006; Lambert et al., 2015).

In 2005, water use for irrigating crops in the United States was about 128 billion gallons per day (U.S. Geological Survey, 2015), falling to 115 billion gallons per day by 2010 (Maupin et al., 2014). Reductions in water use for irrigation can be attributed to more precise application practices. Drip irrigation, for instance, can reduce water use by 20–30% compared to sprinkler-type irrigation systems (EPA, 2016b). However, the adoption of these systems is slowed by water pumping prices, crop type and returns, and land characteristics (Green et al., 1996). In addition, low-pressure systems (eg, drip technology) are more likely to be installed on perennial crops than on annual crops (Schoengold and Sunding, 2014). To make adoption more feasible, drip irrigation may need to be combined with other technologies. For example, drip irrigation systems can be coupled with soil moisture sensors, which determine the soil moisture required by various crops, to reduce water use.

A primary driver for the adoption of any technology is cost. Farmers investing in drip irrigation must consider the large fixed costs (Schoengold and Sunding, 2014). Other factors affecting adoption include farm size, operator education, percentage of income from production, characteristics of sensor-based irrigation systems, reductions in production loss, improvements in product quality, irrigation efficiency, and irrigation management (Lichtenberg et al., 2015). Low expected input prices positively affect the likelihood of a producer adopting precision irrigation technology, while increased variability in profits and risk aversion negatively affect adoption (Schoengold and Sunding, 2014). As mentioned, the size of an agricultural operation can impact the likelihood of adoption. Large grain and oilseed farms in the Midwest and Great Plains are more aware of precision technologies (Daberkow and McBride, 2003).

The US Department of Agriculture's (USDA) Natural Resources Conservation Service offers cost-share for irrigation technologies. The Conservation Stewardship Program will help farmers install more efficient irrigation systems or provide cost-share assistance to farmers to begin or continue the use of variable rate technology. The payment is limited to $40,000 per year and may not exceed $200,000 over the period 2014–18. The Agricultural Management Assistance Program also provides cost-share assistance to construct windbreaks and to improve irrigation efficiency, water quality, and conservation through the use of conservation practices that control soil erosion in states with low participation in federal crop insurance (USDA, NRCS, 2014c).

## 4.2 Residue Management Strategies

Half the sediment that reaches waterways comes from soil erosion (NRC, 2008). This can negatively impact water quality. An increased demand for the production of biofuel crops could also mean a reduction in the amount of land enrolled in environmental programs such as the USDA's Conservation Reserve Program. These negative indirect impacts indicate the need for producers to consider the use of residue management strategies when using agricultural residues as an input for biofuels.

When putting together a residue management strategy, there are multiple practices that can be incorporated to limit or offset the impacts of residue removal. Some estimates suggest that the removal of 30% of residues has little to no impact on runoff or soil loss under no-till scenarios (Andrews, 2006; Lindstrom, 1986). This suggests conversion to a no-till operation, where the soil surface is not disturbed by tillage implements or cultivation except for planting and fertilization. The main benefits are protection from wind and water erosion. Cover crops, grown in rotation between regular cash crop production periods to provide soil protection and improvements, can be used as an alternative to or in conjunction with these practices (Sullivan, 2003). Only harvesting a portion of the available crop residue is another strategy (Anand et al., 2011). These practices can be used alone or in a suite of practices.

The adoption of these residue management practices will have an impact on water use, soil productivity, and water quality. By leaving some crop residue on a field's surface, a producer will avoid decreases in soil organic matter and protect the soil surface from wind and water erosion (USDA, NRCS, 2014a,b). Organic matter has a "sponge-like" characteristic that allows it to hold up to 90% of its weight in water being released later for crop use (USDA, NRCS, 2014a,b). A case study by Anand et al. (2011) indicates that it may be possible to balance economic returns and ecosystem services when harvesting stover for biofuels. Using data from research farms in Minnesota and assuming a biomass price of $50 per dry ton, they estimate a farmer can maximize profit at $164.73 per acre by adopting no-till and harvesting 81% of available corn stover. Particularly noteworthy is that this maximum profit was attained when supplemental nitrogen was not added. Within each of the scenarios considered, the economic objective was to maximize profits while maintaining minimum residue and soil organic content requirements. These results suggest that it may be possible for farmers to profitably remove biomass with limited environmental impacts.

## 4.3 Water Conservation and Biofuel Production

The amount of water needed to produce a gallon of biofuel varies depending on the feedstock used (NRC, 2008). Considering the top producing states of ethanol and/or biodiesel, the amount of water consumed annually was estimated, assuming full production capacity (Table 4). For example, Iowa, with the highest ethanol production capacity and second highest biodiesel production capacity, would require 16.7 billion gallons of water annually to produce 4.0 billion gallons of ethanol and 285 million gallons of biodiesel. In terms of the United States, 62 billion gallons of water per year would be needed to produce 14 billion gallons of ethanol and 2.1 billion gallons of biodiesel.

A biofuel refinery's location is of importance when considering the impact of biofuels on water. Plants located in the states on the High Plains aquifers (eg, Nebraska and Kansas) consume potentially over 9 billion gallons of water per year (Table 4). While this amount may only represent 1% of the total daily withdrawals for all purposes, it nevertheless puts a strain on water resources that are being pumped at higher rates than the recharge rate (NRC, 2008). Thus the location of the plant poses a concern, particularly in areas where

**TABLE 4** Water Use in Ethanol and Biodiesel Production for Top Producing States in the United States

| | Total Operating Capacity (Millions Gallons/Year)[a] | | Water Consumption (Millions Gallons/Year)[b] | | |
|---|---|---|---|---|---|
| **State** | **Ethanol** | **Biodiesel** | **Ethanol[c]** | **Biodiesel[d]** | **Total** |
| Iowa | 3968 | 285 | 15,872 | 855 | 16,727 |
| Nebraska | 1897 | 3 | 7588 | 9 | 7597 |
| Illinois | 1384 | 196 | 5536 | 588 | 6124 |
| Minnesota | 1129 | 125 | 4516 | 375 | 4891 |
| South Dakota | 1019 | – | 4076 | – | 4076 |
| Indiana | 936 | 120 | 3744 | 360 | 4104 |
| Ohio | 528 | 65 | 2112 | 195 | 2307 |
| Wisconsin | 506 | 29 | 2024 | 87 | 2111 |
| Kansas | 479 | 3 | 1916 | 9 | 1925 |
| North Dakota | 360 | 85 | 1440 | 255 | 1695 |
| Texas | 205 | 315 | 820 | 945 | 1765 |
| Missouri | 256 | 195 | 1024 | 585 | 1609 |
| Washington | – | 107 | – | 321 | 321 |
| Mississippi | – | 85 | – | 255 | 255 |
| US Total | 13,966 | 2087 | 55,864 | 6261 | 62,125 |

[a]*USDA, 2015.*
[b]*NRC, 2008.*
[c]*Estimate based on consumptive water use of 4 gallons of water per gallon of ethanol.*
[d]*Estimate based on overall water use of 3 gallons of water per gallon of biodiesel.*

water is drawn from confined and scarce sources (eg, the Silurian-Devonian, Cambrian-Ordovician, and Ogallala aquifers). A plant with a capacity of 100 million gallons per year requires an amount of water equivalent to the water supply of a town of around 5000 inhabitants (NRC, 2008). Apart from water consumption, wastewater discharge maybe a local concern. Aden (2007) argues that many ethanol plants have little wastewater discharge because they recycle the processed water by using a combination of centrifuges, evaporation, and anaerobic digestion.

Although an ethanol plant may require less water than feedstock production, it can generate a local problem because of water consumption needs. Water has become a barrier to the installation of ethanol plants. For instance, a plant owned by Cargill, Inc. in Pipestone, Minnesota, could not be supplied with the water they needed from the Lincoln Pipestone Rural Water System (Keeney and Muller, 2006). Water used for distillation can be reduced by producing broths highly concentrated with ethanol or through alternative technologies such as pervaporation. During the cooling process, water usage can be reduced further by using forced-air fans (Aden, 2007).

Wu et al. (2009) report that since the 1990s the amount of water needed to produce a gallon of ethanol has decreased. In fact, a database maintained by the state of Minnesota shows that from 1998 to 2005 there was a reduction by 20% of the water used by corn ethanol plants. Water use can be decreased by using techniques that optimize the process, such as capturing the vapor from the dryer for reuse and recycling the broiler condensate to reduce its make-up rate (Wu et al., 2009). Furthermore, modern plants can treat municipal waste water for use in the production, which reduces the need to pump water from the ground. This also makes it suitable to locate plants closer to cities and near wastewater treatment plants (Keeney and Muller, 2006).

## 4.4 Water Conservation Policies

To reduce the impacts of biofuel production on water resources, current policies focus on motivating biofuel production from feedstocks that require less water, promoting adoption of efficient water use technology and conserving environmentally sensitive areas.

### 4.4.1 Agricultural and Land Use Policies

Financial and technical assistance is provided to agricultural producers willing to implement conservation practices through the Environmental Quality Incentive Program (EQIP). Programs that provide financial assistance or payments for agricultural practices that promote water conservation and water use efficiency are especially important, since under 10% of farms with irrigation use advanced on-site water management decision tools. Examples of these tools include sensing devices for soil or plant moisture, services to schedule irrigation, and computer-run crop growth simulation models (Schaible and Aillery, 2012). EQIP financial assistance programs can be coupled with other institutional water management measures, such as groundwater management, water markets and Conservation Reserve Enhancement Program land easements (Schaible and Aillery, 2012).

### 4.4.2 Industry-Related Policy

At the industrial level, refineries require a permit to discharge waste water. Under the National Pollutant Discharge Elimination System firms request permits from the state that allow them to discharge total dissolved solids, acidity, iron, residual chlorine, and total suspended solids (NRC, 2008). The impacts of biofuel production on water could be reduced by implementing subsidies that motivate water reuse or recycling.

## 5. CONCLUSION

Water is scarce. Biofuel production has expanded significantly because of strong government policies aiming for reduced carbon and greenhouse gas emissions, as well as ensuring US energy security. Nevertheless, an increase in biofuel production has an impact on the use of and quality of water. Thus water conservation and protection must be considered when making biofuel production decisions and policy.

The RFS program has stimulated biofuel production by mandating production levels for alternative biofuels to be met by 2022. This has increased the demand for biofuels causing significant impacts in agricultural markets and on natural resources, including water. In addition, there is pressure on the production of alternative feedstocks to serve as an input for advanced biofuel production. The production of these feedstocks and the continuation of biofuel production from traditional feedstocks place pressure on existing water resources as the demand for irrigation continues to increase and agricultural biomass is removed from fields for cellulosic biofuel production. To cope with this increased pressure, a number of strategies for conserving water and increasing water use efficiency were presented.

The production of biofuels from agricultural residuals presents an opportunity to increase water use efficiency at the agricultural level. This production though must be conducted sustainably because agricultural residuals, such as corn stover, have the benefit of protecting the soil, reducing erosion, improving water use efficiency, and reducing runoff from agricultural fields. The temptation to remove all crop cover as a value-added income stream for biofuel production may have adverse impacts on water use and water quality. Further options for water conservation include the use of more efficient irrigation technologies, crop rotations with drought-tolerant feedstock varieties, and policies promoting water conservation strategies. At the industrial level, refineries could be motivated toward the use of waste water and water recycling. Through all these mechanisms, water use and demand may decrease, with the added benefit of further protecting the integrity of our ground and surface water sources.

The impact of biofuel production on water does not imply that biofuel production should be hindered. But it is important to be conscious of its impacts on water supply and quality to promote policies and strategies to guide biofuel feedstock production, biorefinery locations, and the adoption of water conservation practices to protect this scarce natural resource, ensuring its use for future generations.

## ACKNOWLEDGMENTS

Graduate student support and funding for the preparation of this chapter was provided by CNH: Coupled Climate, Cultivation and Culture in the Great Plains: Understanding Water Supply and Water Quality in a Fragile Landscape, NSF, CNH Award #1313815; and Collaborative Research: Land Change in the Cerrado: Ethanol and Sugar Cane Expansion at the Farm and Industry Scale—NSF BCS-1227451.

## REFERENCES


Aden, A., 2007. Water usage for current and future ethanol production. Southwest Hydrology 6 (5), 2–23.

Anand, M., Archer, D.W., Bergtold, J.S., Canales, E., 2011. Balancing feedstock economics and ecosystem services. In: Braun, R., Karlen, D., Johnson, D. (Eds.), Sustainable Feedstocks for Advanced Biofuels: Sustainable Alternative Fuel Feedstock Opportunities Challenges and Roadmaps for Six U.S. Regions. Soil Water Conservation Society, Ankeny, IA.

Andrews, S.S., 2006. Crop Residue Removal for Biomass Energy Production: Effects on Soils and Recommendations. U.S. Department of Agriculture: Natural Resources Conservation Service. http://www.nrcs.usda.gov/Internet/FSE_DOCUMENTS/nrcs142p2_053255.pdf.

Barton, B., Clark, S.E., 2014. Water & Climate Risks Facing U.S. Corn Production: How Companies & Investors Can Cultivate Sustainably. Ceres. http://www.ceres.org/resources/reports/water-and-climate-risks-facing-u.s.corn-production-how-companies-and-investors-can-cultivate-sustainability.

Basso, B., Kendall, A., Hyndman, D.W., 2013. The future of agriculture over the *Ogallala aquifer*: solutions to grow crops more efficiently with limited water. Earth's Future 1, 39–41.

Calvino, M., Messing, J., 2012. Sweet sorghum as a model system for bioenergy crops. Current Opinion in Biotechnology 23, 323–329.

Chiu, Y.-W., Walseth, B., Suh, S., 2009. Water embodied in bioethanol in the United States. Environmental Science and Technology 43, 2688–2692.

Curell, C., 2011. Why Is Soil Water Holding Capacity Important? Michigan State University Extension. http://msue.anr.msu.edu/news/why_is_soil_water_holding_capacity_important.

Daberkow, S.G., McBride, W.D., 2003. Farm and operator characteristics affecting the awareness and adoption of precision agriculture technologies in the U.S. Precision Agriculture 4, 163–177.

De Fraiture, C., Giordano, M., Liao, Y., 2008. Biofuels and implications for agricultural water use: blue impacts of green energy. Water Policy 10 (Suppl.), 67–81.

Dufey, A., Grieg-Gran, M. (Eds.), 2006. Biofuels Production, Trade and Sustainable Development Emerging Issues. International Institute for Environment and Development, London.

Energy Information Administration (EIA), 2013. U.S. Ethanol Production and the Renewable Fuel Standard RIN Bank. http://www.eia.gov/todayinenergy/detail.cfm?id=11551.

Energy Information Administration (EIA), 2016. Table 10.4 Biofuels and Other Fuels Overview. http://www.eia.gov/totalenergy/data/monthly/pdf/sec10_8.pdf.

Elwell, H.A., 1985. An assessment of soil erosion in Zimbabwe. Zimbabwe Science News 19, 27–33.

Environmental Protection Agency (EPA), 2016a. Program Overview for Renewable Fuel Standard Program. http://www.epa.gov/renewable-fuel-standard-program/program-overview-renewable-fuel-standard-program.

Environmental Protection Agency (EPA), 2016b. WaterSense-An EPA Partnership Program. http://www3.epa.gov/watersense/outdoor/tech.html.

Ess, D.R., Morgan, M.T., 2010. The Precision-Farming Guide for Agriculturalists. John Deere Publishing, Davenport, IA.

Evans, R.G., LaRue, J., Stone, K.C., King, B.A., 2013. Adoption of site-specific variable rate sprinkler irrigation systems. Irrigation Science 31 (4), 871–887.

Fingerman, K.R., Torn, M.S., O'Hare, M.H., Kammen, D.M., 2010. Accounting for the water impacts of ethanol production. Environmental Research Letters 5 (1), 1–7.

Fingerman, K.R., 2012. Measuring and Moderating the Water Resource Impact of Biofuel Production and Trade (Ph.D. dissertation). University of California, Berkeley.

Gramig, B.M., Reeling, C.J., Cibin, R., Chaubey, I., 2013. Environmental and economic trade-offs in a watershed when using corn stover for bioenergy. Environmental Science & Technology 47, 1784–1791.

Green, G., Sunding, D., Zilberman, D., Partker, D., Trotter, C., Collup, S., 1996. How does water price affect irrigation technology adoption? California Agriculture 50 (2), 36–40. http://californiaagriculture.ucanr.org/landingpage.cfm?article=ca.v050n02p36&fulltext=yes.

Hay, J.F., n.d. Soybeans a biofuel feedstock. Cropwatch. Institute of Agriculture and Natural Resources. University of Nebraska. http://cropwatch.unl.edu/bioenergy/soybeans.

Herkes, J., 2014. Rapeseed and Canola for Biodiesel Production. Extension. http://articles.extension.org/pages/26629/rapeseed-and-canola-for-biodiesel-production.

Holman, J., Gillen, R., Moyer, J.L., Roberts, T., Harmoney, K., Sloderbeck, P., Dumler, T., Martin, K., Staggenborg, S., Fick, W., Maxwell, S., Thompson, C., 2011. Kansas Switchgrass Production Handbook. Kansas State University Extension Publication MF3018. https://www.bookstore.ksre.ksu.edu/pubs/MF3018.pdf.

Howell, T.A., 2001. Enhancing water use efficiency in irrigated agriculture. Agronomy Journal 93, 281–289.

Keeney, D., Muller, M., 2006. Water Use by Ethanol Plants: Potential Challenges. Institute for Agriculture and Trade Policy, Minneapolis. http://www.iatp.org/files/258_2_89449.pdf.

King, C.W., Webber, M.E., 2008. Water intensity of transportation. Environmental Science & Technology 42, 7866–7872.

Lambert, D.M., Paudel, K.P., Larson, J.A., 2015. Bundled adoption of precision agriculture technologies by cotton producers. Journal of Agricultural and Resource Economics 40, 325–345.

Lichtenberg, E., Majsztrik, J., Saavoss, M., 2015. Grower demand for sensor-controlled irrigation. Water Resources Research 51, 341–358.

Lindstrom, M.J., 1986. Effects of residue harvesting on water runoff, soil erosion, and nutrient loss. Agriculture, Ecosystems and Environment 16, 103–112.

Liu, C., Greene, D.L., 2014. Consumer choice of E85: price sensitivity and cost of limited fuel availability. Transportation Research Record 2454, 20–27.

Luchansky, M.S., Monks, J., 2009. Supply and demand elasticities in the U.S. ethanol fuel market. Energy Economics 31, 403–410.

Maupin, M.A., Kenny, J.F., Hutson, S.S., Lovelace, J.K., Barber, N.L., Linsey, K.S., 2014. Estimated Use of Water in the United States in 2010. Department of the Interior, U.S. Geological Survey Circular 1405, Washington, DC, U.S.

McBratney, A., Whelan, B., Ancev, T., 2005. Future directions of precision agriculture. Precision Agriculture 6, 7–23.

Meyer, S., Paulson, N., 2014. Export market for U.S. ethanol. Farmdoc Daily 4, 43. http://farmdocdaily.illinois.edu/2014/03/export-markets-for-us-ethanol.html.

National Research Council (NRC), 2008. Water Implications of Biofuels Production in the United States. The National Academies Press, Washington, DC.

National Renewable Energy Laboratory (NREL), 2014. The Biofuel Atlas. https://maps.nrel.gov/bioenergyatlas/.

Pannell, D.J., Marshall, G.R., Barr, N., Curtis, A., Vanclay, F., Wilkinson, R., 2006. Understanding and promoting adoption of conservation practices by rural landholders. Australian Journal of Experimental Agriculture 46, 1407–1424.

Pimental, D., Patzek, T.W., 2005. Ethanol production using corn, switchgrass, and wood; biodiesel production using soybean and sunflower. Natural Resources Research 14, 65–76.

Pimentel, D., Harvey, C., Resosudarmo, P., Sinclair, K., Kurz, D., McNair, M., Crist, S., Shpritz, L., Fitton, L., Saffouri, R., Blair, R., 1995. Environmental and economic costs of soil erosion and conservation benefits. Science 267, 1117–1123.

Saballos, A., 2008. Development and utilization of sorghum as a bioenergy crop. In: Vermerris, W. (Ed.), Genetic Improvements of Bioenergy Crops. Springer, New York, NY, pp. 211–248.

Sadler, E.J., Evans, R.G., Stone, K.C., Camp, C.R., 2005. Opportunities for conservation with precision irrigation. Journal of Soil and Water Conservation 60, 371–379.

Sanderson, M.A., Adler, P.R., 2008. Perennial forages as second generation bioenergy crops. International Journal of Molecular Sciences 9, 768–788.

Schaible, G.D., Aillery, M.P., 2012. Water Conservation in Irrigated Agriculture: Trends and Challenges in the Face of Emerging Demands. Department of Agriculture: Economic Research Service, U.S.. http://www.ers.usda.gov/media/884158/eib99.pdf.

Schaible, G., Aillery, M.P., 2015. Irrigation and Water Use: Background. Economics Research Service, U.S. Department of Agriculture. http://www.ers.usda.gov/topics/farm-practices-management/irrigation-water-use/background.aspx.

Schill, S.R., 2012. Sorghum Acres to Grow Due to Drought Tolerance, New Varieties. Ethanol Producer Magazine. http://ethanolproducer.com/articles/9408/sorghum-acres-to-grow-due-to-drought-tolerance-new-varieties.

Schnepf, R., Yacobucci, B.D., 2013. Renewable Fuel Standard: Overview and Issues. Congressional Research Service. https://www.fas.org/sgp/crs/misc/R40155.pdf.

Schnepf, R., 2011. Agriculture-Based Biofuels: Overview and Emerging Issus. Congressional Research Service. https://www.fas.org/sgp/crs/misc/R41282.pdf.

Schoengold, K., Sunding, D.L., 2014. The impact of water price uncertainty on the adoption of precision irrigation systems. Agricultural Economics 45, 729–743.

Service, R.F., 2009. Another biofuels drawback: the demand for irrigation. Science 326, 516–517.

Stephenson, A.L., Dennis, J.S., Scott, S.A., 2008. Improving the sustainability of the production of biodiesel from oilseed rape in the UK. Process Safety and Environmental Protections 86, 427–440.

Stone, K.C., Hunt, P.G., Cantrell, K.B., Ro, K.S., 2010. The potential impacts of biomass feedstock production on water resource availability. Bioresource Technology 101, 2014–2025.

Sullivan, P., 2003. Overview of Cover Crops and Green Manures. Appropriate Technology Transfer for Rural Areas. Available at: www.attra.ncat.org.

U.S. Congress, House of Representatives, December 2007. Energy Independence and Security Act of 2007. Title II-Energy Security through Increased Production of Biofuels; Subtitle A - Renewable Fuel Standard. Government Printing Office.

U.S. Department of Energy (DOE), 2001. Biofuels and Agriculture, a Fact Sheet for Farmers. Oak Ridge National Laboratory. http://bioenergy.ornl.gov.

U.S. Department of Energy (DOE), 2016. Alternative Fuels Data Center. http://www.afdc.energy.gov/.

U.S. Department of Agriculture (USDA), 2015. Crop Production 2014 Summary: January 2015. National Agricultural Statistics Service. ISSN:1057–7823. Online. Available at: http://www.usda.gov/nass/PUBS/TODAYRPT/cropan15.pdf.

U.S. Department of Agriculture (USDA), 2016a. Quick Stats: Corn Acres Planted. http://quickstats.nass.usda.gov/results/08638011-B478-3942-B44F-FA130ADDE283?pivot=short_desc.

U.S. Department of Agriculture (USDA), 2016b. Table 6-Soybean Oil Supply, Disappearance and Share of Biodiesel Use. http://www.ers.usda.gov/datafiles/US_Bioenergy/Feedstocks/table06.xls.

U.S. Department of Agriculture, Foreign Agricultural Service (USDA, FAS), 2015. U.S. Ethanol Exports Rebound in 2014. http://www.fas.usda.gov/data/us-ethanol-exports-rebound-2014.

U.S. Department of Agriculture, Natural Resources Conservation Service (USDA, NRCS), 2014a. Soil Quality Kit-Guides for Educators: Soil Organic Matter. http://www.nrcs.usda.gov/Internet/FSE_DOCUMENTS/nrcs142p2_053264.pdf.

U.S. Department of Agriculture, Natural Resources Conservation Service (USDA, NRCS), 2014b. Conservation Stewardship Program Factsheet. http://www.nrcs.usda.gov/wps/portal/nrcs/main/national/programs/financial/csp/.

U.S. Department of Agriculture, Natural Resources Conservation Service (USDA, NRCS), 2014c. Agricultural Management Assistance Factsheet. http://www.nrcs.usda.gov/wps/portal/nrcs/detail/ri/programs/farmbill/?cid=nrcs144p2_016495.

U.S. Geological Survey (USGS), 2015. Irrigation Water Use. The U.S. Geological Survey Water Science School. http://water.usgs.gov/edu/irdrip.html.

U.S. Geological Survey (USGS), 2016. Infiltration – the Water Cycle. http://water.usgs.gov/edu/watercycleinfiltration.html.

Van Gerpen, J., 2007. Biofuel production. In: Ranalli, P. (Ed.), Improvement of Crop Plants for Industrial End Uses. Springer, New York, NY, pp. 281–290.

Varghese, S., 2007. Biofuels and Global Water Challenges. Institute for Agriculture and Trade Policy, Minneapolis.

Wallander, S., Claassen, R., Nickerson, C., August 2011. The Ethanol Decade: An Expansion of U.S. Corn Production, 2000–2009. U.S. Department of Agriculture – Economic Information Bulletin Number 79. http://www.ers.usda.gov/media/121204/eib79.pdf.

Wright, C.K., Wimberly, M.C., 2013. Recent land use change in the western corn belt threatens grasslands and wetlands. Proceedings of the National Academy of Sciences 110, 4134–4139.

Wu, M., Mintz, M., Wang, M., Arora, S., 2009. Water consumption in the production of ethanol and petroleum gasoline. Environmental Management 44, 981–997.

Zhang, Z., Qui, C., Wetzstein, M., 2010. Blend-wall economics: relaxing U.S. ethanol regulations can lead to increased use of fossil fuels. Energy Policy 38, 3426–3430.

Chapter 2.2.2

# Water Use for Biofuels in Europe

**D. Drabik, T.J. Venus**
*Wageningen University, Wageningen, The Netherlands*

## 1. INTRODUCTION

By providing tax credits and tax exemptions and by introducing minimum blending requirements for biofuels, the United States and the European Union embarked on promoting the use of biofuels (ethanol and biodiesel) in the early 2000s.[1] Many other countries, including Canada and India, followed. The United States and the European Union implemented their biofuel policies with the objective of reducing greenhouse gas emissions and dependence on imported oil in the transport sector; promoting the security of energy supply by increasing domestically produced energy; promoting technological development and innovation; and providing opportunities for employment and regional development in rural and isolated areas.

The European Union is an important producer and consumer of biofuels in the world market. Fig. 1 presents the production of ethanol (the solid line) and biodiesel (the dashed line) in the EU-28 in the period 2002–2013. While the production of both biofuels was increasing up until 2010, biodiesel production has been significantly higher for two main reasons: biodiesel targets were higher compared to the ethanol targets and the consumption of diesel (with which biodiesel is blended) has historically dominated gasoline consumption. Germany, France, and the Netherlands are the leading biodiesel producers (European Biodiesel Board, 2015).

Although a debate on the effects of biofuels on food commodity prices has been going on for a few years, a debate on water use for production of biofuels appears far less intensive. This is striking as biofuel production and water use are intimately connected through agricultural crops that serve as feedstock for first-generation biofuels (ie, biofuel produced from crops that are also used for food production).

More generally, agriculture is the major source of nitrogen pollution of European water bodies, including lakes, rivers, groundwater, and the European seas (European Environment Agency, 2015). More intensive agricultural crop production in the European Union because of higher food commodity prices (caused in part by biofuel policies) and limited land expansion potential aggravates this pollution even more. The agricultural sector also accounts for a large proportion of water use across Europe, particularly in southern

1. Biofuels were produced also before 2000, but at a much smaller scale.

**Competition for Water Resources. http://dx.doi.org/10.1016/B978-0-12-803237-4.00008-2**

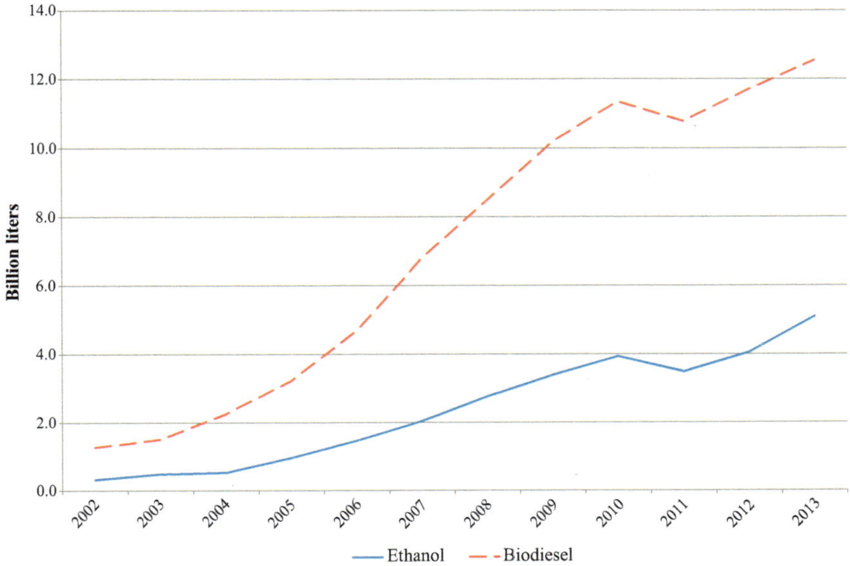


**FIGURE 1** Production of biofuels in the EU-28. Note: Eurostat data in 1000 tonnes of oil equivalent (toe). 1 toe = 39,683,207.2 British thermal units (BTUs), 1 L of ethanol = 20,103.503 BTUs, and 1 L of biodiesel = 31,251.569 BTUs. *Eurostat, 2015b. Primary Production of Renewable Energy by Type. Reproduced from http://ec.europa.eu/eurostat/tgm/refreshTableAction.do?tab=table&plugin=1&pcode=ten00081&language=en.*

countries where the importance of irrigation means that agriculture can account for as much as 80% of total water use in some regions (European Environment Agency, 2009).

Therefore the objective of this chapter is to provide a closer look at the nexus of biofuels and water use in the European Union, especially from an economic point of view. We focus on first-generation biofuels, as these are currently predominantly produced. Since water use to process feedstock into biofuel is comparatively smaller than the amount of water used during the cultivation of biofuel feedstocks, we focus on the latter.

We argue that technical indicators of water efficiency in biofuel production that ignore market-mediated effects (through prices) of biofuel policies can lead to misleading estimates of water consumption for biofuel production. To solve this issue, one needs to calculate not only how much water corresponds to a unit of energy delivered to the market, but also by how much demand for agricultural crops will decline in response to higher food commodity prices.

We find that by determining the dominant biofuel type (biodiesel), EU energy policies essentially determine the domestic water use in biofuel production and also affect water use in the main biofuels (feedstock) import markets. From an economic point of view, water is one of many inputs used to produce biofuel feedstocks and biofuels. Therefore if properly priced, water codetermines the efficiency of biofuel production; that is, if externalities of water use are properly internalized into the price of water, they will also be taken into account as a weight on the water input to influence the cost of biofuel production.

To better understand the driving forces of EU biofuel production, in the next chapter we discuss the main EU biofuel policies.

## 2. EU BIOFUEL POLICIES

The policies governing the production and consumption of biofuels in the European Union are complex. The complexity has three main dimensions. First, biofuel production and consumption are regulated by the Renewable Energy Directive (RED) (Directive 2009/28/EC) and the Fuel Quality Directive (FQD) (Directive 2009/30/EC). Second, EU biofuel policies are shaped by three EU institutions: the Commission, the Parliament, and the Council. In addition, a number of pro- and antibiofuel lobby groups are active in (re) designing biofuel policies. For example, many EU biodiesel producers are associated in the European Biodiesel Board; ePURE represents the European renewable ethanol industry; and Copa-Cogeca, representing European farmers and their cooperatives, supports the production of first-generation biofuels. On the other hand, nongovernmental organizations such as Transport and Environment or Greenpeace are against land-based (ie, first-generation) biofuels. Third, although the EU directives state general objectives to be achieved and principles to be followed at the EU level, the actual implementation of the biofuel legislation differs across the 28 EU Member States (Table 1).

Large-scale biofuel production in the European Union started only after May 2013, when the EU Parliament and the Council passed Directive 2003/30 on the promotion of the use of biofuels for transport. The objectives of this Directive were to replace diesel and gasoline in the transportation sector to contribute to (1) meeting the EU climate change commitments, (2) achieving environmentally friendly security of energy supply, and (3) promoting renewable energy sources. Directive 2003/30 set an indicative target of 2% by 2005 for each Member State for the share of energy coming from biofuels and other renewable fuels in the total energy of fuels used in the transportation sector; the Directive also stipulated a target of 5.75% by 2010.

It is important to notice that the targets in Directive 2003/30 were (and to this date are) expressed as energy shares, as opposed to volumetric shares used in other countries (eg, the United States or Brazil). Most importantly, however, the targets were not binding, which is indicated by Article 4 of the Directive: "Where appropriate, Member States shall report on any exceptional conditions in the supply of crude oil or oil products that have affected the marketing of biofuels and other renewable fuels." This article implies that as long as a Member State was able to explain why a lower energy share of biofuels had been achieved, no consequences followed. Illustrating the nonbinding character of the target, the share of biofuels in total transportation fuels in the European Union reached 1.65% in 2006 and 4.05% in 2010 (USDA, 2010). Furthermore, 22 out of 27 EU Member States failed to achieve their target for 2010 (European Commission, 2013).

Another big milestone in the development of EU biofuel policies was the year 2009 when the RED and the FQD became EU laws. The RED requires (among other things) that by 2020 at least 10% of the total energy consumed in the EU transportation sector comes from renewable sources. Although it is expected that the lion's share of the target will be met by biofuels, other renewable sources of energy (such as renewable electricity) can also be counted. Unlike Directive 2003/30, the RED explicitly uses the term "mandatory target," although it does not specify any enforcement mechanisms.

Although the RED stipulates an overall blend target (ie, ethanol and biodiesel combined, bar the tiny share of other renewable energy sources), each Member State specifies its own trajectory to achieve the overall 10% goal by 2020 and can set ethanol- and biodiesel-specific submandates.

Another important piece of legislation affecting the production and consumption of biofuels in the European Union is the FQD of 2009. The FQD addresses the reduction in

**TABLE 1** Minimum Biofuel Consumption Target in Energy Content for 2014

| | Overall Target (%)[a] | Ethanol Target (%)[a] | Biodiesel Target (%)[a] | Gasoline Consumption (million liters)[b] | Diesel Consumption (million liters)[b] |
|---|---|---|---|---|---|
| France | 7.57 | 7.00 | 7.70 | 619.7 | 2589.4 |
| Poland | 7.10 | | | 336.5 | 722.9 |
| Slovenia | 7.00 | | | 44.6 | 102.5 |
| Sweden | 6.41 | 3.20 | 8.78 | 244.8 | 303.3 |
| Germany | 6.25 | 2.80 | 4.40 | 1617.6 | 2524.3 |
| Finland | 6.00 | | | 128.8 | 196.0 |
| Lithuania | 5.80 | 3.34 | 6.45 | 19.3 | 85.2 |
| Austria | 5.75 | 3.40 | 6.30 | 143.5 | 451.7 |
| Denmark | 5.75 | | | 122.8 | 185.4 |
| Portugal | 5.50 | | | 105.5 | 303.6 |
| Netherlands | 5.50 | 3.50 | 3.50 | 363.8 | 510.3 |
| Belgium | 5.09 | 2.66 | 5.53 | 109.7 | 544.5 |
| Ireland | 4.94 | | | 109.1 | 184.7 |
| Bulgaria | 4.94 | 3.34 | 5.53 | 40.7 | 112.3 |
| Hungary | 4.90 | 4.90 | 4.90 | 109.7 | 164.0 |
| Romania | 4.79 | 3.00 | 5.53 | 116.6 | 280.8 |
| Luxembourg | 4.75 | | | 30.1 | 143.4 |
| Czech Republic | 4.57 | 2.73 | 5.53 | 144.7 | 290.2 |
| Slovakia | 4.50 | 2.73 | 6.27 | 51.7 | 109.5 |
| Italy | 4.50 | | | 772.4 | 1735.2 |
| Malta | 4.50 | | | 6.8 | 7.9 |
| Spain | 4.10 | 3.90 | 4.10 | 429.1 | 1669.1 |
| United Kingdom | 3.90 | | | 1236.8 | 1924.4 |
| Greece | 2.64 | | | 260.6 | 165.4 |
| Croatia | 2.06 | | | 55.3 | 96.1 |

[a]*Biofuels barometer (2014).*
[b]*Eurostat (2015a), consumption data for 2013.*

life cycle greenhouse gas emissions of transportation fuels by 6% by the year 2020 as compared to 2010. With respect to biofuels, it specifies criteria that need to be met for biofuels to count toward the mandatory consumption targets.

Perhaps the most important of these criteria is a requirement that biofuels should save at least 35% of greenhouse gas emissions compared to fossil fuels they are to replace. This threshold increases to 50% on January 1, 2017. Moreover, from January 1, 2018 the saving shall be at least 60% for biofuels produced in plants that started production on or after January 1, 2017. It is important to note, however, that these specified greenhouse gas emissions savings do not take into account carbon emissions from land use change, a topic that gave rise to a heated debate on biofuels in the European Union after 2012.

Moreover, the FQD allows imports of biofuels or biofuel feedstocks only from countries that have ratified important international conventions such as the Convention on International Trade in Endangered Species of Wild Fauna and Flora, the Cartagena Protocol on Biodiversity, or conventions of the International Labor Organization.

The food commodity price booms of 2008 and 2011 and the intensifying "food versus fuel debate" have been an impetus for the reform of the EU biofuel policy. In October 2012, the European Commission proposed to reform the EU biofuel policy (represented by the RED and FQD directives).[2] The Commission has assigned indirect land use change (ILUC) factors to different biofuels but failed to account them for the climate performance of biofuels. Thus the ILUC factors are used only for reporting purposes. In recognition of adverse inflationary effects of first-generation biofuels on food commodity prices, the Commission proposed to cap the use of these biofuels to 5% of energy. Environmentalists, such as Transport and Environment—a Brussels-based environmental organization—opposed this proposal as it did not mean complete abolition of biofuels produced from food crops.

The reshaping of the EU biofuel policy continued in July 2013 when the European Parliament's Environmental Committee voted for the inclusion of the ILUC factors into the RED and for capping all first-generation biofuels at 5.5% of energy. Later in September 2013 the European Parliament voted to cap the first-generation biofuels at 6% and placed a 2.5% minimum requirement to be achieved by 2020 for advanced (third-generation) biofuels from, for example, seaweed or certain types of waste (European Parliament, 2013). In June 2014 the Council of energy ministers decided to cap the use of land-based biofuels to 7% and to put a 0.5% floor on advanced biofuels.[3] After long discussions, the EU Parliament finally approved the Council's proposal on April 14, 2015. These policy developments will have long-term implications for water use in biofuel production in the European Union as they cap the use of land-based biofuels, the production of which requires significant quantities of water.

Besides biofuel policies, water protection regulations also affect the biofuels–water nexus in the European Union. This we discuss next.

## 3. WATER PROTECTION REGULATIONS IN THE EUROPEAN UNION

The Water Framework Directive (WFD) (Directive 2000/60/EC) is the key regulation that protects water resources in the European Union. It is the first directive to consider not only

2. http://ec.europa.eu/clima/policies/transport/fuel/docs/com_2012_595_en.pdf.
3. http://gr2014.eu/sites/default/files/indirect%20land-use%20change_1.pdf.

water quality, but also quantity (Chave, 2001). This is important because biofuel production imposes significant changes in the amount of water used. To improve the quality of natural waters in individual EU Member States, the main objectives of the WFD are to (1) prevent any further deterioration of water bodies, (2) protect and enhance the status of aquatic ecosystems and associated wetlands, (3) promote sustainable water consumption, and (4) contribute to the mitigating effects of floods and droughts. One of the priority issues of the WFD is the adoption of industry-specific measures since in many cases water pollution was caused by specific types of industry that were much more significant in the context of their impact on water quality than others.

The RED, which stipulates mandatory consumption targets for biofuels, also specifies biofuel sustainability criteria that include water use. For example, biofuels are only counted toward the consumption target if they were not made from raw material obtained from wetlands, namely, land that is covered with or saturated by water permanently or for a significant part of the year. In addition, the RED requires each Member State to estimate the biofuel production impact on water resources and water quality (among other indicators) within its territory.

The economic implication of these two directives is that they partially internalize negative externalities related to water use in biofuel production, thus bringing the water price closer to water's social marginal costs. The negative externalities related to water use for biofuels include, for example, over use of water in some areas because of irrigation or pollution of groundwaters with fertilizers because of intensive production of biofuel crops.

Since most water in biofuel production is used during the cultivation of biofuel feedstock, and because a type of feedstock is determined by the type of biofuel, we now look into the main biofuel crops produced in the European Union.

## 4. BIOFUEL TARGETS AND FEEDSTOCK USE IN THE EUROPEAN UNION

Table 1 summarizes the minimum biofuel consumption targets (in energy terms) of selected EU Member States for the year 2014 as reported by the Biofuels barometer. France has the highest target of 7.57 energy percent, while Croatia is at the bottom of the list with 2.06%. Observe that most Member States also specify minimum ethanol and biodiesel submandates. For example, Germany requires that ethanol constitutes at least 2.8% of energy of motor gasoline fuel (ie, gasoline blended with ethanol), and the requirement for biodiesel is 4.4%. Notice, however, that these are minimum requirements since the overall target for Germany is 6.25 energy percent.

Comparing the mandates in the second and third columns of Table 1, we see that of the Member States that have biofuel-specific mandates, a majority (save for the Netherlands and Hungary) favor biodiesel. Because percentages can be misleading if the corresponding bases to which they relate differ, we present the 2013 quantities of gasoline and diesel consumed in the last two columns. (The 2014 Eurostat data were not available at the time of writing.) A higher relative consumption of diesel to gasoline in all EU Member States (except for Greece) puts even more weight on the higher percentage biodiesel submandates. This has an additional effect on the quantity of biodiesel in that it is higher than ethanol. The specification of submandates has important implications for water use as biodiesel feedstocks have different water balances compared to ethanol crops.

Most biofuels currently used in the European Union are derived from crops that can also be used for food production. The main crops processed into first-generation ethanol are wheat, corn, barley, and sugar beet. Wheat is mainly used for ethanol in northwestern Europe, including the United Kingdom. Corn is mainly used in Central Europe and Spain, whereas barley and rye are processed in Germany, Sweden, Poland, and the Baltics. Germany, France, and the Czech Republic derive ethanol also from sugar beet. Wine and wine by-products are important in regions of Italy (USDA, 2014).

Rapeseed oil has been the dominant biodiesel feedstock in the European Union; in 2013 it accounted for 58% of total biodiesel production (USDA, 2014). Palm oil is the second most important biodiesel feedstock used in the Benelux, Spain, Germany, Italy, and Finland. It has gained popularity because of its lower price compared to other feedstocks. However, the use of palm (and soybean) oil in conventional biodiesel is limited because of technical issues related to the iodine value of the fuel.[4] Soybean oil has primarily been used in Spain, France, Italy, and Portugal. After Austria, Denmark, Finland, France, Germany, Ireland, the Netherlands, and the United Kingdom introduced double-counting of biodiesel produced from used cooking oil, the use of this (recycled) feedstock has increased. Because the demand for biodiesel feedstock in the European Union exceeds the supply, the feedstock is imported either in its unprocessed form (soybeans and rapeseed) and is crushed domestically, or it is imported directly as vegetable oil (approximately 1.5 million metric tons) (USDA, 2014).

Fig. 2 depicts the area under the main crops used for biofuel production in the EU-28. Wheat leads the list with almost double the area of barley and triple that of corn. The area under rapeseed cultivation is comparatively smaller and soybean area is close to zero. The relatively small areas of oilseeds compared to ethanol crops explain the excess demand for biodiesel feedstock that needs to be imported. The figure also shows long-term stability of areas under individual crops, which indicates that the growth in food crop commodity prices in 2008 and 2011 did not affect the relative prices among commodities (de Gorter et al., 2015).

The revision of the EU RED has seen a cap on the use of the first-generation biofuels, thus encouraging the introduction of second-generation biofuels. A 2012 proposal of the EU Parliament and the Council[5] suggested that biofuel produced from the following feedstocks be considered at twice their energy content: used cooking oil, animal fats, nonfood cellulosic material, lignocellulosic material except saw logs and veneer logs. In addition, the proposal also lists a set of biofuel feedstocks that are to be counted four times their energy content toward the mandate: algae, biomass fraction of mixed municipal and industrial waste, straw, animal manure and sewage sludge, palm oil mill effluent and empty palm fruit bunches, tall oil pitch, crude glycerin, bagasse, grape marcs and wine lees, nut shells, husks, cobs, bark, branches, leaves, saw dust, and cutter shavings.

In the next section, we investigate how feedstocks for first-generation biofuels perform in terms of water use.

4. Iodine value is an important parameter describing oil, fat, as well as biodiesel characteristics. Heated fuels with a high iodine value tend to polymerize and form deposits on engine nozzles, piston rings and piston ring grooves.

5. http://eur-lex.europa.eu/legal-content/EN/TXT/PDF/?uri=CELEX:52012PC0595&from=EN.

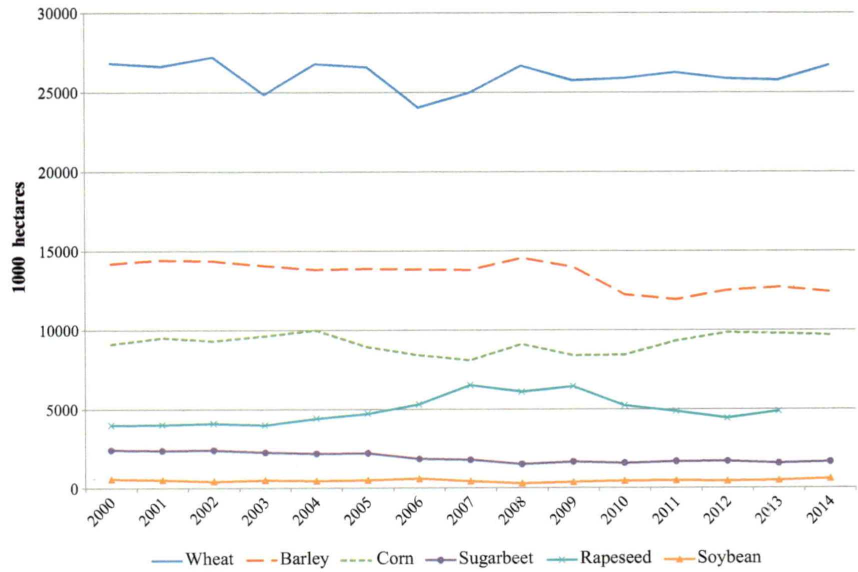


FIGURE 2 Area under the main biofuel feedstocks in the EU-28. *Eurostat, 2015c. Crops Products — Annual Data. Reproduced from http://appsso.eurostat.ec.europa.eu/nui/show.do?dataset=apro_cpp_crop&lang=en.*

## 5. TECHNICAL INDICATORS OF WATER EFFICIENCY IN BIOFUEL PRODUCTION

Most water use related to biofuel production occurs during the cultivation of biofuel feedstock (de Vries et al., 2010). We therefore focus on this stage. We look at two possible criteria that measure water intensity of biofuels and their feedstocks: the virtual water content (VWC) and the water productivity of biofuels (WPB). It should be noted that these technical criteria often used in scientific literature do not take into account commodity market prices and are only partial indicators of water use efficiency as they are unable to reflect other market effects (eg, competition for land). We will discuss an economic evaluation of efficiency of biofuel production in Section 6.

The VWC of a product is defined as the volume of water used to produce the product (eg, a crop), measured at the place where it was actually produced (Chapagain and Hoekstra, 2004, 2007). With respect to biofuels it means the total quantity of water needed to produce a metric ton of a biofuel feedstock (eg, corn or rapeseed). The upper part of Table 2 presents estimates of virtual water contents for the main biofuel feedstocks in selected EU Member States.[6] The VWC varies significantly both across feedstocks and among Member States, reflecting different technological and climatic conditions in cultivation. For feedstocks, sugar beet exhibits the lowest, while rapeseed tends to have the

6. We do not present values for soybeans as their production in the European Union is very low, and more so in individual Member States.

**TABLE 2 Virtual Water Content and Energy From Select Biofuel Feedstocks**

| | Wheat | Corn | Barley | Sugar Beet | Rapeseed |
|---|---|---|---|---|---|
| VWC in $m^3$/ton[a] | | | | | |
| Austria | 981 | 357 | 967 | 60 | 1341 |
| Belgium–Luxembourg | 1168 | 597 | 1237 | 108 | 1841 |
| Czech Republic | 1180 | 564 | 1248 | 93 | 1395 |
| France | 895 | 482 | 886 | 63 | 1390 |
| Germany | 757 | 442 | 826 | 77 | 1128 |
| Greece | 1213 | 706 | 1112 | 121 | NA |
| Hungary | 556 | 666 | 637 | 94 | 539 |
| Italy | 2421 | 530 | 1822 | 117 | 5095 |
| Netherlands | 619 | 408 | 718 | 65 | 1182 |
| Romania | 759 | 1271 | 758 | 190 | 718 |
| Slovakia | 465 | 646 | 584 | 88 | 382 |
| Spain | 1227 | 646 | 1070 | 113 | 3284 |
| Ukraine | 720 | 1362 | 894 | 218 | 664 |
| United Kingdom | 501 | NA | 650 | 56 | 876 |
| | | | | | |
| Energy of biofuel/ton of feedstock[b] (GJ/ton) | 10.17 | 10 | 10.2 | 2.61 | 11.7 |

*VWC*, virtual water content; *NA*, not available.
[a]*Chapagain and Hoekstra (2004).*
[b]*Mekonnen and Hoekstra (2010).*

highest VWC. The pattern is more blurred across Member States, with the VWC in some Member States being a multiple of others (eg, compare wheat in Italy and Slovakia).

The last row of Table 2 presents the gross energy content (ie, unadjusted for the energy input) of biofuels derived from a metric ton of feedstocks. The gross energy content is independent of where the feedstock is produced. Interestingly, the variation in the gross energy content per ton of biofuel feedstocks is much lower than for the VWC. The only exception is sugar beet that yields only 2.61 GJ of ethanol per metric ton. However, this energy "disadvantage" is accompanied by a significantly lower VWC of the crop. To see how much water is needed to produce one GJ of biofuels, one needs to divide the VWC of a crop by the corresponding energy content reported in the last row of Table 2. Then, for example, for corn and sugar beet in Germany we obtain 44.2 $m^3$/GJ (=442/10) and 29.5 $m^3$/GJ (=77/2.61), respectively. Thus, considering only the gross energy yield of a feedstock, the VWC to produce 1 GJ is lower for sugar beet than for corn in Germany.

Although informative, the two measures presented in Table 2 provide little guidance as to what biofuel feedstock could be appropriate (ignoring market prices) if a country has limited availability of water resources to use in biofuel production. It is because the amount of energy used to produce the feedstock is omitted. A more suitable measure is the water productivity of biofuels (WPB). deVries et al. (2010) define this productivity measure as the amount of *net* biofuel energy (ie, deducting the energy needed to produce a biofuel) that is produced using 1 $m^3$ of water lost through evapotranspiration.

de Vries et al. (2010) examine a number of studies to estimate the mean value of the WPB. Their results are summarized in Fig. 3. Biofuel production from oil palm, sweet sorghum, and sugarcane appear relatively water efficient given how much net energy of biofuels is associated with 1 $m^3$ of water used. Intriguingly, sugar beet and rapeseed also perform relatively well. Sugar beet is characterized by a high (fresh) biomass production per volume of water consumed (about double that of sugarcane). However, net energy production of sugar beet ethanol is relatively low because of consumption of large quantities of fossil fuels during processing, while energy required for sugarcane processing is mostly supplied by crop residues (bagasse). Although a metric ton of rapeseed requires significant quantities of water (Table 2), it does at the same time exhibit a favorable net energy yield of 9.1 GJ per ton of processed rapeseed, resulting in a relatively high WPB.

Now we are in a position to evaluate the effects of increased EU biofuel consumption on total water use in EU agriculture. The values in Table 1 suggest that the consumption of biodiesel in the European Union is higher relative to ethanol for two reasons. First, EU Member States mandate higher shares of biodiesel than ethanol, and second, the consumption of diesel in the European Union has historically been greater than the

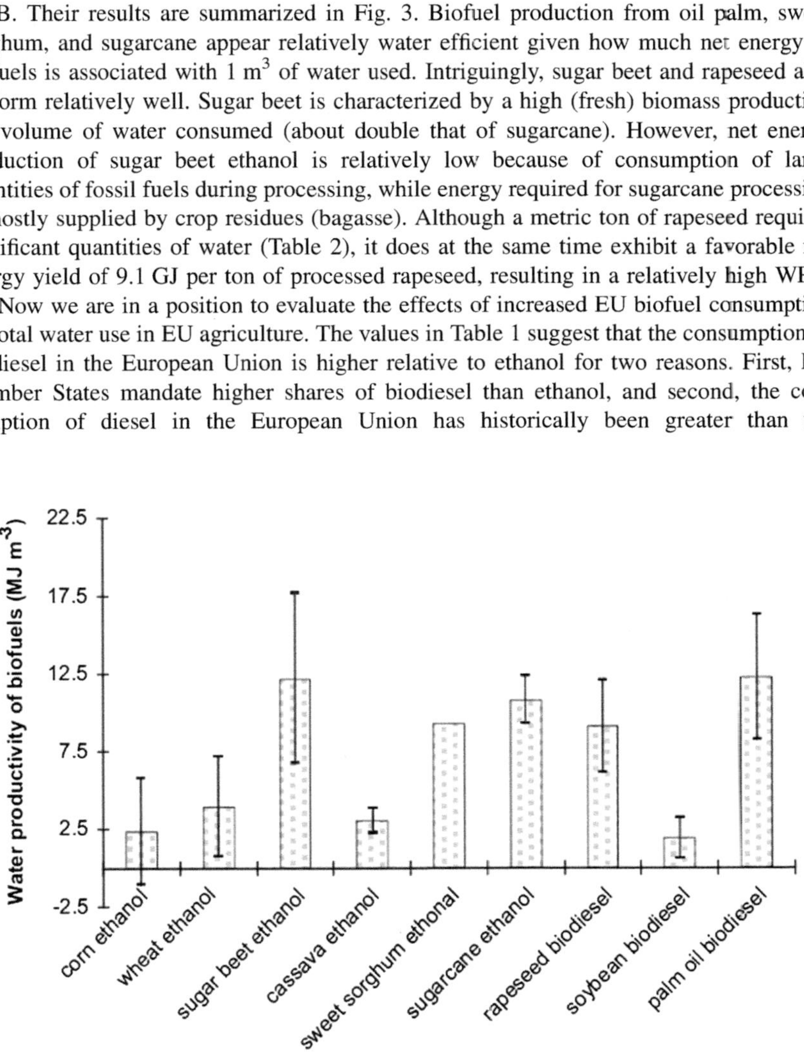


FIGURE 3 Water productivity of biofuels by feedstock. *Modified after de Vries, S.C., van de Ven, G.W.J., van Ittersum, M.K., Giller, K.E., 2010. Resource use efficiency and environmental performance of nine major biofuel crops, processed by first-generation conversion techniques. Biomass and Bioenergy 34, 588–601.*

consumption of gasoline. Combining the observed feedstock use pattern of ethanol (wheat, corn, and sugar beet) and biodiesel (rapeseed and palm oil) with the water productivity of individual feedstocks in Fig. 3, we come to a conclusion that for a given overall biofuel mandate, water use for biofuels decreases with a higher share of biodiesel. The implication of this finding is that if the European Union wants to achieve a 10% share of renewable energy in total transportation energy consumption, and at the same time minimize biofuels-related water use, the blending submandates for biodiesel should increase faster relative to the ethanol submandates.

The earlier conclusion is based on the comparison of the observed biofuel consumption pattern with a counterfactual where ethanol exhibits a higher share. No attention was paid, however, to where the feedstock for the increased biofuel consumption comes from. Fig. 2 shows a stable area under the main biofuel crops in the European Union over time. The stable area of biofuel crops implies unchanged water use for biofuels produced from domestic EU feedstock. However, because of growing biofuel consumption and stable production of domestic feedstock, the difference has to be covered by imports of biofuels or feedstocks (to be processed in biofuels or human consumption). Therefore whether the growing EU biofuels consumption increased or decreased, overall water use depends on the change in the area of biofuels feedstock in the rest of the world.

The imported palm oil is the second most used biodiesel feedstock in the European Union. Because of the favorable water productivity of oil palm relative to other feedstocks (Fig. 3), the additional acreage of oil palm in other countries (mostly Indonesia and Malaysia) could have decreased water use, but only if the crops that otherwise would have been grown in those places were more water-intensive. The net effect is therefore ambiguous because of the EU biofuel policy-induced indirect land use changes (Zilberman et al., 2011; Khanna et al., 2012), and thus cannot be determined a priori without an empirical analysis.

Because these (partial) water efficiency measures do not take into account market prices of biofuels and of their inputs, in the following section we advance a way to incorporate these important market characteristics into the assessment of efficiency of biofuel production.

## 6. A COMPREHENSIVE WAY TO ASSESS THE EFFICIENCY OF BIOFUEL PRODUCTION

To illustrate how to include market effects of biofuel policies to avoid misleading estimates of water consumption for biofuel production, we focus on biodiesel produced from rapeseed oil. Unlike corn and wheat, which are directly processed into ethanol, rapeseed needs to be crushed first, yielding oil and meal. The oil is then processed into biodiesel. Because the two-stage biodiesel production process would complicate our graphical exposition, in the left panel of Fig. 4, we directly present supply of rapeseed oil, $S_{RO}$, that is linked to the underlying rapeseed supply curve. Depicted in the left-hand panel is also the demand for nonbiodiesel oil, $D_{NBRO}$, used, for instance, in human consumption.

If biodiesel were not produced, the intersection of the oil supply and demand curves would determine the oil price denoted by $P_{NB}$, with the subscript NB denoting non-biodiesel. Suppose the oil price increases above $P_{NB}$, for example, because of biodiesel production. Then the quantity of oil supplied exceeds the quantity demanded for non-biodiesel use and the excess supply of oil is diverted to biodiesel production. In Fig. 4 this

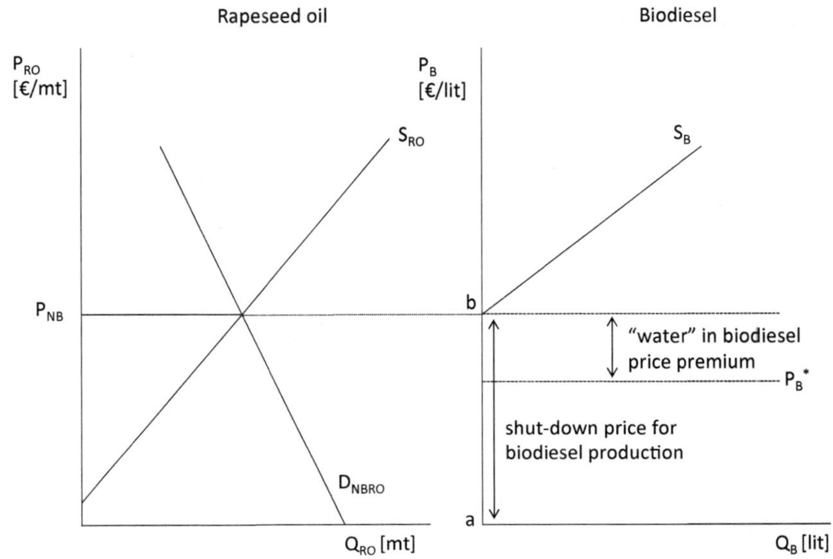


**FIGURE 4** Economic efficiency of biofuel production.

is depicted in the right-hand side panel where the curve $S_B$ is derived as a horizontal difference between the $S_{RO}$ and $D_{NBRO}$ curves in the left-hand panel. Point b on the biodiesel supply curve then corresponds to the price $P_{NB}$ in the left-hand panel. Because the measurement units in both panels differ, the prices and quantities need to be properly converted. Specific equations linking the rapeseed and biodiesel market can be found in Drabik et al. (2014).

One implication of Fig. 4 is that the intercept of the biodiesel supply curve is never at zero, meaning that there is always a positive shutdown price for the industry. Another implication has to do with the relative magnitude of the shutdown price and the free market biodiesel price.

The free market biodiesel price is the price that biodiesel producers would receive if consumers were free to choose a fuel (ie, biodiesel or diesel in our case) based on the number of kilometers a vehicle could travel per unit of a respective fuel, and if no biofuel consumption subsidy (a tax credit or a tax exemption) were provided. Mathematically, this price can be written as (Drabik, 2011)

$$P_B^* = \gamma P_D - (1 - \gamma)t, \tag{1}$$

where $\gamma = 0.91$ denotes kilometers traveled per liter of biodiesel relative to diesel;[7] $P_D$ denotes the diesel market price, and $t$ denotes the fuel tax. The biodiesel price that

7. Calculated as the ratio of the energy content of a liter of biodiesel (31251.6 BTUs, British thermal units) and a liter of diesel (34210.3 BTUs). The actual relative kilometers traveled per liter of both fuels might differ slightly from this ratio because there is not a one-to-one correspondence between energy of fuel and the distance it yields (consider, for example, different driving styles or varying weather conditions during a year).

consumers are willing to pay is lower than the price of diesel for two reasons. First, consumers are willing to pay only 91% of the price of diesel per liter of biodiesel because of fewer kilometers traveled per liter of biodiesel relative to diesel. Second, there is a penalty on blenders because of the volumetric fuel tax. Consumers are willing to pay a fuel tax only on biodiesel that is proportional to its kilometers per liter, $\gamma t$, but blenders have to pay the full tax, $t$. Therefore the difference $(1-\gamma)t$ represents the penalty on biodiesel (reflected in lower biodiesel market prices) because of the volumetric fuel tax. The penalty increases in countries with high fuel taxes and hence makes the production of biodiesel less attractive.

The relative position of the biodiesel free market price and the intercept of the biodiesel supply curve is an empirical issue and depends on the prevailing diesel price and the fuel tax as well as on the relative position of the rapeseed oil supply and demand curves. As Fig. 4 shows, when the free market biodiesel price is below the intercept of the biodiesel supply curve, free market would not support biofuel production. Intuitively, the fossil fuel (diesel) is less expensive than the alternative product (biodiesel).

This has important welfare implications because if biodiesel production does occur because of biofuel policies (eg, the EU Member States' biodiesel targets), then part of the biofuel policy price premium (ie, the difference between the observed and free market biofuel prices) is not effective in increasing the biofuel production; it just fills up the gap between $P_{NB}$ and $P_B^*$. Using the jargon of international economics, de Gorter and Just (2008) term this gap "water" in the biofuel price premium.[8] This means that within the range of "water," a biofuel policy has no effect on feedstock prices. Alternatively, "water" can be thought of as representing the waste of societal resources because diesel (fossil fuel) is less expensive and yet production of more costly biofuel is incentivized through biofuel policies.

Although the economic term "water" (as a measure of policy inefficiency) might be confusing in the discussion of liquid "real" water use for biofuel production, we show that it is useful in explaining why (liquid) water use should not be taken as the key indicator of efficiency of biofuel production.

Consider again the rapeseed oil supply in the left panel of Fig. 4. It is derived directly from the rapeseed supply curve that reflects the (private) marginal production costs associated with rapeseed production. In theory, the marginal cost curve encompasses the competition for land use among crops (eg, wheat or corn vs. rapeseed or soybean); the effects of biofuel policies; and, last but not least, the actual water use in rapeseed production. It is important, however, that water be priced properly so that its true cost is reflected in the social marginal cost of feedstock production. It then follows that water is but one of many components that determine the position of the intercept, and therefore also the level of "water" (ie, the economic term).

Therefore to determine which biofuel is more efficient to produce and from which feedstock, one needs to estimate the level of "water" for every biofuel–feedstock pair and then choose the one with the lowest "water" levels. This is an empirical question left for further research.

8. International economics literature uses the term "water" in an import tariff to represent the difference between bound (ie, the highest permitted) and applied (ie, actual) duties.

## 7. CONCLUSIONS

Water is a key input into production of agricultural crops that are later processed into biofuels. As global consumption of biofuels gradually increases and water becomes scarcer (and more so in different parts of the world), the nexus between biofuels and water becomes more important. In this chapter, we have looked at water use for biofuels in Europe. We have focused on the first-generation biofuels, as these are currently predominantly produced in the European Union, and particularly on the water used during the cultivation of biofuel feedstocks since the amount of water for processing feedstock into biofuel is comparatively smaller.

We find that most EU Member States mandate higher shares of biodiesel relative to ethanol, thus favoring the former. In addition, facing a lower fuel tax, diesel consumption in the European Union has historically been advantaged over gasoline. As a result, more diesel is being consumed in each EU Member State (except for Greece), thus reinforcing the need for biodiesel (as the percentage target applies to a larger base). This implies that by determining which biofuel will be dominant in the European Union, the EU energy policies essentially determine the domestic water use in biofuel production. We show that for a given overall biofuel mandate, water use for biofuels decreases with a higher share of biodiesel.

Because the EU demand for biodiesel is short of supply, many Member States import the biofuel feedstock or biodiesel directly from abroad. This means that EU energy policies also have repercussions for water use in other countries of the world, depending on where biofuels are imported from and the feedstock used. However, since the European Union as a whole is currently consuming only about a half of the 10% target for 2020, and because imports of biofuels (in various forms) play an important role in biofuel consumption, one can expect that EU biofuel consumption will exert even greater pressure on global water use.

Unlike previous scientific literature that determines the efficiency of biofuel production based on partial and mutually disconnected indicators that ignore commodity market prices and market interactions, we stress a holistic economic approach. From an economic point of view, water is one of many inputs used to produce a biofuel feedstock and later a biofuel. Therefore if properly priced (which is likely not the case now), water codetermines the efficiency of biofuel production, but there is no reason to assume that all factors should have the same weights (eg, as in de Vries et al., 2010); the importance of individual inputs is determined by (correct) market prices.

We stress that technical indicators of water efficiency in biofuel production that ignore market-mediated effects (through prices) of biofuel policies can lead to misleading estimates of how much more water is needed because of biofuel production. That is to say, one needs to mechanically calculate not only how much water corresponds to a unit of energy delivered to the market, but also by how much demand for agricultural crops will decline in response to higher food commodity prices.

Finally, dwindling resources of fossil fuels suggest that consumption of biofuels in the European Union will not vanish but is likely to increase over time. So what is the way forward with respect to water use in this scenario? The recent reform of the EU biofuel policies that capped the consumption of first-generation biofuels at 7% of energy is one avenue. Another is improvement of productivity and development of crops with a lower water use; this would improve the efficiency of biofuel production from a water-efficiency perspective. The development of plant cultivars, for example, through a more intensive exploitation of biotechnology, that would require less water, produce higher yields, or be drought resistant would improve the water efficiency of biofuel production as well (eg, Hochman et al., 2008).

## REFERENCES


Biofuels Barometer, 2014. A Study Carried Out by EurObserv'E. http://www.energies-renouvelables.org/observ-er/stat_baro/observ/baro222_en.pdf.

Chapagain, A.K., Hoekstra, A.Y., 2004. Water footprints of nations. In: Volume 1: Main Report. Value of Water Research Report Series. UNESCO-IHE Institute for Water Education, Delft.

Chapagain, A.K., Hoekstra, A.Y., 2007. Water footprints of nations: water use by people as a function of their consumption pattern. Water Resources Management 21, 35–48.

Chave, P., 2001. The EU Water Framework Directive: An Introduction. IWA Publishing, London, UK, 208 pp.

Directive 2000/60/EC of the European Parliament and of the Council of 23 October 2000 Establishing a Framework for Community Action in the Field of Water Policy. http://eur-lex.europa.eu/resource.html?uri=cellar:5c835afb-2ec6-4577-bdf8-756d3d694eeb.0004.02/DOC_1&format=PDF.

Directive 2009/28/EC of the European Parliament and of the Council of 23 April 2009 on the Promotion of the Use of Energy from Renewable Sources and Amending and Subsequently Repealing Directives 2001/77/EC and 2003/30/EC. http://eur-lex.europa.eu/legal-content/EN/TXT/PDF/?uri=CELEX:32009L0028&from=EN.

Directive 2009/30/EC of the European Parliament and of the Council of 23 April 2009 Amending Directive 98/70/EC as Regards the Specification of Petrol, Diesel and Gas-Oil and Introducing a Mechanism to Monitor and Reduce Greenhouse Gas Emissions. http://eur-lex.europa.eu/legal-content/EN/TXT/PDF/?uri=CELEX:32009L0030&from=EN.

Drabik, D., December 2011. The Theory of Biofuel Policy and Food Grain Prices. Working Paper 2011–20. Charles H. Dyson School of Applied Economics and Management, Cornell University.

Drabik, D., de Gorter, H., Timilsina, G.R., 2014. The effect of biodiesel policies on world biodiesel and oilseed prices. Energy Economics 44, 80–88.

European Biodiesel Board, 2015. The EU Biodiesel Industry: 2013 Production by Country. http://www.ebb-eu.org/stats.php.

European Environment Agency, 2009. Water Resources Across Europe—Confronting Water Scarcity and Drought. EEA Report No 2/2009.

European Environment Agency, 2015. SOER 2015—The European Environment—State and Outlook 2015. http://www.eea.europa.eu/soer.

European Commission, March 2013. Renewable Energy Progress Report. http://eur-lex.europa.eu/LexUriServ/LexUriServ.do?uri=COM:2013:0175:FIN:EN:PDF.

European Parliament, 2013. European Parliament Backs Switchover to Advanced Biofuels. http://www.europarl.europa.eu/news/en/news-room/content/20130906IPR18831/html/European-Parliament-backs-switchover-to-advanced-biofuels.

Eurostat, 2015a. Supply, Transformation and Consumption of Oil – Annual Data. http://appsso.eurostat.ec.europa.eu/nui/show.do?dataset=nrg_102a&lang=en.

Eurostat, 2015b. Primary Production of Renewable Energy by Type. http://ec.europa.eu/eurostat/tgm/refreshTableAction.do?tab=table&plugin=1&pcode=ten00081&language=en.

Eurostat, 2015c. Crops Products – Annual Data. http://appsso.eurostat.ec.europa.eu/nui/show.do?dataset=apro_cpp_crop&lang=en.

de Gorter, H., Just, D.R., 2008. 'Water' in the U.S. ethanol tax credit and mandate: implications for rectangular deadweight costs and the corn-oil price relationship. Applied Economic Perspectives and Policy 30 (3), 397–410.

de Gorter, H., Drabik, D., Just, D.R., 2015. The Economics of Biofuel Policies: Impacts on Price Volatility in Grain and Oilseed Markets. Palgrave Studies in Agricultural Economics and Food Policy. Palgrave Macmillan, New York, 316 pp.

Hochman, G., Sexton, S.E., Zilberman, D., 2008. The economics of biofuel policy and biotechnology. Journal of Agricultural & Food Industrial Organization 6 (2). Article 8.

Khanna, M., Crago, C., 2012. Measuring indirect land use change with biofuels: implications for policy. Annual Reviews of Resource Economics 4, 161–184.

Mekonnen, M.M., Hoekstra, A.Y., 2010. The green, blue and grey water footprint of crops and derived crop products. In: Volume 1: Main Report. Research Report Series No. 47. UNESCO-IHE, Institute for Water Education. http://temp.waterfootprint.org/Reports/Report47-Water FootprintCrops-Vol1.pdf.

United State Department of Agriculture (USDA), 2010. EU Annual Biofuels Report 2010. GAIN Report No. NL0019. http://gain.fas.usda.gov/Recent%20GAIN%20Publications/Biofuels%20Annual_The%20Hague_EU-27_6-11-2010.pdf.

United States Department of Agriculture (USDA), 2014. EU-28 Biofuels Annual 2014. GAIN Report Number: NL4025. http://gain.fas.usda.gov/Recent%20GAIN%20Publications/Biofuels%20Annual_The%20Hague_EU-28_7-3-2014.pdf.

de Vries, S.C., van de Ven, G.W.J., van Ittersum, M.K., Giller, K.E., 2010. Resource use efficiency and environmental performance of nine major biofuel crops, processed by first-generation conversion techniques. Biomass and Bioenergy 34, 588–601.

Zilberman, D., Hochman, G., Rajagopal, D., 2011. Indirect land use change: a second best solution to a first class problem. AgBioForum 13 (4).

Chapter 2.2.3

# Water—Energy Nexus and Environmental Aspects of Oil and Gas Production

R. Puls[1], L.D. Sanders[2]

[1]*Oklahoma Water Survey, The University of Oklahoma, Norman, OK, United States;* [2]*Oklahoma State University, Stillwater, OK, United States*

## 1. INTRODUCTION

Water and energy are inextricably linked (Fig. 1). Energy production with modern technology cannot be conducted without water as an input in the process and is also one of the by-products (produced water). Energy is also needed for water acquisition, transport, and treatment. Energy and power production requires water (eg, hydropower and oil and gas extraction). Additionally, water production, processing, and distribution require pumping, conveyance systems, and treatment. The United States is in the midst of an energy boom with no near term end in sight. Domestic production of natural gas has increased dramatically over the past 5 years (Energy Information Administration, 2014). Similarly, domestic production of oil has also increased because of tight oil (shale) production, particularly in North Dakota and Texas. This increase in energy production has resulted in relatively low and stable prices, reduced imports, increased exports, increased jobs, and increased tax revenues. It has also resulted in increased water usage and a myriad of costly

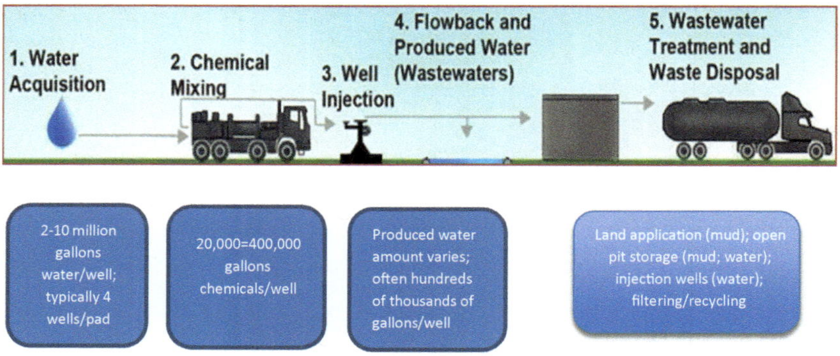


**FIGURE 1** Water and energy production are intimately linked. *Adapted from http://www2.epa.gov/hfstudy.*

Competition for Water Resources. http://dx.doi.org/10.1016/B978-0-12-803237-4.00009-4

problems, especially for local communities. Water use has increased in part because it takes up to 10 million gallons per well in some locations to hydraulically fracture a well over a period of several days (Murray, 2013). While water usage for energy production is significantly less than most uses (agriculture, drinking water, other industrial uses), it is an additional use and can have an adverse impact on local water supplies in areas that are experiencing drought conditions.

Local conflicts may occur where water is withdrawn from small streams, under drought conditions, and where aquifers are already being depleted by irrigation for agriculture. Fig. 1 sets the stage for what is discussed in this chapter. While it is not the intent of the authors to focus on the hydraulic fracturing process itself, the reader will better understand the water-related issues if they have a sense of the process and how important water is to that process.

The technology of hydraulic fracturing and its policy implications go beyond the focus on water. While the primary focus for this chapter is water, some tangential issues will at least be noted and discussed briefly. An outline of the various issues related to hydraulic fracturing is seen in Fig. 2.

## 1.1 Historic Perspective on Boom and Bust

The United States, particularly the western United States, has a long history of boom–bust cycles including the following:

- Gold and silver mining
- Metal mining
- Oil and coal production

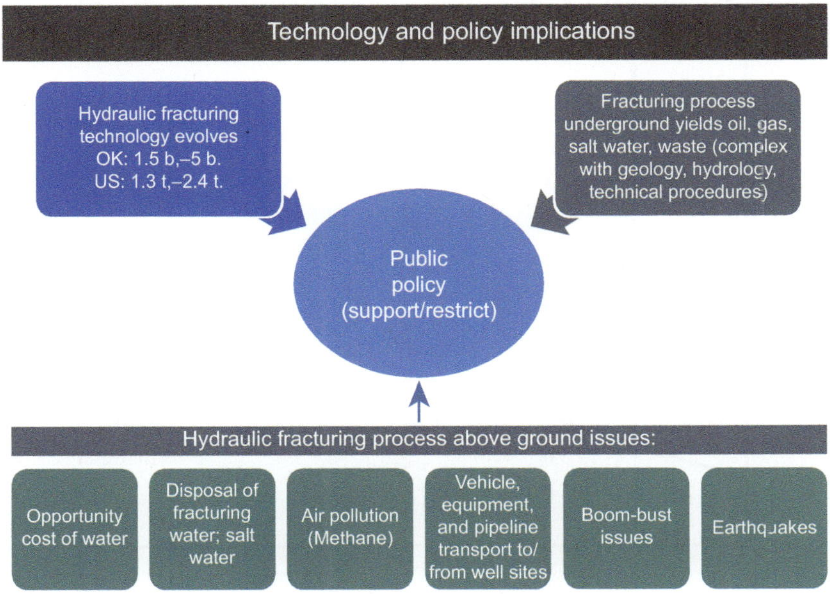


FIGURE 2 Technology and policy implications.

- Potash and other mineral mining
- Real estate speculation
- Unconventional oil and gas production

Historically, the main problem with these cycles is the lack of fundamental planning for sustainable development in communities where these cycles are occurring. They are usually characterized by rapid rise in population and economic activity, increased cost of goods and housing, insufficient infrastructure (water, sewer, schools, businesses, housing, medical facilities, etc.), and potentially increased crime. The "boom" is generally caused by a single industry or resource. When the "boom" is over, the "bust" follows, often as quickly as the "boom" arrived. What is left behind depends on how quickly the community rose to the challenge of providing needed services and infrastructure for ongoing development (Scott et al., 2011).

The implications for oil and gas activity, especially with respect to water use and water quality, become more evident as light is focused on this very real process of an initial burst of economic activity, often welcomed by communities desperate for such, and the inevitable downturn that results from both market glut and exploitation of the limited natural resource. Macke and Gardner (2012) have discussed the boom–bust cycle (also known as "Dutch disease") in three broad phases: (1) preboom without the insertion of the catalytic action; (2) the boom phase, where the action accelerates economic performance above the preboom growth trend; and (3) the boom contraction, beginning at the peak of the boom and continuing through the bust or contraction (Fig. 3). The third phase initially has the contraction or deceleration of economic activity with economic performance still above the preboom growth trend and declining, then the deceleration continues below the preboom growth trend. The difference between the preboom growth trend and the new and lower postboom growth trend captures the lost

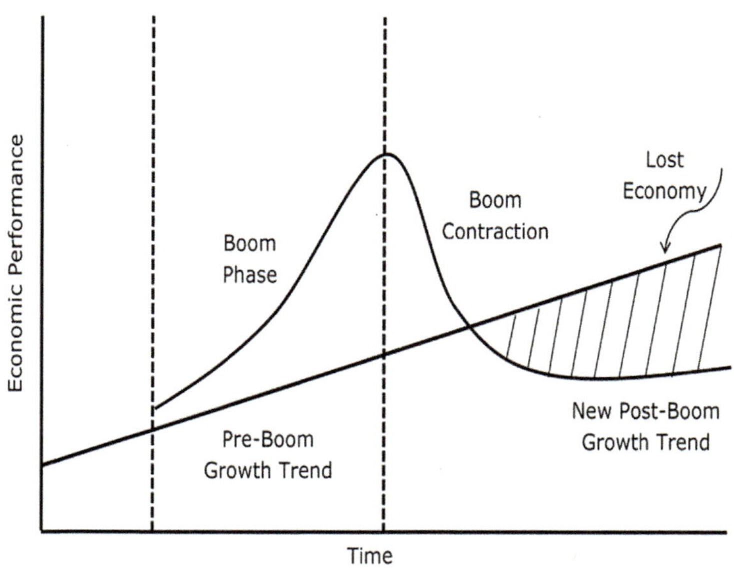


FIGURE 3 Community boom–bust cycle (Macke and Gardner, 2012).

economy that results over the long term. This "lost economy" effect is seldom considered in the planning and development phases by public and private agents who are making long-term investments in infrastructure such as roads, bridges, housing, schools, hospitals, utilities, etc. Thus the community is not only having to carry this over-investment cost (in the form of higher taxes for overbuilt public infrastructure, for example), but also dealing with a labor force that must be retrained for new opportunities to be attracted to the area.

Case after case of natural resource extraction in the oil patch and western United States has generally exhibited the boom–bust cycle to a greater or lesser extent (Weber, 2011; Lindholm, 2012). Of course, the specifics are always unique with respect to duration, expansion, decline, and "smoothness" or "fits-and-starts." During a public meeting coordinated by one of the coauthors of this chapter and in the height of the Oklahoma boom, one otherwise-intelligent Oklahoman sneered at the suggestion that there would be a "bust" following the boom created by hydraulic fracturing. Instead, he insisted this activity was different, and technology would put the economic performance on a long-term upward trend (Sanders et al., 2015). As 2014 came to a close and the oil and natural gas gluts drove prices down, the market was sadly proving the attendee wrong, at least in the near term. Additionally, the specific issues of water use and water quality are of growing concern for the public, public agencies, and private investors. With respect to the boom–bust cycle, the use of fresh water to facilitate hydraulic fracturing diminishes its availability in future years for potential economic activity that may also require cheap, accessible potable water. Also the tandem issues of produced water and reuse water disposal cloud water quality and possibly toxic water disposal further complicates attraction of other potential economic activities during and after the boom. Public and private agents deal with the aftermath of the boom.

## 1.2 Evolution of a Generic Shale Play

Work on a shale play is initiated and leases for mineral rights are obtained. There is initial excitement shared by numerous parties including landowners, local businesses, politicians, oil and gas companies, and supporting industries. A drilling boom begins and there is a rush to hold leases, which may last 3–5 years. Initial drilling identifies "sweet spots" in the shale formation where production is highest and companies focus their activities in these areas (Halliburton). Production rises rapidly but then declines on a per well basis. To maintain overall production increases in an area or play, more wells need to be drilled (Hughes, 2013). To increase production further, lateral lengths may be extended. Gradually, there is a shift to areas outside the "sweet spots" where there is a reduction in production on a per well basis and overall production of the area begins to decline, even with increased drilling.

Finding the "sweet spot" for hydraulic fracturing public policy incorporates the fiscal, physical, and social aspects of success:

- Drilling success in any reservoir is dependent on finding the most prospective areas, or the "sweet spots," and aligning the well bore for maximum borehole exposure to these zones.
- In shale reservoirs this means placing the well in the zones most conducive to fracturing.
- This requires a thorough understanding of the shale gas reservoir characteristics.

- Aiming for the middle is rarely a successful strategy, as shales can have significant variance in thickness and composition.
- For public policy as well, "aiming for the middle" may not be a successful strategy.

As a play develops and becomes more mature, land lease prices tend to rise and overall production costs increase. This is why the industry is constantly looking for new plays and new opportunities for more profit potential. In late 2012 and 2013, with gas prices low, there was a shift from gas-rich shale plays to oil-rich shale plays because these were more profitable because of resource market prices, but the play evolution characteristics are the same for both oil- and gas-rich plays. At what point does a play become uneconomical? Have we discovered or explored all play options? These are largely unanswered questions to date.

## 2. WATER SOURCING

A reasonable question that begs a public policy answer is: will water supplies be sufficient to meet US energy demands in 20 years given other competing uses related to population increases (eg, crop production, drinking water supply, industrial applications)? Populations will increase in many parts of the world and in many regions of the United States, but fresh water will not increase with the possible exception of desalinating nonfresh water resources. Unfortunately, population increases seem to be occurring mostly in regions where water is already stressed (eg, west, southwest United States). Climate change will likely exacerbate this trend. In almost every major economic sector, the question of adequate availability of water is being asked and addressed. Increasingly, golf courses are being encouraged and sometimes forced to use less water and/or marginal waters (eg, nonpotable groundwater, municipal wastewater). Municipalities and states are aggressively pushing more water conservation efforts, especially in water-stressed regions.

When it comes to energy production, the industry tends to be defensive about fresh water usage and points out that oil and gas resource extraction activities generally use less than 1% of all total fresh water consumption (World Resource Institute, 2013). While this is undoubtedly true on a global, national and even regional level, it may not be true at a local level, particularly during droughts. Most discussions do not evaluate this issue at the local level in any great detail, but this is where some impacts have been observed as noted earlier.

Water is the most common and most heavily used fluid in the petroleum industry. Water is produced from virtually every well along with oil and gas. It is used as the base fluid in drilling, completion, and production operations. Well completion itself (ie, hydraulic fracturing) will use most of the water and generally ranges from 3 to 10 million gallons per well. It will be produced throughout the lifetime of the well. Increasingly, it is being recycled and reused in subsequent completion operations. However, fresh water is becoming a scarce commodity in some regions and its value is increasing. This has significant implications for the oil and gas industry in terms of cost, availability, planning of operations, and profitability. Water is typically sourced locally from surface and groundwater. In areas where there is competition for water resources, marginal water sources are being used to some extent. These include brackish groundwater, wastewaters, acid mine drainage, and produced water. To date, there has not been extensive use of these alternative sources, except for produced water. When cost effective, recycling of produced water is becoming more common and states are cooperating with industry to make this easier to do in terms of permits and streamlined processes.

## 2.1 Oklahoma Water Use Trends

Oklahoma stream and aquifer nondomestic water use requires a permit from a state agency. These permits include long-term or permanent permits that are granted for more than a year, and short-term or provisional temporary permits that are granted for 90 days. The latter are the permits usually used by the oil and gas sector for water acquisition for oil and gas operations. These permits can usually be obtained quickly and only require sign-off by the Director of the Oklahoma Water Resources Board (OWRB). Long-term permits require more approval time and can be challenged by other parties.

Fig. 4 shows the annual allocation of the number and amounts allocated for the oil and gas sector with these 90-day permits since 1992. Note the rapid increase in the amounts allocated since 2007. This coincides with the expansion of oil and gas activity in the state and the use of horizontal wells (Fig. 5), requiring on average about 3.3 million gallons per well since 2009.

An analysis of data from the IHS database (industry sponsored oil and gas database) shows a steady increase in the number of horizontal wells completed in Oklahoma since 2000, with a rapid and accelerating increase since 2010.

The state of Oklahoma released the Oklahoma Comprehensive Water Plan (Oklahoma Water Resources Board, 2014) in 2012. It represents Oklahoma's long-range strategy for managing and protecting its surface and groundwater resources for the next 50 years. Results and findings in the plan emphasize demand projections for different state regions and different water usage sectors (eg, irrigation, drinking water supply, thermoelectric power, oil and gas). It identified the oil and gas sector as the sector that would experience the greatest demand growth by 2060. However, this projected increase would still only represent 5% of total water demand compared with slightly more than 2% of the state total at the time of publication of the plan. Nevertheless, current trends in water usage for the oil and gas sector suggest that those estimates might greatly underestimate future demands as current oil and gas water usage permits in 2013 already exceed 2060 demand estimates.

Fig. 6 shows the potential major water deficits and primary reasons through 2060. Central and Southwest Oklahoma face the biggest challenges in water management in the state. There is significant oil and gas activity in the central and west central part of the state. One example of permitted water use for oil and gas activity is Woods County (Fig. 7). Woods County is in the northwest portion of the Central Oklahoma water region. This where the Mississippi Lime formation is being heavily exploited. It occurs in north central Oklahoma and south central Kansas and is an area of the state that is prone to drought and water stress.

Fig. 8 suggests the temporal nature of extraction, with the industry quickly gearing up to exploit the natural resources with increased water input, then moving on. The figure shows acre-feet of water used for oil and gas permits in Canadian County, located in the heart of the Central Oklahoma water region. Rapid expansion and contraction of activity in an area is not atypical. Many factors can contribute to this including movement of resources to plays with greater payoff, reductions in the price of gas or oil, drought or lack of available water, and other economic or social factors.

Another local example of oil and gas sector demands on water resources is in Alfalfa County in north central Oklahoma near the Kansas border. There was a substantial increase in the number of 90-day provisional permits for water supply for oil and gas operations

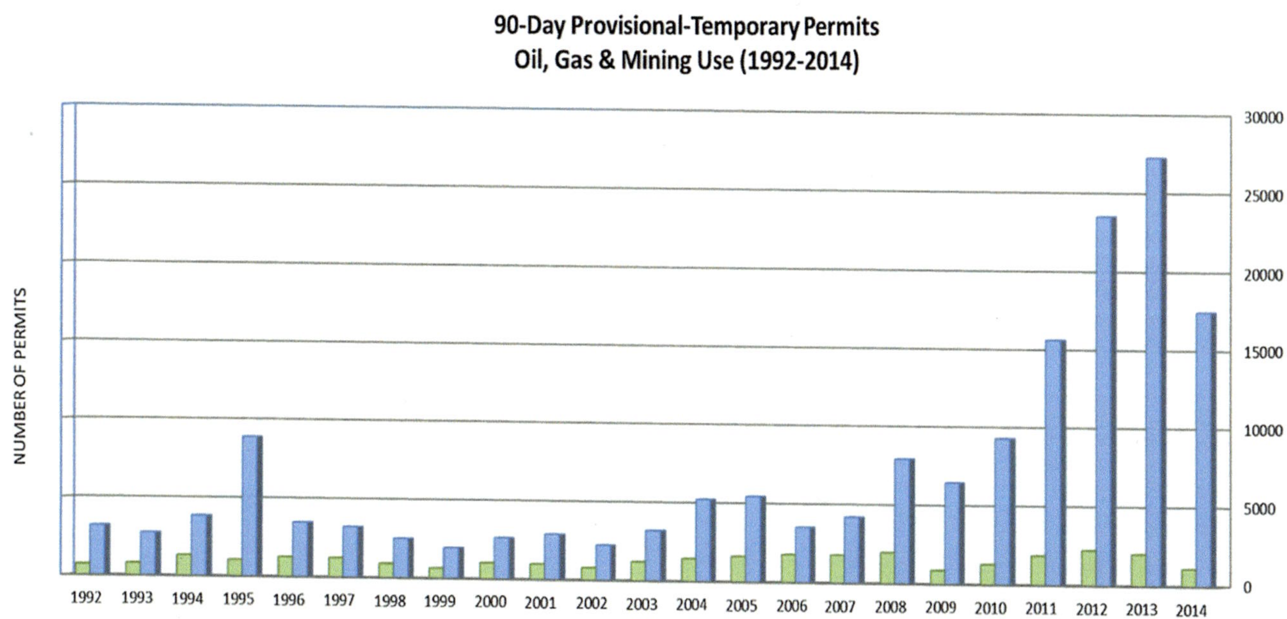


| | 1992 | 1993 | 1994 | 1995 | 1996 | 1997 | 1998 | 1999 | 2000 | 2001 | 2002 | 2003 | 2004 | 2005 | 2006 | 2007 | 2008 | 2009 | 2010 | 2011 | 2012 | 2013 | 2014 |
|---|---|---|---|---|---|---|---|---|---|---|---|---|---|---|---|---|---|---|---|---|---|---|---|
| Number of Permits | 710 | 789 | 1,312 | 1,043 | 1,259 | 1,227 | 904 | 640 | 1,019 | 982 | 793 | 1,206 | 1,449 | 1,628 | 1,794 | 1,792 | 1,993 | 875 | 1,286 | 1,891 | 2,270 | 2,043 | 1,100 |
| Total Amount Permitted | 3,211 | 2,764 | 3,868 | 8,898 | 3,492 | 3,243 | 2,532 | 1,956 | 2,652 | 2,914 | 2,270 | 3,233 | 5,253 | 5,472 | 3,547 | 4,229 | 7,957 | 6,486 | 9,320 | 15,659 | 23,541 | 27,291 | 17,480 |

FIGURE 4 Ninety-day water permits for oil and gas sector since 1992. *Courtesy of Tracy Scopel (Oklahoma Water Resources Board, selected water data, Scopel, T., 2014; Oklahoma Comprehensive Water Plan, 2012).*

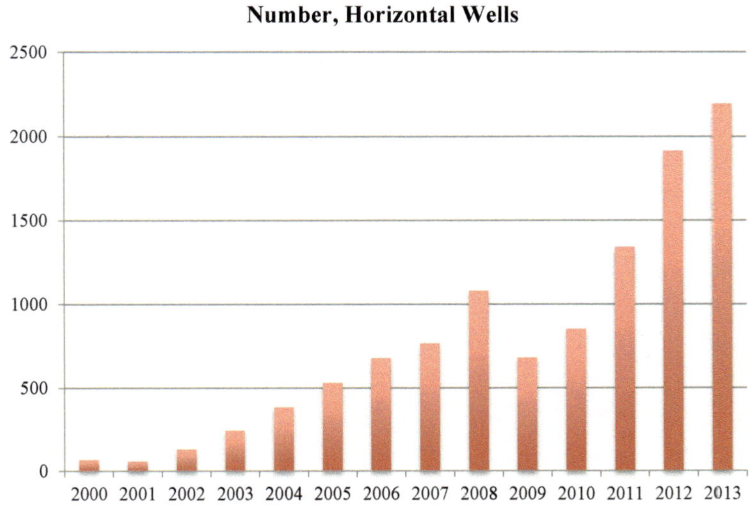


FIGURE 5 Horizontal well completions in Oklahoma since 2000 (IHS database, 2014).

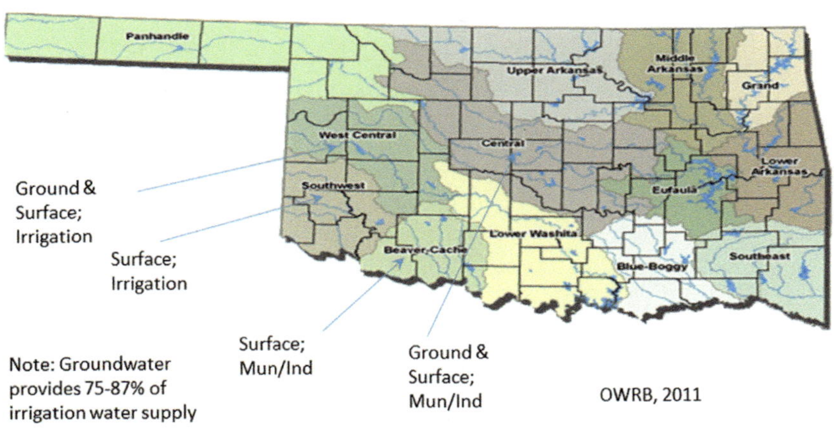


FIGURE 6 Water planning regions, 2012 and potential major deficits through 2060, Oklahoma Comprehensive Water Plan, Oklahoma Water Resources Board.

from 2011 to 2013 (Fig. 9) with the onset of exploitation of the Mississippi Lime formation. A total of 5250 acre-feet was allocated using 90-day permits for oil and gas in Alfalfa County in 2013. The total acre-feet allocated for all oil and gas operations in 2013 for the entire state was slightly more than 30,000 acre-feet. Therefore Alfalfa County represented more than 17% of the state total. The total long-term permits allocated for Alfalfa County for all uses in 2013 was 27,651 acre-feet. Therefore water allocated for oil and gas operations in Alfalfa County represented almost 20% of total usage for all sectors for the county compared with statewide percentage estimates of less than 3%.

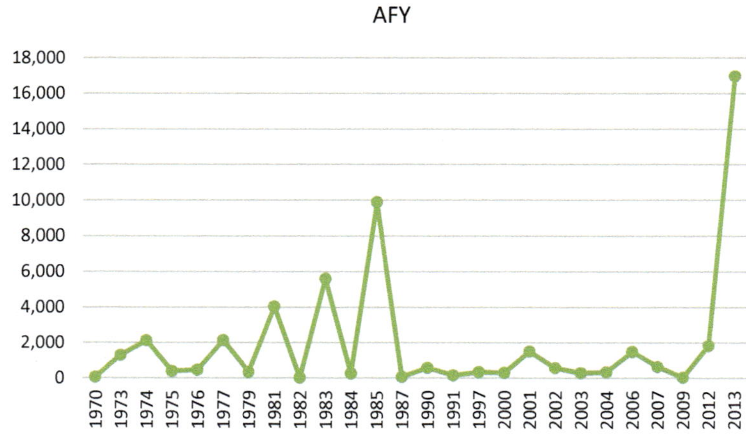


**FIGURE 7** Acre-feet per year (AFY), permitted water use for oil/gas activity in Woods County. *Author's compilation and representation of data from OWRB.*

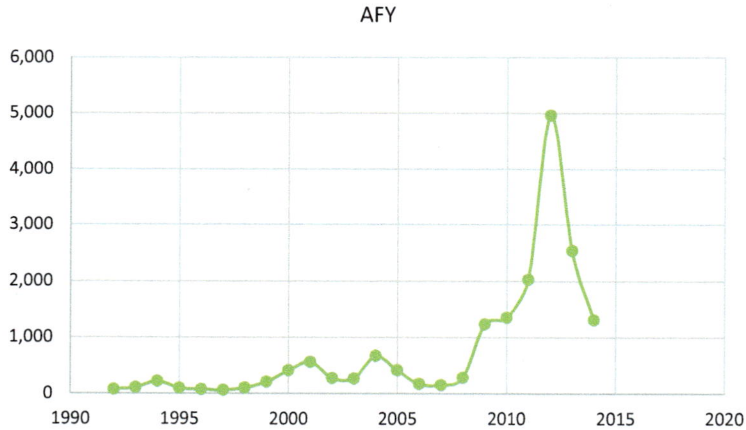


**FIGURE 8** Acre-feet per year (AFY), permitted water use for oil/gas activity in Canadian County. *Author's compilation and representation of data from OWRB.*

## 2.2 Recycling and Reuse

One of the primary means of disposing of produced water from oil and gas operations is underground injection. Some studies have noted an increase in seismic activity in parts of the country where there has been increased underground injection of wastewaters from oil and gas operations (Ellsworth, 2013; van der Elst et al., 2013; Holland, 2011). Fig. 10 shows a correlation between the previous graphs of oil and gas activity in Oklahoma and the incidence of quakes. While correlation does not prove causation, it is suggestive of at least a need for further research. To date, most research does not indicate that hydraulic fracturing causes earthquakes; however, much of the research does indicate that increased

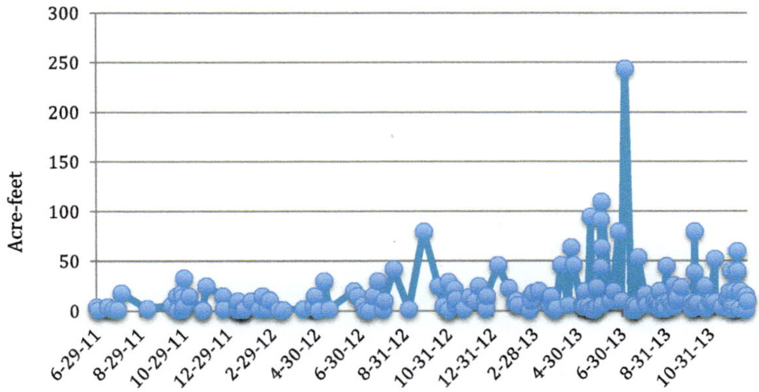


**FIGURE 9** Total acre-feet-year (AFY) allocated for 90 provisional permits for oil and gas in Alfalfa County from 2011 to 2013.

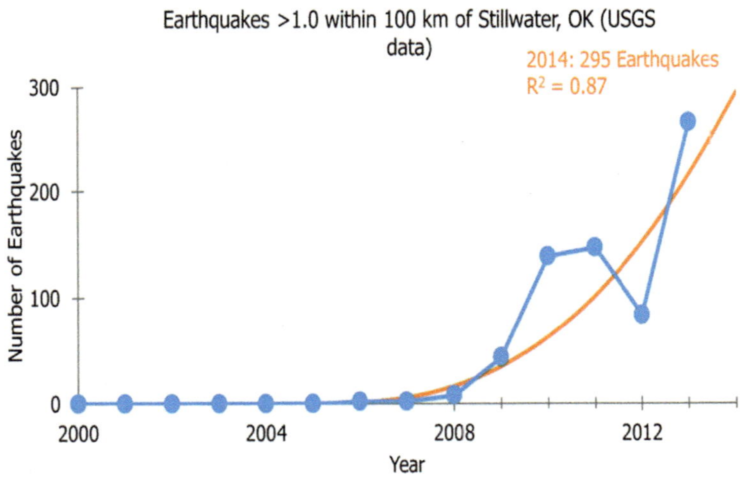


**FIGURE 10** Earthquakes per year near Stillwater, Oklahoma, 2000–2014 (http://earthquake.usgs.gov/research/induced/).

use of injection wells to dispose of produced and used water does result in quakes (USGS, 2014).

If underground injection becomes a less reliable means to dispose of produced water, recycling will increase in importance and industry may be faced with more expensive options for treatment, reuse, and ultimate disposal.

Devon Energy has recently built a 502,000 BBL (21.1 million gallons) capacity water reuse facility in Oklahoma (Fig. 11). It is a state-of-the-art system, with a high density liner and an automated leak detection system. The Oklahoma Corporation Commission cooperated with industry in updating rules to permit the construction of this large facility. It will have water pipelines connected to 36 different sections that will contribute to the completion of 270 wells. The system will reduce truck traffic in the region, reduce fresh

FIGURE 11 Devon recycling impoundment in western Oklahoma.

water usage and reduce operating expenses (Shaffer, 2013). This is a good example of the energy industry working with regulatory authorities and communities to balance environmental stewardship with our need for energy production. Implementation of these types of systems is not trivial. There is a need to do additional research on how to quickly characterize the water and integrate appropriate controls for efficient sediment removal, reduce total dissolved solids, and in some cases remove or degrade organic compounds that may contribute to fouling and reduced performance for subsequent well completions.

## 3. DROUGHT IMPACTS ON OIL AND GAS OPERATIONS

In 2013, a report by CERES, a nonprofit organization that advocates for sustainability leadership among corporations, identified the potential for water stress from hydraulic fracturing in some geographic locations based on some limited data from FracFocus (GWPC, 2014), a public registry for oil and gas operators (Fig. 12). These include Colorado and Texas and to a much lesser extent Oklahoma and Wyoming.

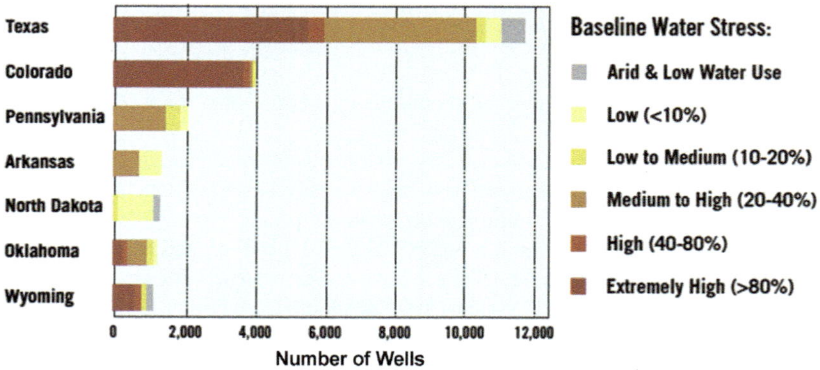


FIGURE 12 Ceres analysis of FracFocus data from Jan. 2011 through Sep. 2012.

Almost 90% of the shale gas and tight oil wells that were hydraulically fractured from January 1, 2011 through September 30, 2012 in Texas (GWPC, 2014) were in areas of medium to extremely high water-stressed regions. In Colorado almost 97% of the wells were developed in high or extremely high water-stressed regions. It should be noted that this does not cover all wells in operation during that timeframe as FracFocus data over this time period were voluntarily submitted by operators and not required. In contrast, less than 2% of the wells developed in Pennsylvania were in high to extremely high water-stressed regions during this same timeframe. Nonetheless, even in Pennsylvania, the Susquehanna River Basin Commission suspended 16 water withdrawal permits in 2012.

During the drought of 2011–2012 that was experienced in the western and southwestern United States, some impacts were noted by the oil and gas industry in terms of water cost increases. Extreme to exceptional drought was present in 95% of the state of Oklahoma (The National Drought Mitigation Center, 2013) (Fig. 13). In addition to the suspension of water withdrawal permits by the Susquehanna River Basin Commission, water auctions went from $9 to $100 per acre-foot in Meade County and water shortages required increased water transport distances in the Eagle Ford play in south Texas. As a result, companies have evaluated contingencies for water supply in times of drought or other potential factors that may limit supply. Many companies have implemented produced water reuse or recycling, particularly in water-stressed areas. Companies are also looking at the use of other nonfresh water sources (eg, brackish and saline water). In addition to drought having negative impacts on water availability, some major aquifers are being depleted at alarming rates as a result of overpumping, mainly because of agricultural demands. A USGS report (USGS, 2013a,b) that assessed the extent and rate of major aquifer depletion in the United States, found that aquifers in peril include the Ogallala in the High Plains of central United States and the Mississippi embayment in the Gulf Coastal Plain. Both of these aquifers touch regions where there is intense hydraulic fracturing activity.

In parts of west Texas and southwest Kansas there have been declines of more than 150 ft in water levels in the Ogallala aquifer primarily because of irrigation practices. However, any continued stress on this aquifer will aggravate current unsustainable water management practices. Figs. 14 and 15 show where 90-day provisional permits for oil and gas were granted in 2013 in Oklahoma and the relative amounts. Many of these locations have experienced recent severe to exceptional drought and are located over water-stressed aquifers (eg, Ogallala, Salt Fork alluvial aquifer). There are more groundwater permits granted in the western part of the state because of lack of available surface water. Intensive water withdrawals of both surface and groundwater have occurred in the north central part of the state. Localized competition for water resources in this area have been steadily increasing with the oil and gas sector claiming about 20% of the total demand.

Water reuse strategies are actively being investigated to minimize fresh water usage and prevent impacts to oil and gas production schedules. The industry is also looking at marginal waters as alternatives to fresh water sources. For example, in Oklahoma, groundwaters that have over 5000 mg/L total dissolved solids do not require a permit for withdrawal of these resources. This would lessen the dependency on fresh water usage and improve a company's social license to operate in water-stressed regions. Another USGS report (USGS, 2013a,b) provides a synthesis of information on the hydrogeology, distribution, and volume of saline groundwater in the midcontinent and south central areas of the United States. Additional research is needed to provide more detail on the capacity and chemistry of these resources.

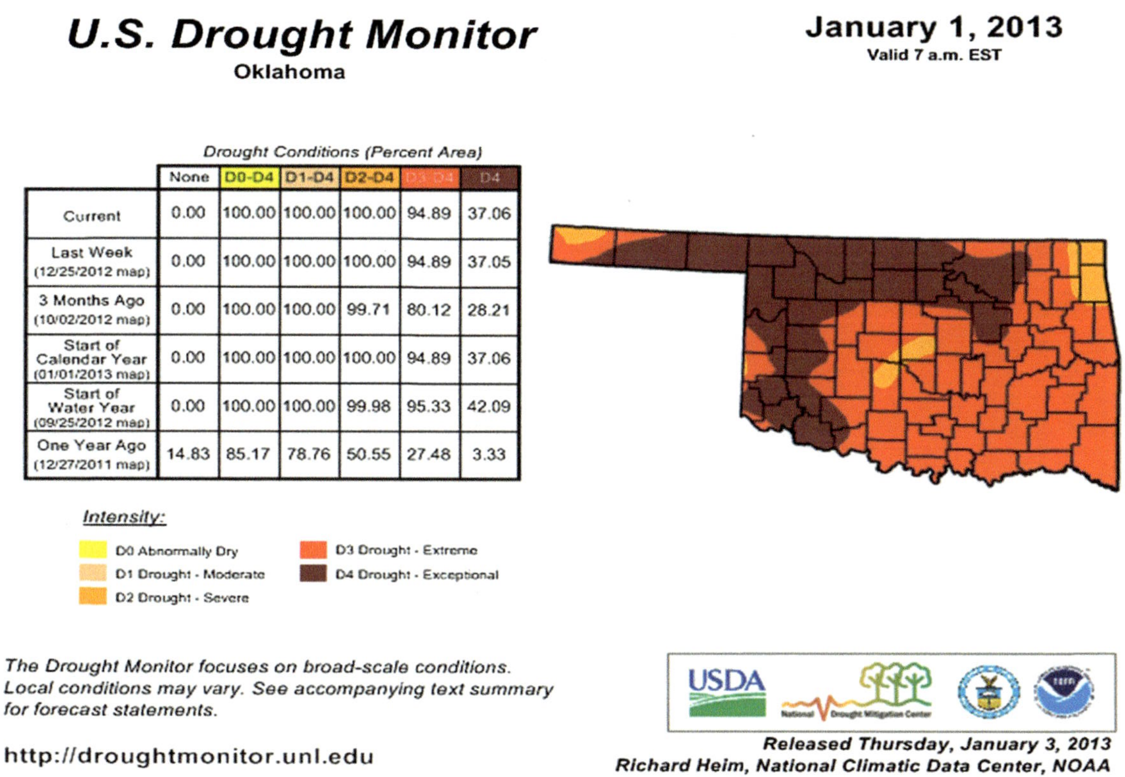


**FIGURE 13** US drought monitor for Oklahoma in Jan. 2013.

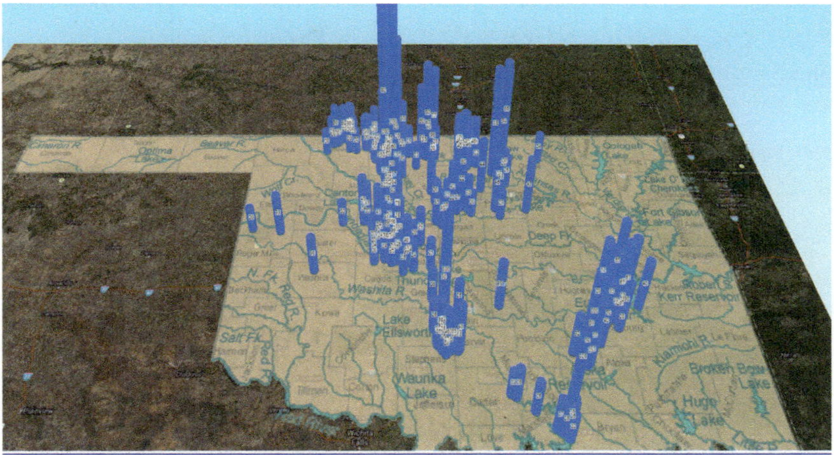

**FIGURE 14** Surface water permits with relative volumes for oil and gas operations in 2013.

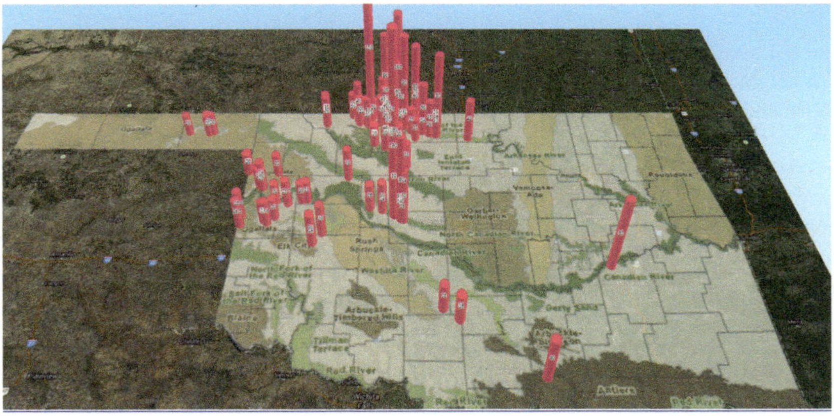

**FIGURE 15** Groundwater permits with relative volumes for oil and gas operations in 2013.

## 4. WATER QUALITY

The dual issues related to water depth are maintaining water sources as noted previously and protecting the quality of fresh groundwater. Whether hydraulic fracturing impacts groundwater by depletion or by contamination is typically a localized issue depending on the water tables and depth of the shale plays. Table 1 shows how variable this issue can be.

Degradation of water quality of fresh water resources will necessarily impact the supply of fresh water and therefore water availability for all uses. Restoration of contaminated groundwater has been shown to be extremely difficult and expensive and in many cases technically impractical. Concerns regarding impacts from oil and gas operations on the quality of water resources, particularly drinking water sources have increased over the last several years (USEPA, 2010a–d), especially in areas where communities

**TABLE 1** Rock Column Thickness by Gas Shale Basin and State in Feet (Ferrell and Sanders, 2013)

| Gas Shale Basin | States | Rock Column Thickness b/w Play and Treatable Water (ft) |
|---|---|---|
| Barnett | TX | 5300–7300 |
| Fayetteville | AR | 500–6500 |
| Haynseville | LA, TX | 10,100–13,100 |
| Marcellus | NY, PA, OH, VA, WV | 2125–7650 |
| Woodford | OK, TX | 5600–10,600 |
| Antrim | MI | 300–1900 |
| New Albany | IL, IN, KY | 100–1600 |

were not used to the presence of conventional oil and gas production. These concerns have been heightened by documented cases of illegal wastewater disposal, well blowouts, stray gas (methane) events, and adverse impacts to private well owners. While the number of such events is small in comparison to the number of wells drilled over this period (>100,000), these events have been highly publicized. The Pennsylvania Department of Environmental Protection (DEP) disclosed that oil and gas operations have damaged Pennsylvania water supplies 209 times since the end of 2007 (Powersource, Pittsburgh Post-Gazette, 2014). However, they did not disclose characteristics of the wells, which companies were responsible, what caused the problems, or what pollutants were found. During that same period more than 20,000 wells were drilled in the state, so this would represent about a 1% rate of impact to water resources. The same article, however, found deficiencies in how DEP maintained and organized its data.

Other states such as Colorado also disclose violations such as spills (Colorado Oil and Gas Commission, 2011). While there continues to be no firm case demonstrating impacts to groundwater from the hydraulic fracturing process itself, there are reported impacts to water resources from spills, wastewater mismanagement and poor production well construction practices (USEPA, 2015). Research by Hildebrand et al. (2015) found elevated levels of 10 different metals as well as the presence of 19 different chemical compounds including BTEX (benzene, toluene, ethyl benzene, and xylenes) compounds associated with hydraulic fracturing in the immediate vicinity of oil and gas operations in the Barnett shale area.

The study also found elevated levels of methanol and ethanol. The researchers noted that these data do not necessarily identify oil and gas activities as the source of contamination. However, they do suggest a need for further monitoring of groundwater quality in this region. There is very little available data on the quality of flowback and produced water from oil and gas operations. This makes it difficult to determine the ultimate fate of pollutants in these wastewaters if they are spilled or leak from holding pits into surface and groundwater.

Drilling muds (mixture of clay and water or oil to lubricate/cool the bit and flush the borehole) are often confused with fracking fluids (mixture of water and other chemicals used to open the rock formation under high pressure to remove oil and natural gas) and complicate the nature of water quality issues. As noted by OSU's Chad Penn (Penn et al., 2014):

> *All wells produce drilling mud as a by-product of drilling. And drilling mud, as opposed to fracking fluids, is often purposely applied to the soil surface for disposal. Drilling mud is used to seal formations, remove/suspend cuttings, lubricate and cool drill bits, control corrosion on the drill stem, and control well bore pressure. The mud is recycled until it cannot be used. Additives vary but may include bentonite clay, barium sulfate, lime, soda ash, lignite, peanut/walnut shells, mica, cellophane, calcium carbonate, plant fibers, and cottonseed hulls. It is either water based or oil based.*

Penn further notes that, when done correctly, the soil spreading process of disposal has few long-term risks. However, incorrect disposal carries such risks as soil salinization, sodic soils, and total petroleum hydrocarbon toxicity to plants. Produced water and flowback water are mostly disposed of by deep injection wells, as noted previously. Occasionally, it is also land applied. Also there are experimental processes to filter and reuse wastewater.

## 5. SUSTAINABILITY OF WATER AND ENERGY RESOURCES

Increasing demands for sources of water, combined with changing land use, population growth, aging infrastructure, and climate change, pose significant threats to our water resources. Failure to manage our waters in an integrated, sustainable manner will limit economic prosperity and jeopardize both human and ecosystem health. Water resource management is typically focused on reallocating water to where and when it is needed. Such a narrow approach may prevent water use for alternatives as needs arise. A new, holistic approach would include the integrated, conjunctive use of surface and groundwater resources and take account of social, economic, policy, and environmental factors.

The drivers for this paradigm shift are emerging risks to future water availability. These include climate change, increasing demand for fresh water because of population increases, pollution of fresh water sources, and energy demands. In the context of the oil and gas industry and its rapid expansion in resource extraction activities in the United States, potential adverse outcomes are increasing costs, delays in production, and potential short-term impacts on water availability in some local communities. Whether water use for oil and gas production is right or wrong is not the question for this chapter. Rather, the question is how does such use fit into an integrated process of achieving societal goals while recognizing physical constraints.

Sustainable principles for energy development include:

- Communication and education
- Water usage efficiency
- Improvements in water quality and water availability monitoring
- Identification and use of alternative water sources
- Water storage to meet varying cyclical demands
- Contingency planning to mitigate shortages
- Economic sustainability of communities

Communication with the public, especially impacted communities, is essential. This involves all stakeholders, not just the energy companies. This will engender trust, alleviate local community concerns, and facilitate the resource extraction process. Transparency must be part of the communication process; in the long run this will help to satisfy the goals and objectives of all parties. Education of the public in a straightforward, unbiased manner is also important.

Water usage efficiency and protection of water quality can be increased through better management practices, hydraulic fracturing where the carrier media is something other than water, improved water conveyance structures that minimize leakage, storage structures that reduce leakage and evaporation, and cooperation and collaboration among oil and gas companies operating in the same play or basin.

Alternative water sources include treated wastewater effluent, brackish groundwater, acid mine drainage, and flowback and produced water from oil and gas production wells. Reuse or recycling of produced water can result in cost savings for industry. Other advantages of using these sources include no competition from agricultural and municipal water uses and the fact that many of these sources do not require permits for usage.

Demands for water vary on an annual basis for certain sectors. Agricultural demand is greatest in the spring and summer months. This is also true for municipal drinking water and irrigation demands. Heating demands are highest in the winter months, but as natural gas becomes more widely used for electricity generation, the demand will be higher in the summer months as well. Water storage when water is more plentiful (eg, spring) can alleviate higher demand periods. This would also be true for increased demands during periods of drought. Aquifer storage and recovery in groundwater can alleviate these shortfalls and minimize evaporative losses associated with surface water storage (eg, reservoirs).

Contingency planning will be an essential piece of any integrative water management strategy. What happens during a drought? Where are alternative water sources if groundwater or surface water withdrawals are restricted? What happens if recycled water becomes unsuitable for use? Alternative options and plans need to be explored and put in place before drilling and completion activities commence.

## REFERENCES

Colorado Oil and Gas Conservation Commission, 2011. Oil and Gas Industry Spills and Releases. Available at: http://dnr.state.co.us/SiteCollectionDocuments/SpillsAndReleases.pdf.

Ellsworth, W.L., 2013. Injection-induced earthquakes. Science 341 (6142). http://dx.doi.org/10.1126/science.1225942.

Energy Information Administration, 2014. Annual Energy Outlook 2014 with Projections to 2040. DOE/EIA-0383.

Van der Elst, N.J., Savage, H., Keranen, K., Abers, G.A., July 12, 2013. Enhanced remote earthquake triggering at fluid-injection sites in the midwestern United States. Science 341 (6142), 164–167. http://dx.doi.org/10.1126/science.1238948.

Ferrell, S.L., Sanders, L., February 2013. Natural Gas Extraction: Issues and Policy Options. White paper for National Agricultural & Rural Development Policy Center.

GWPC, 2014. FracFocus. http://fracfocus.org/.

Halliburton.com. Targeting the Sweet Spot. http://www.halliburton.com/en-US/ps/solutions/unconventional-resources/shale-gas-oil/challenges-solutions/targeting-the-sweet-spot.page?node-id=hgjyd47d.

Hildebrand, Z.L., Carlton Jr., D.D., Fontenot, B.E., Meik, J.M., Walton, J.L., Taylor, J.T., Thacker, J.B., Korlie, S., Shelor, C.P., Henderson, D., Kadio, A.F., Roelke, C.E., Hudak, P.F., Burton, T., Rifai, H.S., Schug, K.A., 2015. A comprehensive analysis of groundwater quality in the Barnett shale region. Environmental Science and Technology 49 (13), 8254–8262.

Holland, A., 2011. Examination of Possibly Induced Seismicity from Hydraulic Fracturing in the Eola Field, Carvin County, Oklahoma. Oklahoma Geological Survey Open-file Report, 2011.

Hughes, J.D., February 2013. Drill, Baby, Drill, Can Unconventional Fuels Usher in a New Era of Energy Abundance. Post Carbon Institute.

Lindholm, M.L., December 18, 2012. The Downsides of Living in an Oil Boomtown. NPR. http://www.npr.org/2012/12/18/167467703/the-downsides-of-living-in-an-oil-boom-town.

Macke, D., Gardner, D., August 27, 2012. Policy Education White Paper- Western North Dakota Energy Project. RUPRI Center for Rural Entrepreneurship. Version. Available online at: http://www.energizingentrepreneurs.org/file_download/717379f4-41ff-4785-8a43-e7d42705a40c.

Murray, K.M., 05/2013. State-scale perspective on water use and production associated with oil and gas operations, Oklahoma, USA. Evironmental Science & Technology 47 (9), 4918–4925. http://dx.doi.org/10.1021/es4000593.

Oklahoma Water Resources Board, selected water data, Scopel, T., 2014. Oklahoma Comprehensive Water Plan, 2012.

Penn, C.J., Whitaker, A., Warren, J.G., August 22, 2014. Surface application of oil-base drilling mud mixed with gypsum, limestone and caliche. Agronomy Journal 106 (5).

Powersource, Pittsburgh Post-Gazette, July 21, 2014. http://powersource.post-gazette.com/powersource/policy-powersource/2014/07/22/DEP-Oil-and-gas-endeavors-have-damaged-water-supply-209-times-since-07/stories/201407220069.

Sanders, L.D., Ferrell, S., Halihan, T., Neurohr, K., Peek, G., Penn, C., Shideler, D., February 2015. Hydraulic Fracturing: Public Policy Issue in Oklahoma. Oklahoma State University.

Scott, C.A., Pierce, S.A., Pasqualetti, M.J., Jones, A.L., Montz, B.E., Hoover, J.H., 2011. Policy and institutional dimensions of the water-energy Nexus. Energy Policy 39, 6622–6630.

Shaffer, D.L., Arias Chavez, L.H., Ben-Sasson, M., Romero-Vargas Castrillon, S., Yip, N.Y., Elimelech, M., 07/2013. Desalination and reuse of high-salinity shale gas produced water: drivers, technologies, and future directions. Environmental Science and Technology. http://dx.doi.org/10.1021/es401966e.

The National Drought Mitigation Center, US Drought Monitor, 2013. http://droughtmonitor.unl.edu/, University of Nebraska at Lincoln.

US Environmental Protection Agency, 2010a. Summary of Public Comments, Binghamton, NewYork, September 13 and 15, 2010. Hydraulic Fracturing Public Informational Meeting. Available at: http://www.epa.gov/hfstudy/publicoutreach.html (accessed 05.09.12.).

US Environmental Protection Agency, 2010b. Summary of Public Comments, Canonsburg, Pennsylvania, July 22, 2010. Hydraulic Fracturing Public Informational Meeting. Available at: http://www.epa.gov/hfstudy/publicoutreach.html (accessed 05.09.12.).

US Environmental Protection Agency, 2010c. Summary of Public Comments, Denver, Colorado, July13, 2010. Hydraulic Fracturing Public Informational Meeting. Available at: http://www.epa.gov/hfstudy/publicoutreach.html (accessed 05.09.12.).

US Environmental Protection Agency, 2010d. Summary of Public Comments, Fort Worth, Texas, July8, 2010. Hydraulic Fracturing Public Information Meeting. Available at: http://www.epa.gov/hfstudy/publicoutreach.html (accessed 05.09.12.).

USGS, 2013a. Groundwater Depletion in the U.S., USGS Scientific Investigations Report 2013–5079.

US Environmental Protection Agency, June 2015. Assessment of the Potential Impacts of Hydraulic Fracturing for Oil and Gas on Drinking Water Resources EPA/600/R-15/047a. www.epa.gov/hfstudy.

USGS, 2013b. Hydrogeology, Distribution, and Volume of Saline Groundwater in the Southern Midcontinent and Adjacent Areas of the United States, Scientific Investigations Report 2013–5017.

USGS, press release, May 12, 2014.

Weber, J.G., 2011. The effects of a natural gas boom on employment and income in Colorado, Texas, and Wyoming. Energy Economics 34 (5), 1580–1588.

World Resource Institute, 2013. http://www.wri.org/our-work/project/aqueduct/aqueduct-atlas.

Chapter 2.2.4

# Water Use for Unconventional Natural Gas Development Within the Susquehanna River Basin

M.K. Shank, J.W. Balay, J. Richenderfer
*Susquehanna River Basin Commission, Harrisburg, PA, United States*

## 1. INTRODUCTION

### 1.1 Emergence of a New Water User

The northeast United States has a long history of hydrocarbon extraction. The nation's first commercial natural gas well was drilled in New York in 1821 and the first oil well was drilled in Pennsylvania in 1859. Since then, over 350,000 oil and gas wells have been drilled within Pennsylvania alone. Recent technological advancements that combined horizontal well drilling and hydraulic fracturing created the unconventional natural gas industry as it currently exists. The unconventional gas industry first began operating within the Susquehanna River Basin (SRB) during 2006–2007. However, it was not until mid-2008 that the industry substantially increased its drilling activities and its need for multiple non-interruptible water sources. Unlike the conventional gas industry (vertical drilling), which used very little water for drilling and development purposes, the unconventional gas industry requires significant quantities of water, typically 4–6 million gallons (Mgal) per well hydraulic fracturing event.

The Susquehanna River Basin Commission (SRBC) is a federal-interstate agency that has regulatory authority over water withdrawals and consumptive water uses within the watershed boundary of the SRB. Because of permitting requirements, industry operators may not begin well construction, drilling, or hydraulic fracturing without SRBC approval of the water use. This requirement allows the SRBC to regulate the industry's individual and cumulative impacts on water resources. This chapter is intended to provide insight into gas development and water use within the SRB through 2013 and also to provide an example of a regulatory response to a new and atypical water user.

**Competition for Water Resources. http://dx.doi.org/10.1016/B978-0-12-803237-4.00010-0**

## 1.2 The Susquehanna River Basin

The SRB is a 27,510 mi$^2$ (71,251 km$^2$) watershed in the northeastern United States, which covers portions of New York, Pennsylvania, and Maryland (Fig. 1). The Susquehanna River is 444 miles in length, making it the largest river lying entirely in the United States that drains into the Atlantic Ocean and the 18th largest river in the United States based on mean discharge (Kammerer, 1990). The Susquehanna is an important tributary to the Chesapeake Bay, which is the largest estuary in the United States. The SRB comprises 43% of the Chesapeake Bay's watershed area and the river provides about one-half of the freshwater flow to the Bay (SRBC, 2013a).

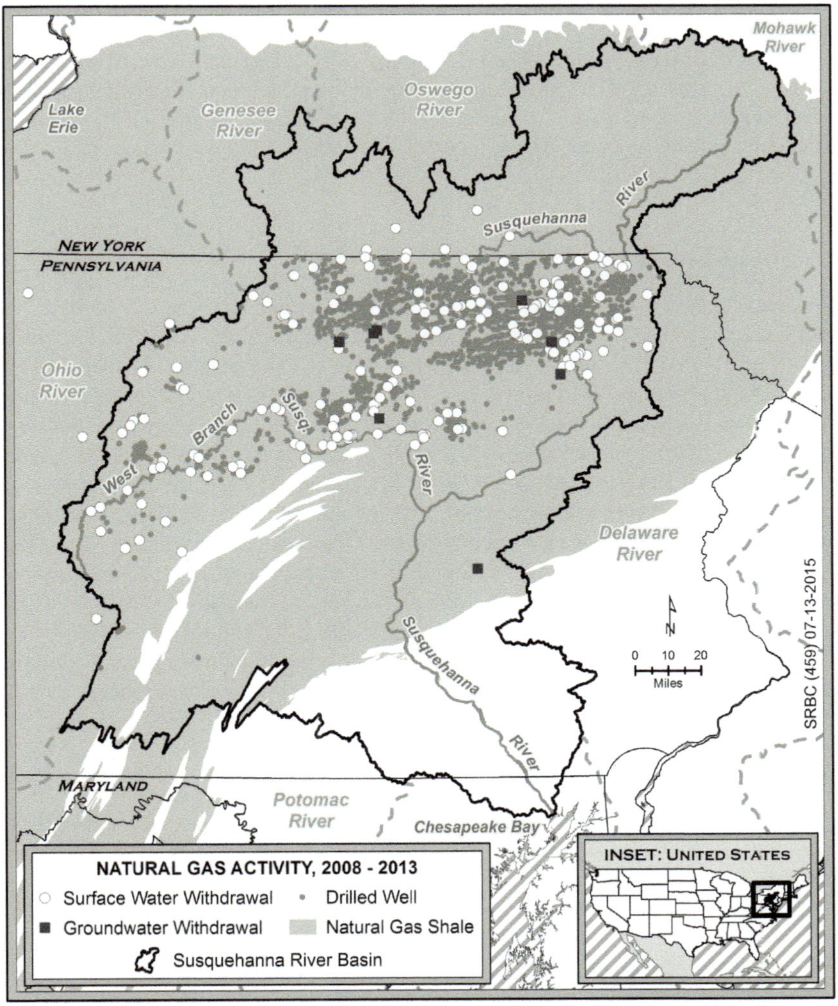


FIGURE 1 Location of drilled unconventional natural gas wells and approved water withdrawals within the Susquehanna River Basin from 2008 to 2013 and underlying gas-containing shale formations in the northeastern United States.

The Appalachian Plateau, Ridge and Valley, and Piedmont physiographic provinces comprise approximately 56, 32, and 11% of the SRB, respectively. Mixed forested and agricultural land uses are found in this relatively rural watershed. The SRB has a human population of more than 4.1 million, with the densest development located in the southern portion in the Piedmont province. Approximately 62.5% of the SRB is forested, while 27.5% is cultivated and 4.2% is developed. About 42 in. of precipitation fall annually on the SRB, with greater than 50 in. in extremely wet years and less than 25 in. in drought years. Changing climate is the suspected cause of increased precipitation in recent decades in the SRB (SRBC, 2013a). Annual minimum and median streamflow have likewise increased considerably post-1970 when compared with historic records (Zhang et al., 2010). The condition of water resources in the SRB is generally good (SRBC, 2013b). Pollution and aquatic habitat alteration persist as a result of large-scale logging activities that peaked in the early 1900s when only 30% of forest cover remained (DePhilip and Moberg, 2010). Abandoned coal mines continue to be a significant source of pollution, causing over 2000 impaired stream miles. Sediment and nutrients, however, are the two largest contributors to stream impairment. These pollutants, coupled with stormwater runoff, are challenging issues caused by urban and suburban development and agricultural practices (SRBC, 2013a).

## 1.3 Shale Gas in the SRB

The Marcellus Shale is the largest gas-containing shale formation in the United States, both in surface area (150,000 mi$^2$) and estimated recoverable gas reserves (141 trillion cubic feet) (Johnson, 2010; USEIA, 2012). Approximately 66% of Pennsylvania is underlain by the Marcellus and 85% of the SRB is underlain by one or more gas-containing shale formations. The most favorable locations for gas development are in the more undeveloped portions of the SRB in northern Pennsylvania, including areas inhabited by sensitive species such as brook trout (Weltman-Fahs and Taylor, 2013; Fig. 1). New York and Maryland state governments currently impose a moratorium on high-volume hydraulic fracturing, thus no unconventional gas development currently occurs within their borders. During early stages of development in 2010, approximately 2 billion ($10^9$) cubic feet (Bcf) of natural gas per day was produced from the Marcellus Shale. Production has expanded greatly, with over 15 Bcf of natural gas produced per day in July 2014, accounting for almost 40% of US shale gas production (USEIA, 2014). The huge increase in gas production in a short amount of time creates a challenge to the environment, and water resources in particular (Rahm et al., 2015).

## 1.4 Terminology Used

The phrase "unconventional natural gas development" herein represents the drilling, casing, cementing, stimulation, and completion of wells undertaken for the purpose of extracting gaseous hydrocarbons from low permeability geologic formations utilizing enhanced drilling, stimulation, or recovery techniques. The word "industry" will be used throughout this document to refer to the unconventional natural gas industry. The "study period" referenced herein is the period from July 1, 2008 to December 31, 2013. "Consumptive water use" is defined as water withdrawn and not returned to the hydrologic cycle of the SRB undiminished in quantity. Well drilling, hydraulic fracturing, and dust control operations are common examples of gas industry consumptive water uses. The term "flowback" is used herein as the return flow of water and formation fluids recovered from the wellbore of a hydrocarbon development well (including unconventional gas wells) following the release of pressures induced as part of the hydraulic fracture stimulation of a target geologic formation, and until the well is placed into production. The return flow of water or formation fluids recovered at the wellhead after the well is placed into production is referred to as "production fluids" or "produced water." Industry water use and well development data referenced herein were obtained from an assessment of unconventional natural gas development occurring within the Susquehanna River Basin from July 2008 through December 2013 (Richenderfer et al., 2016).

### 1.5 Initial Water Needs and Challenges

During the early stages of unconventional gas development within the SRB, which extended from late 2007 through early 2008, the industry had not yet established many water sources that were approved by SRBC. Consequently, during that early period, the industry relied heavily upon water obtained from municipal public water systems to support its hydraulic fracturing operations. As development progressed, the industry began developing a water-sourcing network comprised of approved surface and groundwater sources under the direct control of the industry or independent purveyors and more centrally located to its expanding area of operations than were the public water systems.

For geologic reasons, the industry has been most active in the northcentral and northeastern Pennsylvania portions of the SRB (Fig. 1). The topographic characteristics of this region result in watersheds of smaller sizes when compared to other portions of the SRB, therefore the industry's preference for nearby water sources was initially focused within these smaller watersheds. Typically, these headwater systems have limited water availability, especially during the drier late summer and early fall seasons; are occupied by sensitive yet critically important ecosystems; are generally of very high water quality; and often cannot provide the sustainable water sources the industry seeks for operations.

The industry operates much differently than most regulated water users within the SRB. Most historic, regulated water users exist at a fixed geographic location and use approximately the same amount of water on a routine basis, either daily or seasonally. The gas industry's daily water needs fluctuate widely and routinely migrate over significant distances in relatively short periods of time. Some individual water sources are used at moderate rates by the industry on a daily basis for many months at a time while other water sources are used at varying rates on an infrequent basis. This unpredictable water use pattern proves to be challenging for regulatory agencies responsible for managing water resources.

## 2. EVOLUTION OF REGULATORY PROGRAM

The President of the United States signed the Susquehanna River Basin Compact (SRBC, 1972) into law on December 24, 1970 (US Congress, 1970). The Compact created SRBC, a federal-interstate compact agency comprised of the member jurisdictions of the states of New York, Pennsylvania, and Maryland and the US federal government. SRBC's mission is to enhance public welfare through comprehensive planning, water supply allocation, and management of the water resources of the SRB. Projects involving development of the water resources of the SRB are evaluated in terms of their compatibility with regulatory requirements in 18 Code of Federal Regulations, Parts 801, 806, 807, and 808 and standards set forth in the Comprehensive Plan (SRBC, 2013a).

In general, the SRBC regulates ground and surface water withdrawals of 100,000 gallons per day or more on a 30-day average, consumptive water uses and out-of-basin diversions of 20,000 gallons per day or more on a 30-day average, and all into-basin diversions. SRBC may approve or modify projects, or may deny a project if it determines it is not in the best interest of the conservation, development, management, or control of the SRB's water resources, or is in conflict with the Comprehensive Plan. SRBC does not have a regulatory responsibility in the area of water quality. However, potential water quality impacts of projects are considered in regulatory decisions, and coordination occurs among member jurisdictions to prevent and reduce water pollution and maintain water quality as required by the Comprehensive Plan.

## 2.1 "Gallon One" Rule

SRBC's regulatory thresholds for water withdrawals, consumptive water uses, and diversions were modified during the emergence of the unconventional natural gas industry in the SRB. In 2008, SRBC notified the industry that any amount of water withdrawn or consumptively used to develop wells in shale formations in the SRB will require prior approval (SRBC, 2008). The SRBC regulations allow its executive director to make a determination when water use activities, regardless of the amount of water, have the potential to affect the water resources of the SRB. It was determined that the industry's water use activities could have an adverse, cumulative adverse, or interstate effect on the water resources of the SRB. This action established what has become referred to as the "gallon one" rule. The "gallon one" rule was also enacted to avoid regulatory confusion for the industry. In 2008, the industry did not know if a new well planned for drilling would ultimately be completed as a vertical or horizontal well until drilling began and cuttings were analyzed. Nor did the industry know, beforehand, how much water they would eventually need to successfully fracture the new well. Because these projects would be in violation of SRBC regulations if approval was not obtained prior to initiation, the industry would begin each new well with a high level of uncertainty. To eliminate this confusion, the SRBC made the decision that all unconventional natural gas wells needed regulatory approval, and consequently issued the "gallon one" rule.

## 2.2 Approval by Rule Program

SRBC promulgated new rulemaking in 2008 to help achieve the objectives of the "gallon one rule" without impacting the legitimate development of the SRB's water resources. The rulemaking, which expanded upon existing approval by rule procedures, was tailored to provide a procedure for authorizing and tracking consumptive uses by the gas industry. The previous process was available for use only if the source of water was an approved public water supply system. This expansion allowed gas companies to use the process to seek consumptive use approvals regardless of the water source, including wastewater, mine drainage, and other lesser-quality sources.

At a drilling pad, water is consumptively used for many purposes including well drilling and construction, well completion processes, hydrostatic, geophysical, and other testing, and dust control. SRBC regulates all consumptive use by the natural gas industry on a drilling pad basis through the administrative approval by rule process (SRBC, 2015a). This allows for tracking sources of water transported to and from the site, quantities of water consumptively used, and associated mitigation requirements. The process also allows gas companies to use sources of water previously approved for use at any of their drilling pads and to share sources previously approved for use by another company, as long as access and use agreements are registered with SRBC. This process was intended to encourage water sharing and thereby limit the need for multiple and redundant withdrawals.

## 2.3 Water Withdrawal Approvals and Passby Flow Requirements

Industry applications for the withdrawal of surface or groundwater in any amount are reviewed by SRBC for potential individual or cumulative impacts on water resources and water users. Approvals are issued in the form of dockets that specify maximum instantaneous withdrawal rate, peak day withdrawal amount (for surface water), 30-day average withdrawal (for groundwater), and other project-specific conditions. Water withdrawal approvals issued to the industry are valid for a period of 4 years and subject to additional review prior to renewal. Factors such as foreseeable demand, availability of alternate sources, competing water uses within the watershed, stream classification, stream biology, and other similar factors are considered during the review process. SRBC also incentivizes industry use of lesser-quality water (eg, abandoned mine drainage and municipal wastewater) by decreasing the fees associated with application for withdrawal. This is intended to reduce withdrawal and consumptive use of high-quality sources (SRBC, 2012a).

SRBC requires passby flows for certain surface and groundwater withdrawals. A passby flow is defined as a prescribed streamflow below which withdrawals must cease. Withdrawals that are sufficiently small in rate and quantity that the impacts on streamflows and the ecosystems they support are negligible during all or certain months of the year are considered *de minimis*. Withdrawals, considered individually and cumulatively, that are determined not to be *de minimis* are conditioned with passby flow requirements. This results in the approved withdrawal being interruptible at designated, site-specific, low flow conditions. While streamflow may continue to decline after a withdrawal ceases, passby flow requirements prevent the withdrawal from further exacerbating stressors or impacting other downstream water users during natural low flow periods. Passby flow thresholds are defined within SRBC dockets and are unique to each project location. Water withdrawals approved during the early portion of the study period were conditioned with annual passby flow thresholds set to a percentage of average daily flow (SRBC, 2002).

The Nature Conservancy (TNC) conducted a study entitled *Ecosystem Flow Recommendations for the Susquehanna River Basin* (DePhilip and Moberg, 2010). In the report, TNC presented a set of recommended limitations to flow alteration to protect the species, natural communities, and key ecological processes within the various stream types of the SRB. One of the key findings of the study was that seasonal flow recommendations are preferred to year-round flow recommendations as ecosystem flow needs are naturally seasonal. Based on these recommendations, SRBC adopted a Low Flow Protection Policy in 2012 (SRBC, 2012b), which contains specifications for determining passby flows and conservation releases for approved water withdrawals. Passby flows are currently specified as monthly flow thresholds based on stream drainage area and other special considerations including seasonal uses, instream flow studies, and water quality.

## 2.4 Monitoring and Reporting Requirements

SRBC requires industry withdrawal proposals to include items such as an intake design, a schematic of the withdrawal and associated infrastructure, performance specifications for a pump and flow meter, and a metering plan. Upon approval, projects must install a totalizing meter that records daily water withdrawals, submit photographs of the installation, and certify accuracy of the meter to within 5% of actual flow (SRBC, 2015b). Projects are directed to maintain metering so as to provide an accurate record of withdrawals and certify, once every 5 years, the accuracy of measuring devices. Approved projects are mandated to keep daily records of the withdrawal and well pads receiving the water, and report the data to the SRBC quarterly. Withdrawals with passby flows are required to cease when streamflow, as measured at a local or designated US Geological Survey stream gauge, is equal to or less than the specified passby flow threshold. Such projects are obligated to monitor and record daily data of the gauged stream flow and report the data to the SRBC quarterly or as requested. Groundwater withdrawals are also required to monitor and report groundwater elevation data. Approved consumptive use projects are directed to submit schedules for drilling and hydraulic fracturing wells and post-hydraulic fracturing event reports detailing water use and recovery information.

# 3. NATURAL GAS DEVELOPMENT

## 3.1 Well Development and Associated Infrastructure

Substantial landscape disturbance has accompanied rapid unconventional shale gas development throughout the northeast United States. In the SRB, development mostly occurs in agricultural and forested settings where land disturbance often results. Typically, road construction or widening of existing roads is necessary to facilitate movement of equipment to well pads, which are typically 3–5 acres (1.2–2 ha) in size. When other disturbances associated with well development are included, such as roads, pipelines, and indirect forest impact from new edges, surface disturbance can approach 30 acres (12 ha) per well pad (Johnson, 2010). Unconventional wells are drilled vertically to the depth of the targeted gas-containing shale formation (eg, Marcellus), where lateral drilling begins to follow the contours of the layer. After wells are cased with steel and cement, hydraulic fracturing occurs in the perforated lateral portions of the wells. Water, chemicals, and sand are pumped at high pressures to fracture the shale layer, which facilitates the flow of natural gas into the well. The extracted gas is then collected using gathering lines and, with the help of compressor stations, gas is moved to main transmission pipelines.

In the SRB, the gas industry primarily relies on surface water sources to provide water for hydraulic fracturing. A variety of intake structures are used to withdraw water from streams, rivers, and other surface water bodies in the SRB. For example, floating intakes attached to flexible hoses and deployed over the stream bank are often used for withdrawals of smaller volumes and are advantageous because they are easily removed for maintenance and during periods of ice cover, etc. For larger volume surface water withdrawals, more permanent intakes are buried into the streambank and intakes are fixed to the stream substrate. Pumps are attached to these intake structures and water is withdrawn according to permitted conditions. Water may be stored near the withdrawal location using storage impoundments or tanks. Trucks or pipelines are then used to transport water to well pads where it is stored in impoundments or tanks until it is used in the hydraulic fracturing process. On well pads, flowback water storage impoundments and tanks are also used to collect water used in the hydraulic fracturing processes as it returns to the surface. Flowback water and production fluids are either reused, with or without treatment, for subsequent fracturing events or transported off-site for treatment and/or final disposal with no reuse.

## 3.2 Wells Drilled and Hydraulically Fractured

The industry is required to file a post-hydraulic fracturing report to SRBC for every unconventional gas well stimulated in the SRB. These reports include the date of the hydraulic fracturing event, pressure release date, and quantities and general types of fluids injected and recovered. The types of fluids injected include fresh water, flowback fluids, and production fluids. The well completion reports filed by the industry with the Pennsylvania Department of Environmental Protection (PADEP) and the post-hydraulic fracturing reports submitted by the industry to SRBC were used to compile the number of wells drilled and hydraulically fractured within the SRB by quarter and calendar year. The information pertains only to unconventional natural gas wells located within the Pennsylvania portion of the SRB (Fig. 2A).

Only two wells were reportedly drilled within the SRB in 2005, three wells in 2006, and 14 in 2007. It was not until 2008 that more substantial numbers of gas wells were

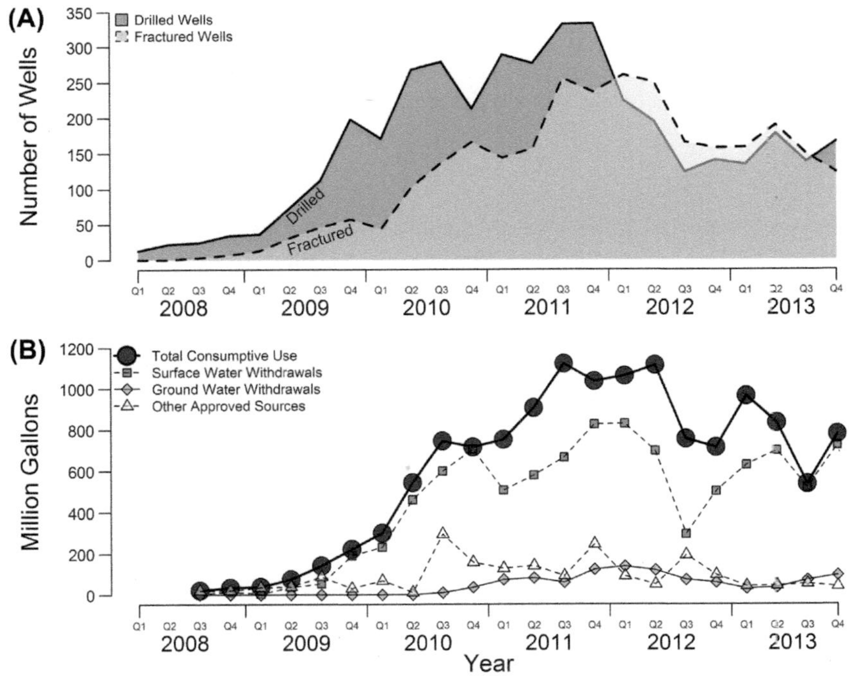


FIGURE 2 Quarterly total of (A) unconventional natural gas wells drilled and hydraulically fractured and (B) total consumptive water use and water withdrawals by quarter from 2008 to 2013 in the Susquehanna River Basin.

permitted and drilled in the SRB, and it was not until 2009 that significant numbers of those wells were hydraulically fractured (Fig. 2A). As of December 31, 2013, the total number of wells drilled and fractured within the SRB was 3995 and 2860, respectively. These numbers suggest that, to date, approximately 70% of the wells drilled were subsequently hydraulically fractured. It is anticipated that a greater percentage of the drilled wells will be fractured as more gathering and transmission pipelines are constructed, and as the price of natural gas rises.

## 4. WATER USE

### 4.1 Pre- and Post-Gas Industry Water Use

A variety of water uses in the SRB were present prior to the emergence of the natural gas industry in 2008. Reported groundwater withdrawals, surface water withdrawals, and consumptive use by category were compared between 2007 and 2013 to assess the significance and uniqueness of the industry's water use. There was no reported water use by the gas industry in 2007. In 2013, total industry withdrawals averaged 0.6 and 7.0 million gallons per day (mgd) of groundwater and surface water, respectively, based on annual averages. The industry represents a relatively minor source of groundwater withdrawals, but surface water withdrawals exceeded those of the mining and recreation sectors in 2013. Reported consumptive use by the industry averaged 8.6 mgd in 2013, which was comparable to that of the public water supply and manufacturing categories (Table 1). There are a number of reasons discussed later that could allow average consumptive use to exceed average total withdrawals in 2013. However, the most likely reason for this discrepancy is the dynamic water storage practices of the industry, which allow water to be withdrawn and held in storage for a period of time before being consumptively used. It is

**TABLE 1** Reported 2007 and 2013 Groundwater Withdrawals, Surface Water Withdrawals, and Consumptive Use in Million Gallons per Day (mgd), by Water Use Category, Regulated by the Susquehanna River Basin Commission

| | Groundwater Withdrawals (mgd) | | Surface Water Withdrawals (mgd) | | Consumptive Use (mgd) | |
|---|---|---|---|---|---|---|
| **Water Use Category** | **2007** | **2013** | **2007** | **2013** | **2007** | **2013** |
| Electric generation | 3.2 | 3.7 | 2267.5 | 2756.1 | 83.2 | 89.2 |
| Public water supply | 52.6 | 52.2 | 23.9 | 70.2 | 11.1 | 8.7 |
| Natural gas | 0.0 | 0.6 | 0.0 | 7.0 | 0.0 | 8.6 |
| Mining | 42.1 | 37.5 | 0.3 | 6.8 | 1.1 | 1.1 |
| Manufacturing | 18.3 | 19.2 | 23.4 | 26.2 | 7.9 | 9.1 |
| Recreation and other | 13.0 | 9.9 | 2.6 | 3.3 | 7.1 | 5.3 |

important to note that while all of the water use by the gas industry is considered consumptive, the same is not true of other industries. For instance, water withdrawn by the public water supply sector is used by consumers and the majority is returned to nearby streams in the form of treated effluent. This is an example of a non-consumptive use. This distinction between consumptive and non-consumptive uses is the reason why other sectors such as public water supply are responsible for much larger volumes of water withdrawals, but their consumptive use is roughly equivalent.

## 4.2 Approvals for Consumptive Water Use on Well Pads

A total of 2249 approvals were issued by SRBC to the industry for the consumptive use of water withdrawn from SRB sources and used on approved pad sites during the study period. There were four counties within the SRB in northern Pennsylvania with the greatest number of approvals by rule: Bradford (699), Susquehanna (400), Tioga (395), and Lycoming (289). Together, these four counties contained approximately 80% of the total approvals by rule issued to the industry by SRBC. Ongoing reviews of the approvals by rule indicate that approximately 60–80% of the well pads for which approvals were issued during the study period resulted in actual pad construction and the drilling of at least one well per pad. The remainder of the approvals either expired over time without pad construction or are currently active and awaiting pad construction and well drilling efforts. Data also indicate that the majority of well pads (77%) contain between one and five wells. Through the end of the study period the maximum number of wells installed on a single pad was 11 (Fig. 3).

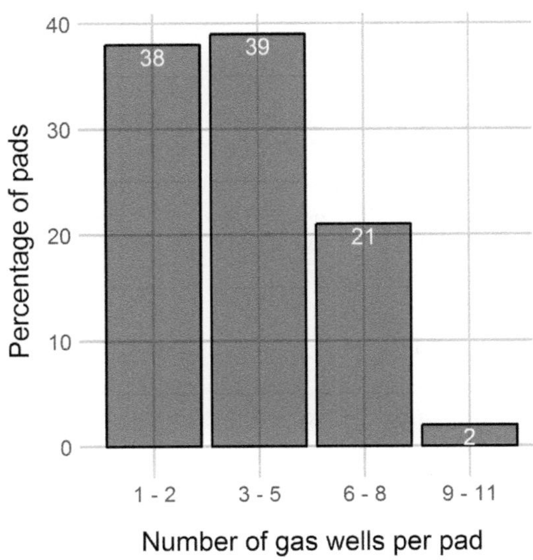


**FIGURE 3** Percentage of unconventional natural gas well pads containing 1 to 11 wells in the Susquehanna River Basin from 2008 to 2013. Exact percentage of each bin appears at the top of each bar.

## 4.3 Total Consumptive Water Use by the Industry

Water used by the industry originated primarily from either surface water or groundwater sources, or a combination of the two. Surface water sources include water withdrawn from streams, creeks, rivers, ponds, and lakes. Groundwater sources include water withdrawn from water wells. Water withdrawn from public water systems can be comprised of a combination of both surface water and groundwater sources.

The total amount of water consumptively used and reported by the industry during the study period was 13.4 billion ($10^9$) gallons (Bgal). This number exceeded the combined total quantities of groundwater withdrawals (998 Mgal), plus surface water withdrawals (11.7 Bgal), plus into-basin diversions (38 Mgal) by approximately 637 Mgal. The difference equates to approximately 4.8% of the total consumptive use of 13.4 Bgal. This difference is possibly attributed to a number of factors, including but not limited to: (1) accuracy of the meters used to measure water quantity at the withdrawal locations; (2) human error during daily monitoring and periodic reporting; (3) capture and use of top-hole water and stormwater; (4) evaporation from storage facilities; and (5) the dynamic water storage practices of the industry. However, it is impossible to entirely discount the possibility that some amount of water from unapproved sources found its way into the industry's water supply system.

On average, the industry consumptively used approximately 6.7 mgd during the study period and 8.6 mgd in 2013 alone. The largest amount of water consumptively used by the industry occurred in the period from the third quarter of 2011 to the second quarter of 2012 (average of 1083 Mgal; 11.8 mgd). This period of peak consumptive use coincided with the period when the largest number of wells were hydraulically fractured (Fig. 2).

### 4.3.1 Groundwater

Eight groundwater withdrawals were approved for industry use during the study period, with none rescinded or expired by the end of 2013. Approximately 998 Mgal, which is 7.4% of the total water consumptively used by the industry during the study period, originated solely from groundwater sources. The majority of this groundwater (774 Mgal) originated at water well fields owned and operated by public water systems or third-party water purveyors approved by SRBC. The balance (224 Mgal) originated at other approved public water systems relying on water wells. Approximately 177 Mgal (18%) of the total 998 Mgal of groundwater used by the industry originated from groundwater wells approved by SRBC and under the direct control or ownership of the industry.

### 4.3.2 Into-Basin Diversions

The diversion of water into the SRB for gas development from the Ohio River Basin during the study period was approximately 38 Mgal. This quantity constitutes only 0.3% of the total amount of water consumptively used by the industry during the study period.

### 4.3.3 Surface Water

A total of 9.76 Bgal of surface water was withdrawn from waterways within the SRB and consumptively used by the industry during the study period. Approximately 70% of the approved surface water withdrawals for the industry include site-specific passby thresholds below which the withdrawal must cease. An additional 1.97 Bgal of water was withdrawn from public water systems composed of varying portions of both surface water and groundwater sources. Together, these two major sources of water comprised approximately 88% of the total amount of water consumptively used by the industry.

From the third quarter of 2008 through the third quarter of 2009, approximately 60–90% of water consumptively used by the industry originated at public water systems. These public systems relied upon both surface water and groundwater sources to meet their overall water demands. Beginning in the fourth quarter of 2009 and extending through the fourth quarter of 2012, the primary sources of water for the industry transitioned from public systems to surface water withdrawals developed and controlled by gas companies or private third-party water purveyors. By the first quarter of 2013, the amount of water taken from public water systems and consumptively used by the industry had dropped below 5% of total water used. It remained below 10% throughout the remainder of 2013 (Fig. 2B).

The maximum average daily water withdrawal rate calculated on a quarterly basis for the industry was 10.7 mgd and occurred during the first quarter of 2012. Calendar year 2012 also had the greatest annual number of wells fractured at 836, followed closely by 2011, which had 794 wells fractured during that year. During calendar year 2013, the total number of wells hydraulically fractured dropped to 623 wells. The correlation between wells fractured and total consumptive use/water withdrawals by quarter is notable (Fig. 2).

During the study period there were a total of 222 surface water withdrawals approved by SRBC for use by the industry. Of that total, 28 approvals were rescinded for various administrative reasons, 35 approvals expired and were subsequently renewed, and 58 approvals expired and were not renewed as of December 2013. The approved surface water withdrawals were located within 61 individual watersheds ranging in size from 0.5 to 10,539 mi$^2$. The number of withdrawals approved in watersheds ranged from one to two in 44 watersheds to 25 withdrawals in the North and West Branch of the Susquehanna River (Fig. 4), the two largest watersheds identified.

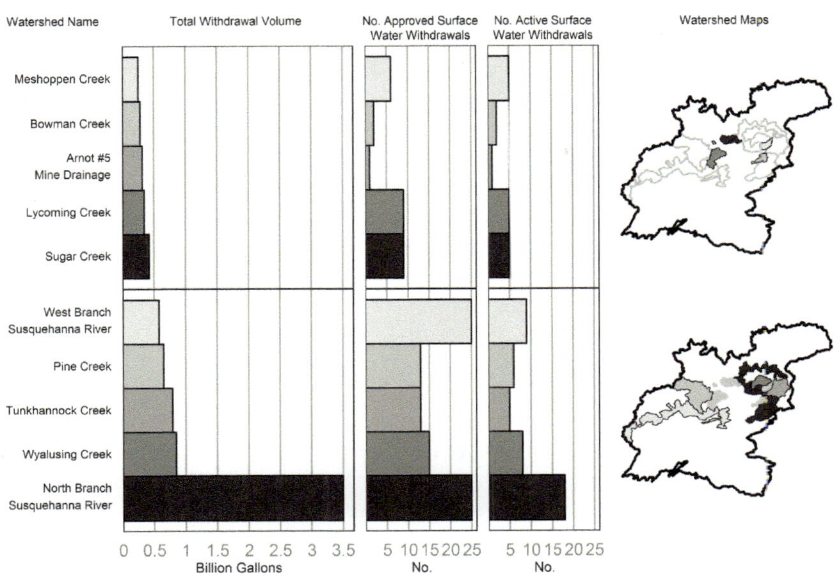


**FIGURE 4** Cumulative surface water withdrawals and number of approved and active surface water withdrawals during 2008–2013 from 10 most used watersheds in the Susquehanna River Basin.

A total of 114 out of the 222 surface water withdrawals approved were never actively used. The remaining 108 (48%) withdrawals were actively used during the study period, meaning that water was withdrawn for at least one day. Of the 61 watersheds containing at least one approved withdrawal, withdrawals were only active in 39 of those watersheds. Of these 39 watersheds, most (27) contained one to two active withdrawals, while the North Branch Susquehanna River watershed contained 18 active withdrawals (Fig. 4).

Of the 108 actively used withdrawals, 64 were located within the same 10 watersheds. These 10 watersheds accounted for over 82% of the 9.7 Bgal of surface water withdrawn by the industry during the study period (Fig. 4). Withdrawals from the top five watersheds (North Branch Susquehanna River, Wyalusing Creek, Tunkhannock Creek, Pine Creek, and West Branch Susquehanna River) constituted approximately 66% of total surface water withdrawals. Together, the North Branch and West Branch of the Susquehanna River supplied approximately 45% of the total surface water withdrawn by the industry (Fig. 4).

Of the available surface water withdrawals, 37 were from lesser-quality waters, including three discharges from water treatment plants and 34 mine drainage waters associated with past coal mining activities. Thirteen of the lesser-quality water sites were actively used by the industry during the study period, with a total of approximately 865 Mgal of lesser-quality water withdrawn and used.

Of the 222 approved surface water withdrawals for the industry during the study period, 212 had measurable drainage areas. Ten projects, including quarry pits and ponds, were excluded since they did not have clearly defined drainage areas. Drainage areas were divided into five size classes for frequency analyses. Results indicate that 23 (10.8%) of the 212 withdrawals were located within 0–10 mi$^2$ watersheds. Additionally, 48 withdrawals (22.6%) were located within watersheds 10–50 mi$^2$, 60 (28.3%) within 50–200 mi$^2$, 34 (16.0%) within 200–1000 mi$^2$, and 47 withdrawals (22.2%) were located within watersheds larger than 1000 mi$^2$ (Fig. 5). The siting of the majority of approved surface water withdrawals (154; 72.6%) in watersheds less than 500 mi$^2$ is believed to be the result of several factors. First, primarily because of productivity of shale formations, the industry is most active in the Appalachian Plateau and Allegheny Front physiographic provinces of the northcentral Pennsylvania portion of the SRB. The geomorphic characteristics of those areas produce relatively steep mountainous terrain resulting in localized watersheds of smaller sizes when compared to other physiographic provinces within the SRB. The relatively steep terrain also creates a preference by the industry for siting well pads on hilltops, which are commonly located in smaller watersheds. The industry's effort to minimize the transport distances between well pads and water sources results in its preference for seeking water sources in nearby, smaller watersheds.

## 4.4 Water Use for Hydraulic Fracturing

A total of 2860 gas wells were hydraulically fractured within the SRB during the study period. Approximately 12.9 Bgal (96%) of the water withdrawn by the industry during that period was consumptively used in the hydraulic fracturing process. The remaining 0.5 Bgal (4%) of the water was consumptively used for other activities at the drilling pads such as well drilling, preparation of drilling muds and grout, dust control, maintenance operations, and site reclamation. Reports from individual hydraulic fracturing events performed during the study period indicate that the industry consumed an overall average of 4.3 Mgal of water per well. Of that 4.3 Mgal of water used during the average fracturing event,

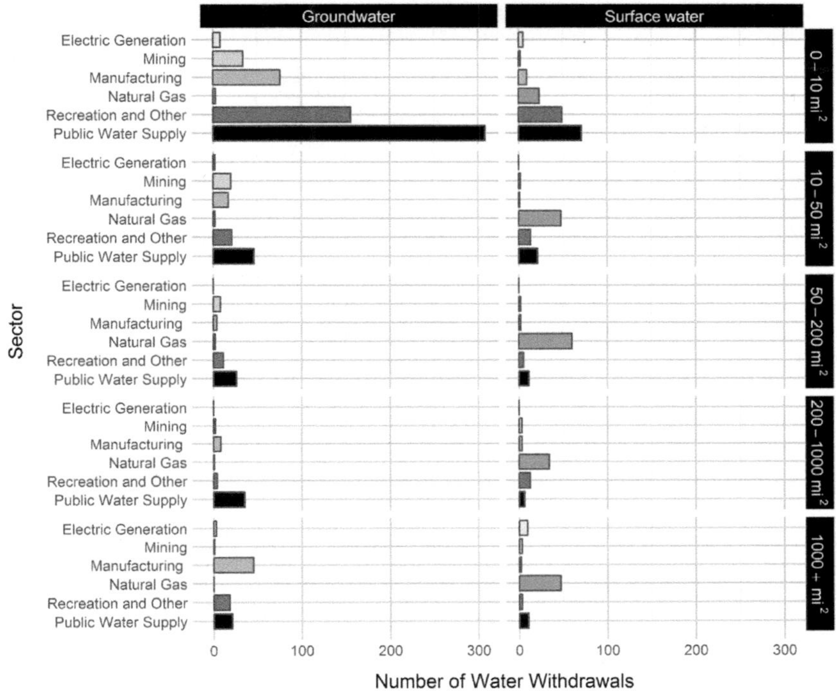


FIGURE 5 Histogram of number of approved groundwater and surface water withdrawals, by drainage size, in the Susquehanna River Basin from 2008 to 2013.

3.6 Mgal (84%) was comprised of fresh water and 0.7 Mgal (16%) was comprised of reused flowback fluids.

The average amount of water used per fracturing event was relatively low in the second half of 2008 and the first quarter of 2009, ranging from 1.6 to 2.1 Mgal per event. These relatively lower amounts of water used per event during this early period were believed to be primarily because of smaller exploration companies performing limited fractures on vertical wells and "toe fractures" on shorter laterals in horizontal wells to secure land leases with property owners. The shorter laterals were also used by the exploration companies to test the productivity of the target formations and prove the resource. As the industry transitioned from the exploratory phase to the production phase, companies started drilling longer laterals to achieve better gas recovery from the shale formations and to access more of the formation from a single well. This led to a more systematic development of the resource overall and an increase in the amounts of water used per fracturing event. From the third quarter of 2010 through the fourth quarter of 2012, the amount of water used held relatively steady at 4.3–4.8 Mgal per fracture event. During 2013, the industry started lengthening the laterals and the average amount of water used increased to approximately 5.1–6.5 Mgal per fracturing event.

During the second quarter of 2009, the industry began reusing flowback fluids in subsequent fracturing events in a more concerted manner. The amount of flowback used in fracturing events increased on an annual basis from 2009 through 2013. This increased reuse of flowback reflects the value of these fluids in subsequent fracturing events and represents a reduction in the amounts of fresh water needed for subsequent fracturing events. The reuse also resulted in a reduction in the amount of waste fluids requiring disposal or treatment.

Post-hydraulic fracture data from the study period indicate that the average amount of flowback recovered from the wellbore of stimulated wells within the first 30 days following the release of pressures ranged from a low of approximately 5% to a high of approximately 12%, with a long-term average of approximately 10%. Therefore, given the average of 4.3 Mgal of water used per fracturing event, the amount of flowback from each stimulated well ranged from approximately 0.2 Mgal (at 5%) to 0.5 Mgal (at 12%). Using the long-term average flowback recovery rate of 10% per fracturing event, with an average of 4.3 Mgal of water used per well fracturing event for 2860 wells fractured, indicated that approximately 1.2 Bgal of flowback fluids were recovered from wells during the study period. This flowback water was either: (1) reused for subsequent fracturing events without treatment; (2) treated on the originating pad and reused for additional fracturing; (3) transported to an off-site facility for treatment and then back for subsequent fracturing; or (4) transported off-site for treatment and/or final disposal with no reuse.

Information from PADEP files indicate that approximately 99% of this flowback was reused by the gas industry. The remainder was transported to deep injection wells, landfills, or treated and discharged into surface waters. Information taken from PADEP files also indicated that approximately 86% of all produced fluids from wells located within the SRB were reused by the industry, with only 14% of produced fluids destined for final disposal including deep injection wells (PADEP, 2014).

## 5. COMPETITION FOR WATER RESOURCES IN THE SRB

Riparian water rights govern water use in the northeast United States (and SRB), not prior appropriation water rights as in the western United States. Because of the directive of the riparian doctrine and the SRBC Compact, the emergence of the gas industry as a legitimate water use must be considered equally as important as traditional and established water uses in the SRB. When compared with 2007, water use in 2013 remained generally unchanged in all sectors other than the gas industry (Table 1). This suggests that water use by the gas industry, which increased over the same period, has not negatively impacted established water users. This may be because of the geographic area occupied by the gas industry. The northern Pennsylvania portion of the SRB, where the industry is most active, is rural and/or undeveloped with relatively little existing water use. As such, there is little evidence for competition for water resources between the gas industry and existing anthropogenic water users in the SRB.

The primary competition for water resources has occurred between the industry and aquatic ecosystems in the small watersheds with characteristically low water yield in the northern Pennsylvania portion of the SRB. This challenge was met with regulatory changes implemented by SRBC intended to be protective of water resources while allowing for sustainable development. Desktop environmental screenings and on-site aquatic resource surveys of proposed withdrawal locations have proven to be valuable tools to inventory current conditions at sites where data are lacking, which allows for more effective technical reviews of proposed withdrawals. Limits on withdrawal magnitude and examination of cumulative withdrawals in watersheds have provided insight regarding the suitability of watersheds to accommodate newly proposed water use. Passby flow requirements have also been especially useful in maintaining adequate instream flows and not exacerbating natural low flow conditions. Additionally, monitoring and reporting requirements for industry withdrawals and consumptive use have allowed managers to accurately quantify and examine trends in industry water use. Lastly, the active compliance program implemented by SRBC to ensure adherence to withdrawal conditions such as passby flows has proven to be an effective regulatory function.

Secondary competition has arisen within the gas industry, specifically between different operating companies. The small watersheds in the northern Pennsylvania portion of the SRB constitute interruptible water sources, as many withdrawals are conditioned with passby flows. In response, the industry has developed a dynamic water storage and distribution system, allowing water to be obtained when conditions allow and saved for later use. SRBC has enacted regulatory changes intended to allow sharing of water sources between companies, to decrease the need for multiple and redundant withdrawals in individual watersheds. The term of approval for industry withdrawals has also been maintained at 4 years, which allows more frequent revisiting of projects and examination of water use to better align allotted quantities with actual use.

## 6. RELEVANT STUDIES AND APPLIED RESEARCH

Significant effort has been directed toward researching the potentially adverse impacts of gas development, especially research specifically targeting water resources. Of concern is potential groundwater contamination caused by methane migration from improperly cased wells and vertical migration through geologic fractures (eg, Osborn et al., 2011). Impacts to surface waters are also of concern. Sedimentation from land disturbance and contamination from produced waters high in salts, metals, and radioactivity are threats that have been evaluated in gas plays (eg, Olmstead et al., 2013; Brantley et al., 2014; Hintz and Steffy, 2015). Attention has also been paid to trends in environmental violations by gas drillers and the effectiveness of regulations levied by state agencies (eg, Rahm et al., 2015).

To evaluate the impacts of industry surface water withdrawals on stream ecology, data were collected at gas industry surface water withdrawals in three stream types in the SRB (Shank and Stauffer, 2015). The study concluded that landscape characteristics of watersheds better explained variation of fish and macroinvertebrate assemblages than did water withdrawal intensity at study sites. It should be noted that this study was conducted during a relatively early stage in what is expected to be the long-term development of shale gas. Additionally, this research examined water withdrawals regulated by SRBC's previous regulatory scheme of year-round passby flow thresholds set to a percent of average daily flow (ADF). Findings indicated that the largest withdrawals relative to stream size observed in this study were from headwater streams, which averaged 6.8% of ADF daily. These large withdrawals in small watersheds have a greater potential for impacts (Shank and Stauffer, 2015). A study using theoretical withdrawal scenarios in the Ohio River Basin portion of Pennsylvania highlighted the importance of longer streamflow gauge records when determining passby flows. The results suggested that, when using seasonal/monthly percent exceedance flow statistics as passby thresholds, the period of record required to accurately estimate flows at ungauged sites is substantially longer than what is needed for static, year-round ADF thresholds (Mitchell et al., 2014). SRBC initiated research to evaluate the predictive accuracy associated with using reference US Geological Survey stream gauges to estimate passby flow conditions at ungauged water withdrawal sites. Correlation analyses between on-site flow measurements and concurrent streamflow records from selected reference gauges were conducted at 18 surface water withdrawal sites during low and base flow conditions. Preliminary results suggest high correlation between on-site flow measurements and references gauges at the majority of ungauged sites, with the exception of one located in a unique hydrogeologic spring setting (Liu et al., 2016).

SRBC conducted a Cumulative Water Use and Availability Study to evaluate the potential cumulative impact of consumptive use within the SRB. The study entailed computing existing and projected consumptive use, determining water capacity at varying spatial scales, developing a GIS-based water availability tool, and evaluating alternatives for mitigating potential impacts. To integrate sustainable limits of water development and low flow protection criteria in existing SRBC policies and plans, a water capacity threshold based on the low flow margin of safety method (Domber et al., 2013) was defined for assessing water availability. The assessment tool will be instrumental in identifying water-limited areas and evaluating various management measures and their effects on water availability (Balay et al., 2016).

## 7. LESSONS LEARNED AND FUTURE OUTLOOK

There were several important lessons learned from unconventional natural gas development in the SRB. The ability to move swiftly to establish a regulatory framework to accommodate this unique water use, and remain flexible to implement adaptive management measures as development evolved, proved effective in attaining regulatory compliance for this new industry. The highly mobile and decentralized nature of the industry's water use poses a significant challenge for enacting appropriate oversight. An overarching goal in the SRB was to ensure that all water use came from a source that was approved or otherwise recognized as appropriate. It was critical to institute administration, tracking, and reporting of water use in such a way that water movement was accountable, while providing flexibility for the transport and reuse of water by the industry. It is advantageous to plan out supporting water infrastructure prior to initiation of gas development to optimize siting of water sources, transportation and distribution routes, and storage facilities. Doing so helps identify sustainable supplies, avoid redundant sources, and minimize environmental impacts. The industry's presence in more remote, headwater settings introduces concerns not typically associated with conventional energy development such as the effects of habitat fragmentation, land use and disturbance, and overallocation of water resources in small watersheds. More protective standards are necessary to ensure proper management in sensitive headwater settings.

The industry's water use pattern is unique in that it exhibits a relatively low frequency and duration, but high magnitude, of water use. This introduces uncertainty into actual water use accounting and long-term planning. It is unrealistic to assume that all recognized sources are used routinely at the maximum capacity, but it is very difficult to predict the timing and quantity of water use. This makes planning for the industry's water demand, without overallocating and unnecessarily constraining the water available for other economic development, a challenging task. Incentivizing sharing of sources, use of lesser-quality waters, and recycling/reuse are important management strategies for reducing water demand, system stress, and waste streams. Protective conditions such as limiting withdrawals to viable rates and imposing passby flow requirements are critical to ensuring sustainable water development and avoiding impacts to competing instream uses, particularly during times of low flow and drought. Integrating an active compliance and enforcement program, with mandatory reporting requirements and routine field inspections, is essential to ensuring adherence to water use regulations and avoidance of environmental harm.

The potential for future natural gas development in New York, and presence of numerous other tight shale formations in the SRB, suggests that natural gas development in the SRB is likely to continue well into the future. As the water source, distribution, and storage network matures, competition for additional water sources is anticipated to stabilize. Increasing trends in length of well laterals, and associated volume of water used per hydraulic fracturing event, could result in increased future water demands. However, concentration of water use on fewer hydraulic fracturing events could provide an opportunity to reduce competition through a focus on larger events more spaced out temporally, as opposed to more frequent smaller events occurring in quicker succession. Experience suggests that water use for unconventional natural gas development in the SRB can be accommodated without impacts to competing users, with appropriate regulatory oversight and protective conditions in place. As development continues in the future, ongoing evaluation of industry trends and adaptive management will be critical to striking an appropriate balance between energy development and water resources management in the SRB.

## ACKNOWLEDGMENTS

We extend our thanks to Andrew Dehoff, Andrew Gavin, and Erin Lynam for contributing critical reviews of this manuscript. Bret Wagner, Jeff Zimmerman, and Can Liu lent their expertise in generation and presentation of the data contained within.

## REFERENCES

Balay, J., Zhang, Z., Zimmerman Jr., J., MaCoy, P., Frank, C., Markowitz, G., Liu, C., 2016. Cumulative Water Use and Availability Study for the Susquehanna River Basin. Susquehanna River Basin Commission (Publication Number 303), Harrisburg, Pennsylvania.

Brantley, S.L., Yoxtheimer, D., Arjmand, S., Grieve, P., Vidic, R., Pollack, J., Llewellyn, G.T., Abad, J., Simon, C., 2014. Water resource impacts during unconventional shale gas development: the Pennsylvania experience. International Journal of Coal Geology 126, 140–156.

DePhilip, M., Moberg, T., 2010. Ecosystem Flow Recommendations for the Susquehanna River Basin. The Nature Conservancy, Harrisburg, Pennsylvania.

Domber, S., Snook, I., Hoffman, J.L., 2013. Using the Stream Low Flow Margin Method to Assess Water Availability in New Jersey's Water-Table-Aquifer Systems. New Jersey Geological and Water Survey, Technical Memorandum, 13–3.

Hintz, D., Steffy, L., 2015. Remote Water Quality Monitoring Network Data Report of Baseline Conditions for 2010–2013. Susquehanna River Basin Commission (Publication Number 297), Harrisburg, Pennsylvania.

Johnson, N., 2010. Pennsylvania Energy Impacts Assessment. The Nature Conservancy.

Kammerer, J.C., 1990. Largest rivers in the United States. U.S. Geological Survey Fact Sheet. Open File Report 87–242.

Liu, C., Zhang, Z., Balay, J., 2016. Evaluation of Passby Flow Reference Gage Determinations in the Susquehanna River Basin (Publication in Preparation). Susquehanna River Basin Commission, Harrisburg, Pennsylvania.

Mitchell, A.L., Small, M., Casman, E.A., 2014. Surface water withdrawals for Marcellus Shale gas development: performance of alternative regulatory approaches in the upper Ohio River basin. Environmental Science and Technology 47, 12669–12678.

Olmstead, S.M., Muehlenbachs, L.A., Shih, J.S., Chu, Z., Krupnick, A.J., 2013. Shale gas development impacts on surface water quality in Pennsylvania. Proceedings of the National Academy of Sciences of the United States of America 110, 4962–4967.

Osborn, S.G., Vengosh, A., Warner, N.R., Jackson, R.B., 2011. Methane contamination of drinking water accompanying gas-well drilling and hydraulic fracturing. Proceedings of the National Academy of Sciences of the United States of America 108, 8172–8176.

Pennsylvania Department of Environmental Protection (PADEP), 2014. Oil and Gas Reporting Website. www.paoilandgasreporting.state.pa.us.

Rahm, B.G., Vedachalam, S., Bertoia, L.R., Mehta, D., Vanka, V.S., Riha, S.J., 2015. Shale gas operator violations in the Marcellus and what they tell us about water resource risks. Energy Policy 82, 1–11.

Richenderfer, J.L., Wagner, B., Shank, M.K., Balay, J., Hintz, D., Hoffman, J., Ballaron, P.B., McFeaters, S., Zimmerman Jr., J., 2016. Water Use Associated with Natural Gas Shale Development: An Assessment of Activities Managed by the Susquehanna River Basin Commission through December 2013. Susquehanna River Basin Commission (Publication Number 299), Harrisburg, Pennsylvania.

Shank, M.K., Stauffer Jr., J.R., 2015. Land use and surface water withdrawal effects on fish and macroinvertebrate assemblages in the Susquehanna River basin, USA. Journal of Freshwater Ecology 30 (2), 229–248.

Susquehanna River Basin Commission (SRBC), 1972. Susquehanna River Basin Compact.

SRBC, 2002. Policy No. 2003–01: Guidelines for Using and Determining Passby Flows and Conservation Releases for Surface-Water and Ground-Water Withdrawal Approvals. SRBC, Harrisburg, Pennsylvania.

SRBC, 2008. Press Release: SRBC to Require All Natural Gas Well Development Projects in Susquehanna Basin to Obtain Prior Approval Regardless of Amount of Water Used. Released August 15, 2008.

SRBC, 2012a. Resolution Number 2012–01: Use and Reuse of Lesser Quality Water.

SRBC, 2012b. Policy Number 2012–01: Low Flow Protection Policy Related to Withdrawal Approvals.

SRBC, 2013a. Comprehensive Plan for the Water Resources of the Susquehanna River Basin.

SRBC, 2013b. State of the Susquehanna 2013 Report.

SRBC, 2015a. Project Review Regulations, 18 Code of Federal Regulations Part 806.22(f).

SRBC, 2015b. Project Review Regulations, 18 Code of Federal Regulations Part 806.30.

United States Congress, 1970. Public Law 91–575, Susquehanna River Basin Compact.

U.S. Energy Information Administration (USEIA), 2012. Annual Energy Outlook 2012. Released June 2012. http://www.eia.gov/forecasts/aeo/pdf/0383(2012).pdf.

USEIA, 2014. Marcellus Region Production Continues Growth. Released August 05, 2014. http://www.eia.gov/todayinenergy/detail.cfm?id=17411#.

Weltman-Fahs, M., Taylor, J.M., 2013. Hydraulic fracturing and brook trout habitat in the Marcellus Shale region: potential impacts and research needs. Fisheries 38, 4–15.

Zhang, Z., Dehoff, A.D., Pody, R.D., Balay, J.W., 2010. Detection of streamflow change in the Susquehanna River basin. Water Resources Management 24, 1947–1964.

Chapter 2.2.5

# Water Use for Unconventional Gas Production in the European Union

L. Reins
*KU Leuven University, Leuven, Belgium*

## 1. INTRODUCTION

The management and regulation of shale gas and its impacts on water resources has been identified as an important action area in the European Union by various stakeholders, such as industry actors, as well as European Commission representatives (Barton, 2013). A European Parliament Resolution on environmental impacts of unconventional gas activities has further called for "compulsory water management plans by operators, in cooperation with the drinking water companies and the competent authorities; [the Parliament] stresses, however, that existing treatment plants are ill-equipped to treat hydraulic fracturing waste water and may be discharging pollutants into rivers and streams; [the Parliament] considers, to this end, that a full assessment of all the relevant water treatment plants in the Member States concerned should be carried out by the competent authorities" (European Parliament, 2012). The importance of this is also recognized through the fact that the Joint Research Center (JRC) of the European Commission has issued a special study on the Assessment of Land and Water Use Scenarios for Shale Gas Development in Poland and Germany (European Parliament, 2012). However, the exact impacts of fracturing activities are still subject to uncertainty. This is also reflected by the fact that the JRC study employed three different water use scenarios, as it is still uncertain whether and to which extent the shale gas technology will break through. Accordingly, low, average, and high rate of activity based on the parameters "different fracks per 10 years," "recycling," and "water consumption per well" has been used (Lavalle, 2013). Estimates from the International Energy Agency predict a high development scenario of 50,000 wells in the Union until 2035, resulting in increased challenges for water and land use in the Union's territory (Lavalle, 2013). Indeed, the shale gas development process illustrates the overall water issues along the value or process chain very well. As the JRC establishes, "the main environmental concerns associated with shale gas fracturing today are due to the usage of water: the high volumes of water used and lost underground, the need to process

**Competition for Water Resources. http://dx.doi.org/10.1016/B978-0-12-803237-4.00011-2**

flowbacks, the potential contamination of aquifers by leaks of chemicals employed in the fracturing fluids, etc." (Gandossi, 2014).

Whereas in the European Union no commercial extraction of shale gas is practiced as yet, it is understood from the United States' trial and error and laissez faire approach to regulation that this is not an option in the Union, especially because of the large public opposition. Hence an ex-ante regulatory regime is necessary. As a result of these uncertainties, the management and regulatory framework in the Union also has been uncertain for some time. The regulation of shale gas and its impacts on water resources is generally covered by the basic energy and environmental legislation. An assessment of the regulatory regime has, however, concluded that the "existing EU legislation is not fully equipped to tackle the resulting environmental impacts and risks (eg, surface and ground water contamination, air emissions including greenhouse gas emission)" (Impact Assessment, 2014). As a reaction to this and to clarify the existing framework, the Commission published a nonbinding Recommendation establishing minimum principles for the regulation of shale gas,[1] the only specific regulatory instrument on shale gas at the European level.

This chapter assesses the applicable regulatory framework for shale gas resources at an EU level. In particular, it will first establish the state of shale gas development in the Union and then briefly explain the potential impacts and regulatory challenges. The chapter will then analyze the regulatory framework applicable to water resources. More precisely, it will look at four pressing challenges throughout the shale gas value chain: the regulatory requirements applicable to shale gas development prior to operation, the regulation of surface water issues, the framework applicable to underground injection and groundwater, and wastewater management. A conclusion will establish the way forward.

## 2. SHALE GAS IN THE EUROPEAN UNION

Estimates of the US Energy Information Administration (EIA) for shale gas resources from June 2013 predict that the European Union in total has the potential of 13,309 billion cubic meters of technically recoverable shale gas resources (US Energy Information Administration, 2013). Poland and France, with a predicted 4.19 and 3.87 billion cubic meters of technically recoverable resources respectively, hold the greatest amount of resources. Romania (1.44 billion cubic meters), Denmark (900 million cubic meters), the United Kingdom and the Netherlands (both 730 million cubic meters), as well as Germany and Bulgaria (both 481 million cubic meters)[2] are the remaining Member States with predicted technically important amounts of shale gas resources (US Energy Information Administration, 2013). Two years after the EIA's

1. Commission Recommendation from 22 January 2014 on minimum principles for the exploration and production of hydrocarbons (such as shale gas) using high-volume hydraulic fracturing, OJ [2014] L39/72 ("Recommendation").
2. The estimates from the German authorities from 2012, however, mention recoverable resources of 1.3 billion cubic meters. Bundesanstalt für Geowissenschaften und Rohstoffe, *Abschätzung des Erdgaspotenzials aus dichtem Tongesteinen (Schiefergas) in Deutschland* (2012), at 31. Estimates thus differ considerably. See note 4 for further explanation.

estimates were published, however, it seems that the technically recoverable resources in most Member States are significantly below these estimates.[3]

Prospection and explorations are under way in several EU Member States, even if shale gas activity is not pursued in a commercial manner as yet in Europe. Despite the European Union no longer being listed among the top 10 key global players with technically recoverable resources in recent studies, the amount of resources is still immense.[4] Emerging citizen initiatives in several Member States[5] show public concern regarding environmental, climate change, and health-related issues, with notable public opposition leading to the adoption of moratoria and bans in some states (Fleming, 2013).

The political context could not be more diverse among the Member States. Some Member States are overall in favor of shale gas extraction, such as Poland, Hungary, and Lithuania. A study found that "in general, these countries' political and legal environments present a fairly stable atmosphere in which investment risk is not a primary concern" (KPMG, 2012) regarding unconventional gas activities. A reason for the strong political interest in shale in Eastern European countries is the fact that these states are more dependent on energy imports from third (non- EU) countries, especially Russia.

Poland, with the largest share of shale gas resources in Europe, is taking the lead role within the European Union to enhance the shale gas exploitation process and to become a leading commercial undertaking in the future (Orlen, 2010). For this reason, Poland is often considered a "test case for European shale gas development," which is going to determine the further process in the European Union (Meiβner, 2011). According to the Ministry of the Environment, Poland has granted at least 109 concessions for shale gas, covering a minimum area of 88,000 $km^2$ with 64 exploration wells. In nine of these wells, hydraulic fracturing technology is used, and a further 11 wells include horizontal drilling operations as of April 2015 (Wagrodzka, 2013).

In the United Kingdom, to date, one well has been hydraulically fractured by Cuadrilla, one of the biggest oil and gas exploration companies in the United Kingdom. A temporary

---

3. See, for example, A.E. Mihalache, "No shale gas in Eastern Europe, after all: implications of Chevron's exit from Romania," *Energy Post* 09.04.2015, available at http://www.energypost.eu/shale-gas-eastern-europe-implications-chevrons-exit-romania/. This is mostly because of differences in the measuring methods used by the various institutions, but also because the exact amount of technically recoverable resources can only be assessed during exploration. See also I.J. Andrews, *The Carboniferous Bowland Shale gas study: geology and resource estimation* (London: British Geological Survey for DECC, 2013), at 5f and 10f, as well as DECC, "Resources vs Reserves—what do estimates of shale gas mean?", 27.07.2013, available at https://www.gov.uk/government/uploads/system/uploads/attachment_data/file/256358/Publication_Resources_vs_Reserves_June13.pdf. See also C. McGlade, J. Speirs, and S. Sorrell, "Methods of estimating shale gas resources—comparison, evaluation and implications," 54 *Energy* 59 (2013), 116–125.
4. This was still the case in the International Energy Agency, *World Shale Gas Resources: An Initial Assessment of 14 Regions Outside the United States*, available at http://www.eia.gov/analysis/studies/worldshalegas/; however, this is no longer in the updated version. In that version, the countries of China, Argentina, Algeria, the United States, Canada, Mexico, Australia, South Africa, Russia, and Brazil form the top 10.
5. For example, "Gegen Gasbohren" in Germany, available at http://www.gegen-gasbohren.de/, "Fracturing Free Ireland" in Ireland, available at http://frackingfreeireland.org/, and the Your Voice Consultation at European Union level, available at http://ec.europa.eu/yourvoice/ipm/forms/dispatch?form=SHALEGAS.

moratorium was in place from spring 2011 until the end of 2012 after the occurrence of seismic events. The United Kingdom, after Poland, is one of the European Member States with a very proactive attitude towards shale gas, having passed several government programs promoting the activity (Petroff, 2013; N.N., 2013).

Further Member States such as Germany are still assessing the pros and cons of potential exploitation (National Gas Europe). In the Länder of Lower Saxony, North Rhine-Westphalia, and Thuringia, experimental drilling sites have been installed. However, the German authorities take into account the scientific uncertainties and lack of knowledge regarding the exact impacts on environmental resources (Umweltbundesamt, 2011). After the elections of Sep. 2013, the new Grand Coalition included a section on "Fracking" in its Coalition Agreement. The section stresses the "enormous risk potential" (CDU et al., 2013) of the technology and that the effects on humans, nature, and the environment are not sufficiently scientifically verified as of yet. Accordingly, the approval of any shale gas activity can only be granted if the necessary data on which to base an assessment are available, and if it can be determined beyond all doubt that the technology does not have any adverse effect on water quality, and is thus in line with the "duty of care principle" ("*Besorgnisgrundsatz*") of the Federal Water Act. The coalition will, with the involvement of the Länder, the scientific community, and industry representatives, develop which specific findings the exploration undertakings have to provide to eliminate gaps in knowledge and to provide a sufficient basis for possible subsequent steps (CDU et al., 2013). The agreement does not include any new findings; it essentially confirms the state of the German practice and its careful attitude in this regard. In April 2015, after considerable toing and froing, the German government presented its draft legislation on the issue, allowing fracturing under strict limitations (Bundesrat, 2015). The draft is still pending.

Bulgaria and France imposed a ban on the exploration and extraction of hydrocarbons using hydraulic fracturing. Bulgaria imposed the ban on grounds of public opposition in 2012 (Shale Gas Europe), following a similar action by France in 2011.[6] In France, the government had to defend its ban in court: the US company Schuepbach Energy challenged the ban before the lower administrative court at Cergy-Pontoise, which referred the matter to the Conseil d'Etat, which in turn transferred it to the French Constitutional Court. Schuepbach Energy held two exploration licenses in the south of France, which were suspended after the ban in 2011 (Van Calster, 2013). The court upheld the decision (Cournil, 2013; Martor).

Because of the strong public debate and divergent approaches in the European Member States, the European Commission undertook a public consultation on "Unconventional fossil fuels (eg, shale gas) in Europe" from December 2012 to March 2013. The majority of respondents were private individuals, but other stakeholders including companies, NGOs, industry associations, the academic sector, and governmental bodies also took part. The results were published at the beginning of June 2013. One key finding of the stakeholder participation was that the "large majority of respondents agree on the lack of adequate legislation, the need for public information, the lack of public acceptance of unconventional fossil fuels (eg, shale gas)" and that "doing nothing at the EU level is the least favored option" (European Commission, 2013). Around 60% of the respondents are generally in favor of the development of unconventional fossil fuels such as shale gas if proper health and environmental safeguards

6. Act 2011–835 of 13 July 2011 (*Loi n° 2011–835 du 13 juillet 2011 visant à interdire l'exploration et l'exploitation des mines d'hydrocarbures liquides ou gazeux par fracturation hydraulique et à abroger les permis exclusifs de recherches comportant des projets ayant recours à cette technique*).

are in place (European Commission, 2013). Key potential benefits associated with the activity are the avoidance of increasing the EU's energy import dependency, the strengthening of its negotiating position towards external energy suppliers, increasing the diversity of the EU energy mix, and the creation of employment (European Commission, 2013).

## 3. POTENTIAL IMPACTS ON WATER RESOURCES

It is well known that the extraction of shale gas is associated with impacts on the environment (Bailey, 2010). Extraction can adversely affect various aspects of the environment, such as water resources, ecosystems and wildlife, air quality, noise levels, and the seismicity of rock formations, and can cause visual impacts on the site (Department of Environmental Conservation, New York State, 2009). Regarding water resources, concerns relate to the "high usage of water, methane infiltration in aquifers, aquifer contamination, extended surface footprint, induced local seismicity, etc." (Gandossi, 2014).

Concerns regarding water resources relate to the fact that, depending on the individual rock formation, the hydraulic fracturing process can require between 3 and 4 million gallons of fresh water per well (Commission Communication, 2014; Impact Assessment, 2014). This arguably can have significant impacts on the local water capacity and lead to an overuse of local water resources (Commission Communication, 2014; Impact Assessment, 2014). Associated concerns are adverse impact of water withdrawals, wastewater treatment and disposal, as well as impacts on the local drinking water supply (Cady, 2010a; Department of Environmental Conservation, 2009). Furthermore, the use of water and chemicals during the drilling and fracturing phase can lead to water pollution through flowback[7] and produced water,[8] as well as drilling fluids.[9] Additional concerns relate to the pollution of surface and groundwater caused by surface spills and leaks in pits (Cady, 2010b). Depending on the location and surroundings of the shale, groundwater usage can range from 45% to 90% (Cady, 2010a). In addition, it is estimated that between 25% and 90% of the fracturing fluids that are injected into the subsoil remain underground. The fracturing process can result in large amounts of wastewater, either flowback water from the treatment itself or produced water, which is concentrated subterranean saltwater from the rock formation brought to the surface during the drilling process (Cady, 2010a; Arthur, 2008). The flowback water that returns to the surface after the treatment is mixed with the fracturing fluids can be contaminated with chemicals (Cady, 2010a).

A connected problem is the handling of waste and drilling water. Depending on the individual shale, three treatment options are possible: underground injection, treatment, and discharge or recycling (Baily, 2010). Concerns regarding underground injection are the contamination of groundwater caused by leakage of those pits or spills, as well as the destruction of flora and fauna on the surface (Cady, 2010a) and increased seismicity

7. The US EPA (2016) explains flowback water as "after the hydraulic fracturing procedure is completed and pressure is released, the direction of fluid flow reverses, and water and excess proppant flow up through the wellbore to the surface. The water that returns to the surface is commonly referred to as 'flowback'."
8. Explained by the US EPA (2016) as "after the drilling and fracturing of the well are completed, water is produced along with the natural gas. Some of this water is returned fracturing fluid and some is natural formation water. These produced waters move back through the wellhead with the gas."
9. The term drilling fluids refers to the fluids injected during the drilling process.

(Ellsworth, 2013). In case of off-site treatment, the produced water is transported to either municipal wastewater treatment plants or commercial treatment facilities (Arthur, 2008). Accidents and spills caused by transport on highways and through residential areas are related concerns (Cady, 2010a). The third method, to recycle the water, is an option for drilling water as well as wastewater, even if recycling for this kind of water is more challenging because of the high degree of corrosiveness and contamination. With regard to the amount of water needed for the operation, this method seems to be reasonable and appropriate, but its use is still limited because of inefficiency, the limited amount of recycling capacity, and economic reasons, as the disposal of drilling water in pits is less expensive (Cady, 2010a).

## 4. THE REGULATION OF WATER USE IN THE CASE OF UNCONVENTIONAL GAS[10]

Besides the nonbinding Recommendation outlining minimum principles for the exploration and production of hydrocarbons of 2014, there is no specific shale gas act enacted within the European Union. The regulation thus falls under general acts of environmental and/or (conventional) energy law.[11] In 2011, the European Parliament released a study listing 10 pieces of the most relevant legislation applicable to shale gas extraction activities (Tomescu), as well as a further 36 relevant EU Directives applicable to the regulation associated with the activity, including legislation on water, protection of the environment, safety at work, radiation protection, waste, and chemicals (DG for Internal Policies European Parliament, 2011). The key pieces of legislation are the Mining Waste Directive,[12] the Ambient Air Quality Directive,[13] the Seveso II Directive,[14] the EIA Directive,[15]

---

10. The current section is based on L. Reins, "The shale gas extraction process and its impacts on water resources," 20 *Review of European Community & International Environmental Law* 3 (2011), 300–312.
11. Article 288 TFEU establishes that "[t]o exercise the Union's competences, the institutions shall adopt regulations, directives, decisions, recommendations and opinions. A regulation shall have general application. It shall be binding in its entirety and directly applicable in all Member States. A directive shall be binding, as to the result to be achieved, upon each Member State to which it is addressed, but shall leave to the national authorities the choice of form and methods. A decision shall be binding in its entirety. A decision which specifies those to whom it is addressed shall be binding only on them. Recommendations and opinions shall have no binding force."
12. Directive 2006/21/EC on the management of waste from extractive industries and amending Directive 2004/35/EC, [2006] OJ L 102/15.
13. Directive 2008/50/EC of 21 May 2008 on ambient air quality and cleaner air for Europe, [2008] OJ L 152/1.
14. Directive 96/82/EC of 9 December 1996 on the control of major-accident hazards involving dangerous substances, [2007] OJ L 10/13, as amended.
15. Directive 2011/92/EU of 13 December 2011 on the assessment of the effects of certain public and private projects on the environment, [2012] OJ L 26/1.

the REACH Regulation,[16] the Habitats and Birds Directive creating the NATURA 2000 network,[17] the Water Framework,[18] and the Groundwater Directive.[19]

The regulation of shale gas and its impacts on water resources is subject to the general EU water legislation, namely, the Drinking Water Directive,[20] the Water Framework Directive,[21] the Groundwater Directive,[22] the Waste Directive,[23] and the Mining Waste Directive.[24] In addition, the Recommendation includes several provisions addressing the use of water for shale gas exploration and extraction. It aims at "ensuring that the public health, climate and environment are safeguarded, resources are used efficiently, and the public is informed" (Section 1 of the Recommendation). To achieve this, water use requirements are included throughout the entire regulatory lifecycle of the activity. The following sections analyze these in more detail.

## 4.1 Regulatory Requirements Prior to Operation

The Recommendation puts special emphasis on the site selection and construction phase, as this aspect of the shale gas value chain is currently unaddressed by the existing (water) legislation. Accordingly, at the strategic planning and environmental assessment stage, Member States are required to establish restrictions, minimum distances, and clear rules on shale gas activities in sensitive areas, such as in protected, flood- or seismic-prone areas, as well as minimum depth limitations between the fracturing area and groundwater resources. The risk assessment carried out at the shale gas development selection stage should further include an assessment of the "changing behavior of the target formation, geological layers separating the reservoir from groundwater and existing wells or other manmade structures exposed to the high injection pressures used in high-volume hydraulic fracturing and the volumes of fluids injected; [and] respect a minimum vertical separation distance between the zone to be fractured and groundwater" (Section 5.3 of the Recommendation).

Only if the risk assessment confirms that the fracturing activity will not "result in a direct discharge of pollutants into groundwater and that no damage is caused to other activities around the installation"(Section 5.4 of the Recommendation) can a site be selected. In a next step, a baseline study needs to be carried out determining the "(a) quality

16. Regulation No. 1907/2006 of 18 December 2006 concerning the Registration, Evaluation, Authorisation and Restriction of Chemicals (REACH), establishing a European Chemicals Agency, amending certain Directives, [2006] OJ L 396/1.
17. Directive 92/43/EEC of 21 May 1992 on the conservation of natural habitats and of wild fauna and flora, [1992] OJ L 206/7, as amended and Directive 2009/147/EC of 30 November 2009 on the conservation of wild birds, [2010] OJ L 20/7.
18. Directive 2000/60/EC of 23 October 2000 establishing a framework for Community action in the field of water policy, [2000] OJ L 327/1.
19. Directive 2006/118/EC on the protection of groundwater against pollution and deterioration, [2006] OJ L372/49.
20. Directive 98/83/EC on the quality of water intended for human consumption, [1998] OJ L 330/32.
21. Directive 2000/60/EC establishing a framework for Community action in the field of water policy, [2000] OJ L 327/1.
22. Directive 2006/118/EC on the protection of groundwater against pollution and deterioration, [2006] OJ L372/49.
23. Directive 2008/98/EC on waste and repealing certain Directives, [2008] OJ L 312/3.
24. Directive 2006/21/EC on the management of waste from extractive industries and amending Directive 2004/35/EC, [2006] OJ L 102/15.

and flow characteristics of surface and ground water; (b) water quality at drinking water abstraction points; [and] presence of methane and other volatile organic compounds in water" (Section 6.2 of the Recommendation). The installation and construction of the infrastructure needed for shale gas development needs to be built in a "way that prevents possible surface leaks and spills to soil, water or air" (Section 7 of the Recommendation). The requirements prior to operation are very detailed, taking into account that there is no specific legislation applicable to this stage of development at the European level. Regarding conventional extraction, the requirements are largely established at the Member State level. Even if detailed in nature, the requirements are to a large extent narrative. It is still at the discretion of the Member States as to how to carry out, for example, baseline studies, leaving a broad room for maneuver.

## 4.2 Regulation of Surface Water

In its Preamble, the Recommendation clarifies that the Water Framework Directive "requires the operator to obtain authorisation for water abstraction and prohibits the direct discharge of pollutants into groundwater." The Water Framework Directive constitutes the EU framework for water protection and management. The Directive sets the requirement for surface waters to achieve "good ecological status" (European Commission, 2010). At the European level, the Water Framework Directive builds the Community framework for water protection and management. Based on river basins and districts, the Member States adopt management plans and programs and are required to revise them every 6 years. These management plans have the aim of preventing and reducing pollution from discharges and emissions of hazardous substances, ensuring a balance between groundwater abstraction, promoting sustainable water use (Article 1 WFD), and of achieving the ultimate "good ecological and chemical status for all Community waters by 2015." The European Court of Justice recently clarified in this regard that this objective is not merely a management-planning objective but applies to each project authorization procedure.[25] The goal is thereby twofold and has to be understood as an obligation to prevent deterioration and an obligation to enhance (ECJ, at 39). Consequently, a project authorization has to be rejected if it may cause deterioration to the status of the surface water. Thereby it is sufficient if the deterioration affects one of the water quality elements included in Annex V of the Directive in a way that it falls by one water quality class (ECJ, at 69). Even though this case did not deal with shale gas projects, it is certainly also important for the latter, in the sense that an authorization for these might be rejected if they result in deterioration to the surface water status.

The Water Framework Directive is applicable to all inland surface waters, groundwater, transitional waters, and coastal waters. For the shale gas process, the provisions on surface waters and groundwater are most important. Concretely, the requirement for surface waters is the achievement of a "good ecological status" according to Annex V of the Directive and "good chemical status," which means compliance with all EU chemical standards. Article 11(j) establishes a general prohibition against discharging pollutants directly into groundwater. However, an exception is included for hydrocarbons. Under the exemption, specific conditions for the "injection of water containing substances

25. ECJ, Case C-461/13, *Bund für Umwelt und Naturschutz Deutschland* [2015] ECLI:EU:C:2015:433, at 43.

resulting from the operations for exploration and extraction of hydrocarbons" are authorized, provided that "such discharges do not compromise the achievement of the environmental objectives established for that body of groundwater" (Article 11). In other words, the discharge of wastewater resulting from the extraction of shale gas into groundwater is permitted if it does not lead to an immediate failure to meet other environmental goals established in the individual management plan. The exception in the Water Framework Directive is similar to the exception in the US Safe Drinking Water Act that excludes underground injections.[26] However, the provision in the European Directive is not as strict, having added the condition that any such measures shall not go against "environmental objectives" in general; although in practice the inclusion of that condition does not seem to be relevant.

Moreover, one of the Directives that has been integrated into the Water Framework Directive was the Dangerous Substances in Water Directive.[27] Articles 16 and 22 of the Water Framework Directive establish provisions for the discharge of dangerous substances. However, to date no such substances are included in the shale gas fracking fluids, and the provisions do not refer to unconventional gas resources.

Furthermore, the Water Framework Directive contains important provisions relating to hazards by floods and establishes together with Directive 2007/60/EC on the assessment and management of flood risks ("Floods Directive") the framework of flooding and flood risk regulation in Europe. In case of a flooding of the shale gas exploitation site and wells, these Directives apply. Not having any specific and important provisions, flooding management is only secondary in this context.

The Drinking Water Directive from 1998 outlines the essential quality standards that water must meet to be declared as drinking water. The Directive is aimed at the protection of "human health from the adverse effects of any contamination of water intended for human consumption by ensuring that it is wholesome and clean" (Article 1). The Directive sets quality standards (Article 5) and minimum requirements (Article 4 in conjunction with Annex I parts A and B) for drinking water. Moreover, it establishes a monitoring system where Member States have to disclose information regarding water quality to consumers (Article 13). However, the Directive does not contain any specific rules concerning measures on exploiting unconventional gas and the possible influences on drinking water quality, but rather contains more general provisions on impacts on drinking water. Derogations from the overall aim of the Directive is included in Article 9. These possible derogations are only granted under exceptional circumstances and if there is no danger to human health. In addition, the derogations have to be as short in time as possible.[28] This provision serves as a temporary opt-out mechanism for Member States from the chemical quality standards specified in Annex I. However, the shale gas extraction process is unlikely to be considered an "exceptional circumstance" as it is a standard industry practice that was intended and planned in advance. Also, as explained earlier, the chemicals and the high concentration of saltwater from the subterranean rock formations brought to the

26. See for further discussion Chapter 2.2.4 of this volume.

27. Directive 76/464/EEC—Water pollution by discharges of certain dangerous substances, codified as Directive 2006/11/EC of the European Parliament and of the Council of 15 February 2006 on pollution caused by certain dangerous substances discharged into the aquatic environment of the Community, [1976] OJ L 129/23.

28. The maximum is 3 years, unless the Commission extends it for another period of 3 years (Article 9 Directive 98/83/EC).

surface in the drilling process (Arthur, 2008) would likely be considered a threat to human health. Additionally, to be a commercially profitable activity, shale gas extraction generally must produce gas for at least 3 years.

The Recommendation further clarifies that project-specific water management plans should be established to guarantee the efficient and responsible use of water sources during the entire project. The plan should allow tracing back the individual water flows and take into account the changes in water availability (Sections 9.2 and 9.3 of the Recommendation). As part of the monitoring requirements, in addition to the baseline study, the water volume per well used for the fracturing activity, as well as the composition of the fluids used, shall be monitored (Section 11.3(a) and (b) of the Recommendation).

## 4.3 Regulation of Underground Injection and Groundwater

The regulatory framework for underground injection consists of the Groundwater Directive, the Mining Waste Directive, and the Water Framework Directive. In addition, the Recommendation on shale gas contains provisions that clarify the applicability of these Directives. It clarifies that operators are required to "put in place measures that prevent or limit the input of pollutants into groundwater" (Preamble 7 of the Recommendation). This clarification is indeed not self-evident as earlier studies revealed that Member States have different interpretations of the applicability and content of the provisions of the Directive (Ballesteros et al., 2013; Commission Communication, 2014).

The Groundwater Directive of 2006 completes the Water Framework Directive, especially the requirements established in Article 17. It requires establishing groundwater quality standards to achieve compliance with the "good chemical and ecological status criteria." It also introduces measures to prevent or limit inputs of pollutants into groundwater and a requirement to carry out pollution trend studies, taking into account local conditions (Article 1 of the Directive). There are no specific provisions relating to hydrocarbons or unconventional gas. Thus for the shale gas exploitation process the only requirement is not to exceed the defined maximum concentrations of chemicals and pollutants in groundwater defined in the individual area (Meiβner, 2011).

The Mining Waste Directive provides for "measures [...] to prevent or reduce [...] adverse effects on the environment, in particular water, air, soil, fauna and flora and landscape [resulting from] the management of waste from the extractive industries" (Article 1). According to the definitions outlined in Article 3 of the Directive, "extractive industries" include "all establishments and undertakings in surface or underground extraction of mineral resources for commercial purposes, including extraction by drilling boreholes" (Article 3(6)). Thus, at first sight, it appears to cover the extraction of shale gas. The Directive applies to all waste from the prospecting, extraction, treatment, and storage of mineral resources (Article 2). However, it is important to note that the regulation of the "injection of water and re-injection of pumped groundwater as defined in [...] Article 11(3)(j) of Directive 2000/60/EC" (Article 2(2)(c)) is excluded from the Mining Waste Directive.

According to Article 11(3)(j) of the Water Framework Directive, injection might be allowed by the Member States subject to specific conditions. In other words, the injection process of wastewater resulting from the drilling process of shale gas is excluded, and thus subject only to individual Member States' regulations. This is a loophole and the Member States do not have a common interpretation of the applicability of this provision to shale gas activities (Ballesteros et al., 2013). The Recommendation tries to address this issue.

It includes provisions to establish minimum depth limitations between the area to be fractured and groundwater (3.2), as well as provisions requiring a risk assessment on the changing behavior of groundwater reservoirs (5.3(b)). Direct discharge of pollutants into groundwater is prohibited (5.4) and water management plans for the entire project chain are needed (9.2(a)). In case of "loss of the well integrity or if pollutants are accidentally discharged into groundwater," the operation has to be stopped and necessary remedial actions need to be taken immediately.

## 4.4 Wastewater Management

Stakeholders, such as the International Oil and Gas Producers Association, asked the European institutions to assess and clarify the "legal status of flowback water as to whether or not it is considered as waste when re-used in other fracturing operations" (Impact Assessment, 2014) and thus the applicability of the Mining Waste Directive as well as the general Waste Directive. The Mining Waste Directive is applicable to waste resulting from the shale gas exploitation process outside the injection process itself. The operator of the "extractive industry site" has to implement a "major-accident prevention and information" plan (Article 6 Mining Waste Directive), a monitoring regime (Article 11 Mining Waste Directive), as well as a waste management plan. The latter has to be reviewed on a 5-year basis to prevent or reduce the generation of waste and its harmful nature and to encourage the short- and long-term safe disposal of waste (Article 5 Mining Waste Directive). None of the plans and provisions of the directive contain specific measures on waste produced by unconventional gas resources. However, the special characteristics of the waste produced by unconventional gas extraction processes, like the shale gas process, need a specific regulation, as the waste, especially the polluted water, is more toxic and dangerous and above all hard to capture and collect (Meiβner, 2011).

The European Hazardous Waste Directive[29] used to lay down the framework for the management, recovery, and correct disposal of hazardous waste. However, the Directive has been repealed by Directive 2008/98/EC on waste, with effect from December 2010. Hence the treatment of hazardous waste falls under the section on hazardous waste in the general Waste Directive 2008/98/EC.

The exact composition of the chemicals used in the fracking process and their proportions is generally considered to be a confidential trade secret (Howell, 2009). Therefore undertakings are not required to disclose the exact composition of the materials used. Even if some of the disclosed components used in the United States are classified as hazardous, the entire fracking fluid mixture is not considered hazardous waste (Hanlon, 2011). The same is true in the European Union, where fracking fluid does not fall under the definition of "hazardous waste" as defined in the European Waste Directive. However, this might change in the future. The properties that render waste hazardous are currently under review, so any given component might be listed as "hazardous". Consequently, the regulation of hazardous waste under the Waste Directive is not completely irrelevant to the shale gas extraction process. In fact, the entire Waste Directive does not provide for specific provisions on the treatment of (hazardous) waste resulting from the unconventional gas extraction processes. Articles 17–20 lay out stricter rules applying to hazardous waste, transportation, packing, labeling (Articles 17 and 19), monitoring, the mixing of hazardous substances (Article 18), and permit exemptions.

29. Council Directive 91/689/EEC of 12 December 1991 on hazardous waste, [1991] OJ L 377/20.

Also control obligations "from cradle to grave" are aimed at safe and controlled management (European Commission, 2011). Regarding the high degree of contamination and concentration of chemicals in fracking wastewaters, it would be a reasonable step to classify the fracking fluid mixture as "hazardous" to make the entire fracking wastewaters subject to stricter hazardous waste regulation, unless the operator shows that a nonhazardous activity is carried out, and thus to guarantee safe and controlled disposal.

The Recommendation in Section 10.2 further encourages research into "fracturing techniques that minimize water consumption and waste streams and do not use hazardous chemical substances." In addition, operators are encouraged to disclose and disseminate information to the general public on the chemical substances and volumes of water used for the fracturing activities. The Recommendation foresees as a means for disclosure the "names and Chemical Abstracts Service (CAS) numbers of all substances and to include a safety data sheet, if available, and the substance's maximum concentration in the fracturing fluid" (Section 15(a) of the Recommendation).

And indeed, even if water-based fracturing is currently the most used technique, "other methods for fracturing [...] exist that do not make use of water-based fluids (for instance, explosive fracturing, dynamic loading, etc.), or that make use of fluids other than water" (Gandossi, 2014). For example, "foam technologies, [...] offer an alternative to reduce the amount of water used in shale gas stimulation. These are available across the industry" (Gandossi, 2014). However, these are to date "not extensively applied due to performance considerations" (Gandossi, 2014).

## 5. CONCLUSION

From the analysis of the regulatory requirements and its applicability to water resources along the shale gas value chain it can be concluded that the Recommendation tries to close existing gaps and uncertainties in the legislative framework. As such, the Recommendation is well drafted. However, it is striking, that the Recommendation does not establish specific regulatory requirements regarding water resources or the closure and postclosure phase (Section 14 of the Recommendation). The establishment of a coherent ex-ante approach requires that no element is left out, even if practical examples are still missing. The Commission itself stresses the need for "robust and clear rules to accompany shale gas developments to ensure that negative impacts can be reduced and risks can be managed" (Commission Communication, 2014). The question, however, is whether this can be guaranteed through nonbinding guidelines. As mentioned by the Commission, the important threshold is that "if fully applied...," it is expected that the Recommendation "...would contribute to enabling shale gas activities while ensuring that climate and environmental safeguards are in place" (Commission Communication, 2014). However, it is inherent to the very nature of a "Recommendation,"[30] that the provisions included are not binding on the Member States to be applied in practice, but merely are an invitation to do so.

At the EU level, the entire shale gas regime is still some steps behind the American regime. The efforts of the Union institutions and those of Member States such as the United Kingdom and Germany show that a laissez-faire approach, as has been the governing

30. Article 288 TFEU states that "Recommendations and opinions shall have no binding force."

principle at the start of shale gas activities in the United States, is not what seems to work best and is the policy preference in the European Union. As such, law and policymakers learned from the United States experience and realized that ex-ante regulation is necessary. However, the regulation of the "shale gas value chain" is still in its infancy. Former Energy Commissioner Günther Oettinger has stated that: "European common standards [will guarantee] a high level of safety and security and quality for environmental interests" (Agence France Presse, 2011). One scholar sums up correctly that: "[T]he laws and regulations covering oil and gas exploration and development in Western Europe are underdeveloped in terms of managing unconventional gas; so much so that current regulations do not even make reference to these types of resources. This lack of regulatory framework may create a real obstacle to Europe's unconventional gas exploration." (Petersen, 2014). This also holds true of the regulation of water resources, as binding measures on the issue are still missing. The European Recommendation is an initial starting point, rather than a definitive answer: it remains to be seen how the Member States will implement these recommendations in practice, and whether such minimum principles are enough to regulate an activity that is associated with so many uncertainties and faces public opposition and political debate in nearly all Member States. Regardless of the many black letters and the very narrative character of the Recommendation, it attempts to prevent a race to the bottom of environmental standards and creates legal clarity, in the way a nonbinding instrument can do so. Along with the development of practical experience with the technology, regulation might indeed evolve "on the go." So far, it has been argued that the Recommendation has not had a significant effect on the Member States policy and legislative approaches on shale gas in general (Molyneux, 2014), and more specifically on water resources.

## REFERENCES

Agence France-Presse, 09.09.2011. Brussels Seeks EU Shale Gas Rules. France 24. Available at: http://www.france24.com/en/20110909-brussels-seeks-eushale-gas-rules-oettinger.

Arthur, J.D., Langhus, B., Alleman, D., 2008. An Overview of Modern Shale Gas Development in the United States. ALL Consulting.

Bailey, A.J., 2010. The Fayetteville Shale Play and the need to rethink environmental regulation of oil and gas development in Arkansas. 63 Arkansas Law Review 815–848.

Ballesteros, M., Pelsy, F., Reins, L., 2013. Regulatory Provisions Governing Key Aspects of Unconventional Gas Extraction in Selected Member States. Milieu Ltd. Available at: http://ec.europa.eu/environment/integration/energy/pdf/Final%20Report%2024072013.pdf

Barton, B., 04.11.2013. Why Is Europe Failing to Take the Energy-Water Connection Seriously? The Guardian. Available at: http://www.theguardian.com/sustainable-business/europe-failing-energy-water-connection.

Bundesrat, 01.04.2015. Gesetzentwurf der Bundesregierung, Entwurf eines Gesetzes zur Änderung wasser- und naturschutzrechtlicher Vorschriften zur Untersagung und zur Risikominimierung bei den Verfahren der Fracking-Technologie. Drucksache 143/15.

Cady, M.R., 2010a. Drilling into the issues: a critical analysis of urban drilling's legal, environmental and regulatory implications. 16 Texas Wesleyan Law Review 138f.

Cady, M.R., 2010b. Drilling into the issues: a critical analysis of urban drilling's legal, environmental and regulatory implications. 16 Texas Wesleyan Law Review 1, 127–158.

CDU, CSU, SPD, 2013. Deutschlands Zukunft gestalten – Koalitionsvertrag zwischen CDU, CSU und SPD, 18. Legislaturperiode, at 44. Available at: https://www.cdu.de/sites/default/files/media/dokumente/koalitionsvertrag.pdf.

Cournil, C., 2013. Adoption of legislation on Shale gas in France: hesitation and/or progress? 22 European Energy and Environmental Law Review 4, 141–151.

Department of Environmental Conservation, New York State, Chapter 6, Potential Environmental Impacts in: Draft Supplemental Generic Environmental Impact Statement on the Oil, Gas and Solution Mining Regulatory Program- Well Permit Issuance for Horizontal Drilling and High-Volume Hydraulic Fracturing to Develop the Marcellus Shale and Other Low-Permeability Gas Reservoirs, 2009.

DG for Internal Policies European Parliament, 2011. Impacts of Shale Gas and Shale Oil Extraction on the Environment and on Human Health. European Parliament, Brussels.

Ellsworth, W.L., 2013. Injection-induced earthquakes. Science 341 (6142), 1225942.

European Commission, 2013. In: Presentation of the Stakeholder Participation on Unconventional Fossil Fuels (e.g. Shale Gas) in Europe. Available at: http://ec.europa.eu/environment/integration/energy/pdf/Presentation_07062013.pdf.

European Commission, 2010. Water Protection and Management. Available at: http://europa.eu/legislation_summaries/environment/water_protection_management/l28002b_en.htm.

European Commission, 2011. Waste: Hazardous Waste. Available at: http://ec.europa.eu/environment/waste/hazardous_index.htm.

European Commission, 2014. Staff Working Document, Impact Assessment, Exploration and Production of Hydrocarbons (Such as Shale Gas) Using High Volume, Hydraulic Fracturing in the EU. Brussels, 22.1.2014, SWD, p. 21 (Impact Assessment).

European Commission, 2014. Communication on the exploration and production of hydrocarbons (such as shale gas) using high volume hydraulic fracturing in the EU. Brussels, 17.3.2014, COM, p. 23.

European Parliament, 2012. Resolution of 21 November 2012 on the Environmental Impacts of Shale Gas and Shale Oil Extraction Activities (2011/2308(INI)).

Fleming, R., 2013. Shale gas – a comparison of European moratoria. 22 European Energy and Environmental Law Review 1, 12–32.

Gandossi, L., 2014. An Overview of Hydraulic Fracturing and Other Formation Stimulation Technologies for Shale Gas Production. JRC.

Hanlon, J., 2011. Natural Gas Drilling in the Marcellus Shale NPDES Program Frequently Asked Questions. EPA.

Howell, K., 4.11.2009. More Oversight Sought for Hydraulic Fracturing. New York Times. Available at: http://www.nytimes.com/gwire/2009/11/04/04greenwire-more-oversight-soughtforhydraulic-fracturing-35961.html.

KPMG, 2012. Central and Eastern European Shale Gas Outlook.

Lavalle, C., Baranzelli, C., Vandecasteele, I., Barranco, R., Rivero, I., Sala, S., Ballesta, P., Borowiak, A., Field, R., De Roo, A., Burek, P., Gawlik, B., Pelletier, N., 2013. Spatially-Resolved Assessment of Land and Water Use Scenarios for Shale Gas Development: Poland and Germany. JRC.

B. Martor. France: Evolutions in the Legal Framework for Shale Oil and Gas. Available at: http://www.shale-gas-information-platform.org/categories/legislation/expert-articles/martor-article.html#c2405S.

Meißner, F., Naumenko, D., 2011. Non-Conventional Gas Regulation in Europe: Implications for Ukraine. Institute for Economic Research and Policy Consulting, Berlin/Kyiv.

Molyneux, C.G., Tosoni, L., 2014. What's in a name? legal implications of the EU recommendation on Shale gas. 44 Environmental Law Institute 5, 10351–10353.

N.N., 19.07.2013. UK outlines proposed tax break for Shale gas. Oil&Gas Journal. Available at: http://www.ogj.com/articles/2013/07/uk-outlines-proposed-tax-break-for-shale-gas.html.

National Gas Europe. Germany's Shale Gas Potential Threatened by Environmental Opposition. Available at: http://www.naturalgaseurope.com/germanys-shale-gas-potential-threatened.

Orlen, S., 2010. Gas-Basic Information.

Petersen, A., 2014. Unconventional Gas in Europe: Opportunities and Risks. Available at: http://www.naturalgaseurope.com/unconventional-gas-in-europe-opportunities-risks.

Petroff, A., 19.07.2013. UK Plans Big Tax Breaks for Shale Gas. CCN Money. Available at: http://money.cnn.com/2013/07/19/news/economy/fracturing-uk-taxes/index.html.

Shale Gas Europe. Bulgaria. Available at: http://www.shalegas-europe.eu/shale-gas-explained/shale-gas-and-europe/bulgaria/.

Tomescu, M., Unconventional/Shale Gas & the Environment. Available at: http://www.env-health.org/IMG/pdf/111007_Shale_Gas_NGO_meeting_EU_Commission_DG_Envt_Mihai_Tomescu.pdf.

Umweltbundesamt. Einschätzung der Schiefergasförderung in Deutschland, 2011.

US Energy Information Administration, 2013. Technically Recoverable Shale Oil and Shale Gas Resources: An Assessment of 137 Shale Formations in 41 Countries Outside the United States. Available at: http://www.eia.gov/analysis/studies/worldshalegas/.

US EPA, 2016. The Hydraulic Fracturing Water Cycle. Available at: http://www.epa.gov/hfstudy/hydraulic-fracturing-water-cycle#ftn6.

Van Calster., G., 2013. Fracking Litigation in France – Reference to the Constitutional Court. Available at: http://gavclaw.com/2013/07/29/fracking-litigation-in-france-reference-to-the-constitutional-court/.

Wagrodzka, M., 2013. Update on Hydrocarbon Law and Projects of Shale Gas Resources (American-Polish Energy Roundtable). Available at: http://www.usea.org/sites/default/files/event-/Shale_Gas_2_Ministry_of_Environment_Marta_Wagrodzka.pdf. This presentation is updated with information obtained during a meeting of the European Science and Technology Network on Unconventional Hydrocarbon Extraction Working Group 1 (WG 1): Exploration, Demonstration and Production Projects in the EU at 23.2.2015.

Chapter 2.2.6

# Water for Electricity Generation in the United States

M.E. Webber
*The University of Texas at Austin, Austin, TX, United States*

## 1. INTRODUCTION

As noted by others and elsewhere in this book, energy and water are interconnected in many ways (Sanders, 2015). One important factor of the relationship between energy and water are the extensive water needs by the power sector. These needs include direct water use (eg, at hydroelectric power plants) and indirect water use (such as the water for cooling thermoelectric power plants, producing fuels, and transporting fuels). In the United States, which has very high electricity consumption per capita, this situation is particularly acute. In addition, the per capita water use in the United States is relatively high because it has extensive industrial activity and is a major agricultural producer. Thus the power sector is at risk of facing severe competition for water resources with other sectors in society, and vice versa.

Withdrawals were 355 billion gallons per day in 2010 in the United States, which was greater than 1100 gallons per person per day (USGS, 2014). Of that, the thermoelectric power sector was responsible for 161 billion gallons per day, or 45% of total withdrawals (USGS, 2014). This water is withdrawn from a mixture of sources (groundwater and surface water) from a variety of source qualities (including saline and fresh) for a mix of end uses (Fig. 1).

While the daily withdrawals for the power sector are higher than for agriculture in the United States, globally, agriculture is a much larger cause of withdrawals. While there is significant water consumption via evapotranspiration for photosynthesis in the agricultural sector and evaporation for cooling in the power sector, most of the water that is withdrawn for power plant cooling is returned. Overall, while the power sector is responsible for the greatest volume of water withdrawals in the United States, the agricultural sector has the greatest water consumption (USGS, 1998).

Different aspects of water for power are illustrated in Fig. 2. For example, as shown in the upper right portion of the figure, rain that is collected into a catchment by a dam can be used directly to generate power by use of hydroelectric turbines. Further downstream, water provides electric power indirectly by cooling a thermoelectric power plant. The pumps that withdraw and circulate the water for cooling require some of the energy produced on-site, reducing the output of the power plant. Water is also used for fuels

**Competition for Water Resources. http://dx.doi.org/10.1016/B978-0-12-803237-4.00012-4**

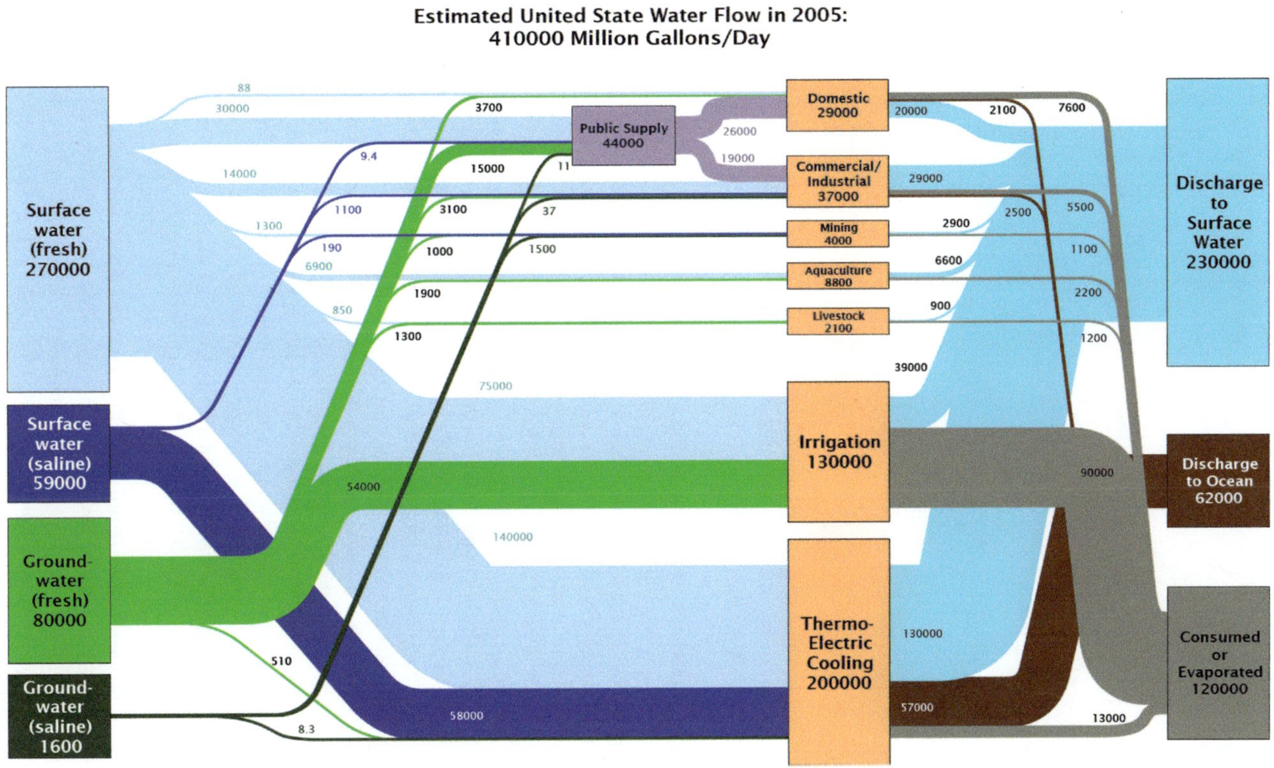


FIGURE 1 US water withdrawals in 2005 were 410 billion gallons per day. While cooling of the power sector is responsible for over half of daily water withdrawals, primarily from surface water, it is responsible for a small fraction of overall consumption (LLNL, 2011).

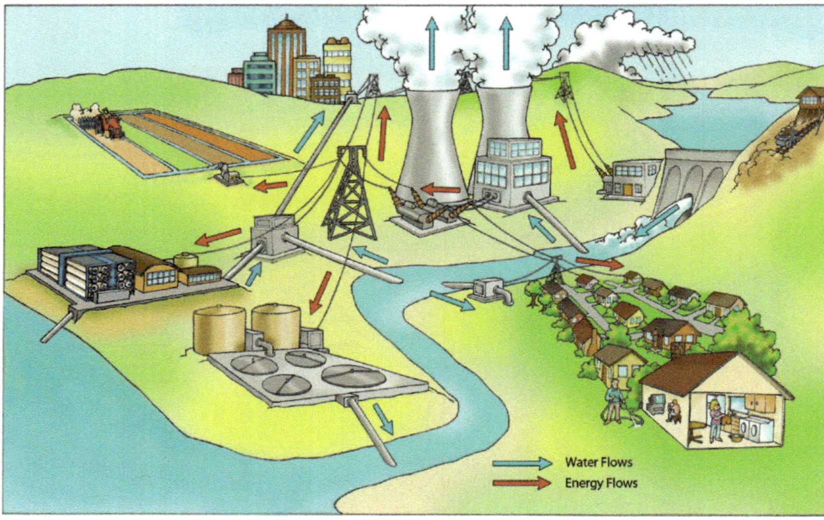


**FIGURE 2** This illustration graphically depicts many elements of the energy water nexus. *Red arrows* denote flows of energy, and *blue arrows* denote flows of water. *Courtesy of the Electric Power Research Institute (EPRI), Courtesy of Bob Goldstein, 2006.*

production, and in particular for the extraction of fuels such as uranium, coal, oil, and gas. A mine shown in the upper-right corner of the illustration presumably uses water as part of its production process (for washing or leaching minerals), and has potential water quality impacts from runoff, as shown with the cascading dirt below the mining equipment.

Water is also used extensively for transportation of fuels for the power sector. In particular, barges are used to transport coal to power plants along major rivers. There are 11,000 miles of inland waterways that move 0.5 million ton-miles of freight annually (compared with 1.7 million ton-miles of freight for the United States' 141,000 miles of railroads) (TEDB, 2012; FHA, 2011). While waterborne transportation is very efficient, droughts or floods that disrupt barge movement can be impactful on the power sector.

In parallel, the water industry uses significant amounts of energy for moving, pumping, treating, and heating water. A water treatment plant at the far left side requires energy to remove impurities of the water via reverse osmosis using coiled, concentric membranes in tubes lying horizontally. The wastewater treatment plant at the bottom center of the figure is there for a reason: namely, wastewater plants are usually placed at the lowest-altitude location so that the solid-laden flows of sewage can flow by gravity to the plant. There, energy is used to clean up the flows to a suitable standard that meets goals for maintaining a healthy ecosystem, after which the water is returned to the watershed so that the next user downstream has that water available for other purposes. In addition, energy-intensive pumps are used to move water uphill from the river and water treatment plant to the urban area at the top of the figure and to the residential area on the right-hand side. Once water arrives at the home or business, additional energy is invested to heat, filter, treat, chill, or pressurize the water so that it is useful for other purposes. While some energy for water is in the form of natural gas, especially for residential water heating, a vast preponderance of energy requirements for the municipal water systems is in the form of electricity for pumps and blowers. This means the water sector represents a nontrivial amount of the demand for electricity from the power sector.

## 2. DIRECT WATER USE FOR POWER GENERATION: HYDROELECTRIC DAMS

Hydropower is a power generation technology that does not use steam boilers. Hydroelectricity provides the largest share of non-thermoelectric generation worldwide, accounting for over 16% of generation. In the United States, hydroelectricity makes up the largest portion of renewable energy consumption, responsible for over 2% of annual energy consumption of all types, and more than 5% of electricity production (EIA, 2016). The water use implications of hydroelectric power differ significantly from thermoelectric generation because it does not withdraw or consume water for cooling. Instead, hydroelectric facilities use the force of gravity to pass water through turbines to generate electricity.

The design is pretty straightforward: a dam is built to create a large reservoir of water with a significant elevation differential (Fig. 3). The elevation difference between the water behind the dam and the river downstream of the dam creates potential energy that can be converted to mechanical energy (from rotating turbines) that can be converted to electrical energy (from the generator).

The power output of the system is a function of height differences through which the water falls and the volumetric flow rate. Large volumes of water falling long distances at a high rate of speed generate a lot of power. The power output from a hydroelectric dam can be described as $P \propto 10 \times H \times Q$ (kW), where $H$ is the head (m), or the height difference through which the water falls from behind the dam to the stream beyond it, and $Q$ is the volumetric flow rate ($m^3/s$). This means the power output from a hydroelectric facility varies with the water availability, which changes by season and year to year based on broader meteorological and climatic conditions.

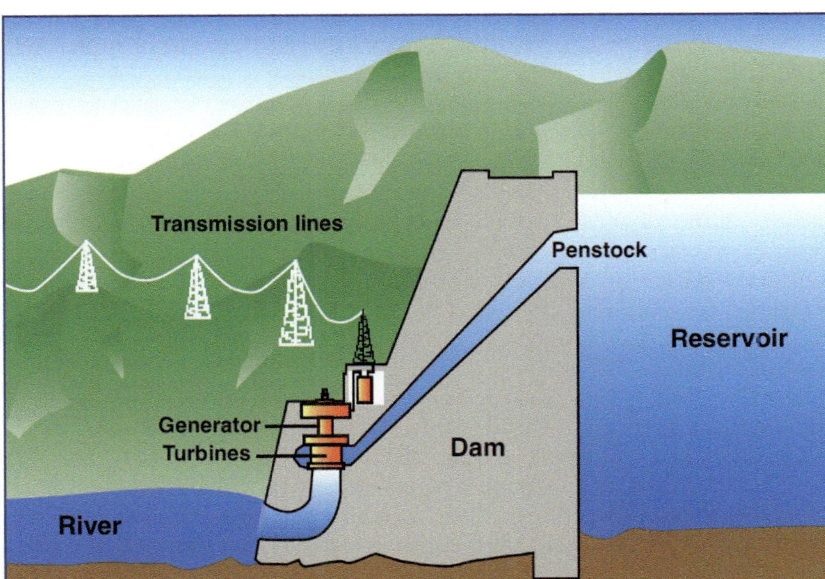


FIGURE 3 Hydroelectric power plants use a height differential of flowing water created by a dam to generate electricity (USACE, 2001).

To make power, water falls through the curved blades, forcing them to rotate a shaft that is attached to a generator. The process is very simple, and consequently dams are highly efficient: they usually achieve a 90% or higher conversion efficiency from potential energy (from the elevation of the water) to electrical energy (USACE, 2001). This performance is much better than the 30–40% efficiency that is typical for conventional steam cycle power plants, and still 1.5 times better than the state-of-the-art natural gas combined cycle power plants that achieve 60% efficiency. One of the key advantages is that solar power does the heavy lifting for us, raising water to high elevations from evaporation of the oceans, after which we can harness the free force of gravity on the way back down.

Hydroelectric power plants can be absolutely massive, both in area and power generation. The largest power plant in the United States is the Grand Coulee Dam along the Columbia River, topping out at over 6 GW, or approximately the size of six nuclear power plants. The Hoover Dam is only 2 GW by comparison. In regions with abundant water, dams can be used for baseload power. But hydroelectric power plants also have the ability to be quickly turned on and off, which gives them great operational flexibility. This means they can be used to meet peak load or to firm up the power grid. In addition, dams are black-start rated, which means they can start operation even if the grid is down, which helps bring the grid back to full power.

And, at the same time, because there is no combustion at the point of use, hydroelectric power is relatively clean since no greenhouse gas emissions (GHGs) are released during power generation. However, the large reservoirs release significant amounts of methane from the anaerobic decomposition of organic biomatter that was flooded during the filling of the reservoir (Li and Lue, 2012). Furthermore, GHGs are released because of the energy consumed during construction of the dam.

Because the construction of large dams has such a large impact on ecosystems, building new ones is contentious in most OECD countries. Therefore efforts to identify opportunities for increasing hydropower generation have focused on smaller-scale opportunities ("small hydro") or improved efficiency and expansion of hydropower at existing facilities through uprating processes (Sternberg, 2010; Carlton, 2009).

Although hydropower does not require water for cooling like thermal generation, it is often considered a highly water consumptive technology because of the large volumes of water evaporated from the surface of reservoirs behind dams. Hydroelectric dams are associated with a significant amount of water consumption for power generation primarily because the increased surface area of manmade reservoirs beyond the nominal run-of-river accelerates the evaporation rates from river basins (Torcellini et al., 2003). Notably, the estimates for this increased evaporation depend significantly on regional location. Furthermore, whether all the evaporation should be attributed to power generation is not clear, because reservoirs serve multiple purposes, including water storage, flood control, irrigation, navigation, and recreation.

As will be discussed, thermoelectric power plants, which use water for cooling, have thermal impacts on the water, heating the water up before it is returned to the source. Hydropower facilities can also cause temperature impacts, though typically in a reversed fashion. Usually, temperatures are relatively uniform for a shallow, free-flowing river. But, after a reservoir is built, the temperature can vary significantly from the water surface (relatively warm) to the bottom of the water column (relatively cold). Because the water flowing through the turbines comes from a part of the reservoir that is not the surface, it exits at a lower temperature than the temperatures to which the native river species are adapted (Hayes et al., 2006). Native river species must often migrate upstream of the dam to reach normal conditions or move downstream until temperatures stabilize. At the same time, some surface-release dams in Australia release water at a higher temperature to downstream locations (Lugg and Copeland, 2014). Dams also interfere with fish migration patterns, which have been solved in some places with fish ladders.

## 3. INDIRECT WATER USE FOR POWER GENERATION: THERMOELECTRIC POWER PLANTS

In addition to direct power generation, water indirectly enables power generation through the cooling it provides for thermoelectric power plants operating on the steam cycle (also known as the Rankine cycle). Thermoelectric power plants use heat to make power and are responsible for more than 90% of the electricity generated in the United States (approximately 3500 of the 4000 million MWh generated annually). Most of those power plants, satisfying 75% of power needs, use the steam cycle, which requires extensive cooling. Power plants also use water for fuels production at the mine or point of extraction and for emissions control at the power plant. The mining sector, which includes the extractive industries for fuels production, requires another 4 billion gallons per day, and the industrial sector, which includes refineries and other facilities for upgrading fuels, is responsible for another 14 billion gallons per day of withdrawals in the United States (USGS, 2014).

As noted earlier, the power sector is the largest cause of water withdrawals, but the agricultural sector is the greatest consumer of water. Such a phenomenon is because most of the water that is withdrawn for power plants is returned to the source, though with a different quality (primarily at a different temperature). The power sector primarily withdraws surface water, though in some locations it also withdraws groundwater. Of the surface water, about a third is saline water. Most of the saline water withdrawn is for cooling power plants located on the coast (though some power plants use brackish groundwater for cooling).

Across the thermoelectric power sector nationally, approximately 15 gallons of water are withdrawn and less than 1 gallon is consumed for every kilowatt-hour of electricity that is generated. Hydroelectric dams are associated with nearly 20 gallons of water consumed per kilowatt-hour primarily because the increased surface area of manmade reservoirs beyond the nominal run-of-river accelerates the evaporation rates from river basins (Torcellini et al., 2003).

The amount of water that is withdrawn and consumed by thermal power plants is driven primarily by a mix of factors:

- **fuel**: coal, natural gas, biomass, oil, nuclear, solar thermal;
- **power cycle**: Rankine (steam) cycle, Brayton (open, simple, or combustion) cycle, combined cycle;
- **cooling technology**: open-loop wet cooling, pond cooling, closed-loop (recirculating) wet cooling, hybrid wet–dry cooling, dry cooling;
- **meteorological conditions**: temperature, humidity, wind speed.

The Rankine cycle, named after famed thermodynamicist William Rankine, is also known as the steam cycle. It uses heat to create steam that drives a turbine that spins a generator to make electricity. The steam cycle is used to generate approximately 75% of all power in the United States. A key step in the steam cycle is cooling to condense the steam into liquid water so that it can be used again in a continuous loop. This cooling can be accomplished with a variety of fluids, but because of water's high heat capacity, relative abundance, and widespread distribution, it is the world's most common coolant.

Other power cycles include the Brayton cycle, which is also known as the open cycle, simple cycle, or combustion turbine. These systems often use turbines that are billed as "aeroderivatives" because of their lineage with turbines that are used for airplane propulsion. A combined cycle is so named because it combines a Rankine cycle and a Brayton cycle to operate at higher efficiency.

The three most prevalent cooling methods are open-loop, closed-loop, and air cooling (Fig. 4). Hybrid wet−dry systems also exist, but are not widely implemented. Meteorological conditions such as prevailing temperature, humidity, wind speed, etc. are also important because they affect overall plant efficiency and the cooling effectiveness of the atmospheric and water-based heat sinks. For values on water withdrawal and consumption by power plants, see Table 1 for a typical breakdown by power cycle, fuel, and cooling type. Water is also needed for the production of the fuels.

Open-loop, or once-through, cooling withdraws large volumes of surface water, fresh and saline, for one-time use and returns nearly all the water to the source with little of the overall water being consumed because of evaporation. While open-loop cooling is energy efficient and low in infrastructure and operational costs, the discharged water is warmer than ambient water, causing thermal pollution, which can kill fish and harm aquatic ecosystems. Thus environmental agencies regulate discharge temperatures, taking into account a water body's heat dissipation capacity.

Closed-loop cooling requires less water withdrawal, since the water is recirculated through use of cooling towers or evaporation ponds (which are reservoirs dedicated for power plant cooling). However, since the cooling is essentially achieved through evaporation, closed-loop cooling causes higher water consumption. The alternative, air cooling, does not require water, but instead cools by use of fans that move air over a radiator similar to that in automobiles. However, power plant efficiency for air cooling is lower, upfront capital costs are higher, and real estate requirements are sometimes larger, often making this option less attractive economically unless water resources are scarce.

Even though power plants return most of the water they withdraw, the need for such large amounts of water at the right temperature for cooling introduces vulnerabilities for the power plants. If a severe drought or heat wave reduces the availability of water or restricts its effectiveness for cooling because of heat transfer inhibitions or thermal pollution limits, the fact that the power plant consumes so little water becomes less important than the fact that it needs the water in the first place.

Power plants constructed over 50 years ago almost exclusively used open-loop cooling designs, which have very high water withdrawals. When these power plants were built, water was perceived as abundant and environmental regulations were practically nonexistent. During the 1960s and 1970s, environmental concerns about water increased, kicking off an era of regulatory pressure to reduce water use at power plants.

They key legislation was the Clean Water Act (CWA), which according to the Environmental Protection Agency (EPA) "…establishe[d] the basic structure for regulating discharges of pollutants into the waters of the United States and regulating quality standards for surface waters" (EPA CWA Summary, EPA CWA History). The Federal Water Pollution Control Act of 1948 served as the basis for the regulatory framework that later became the CWA in popular parlance in 1972 after significant reorganization and

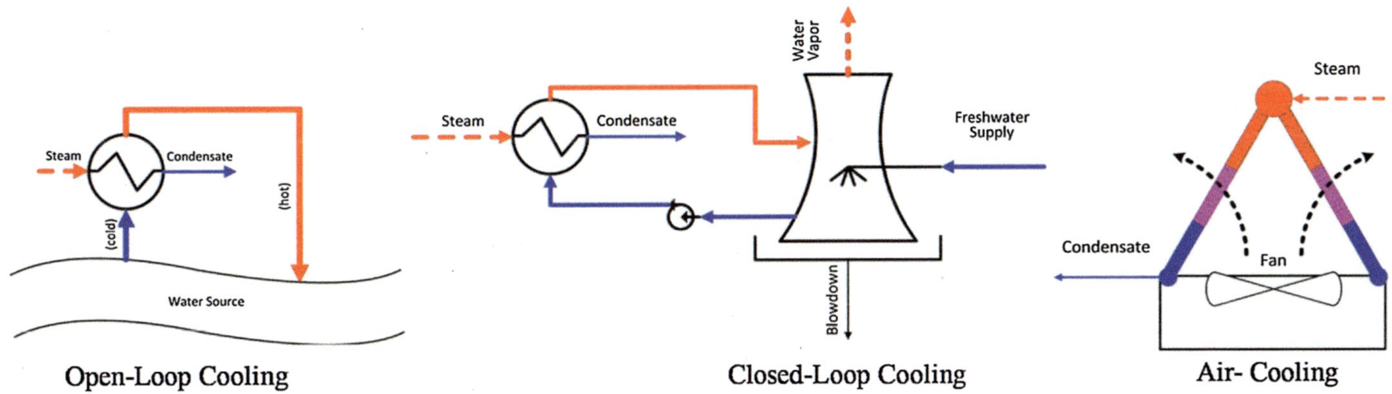


FIGURE 4 There are three basic cooling methods: open-loop, closed-loop, and air cooling. *Courtesy of Stillwell, A.S., 2010. Energy-Water Nexus in Texas (Master's thesis). The University of Texas at Austin.*

**TABLE 1** Water Use at Power Plants Varies by Fuel, Power Cycle and Cooling Technology (Typical Values are Listed) (Stillwell et al., 2011)

| | Cooling Technologies | | | |
|---|---|---|---|---|
| | Closed-Loop (Cooling Tower) | | Open-Loop (Once-Through) | |
| Fuels and Power Cycles | Withdrawals (gal/kWh) | Consumption (gal/kWh) | Withdrawals (gal/kWh) | Consumption (gal/kWh) |
| Concentrating Solar Power | 0.8 | 0.8 | – | – |
| Nuclear | 1.0 | 0.7 | 42.5 | 0.4 |
| Coal/Natural gas (steam cycle) | 0.5 | 0.5 | 35.0 | 0.3 |
| Natural gas (combined cycle) | 0.23 | 0.18 | 13.8 | 0.1 |
| Natural gas (open cycle) | Negligible | Negligible | Negligible | Negligible |
| Solar PV | Negligible | Negligible | Negligible | Negligible |
| Wind | Negligible | Negligible | Negligible | Negligible |

expansion. The CWA gives the EPA authority to implement pollution control programs, including the establishment of wastewater standards for industry and water quality standards for surface waters.

The CWA outlawed the unpermitted discharge of any pollutant from a point source into navigable waters, which led to the creation of the EPA's National Pollutant Discharge Elimination System (NPDES) permit program to control discharges. Point sources (ie, the discrete locations such as pipes or manmade ditches) are regulated by the CWA. While homes do not generally need an NPDES permit for their wastewater flows into the sewers or septic systems, industrial, municipal, and other facilities must obtain permits for their discharges that go to surface waters. In this way, the CWA regulates discharges from power plants. They also regulate intake requirements.

Power plants built since then have almost exclusively used closed-loop designs with cooling towers as a way to serve many environmental interests by greatly reducing the entrainment (fish and aquatic organisms are withdrawn from the environment into the power plant facility) and impingement (fish and aquatic organisms are pinned against water intake screens) of aquatic wildlife. Doing so meant that water withdrawals have decreased in response to §316(b) of the CWA passed in 1972.

They also prevent the artificial heating of aquatic environments, which is a form of thermal pollution and is regulated by §316(a) of the CWA. Conventional wisdom concludes that cooling towers are less impactful than open-loop cooling systems because they withdraw less water, even though cooling towers consume more water, as noted earlier.

In the first decade of the 21st century, 43% of US thermoelectric power plants were large power facilities with generation capacities of over 100 MW. Of these large power plants, 42% used wet-recirculating cooling towers (ie, closed-loop) and 14.5% used cooling reservoirs. The remaining 43% of these large power plants used once-through cooling, and just under 1% use dry cooling (King et al., 2013). Most of those plants with once-through cooling systems were built before the CWA was enacted or were grandfathered in once the legislation was passed. Many of them are also the same plants that were built before strict emissions controls. This means most of them are decades old and are simultaneously dirty and thirsty (except for the ones that added scrubbers) and whether they are shut down in exchange for newer, cleaner, leaner plants remains a hotly contested public policy debate.

Moving forward, new hybrid and dry systems might see greater implementation because of looming regulatory requirements and competition for water. For example, the California State Lands Commission proposed a moratorium on construction of new power plants with open-loop cooling systems, which clashes with separate efforts to push power plants to coastal regions where open-loop cooling can use seawater to spare inland freshwater (CASLC, 2006). Coastal water has higher performance benefits because it is at a relatively lower temperature, which improves efficiency of the power plant. However, environmental concerns about oceanic wildlife are in direct conflict with environmental concerns about inland freshwater supply.

As noted earlier, more water-efficient cooling technologies exist; however, these systems have drawbacks. Dry-cooled systems withdraw and consume less than 10% of the water of wet-cooled systems. However, dry-cooling systems have higher capital costs and reduce overall efficiency of the plant, which increases costs and emissions per unit of electricity generated. Because the heat capacity of air is so much lower than water, much more air has to be moved to achieve the same cooling as with water. This means much larger facilities to create the larger cooling surfaces in dry-cooling systems, which dramatically increases capital costs. Furthermore, a power plant with dry cooling can experience a 1% loss in efficiency for each 1°F increase of the condenser, limiting power generation based on ambient air temperatures (Kutscher et al., 2006).

Because they include both closed-loop wet and dry cooling, hybrid wet–dry-cooling systems provide a compromise between wet- and dry-cooling systems. Thus hybrid wet–dry-cooling systems can have low water consumption for much of the year by operating primarily in dry mode, but have the flexibility to operate more efficiently in wet mode during the hottest times of the year. Unfortunately, water resources are typically less available during these peak demand times. Although dry- and hybrid-cooling systems are proven technologies, low water prices and senior water rights for power generators usually prevent them from being economically competitive designs. However, in water-constrained regions where water is not available for cooling, dry cooling is often the only alternative. In such cases, the upfront capital costs and parasitic efficiency loads are more readily justifiable.

## 4. WATER FOR POWER GENERATION: NONHYDRO RENEWABLE ENERGY

In addition to the water needs for hydroelectric power and for cooling conventional thermoelectric power plants fueled by coal, natural gas, and nuclear, the other forms of renewable power—solar, wind, geothermal, and biomass—also need water for their operation. The range of water they need varies dramatically.

Renewable electricity technologies such as wind turbines and solar photovoltaic (PV) panels do not use thermoelectric processes and thus have no need for cooling water. They need small volumes of water for manufacturing components (eg, at the steel mill where the turbine parts are fabricated, or the semiconductor fab where the solar panels are printed) and cleaning equipment, but other than that the water needs are minimal. Other forms of renewable technologies such as concentrating solar power (CSP) configurations, enhanced geothermal systems, and biomass-powered plants use thermoelectric processes (as before) to convert heat into electricity, and thus also need water for cooling.

By conventional wisdom, CSP systems are preferred over solar PV by utilities because they can achieve large scale and integrate readily with thermal storage technologies (such as heat exchangers with molten salts that can save daytime heat for nighttime power generation) and natural gas turbines that allow facilities to more consistently produce electricity during the day and into the night hours. Because CSP plants operate at lower temperatures than fossil- and nuclear-powered plants, their steam cycles are less efficient. As a result, they require more cooling water per unit of electricity generated than their fossil-fueled peers. Furthermore, areas that provide the best solar insolation for CSP are typically dry and hot, which limits large-scale use of wet-cooling technologies because of water resource scarcity (eg, the first major installations are in Granada, Spain, and in Arizona).

Although dry-cooling technology can be coupled to CSP, doing so introduces parasitic efficiency losses, particularly on hot days. Nonetheless, some CSP companies have committed to dry cooling to avoid the political, availability, and environmental barriers because of concerns over water issues. These new systems demonstrate the feasibility of dry-cooling technology for large-scale systems and might be indicators of a new trend in electricity.

Geothermal power plants utilize naturally occurring heat below ground (especially in locations near volcanic activity, such as Iceland and the mountainous west of the United States) to create steam and generate electricity. However, the majority of the global geothermal resource is deep, dry, hot rock that does not contain adequate water to recover the embedded thermal energy necessary to run steam-powered turbines. Consequently, water has to be added. Enhanced geothermal systems exploit the dry, hot rock by injecting large volumes of water into fractured rock from an external water supply. The injected water absorbs the geothermal heat and is pumped to the surface to power the steam cycle. The same water volume is then injected back into the rock to form a closed-loop system. However, because of lower operating temperatures and losses during the round trip from the surface, geothermal systems need more water than nuclear or CSP.

Electricity generation from burning biomass requires the use of similar amounts of cooling water as coal- and nuclear-fueled thermoelectric facilities, because the power generation process is very similar. However, in addition to the water needed to cool the power plant, water is also needed to grow the fuel.

## 4.1 Water Constraints on the Power Sector

One of the main consequences of how much the power sector depends on water is that water constraints can become energy constraints. Water can be too hot, too cold, too abundant, or too scarce for full operation of power plants, leaving a sweet spot where the water is just right (Fig. 5). This means heat waves, freezes, droughts, and floods all cause problems for power plants.

**FIGURE 5** The power sector needs water to be at just the right abundance and temperature to function at full power.

### *4.1.1 Heat Waves*

Heat waves are relevant in two different ways: they reduce performance because of hampered power plant efficiency and they put power plants at risk of violating thermal pollution limits. The efficiency of a power plant is proportional to the ratio of absolute hot and cold temperatures between which the power plant operates. For example, a coal-fired power plant might have a combustion chamber that operates at 2700°F (1500°C) with a cooling system that sheds heat to river water at 58°F (or 14°C). The Carnot efficiency, or maximum possible efficiency for such a system, is described as:

$$\eta = 1 - \frac{T_{\mathrm{cold}}}{T_{\mathrm{hot}}}$$

where $T_{\mathrm{hot}}$ is the high temperature in the system (the combustion chamber) and $T_{\mathrm{cold}}$ is the temperature sink to which waste heat is rejected (ie, the cooling water or the atmosphere, or some combination of both). This equation is only for absolute temperatures (kelvin, K, or degrees Rankine, °R), as opposed to relative temperatures (Celsius, °C, or Fahrenheit, °F).[1] Thus, for the situation just described, the overall efficiency of the coal-fired power plant is:

$$\eta = 1 - \frac{T_{\mathrm{cold}}}{T_{\mathrm{hot}}} = 1 - \frac{14 + 273}{1500 + 273} = 84\%$$

Note that while the ideal efficiency (or the highest possible efficiency) for power plants operating between those temperatures is 84%, actual power plants induce significant

1. To adjust from relative to absolute temperatures, use the following relationships: (1) T (K) = T (°C) + 273, and (2) T (°R) = T (°F) + 460.

losses, achieving a typical real-world efficiency of 30–40%. For the same operating conditions in the combustor, but with a heat wave present, the river temperatures might be 78°F (25°C) instead of 68°F (20°C). For that situation, the overall efficiency is:

$$\eta = 1 - \frac{T_{\text{cold}}}{T_{\text{hot}}} = 1 - \frac{25 + 273}{1500 + 273} = 83\%$$

In other words, the efficiency of the power plant drops a percentage point just for the case where the water is hotter from a heat wave.

In addition to the performance problems, there are also environmental regulations that are relevant. As was discussed earlier, power plant cooling structures are regulated by thermal pollution standards from §316(a) of the 1972 CWA. These standards are often implemented with a maximum exit temperature that is allowable for the return water from a power plant's cooling system. Typical exit temperatures are around 104–112°F (40–44°C). Those thermal pollution limits are designed to protect the aquatic environment.

When a heat wave occurs, the prevailing temperature of the cooling water will increase above normal. Doing so reduces the overall performance of the power plant as noted earlier. But it also puts the power plant at risk of needing to turn down its output. When the temperature difference between the exit threshold and the inlet (which is the same as the river temperature) is smaller, then the power plant can extract less cooling from the water. The key here is that the thermal pollution limits are a fixed threshold, as opposed to a differential. Thus, as the river temperatures increase, there is less cooling available. And ultimately, when heat waves are extreme enough, power plants will need to draw down their power output.

Such extreme heat waves can happen. The 2003 heat wave in France was one prominent example. That heat wave stretched across Europe, but its greatest impact was in France. In many places in France, temperatures were 10°C (18°F) hotter in 2003 than in 2001 (NASA, 2003). This was a killer heat wave, responsible for tens of thousands of deaths. In particular, elderly women were vulnerable. Consequently, the demand for power was spiking as people turned on their air conditioners to avoid dying from the heat.

At the same time that the demand for power was spiking, the French nuclear fleet had to draw down its output to avoid violating the thermal pollution limits. Because the French power sector has a very high contribution from nuclear power with open-loop cooling systems (which are very water intensive) and because the French nuclear plants are often sited on inland rivers (37 of their 58 nuclear power plants are situated on rivers) (Godoy, 2006), they are doubly exposed to heat waves. By contrast, ocean water maintains a much more stable and cooler temperature compared with inland rivers and lakes.

Because of the combination of the higher energy demand and higher river temperatures, 16–17 of France's 58 nuclear reactors were at risk of violating the thermal pollution limits and had to reduce their output or turn off completely (Poumadere et al., 2005; Forster and Lilliestem, 2010). The rivers were too hot and the water levels too low to guarantee adequate cooling of nuclear power plants, putting the entire system at risk of failure. Ultimately, Électricité de France, the main power provider in France, requested exemption from its operational limits and cut its power exports in half to keep the power system operating (Poumadere et al., 2005). After those temporary exemptions were granted, nuclear facilities were allowed to operate but at reduced capacity. Consequently, the nuclear power fleet had to dial back its output by up to 15% for 5 weeks (Hightower and Pierce, 2008). That is, just as demand for electricity was spiking for air conditioning in response to the heat and with life hanging in the balance, power supplies were being cut back because of the very same heat wave.

In other words, France, and many other regulatory bodies in Europe, "overrode their environmental laws and allowed for higher waste water temperatures" (Forster and Lilliestem, 2010). When making the choice between impinging the environment through overheating of the ecosystem and saving human lives, the latter was the clear selection.

While the heat wave in France was certainly not a normal situation, the fear with climate change is that those killer heat waves will become the new normal. And the phenomenon is not restricted to France: in May 2012, the Illinois Environmental Protection Agency granted Exelon's Quad Cities nuclear power plant a "provisional variance from the National Pollutant Discharge Elimination System (NPDES) water discharge permitted temperature limits, due to recent unseasonably warm weather conditions" (IEPA, 2012). Just like in France, the thermal pollution limits were temporarily voided during a heat wave to keep the power on.

And while ocean water is generally cooler and more steady with its temperature than inland rivers, even seaside power plants are at risk of shutting down from heat waves. A few months after the Quad Cities plant needed a variance, the Millstone Power Station, a nuclear power plant in Connecticut on the Long Island Sound, shut down for 2 weeks because of overly hot seawater, which exceed the 75°F (24°C) temperature limits for the cooling water (AP, 2012). As before, the authorities did their best to shunt aside environmental regulations to keep the power on, but despite an emergency license amendment from the US Nuclear Regulatory Commission (NRC) to show a lower temperature by averaging different measurements, the safety threshold was violated, triggering a full shutdown (AP, 2012).

### *4.1.2 Freezes*

While water can be too hot—a problem that has affected power plants in many places and might be more prevalent with global climate change—water can also be too cold. It is known that weatherization and preparing for cold storms includes steps such as draining water from nonessential water systems (to avoid damage from ice formation), and confirming that the water in essential water systems circulates, which reduces the likelihood of a freeze (FERC, 2011). So, in effect, power plant operators know that freezing water poses a risk to reliability. But that does not mean they have taken appropriate action.

In one dramatic incident, Texas endured a statewide freeze in February 2011. That cold front caused water to freeze in some instrumentation pipes or switches, tripping some large coal plants offline, which triggered a whole cascading series of power plant failures elsewhere, ultimately leading to rolling blackouts across the state (Souder, 2011). On one day, six more coal-fired units at four locations went offline.

Because the natural gas demand was at a record high to meet the heating requirements for homes and businesses, backup power plants fueled by natural gas did not have enough gas available on line. Ultimately, more than 50 gas plants were not able to work. Compounding things, many gas-producing wells using electrical equipment and some gas compressors along pipelines use electrical pumps, so when the power went out, they could not keep up with the demand to keep the pipelines filled with gas, causing pressures to drop and straining the system further (Galbraith, 2011).

Ultimately, over a 4-day span of time between February 1 and 4, 2011, more than 200 individual generating units within the Texas grid experienced an outage, were derated, or failed to start (FERC, 2011). All of these events combined triggered statewide rolling blackouts: since enough power could not be produced to meet demand, demand had to be cut across the state. The grid operator started turning off the power to different parts of the state one after the other. Over 3 million meters (and there are usually several consumers served per meter) lost their power, and a total of 4000 MW of capacity was shed (FERC, 2011).

The after-action report that investigated the blackout episodes concluded that a vast preponderance of the failures were from frozen water in sensing lines, equipment, and frozen water lines (FERC, 2011). Making things worse, natural gas production dropped, which made it hard for gas-fired power plants to come online.

So a set of small frozen water pipes tripped major power plants and some gas wells clogged with frozen water reduced gas pressures in the pipelines making it hard for gas plants to pick up the slack, ultimately shutting power down for at least 3 million people. Essentially, little pipes frozen with water brought the whole grid down.

Freezes can also interrupt the power sector by inhibiting fuels delivery. During the polar vortices of 2014, Lake Superior froze, preventing delivery of coal by ship to power plants along the shore. Consequently, emergency shipments by truck were implemented from one neighboring utility to another to keep the power plants operating.

### 4.1.3 Droughts

In addition to water temperature affecting the energy sector, the abundance of water also matters. In particular, water scarcity from extended drought can be a problem. Though hydroelectric power is attractive for many reasons, it is least reliable during droughts when the need to use water for other purposes such as drinking and irrigation might take precedence over hydroelectricity. For example, low water levels in hydroelectric reservoirs can force power plants to dial back or turn off. Nearly 60% of US hydroelectricity is generated in California, Oregon, and Washington alone, making the power supply vulnerable to regional changes in water availability.

This region is also particularly sensitive to climate change: as the climate warms up, the snowmelt and precipitation patterns are distorted in ways that are detrimental. And the cumulative impact of the changes is nonlinear. Quoting a National Oceanic and Atmospheric Administration report from 2009 on the impacts of climate change, it is noted that:

> *Hydroelectric generation is very sensitive to changes in precipitation and river discharge. For example, every 1 percent decrease in precipitation results in a 2 to 3 percent drop in streamflow; every 1 percent decrease in streamflow in the Colorado River Basin results in a 3 percent drop in power generation. Such magnifying sensitivities occur because water flows through multiple power plants in a river basin.*
>
> NOAA (2009)

For a large basin like that of the Colorado River, small changes in precipitation cause major droughts, which in turn can dramatically reduce power output from a whole chain of hydroelectric dams. At the same time, many millions of people depend on that basin's water for irrigation, drinking, commercial activity, industrial process, and, of course, for power production. And things look like they might get worse.

For example, without a change in water usage patterns, Lakes Mead and Powell along the Colorado River, which are used for hydroelectric power and municipal supply, are projected to have a 50% chance of running dry by 2021 (Scripps, 2008). In the American southwest, reduced hydroelectricity generation because of water scarcity is a way of life (Gertner, 2007).

Ultimately, reduced hydroelectric power output can strain the entire grid. The Energy Information Administration's monthly update in April 2012 raised an alarm that the California grid would face a number of challenges in the summer, partly because of the lower-than-normal snowpack in the Sierra Nevada mountain range, which was expected to reduce hydroelectric output (EIA, 2012). Then, in July 2012, the grid did come crashing down—in India—because of strains induced by reduced hydroelectric output. This event was the largest blackout in history, affecting 600 million people (Gottipati, 2012). Drought, triggered by the later-than-normal monsoon season, played a role in two ways (Yardley and Harris, 2012). First, because of reduced rainfall, the need for electricity from farmers to pump water for irrigation was higher than normal, straining the grid because of the higher demand for power. Second, there was reduced hydroelectric power, making it harder for the grid to meet demand. Ultimately, a population larger than all of Europe and twice as large as the United States was plunged into darkness, with railways and other critical services brought to a sudden halt. So drought can trigger massive, cascading blackouts across a nation.

The problem of water scarcity is not just limited to hydroelectric reservoirs: it also affects thermal plants that need vast amounts of cooling water. This problem was particularly acute during the extensive drought in 2007 and 2008 in the southeastern United States. Unlike the problems of heat waves and thermal pollution, which might cause a power plant to reduce its power output, if water levels drop below the physical location of water intake pipes, then there is a hard stop where the power plant quits operating completely. During the southeastern drought, lake levels came within 3.5 feet of the water intake pipe for the cooling system at the Harris reactor near Raleigh, North Carolina, and within 1 foot of the intake level needed for one of the backup systems at the McGuire plant near Charlotte (AP, 2008). And it is difficult to expand, lower, or lengthen the water intake pipes so that they are less exposed to falling lake levels. According to one Associated Press report:

> *Extending or lowering the intake pipes is not as simple as it sounds and wouldn't necessarily solve the problem. The pipes are usually made of concrete, can be up to 18 feet in diameter and can extend up to a mile. Modifications to the pipes and pump systems, and their required backups, can cost millions and take several months. If the changes are extensive, they require an NRC review that itself can take months or longer. Even if a quick extension were possible, the pipes can only go so low. It they are put too close to the bottom of a drought-shrunken lake or river, they can suck up sediment, fish and other debris that could clog the system.*
>
> AP (2008)

Lake Lanier, a key reservoir in northern Georgia that provides drinking, irrigation, and cooling water, reached its record low level in December 2007 (Lanier, 2012). Consequently, power plants near Atlanta during the winter 2008 drought were within days of shutting off because the vast amounts of cooling water were at risk from diversion for other priorities such as municipal use for drinking water (Mungin, 2007). Overall, 24 of the United States' 100+ nuclear reactors are sited in the region that endured that drought (AP, 2008).

And when drought and heat waves are combined, as they often are, the problems are compounded. For the same heat wave in France in 2003 described earlier, there was a simultaneous drought that caused France to lose 20% of its hydropower capacity (Hightower and Pierce, 2008). France gets 78% of its electricity from nuclear power and approximately 11% from hydropower, which made this drawback hard to accommodate (MEEDDM, 2011). The same problems happen in the United States. According to reporting in *The Christian Science Monitor*:

> *In August 2007, the Tennessee Valley Authority shut down one of three reactors at its Browns Ferry nuclear plant. The plant draws from and returns its cooling water to the Tennessee River. Drought shrank water levels in the river. And the hottest temperatures in some 50 years (plus water warmed by power plants upstream) heated the available water to levels that would have topped permissible limits at the outflow pipe if the plant had continued to run all three reactors. Consumers already were bracing for rate increases because the low flow in the Tennessee's tributaries had cut the amount of juice hydropower plants in the system could produce.*
>
> Mungin (2007)

The drought lowered the water levels in the rivers, which made them more vulnerable to increased temperatures from the heat wave, causing a violation of the thermal pollution limits that were described earlier. The power problems were exacerbated by the fact that the same drought that caused problems for the thermal plants also reduced output from the hydroelectric power plants, causing upward pressure on power prices.

Drought causes other problems for the power sector. For example, severe drought in 2012 threatened a 200-mile stretch of the Mississippi River from St. Louis to Cairo, Illinois, with the risk of being shut off for barge traffic in early 2013 (Keen, 2012). Because of the severe drought in 2012, the Army Corps of Engineers decided to save water for summer irrigation in its upstream reservoirs. Consequently, releases were reduced, diminishing flow to the Missouri River, which subsequently reduced the flow into the Mississippi River. If water levels get too low, barges heavily laden with goods would not be able to travel safely. Those barges carry freight such as fertilizers, salt (for winter road treatments), and agricultural products. They also carry a lot of coal to Midwestern power plants. Thus a water scarcity event threatens the supply chain for the energy industry in yet another way.

### 4.1.4 Floods

In addition to water being too hot, too cold, and too scarce, it can also be too abundant. It is hard to imagine that water abundance can be a problem, given the prior discussion about the risks that droughts pose to energy production. But too much water puts power plants at risk of damage. For example, a Nebraska nuclear power plant nearly shut down because of flooding of the Missouri River in June 2011. The floodwaters came within a few feet of cresting over the flood walls surrounding the facility, which would have triggered a shutdown to prevent safety risks. And, for the Fukushima incident in 2011, it was not the earthquake that caused the biggest problems. Rather, it was the tsunami and the subsequent flooding that triggered the meltdowns and explosions within the nuclear power plants. Bringing all these points together reveals that water has to be just right for the power sector to work with optimal performance.

## 5. POTENTIAL SOLUTIONS

There are a range of solutions available for mitigating the power sector's vulnerabilities because of its dependence on water. A few of those are discussed here:

- **Fuel switching and water switching**: We can choose to switch the fuel sources we use (from water-intensive to water-lean options) and switch the water sources (to gray-water, effluent, brackish, or saline sources) to avoid competition with freshwater.
- **Advanced technologies**: We can also invest in developing and implementing advanced technologies that are water lean.
- **Cross-sectoral production**: We can look for cross-sectoral benefits, whereby we use the water sector to produce power or the power sector to produce water.

In some cases these solutions are already being implemented somewhere worldwide and just need to get broader implementation. In other cases, much innovation remains to be done. Fuel switching is straightforward. In this case, investments can be made in the power sector to select less water-intensive fuels. Rather than conventional nuclear, coal, or natural gas operating on the steam cycle, water-lean options such as natural gas combined cycle, wind, and solar PV could be used instead. For example, in Texas, displacing legacy coal plants with new, advanced natural gas combined cycle power plants would reduce average water consumption from 0.61 gallons per kWh to 0.25 gallons per kWh (Grubert et al., 2012). That means fuel switching reduces the carbon and water intensity simultaneously. The water and carbon savings would be even more pronounced by displacing to wind and solar PV, though those options have important performance differences that would need to be accommodated.

In addition to fuel switching, power plants can also switch their water sources (Tidwell et al., 2014). Rather than using freshwater, they can use treated effluent or saline sources. Out of more than 1000 power plants in the United States, a few dozen of them already use reclaimed water for their coolant. The most prominent example is the Palo Verde nuclear generating station in Arizona. Conducting a spatial analysis in Texas, it was found that a significant fraction of major power plants are located within 25 miles of a large municipal source of treated effluent (Stillwell and Webber, 2014). The effluent is sufficient in total to meet the total consumptive needs of the power sector. Applying that approach more broadly would have significant impact nationwide while also providing a customer for the effluent, which helps cover the infrastructure investment costs.

Beyond switching fuels or water sources, cooling technologies can be switched (Tidwell et al., 2014). Because power plants have prioritized closed-loop cooling over open-loop cooling, total water withdrawals for the power sector have leveled and dropped. These technologies can be improved further with implements such as better heat exchangers, sophisticated materials and coatings, and different configurations. In particular, dry-cooling and hybrid-cooling systems are gaining traction in water-scarce locations. Because the power plant can keep working even during drought, some economic analysis has deduced that the costs of the dry-cooling systems can be recovered during drought events (Stillwell and Webber, 2013).

There are also opportunities for integrated design to achieve cross-sectoral solutions. For example, the power sector can be used to generate water. In particular, variable renewables such as wind and solar energy can be used to desalinate brackish groundwater (Clayton, 2014a,b). Doing so accomplishes two goals. First, it increases water supply in a way that does not exacerbate carbon emissions. Second, the water treatment system can be dialed up and down to follow the supply of renewable electricity. In that way, desalination serves as a proxy for energy storage, and thereby offers ancillary services and reliability benefits to the grid.

Beyond these three categories of solutions, there are many other opportunities to solve the various vulnerabilities of the power sector induced by water availability.

## 6. CONCLUSIONS

The power sector in the United States is a major user of water primarily for cooling thermoelectric power plants. Those water needs are determined by the fuel, power cycle (steam cycle, combined cycle, etc.), cooling technology, and prevailing climatic conditions. In addition, water is used to generate power at dams, produce fuels, transport fuels, and operate emissions controls on-site. Because of the power sector's dependence on water, it is vulnerable to water constraints. Namely, if water is too scarce, abundant, hot, or cold, it induces strain on the power sector that can lead to blackouts. However, there are different solutions available, including source switching (either switching to different fuels or source of cooling water), advanced technologies, and cross-sectoral integration.

## REFERENCES


Associated Press, August 27, 2012. Conn. Nuclear Plant Unit Closed Due to Too-warm Seawater Reopens. Associated Press via Christian Science Monitor.

Associated Press (AP), January 23, 2008. Drought Could Shut Down Nuclear Power Plants. Associated Press via MSNBC.

Carlton, J., August 21, 2009. Deep in the wilderness, power companies wade in. Wall Street Journal.

CASLC, 2006. Proposed Resolution by the California State Lands Commission Regarding Once-through Cooling in California Power Plants. Technical report. California State Lands Commission.

Clayton, M.E., Stillwell, A.S., Webber, M.E., 2014a. Implementation of brackish groundwater desalination using wind-generated electricity: a case study of the energy-water nexus in Texas (special issue The Energy Sustainability Nexus). Sustainability 6, 758–778 (2014).

Clayton, M.E., Kjellsson, J.B., Webber, M.E., November 2014b. Wind-Solar-Desalination: How Integrated Systems Can Solve Our Water and Energy Issues. Earth Magazine.

Electric Power Research Institute (EPRI), 2006. Courtesy of Bob Goldstein.

Energy Information Administration (EIA), April 2012. Electricity Monthly Update. Technical report. U.S. Department of Energy, June 2012.

Energy Information Administration (EIA), January 12, 2016. Short Term Outlook. U.D. Department of Energy.

Federal Energy Regulatory Commission (FERC), August 2011. Outages and Curtailments During the Southwest Cold Weather Event of February 1–5, 2011. Technical report.

Federal Highway Administration (FHA), 2011. Freight Facts and Figures 2011. Technical Report FHWA-HOP-12-002. U.S. Department of Transportation.

Forster, H., Lilliestam, J., 2010. Modeling thermoelectric power generation in view of climate change. Regional Environmental Change 10.

Galbraith, K., February 3, 2011. The Rolling Chain of Events Behind Texas Blackouts. Texas Tribune.

Gertner, J., October 21, 2007. The Future Is Drying Up. New York Times.

Godoy, J., July 28, 2006. European Heat Wave Shows Limits of Nuclear Energy. Common Dreams.

Gottipati, S., July 31, 2012. India Endures World's Largest Blackout. New York Times.

Grubert, E.A., Beach, F.C., Webber, M.E., 2012. Can switching fuels save water? a life cycle quantification of freshwater consumption for Texas coal- and natural gas-fired electricity. Environmental Research Letters 7.

Hayes, D.B., Dodd, H., Lessard, JoA., 2006. Effects of small dams on cold water stream fish communities. In: American Fisheries Society Symposium.

Hightower, M., Pierce, S.A., March 20, 2008. The energy challenge. Nature 452.

Illinois Environmental Protection Agency (IEPA), May 25, 2012. Illinois EPA Grants Exelon Quad Cities Station Provisional Variance from Discharge Requirements. Press Release.

Keen, J., November 2012. Low River May Paralyze Commerce. USA Today, pp. 23–25.

King, C.W., Stillwell, A.S., Sanders, K.T., Webber, M.E., 2013. Coherence between water and energy policies. Natural Resources Journal 53.

Kutscher, C., et al., 2006. Hybrid wet-dry cooling for power plants. In: Parabolic Trough Technology Workshop. National Renewable Energy Laboratory, Incline Village, NV.

Lanier, L., 2012. Lake Lanier Water Levels. http://www.lakelanier.com/about/water-levels/.

Lawrence Livermore National Laboratory (LLNL), U.S. Department of Energy, 2011. Estimated Water Flows in 2005. LLNL-TR-475772.

Li, S., Lue, X.X., 2012. Uncertainties of carbon emission from hydroelectric reservoirs. Natural Hazards 62, 1343–1345.

Lugg, A., Copeland, C., 2014. Review of cold water pollution in the Murray—Darling basin and the impacts on fish communities. Ecological Management and Restoration 15, 71–79.

Ministére de L'ecologie, du D'eveloppement Durable et de L'Energie, La description du parc 'electronucl'eaire francais. Updated March 10, 2011 and Accessed November 22, 2012, http://www.developpement-durable.gouv.fr/La-production-d-electricite.html

Mungin, L., December 20, 2007. Two off-line power plants help region hit water goal. Atlanta Journal-Constitution.

NASA, 2003. National Aeronautics and Space Administration, Earth Observatory. Image courtesy Reto Stockli and Robert Simmon, based upon data provided by the MODIS L and Science Team. http://earthobservatory.nasa.gov/IOTD/view.php?id=3714.

Poumadere, M., Mays, C., Mer, S.L., Blong, R., 2005. The 2003 heat wave in France: dangerous climate change here and now. Risk Analysis 25 (6).

Sanders, K.T., 2015. Critical review: uncharted waters? The future of the electricity-water nexus. Environmental Science and Technology 49, 51–66.

Scripps Institute of Oceanography, February 12, 2008. Lake Mead Could Be Dry by 2021. Press Release. http://scrippsnews.ucsd.edu/Releases/?releaseID=876.

Souder, E., Gwynne, S., Jacobson, G., February 6, 2011. Freeze Knocked Out Coal Plants and Natural Gas Supplies, Leading to Blackouts. Dallas Morning News.

Sternberg, R., 2010. Hydropower's future, the environment, and global electricity systems. Renewable and Sustainable Energy Reviews 14.

Stillwell, A.S., 2010. Energy-Water Nexus in Texas (Master's thesis). The University of Texas at Austin.

Stillwell, A.S., Webber, M.E., 2014. Geographic, technologic and economic analysis of using reclaimed water for thermoelectric power plant cooling. Environmental Science and Technology 48, 4588–4595.

Stillwell, A.S., Webber, M.E., 2013. Novel methodology for evaluating economic feasibility of low-water cooling technology retrofits at power plants. Water Policy 15, 292–308.

Stillwell, A.S., King, C.W., Webber, M.E., Duncan, I.J., Hardberger, A., 2011. The energy-water nexus in Texas (Special Feature on The Energy-Water Nexus: Managing the Links Between Energy and Water for a Sustainable Future). Ecology and Society 16 (1).

Tidwell, V.C., et al., 2014. Transitioning to zero freshwater withdrawal in the U.S. for thermoelectric generation. Applied Energy 131, 508–516.

Torcellini, P., Long, N., Judkoff, R., 2003. Consumptive Water Use for U.S. Power Production. Technical report, NREL-TP-550-33905. National Renewable Laboratory, U.S. Department of Energy.

Transportation Energy Data Book (TEDB), 2012. U.S. Department of Energy.

U.S. Army Corps of Engineers (USACE), 2001. Hydropower Value to the Nation. Technical report. Fall.

U.S. Environmental Protection Agency (EPA), 2012a. History of the Clean Water Act. U.S. Department of the Interior. http://www.epa.gov/lawsregs/laws/cwahistory.html.

U.S. Environmental Protection Agency (EPA), 2012b. Summary of the Clean Water Act, 33 U.S.C. §1251 et seq. (1972). Department of the Interior. http://www.epa.gov/lawsregs/laws/cwa.html.

U.S. National Oceanic and Atmospheric Administration (NOAA), 2009. Global Climate Change Impacts in the United States. Cambridge University Press.

United States Geological Survey (USGS), 2014. Estimated use of water in the United States in 2010. Circular 1405.

United States Geological Survey (USGS), 1998. Estimated use of water in the United States in 1995. Circular 1200.

Yardley, J., Harris, G., July 31, 2012. 2nd Day of Power Failures Cripples Wide Swath of India. New York Times.

Chapter 2.2.7

# Long-Term Water and Energy Issues in European Power Systems

**N. Maïzi, S. Bouckaert, E. Assoumou**
*MINES ParisTech, PSL Research University, Sophia Antipolis, France*

## 1. INTRODUCTION

Water and electricity are both identified as strategic resources in Europe, and both sectors have to cope with numerous challenges.

### 1.1 Finding the Adequate Portfolio for a Low-Carbon System

Triggered by population growth and improved living conditions, the global demand for electricity increased sharply by 47.5% between 1990 and 2010, 22.5% in Europe. This thirst for electricity has environmental downsides. With a 40% share of coal plants and subject to growing demand, the power sector is the first and fastest-growing contributor to energy-related greenhouse gas (GHG) emissions worldwide. Despite increasing use of renewables globally, power sector-related $CO_2$ emissions rose by 80% between 1990 (the reference year for the Kyoto protocol) and 2011 (IEA, 2013). In Europe, coal is still the largest source of emissions, reaching 68.2% in 2008 (IEA, 2011a), with Germany, the United Kingdom, Poland, and Italy accounting for 54% of total European emissions from electricity in 2008. The European generation mix is also characterized by a large share of nuclear (around 25%) and a rapidly increasing share of gas.

The European roadmap to 2050 has made the decarbonization of electricity a key component of its global energy policy including a high decarbonization target for the electricity sector. The European Union has set penetration targets for renewable energy sources: an objective of 20% renewable energy in final energy consumption by 2020 (EU, 2009) recently extended to 27% by 2030 (EU, 2014). Thanks to constant innovation and research efforts, technical solutions have been proposed for a low-carbon electric system but their interplay remains complex. Coal or gas thermal power plants equipped with carbon capture and storage (CCS), nuclear power, and biomass thermal plants are candidate solutions to mitigate the power sector's contribution to climate change, but they raise water and security concerns. Wind and solar power are alternative

**Competition for Water Resources. http://dx.doi.org/10.1016/B978-0-12-803237-4.00013-6**

renewable candidates but their intermittent nature points to reliability issues for achieving high penetration, while concentrated solar power plants also raise water issues.

## 1.2 Heterogeneous Supply Mix

Beyond the European strategy, at a country level, energy policies still rely on national decisions with the result that power systems are highly diverse (Fig. 1). Although renewable integration programs are foreseen in most of these countries thanks to strong policy support measures, they face difficulties in reaching their targets. Nonhydro renewables are dominated by wind generation, which is mainly concentrated in Germany and Spain. Some countries (Germany, Sweden, Switzerland, Italy) are reducing or phasing out nuclear generation.

Added to this diversity in power generation structures, size is also highly heterogeneous, as illustrated in Fig. 1: 23% of the generation capacities per country are less than 20 GW and 42% are less than 50 GW. Regardless of their size, some systems concentrate significant shares of generation capacity, such as hydropower: European hydropower capacity reaches 27% for systems less than 20 GW and 61% for systems under 50 GW. The numerous challenges facing power systems crucially involve water, as the two resources are closely interconnected: energy is needed for pumping, transporting, and treating water, and water is present throughout the energy chain in extraction processes, cooling systems, hydropower, and emission control processes like carbon capture and flue gas desulfurization (Voinov and Cardwell, 2009).

In light of this (greening the system with heterogeneous mixes), when considering power issues, concerns should ideally be shared with water experts. For Europe, these growing concerns regarding water availability and sustainability (UN, 2006, 2009, 2014) are tackled

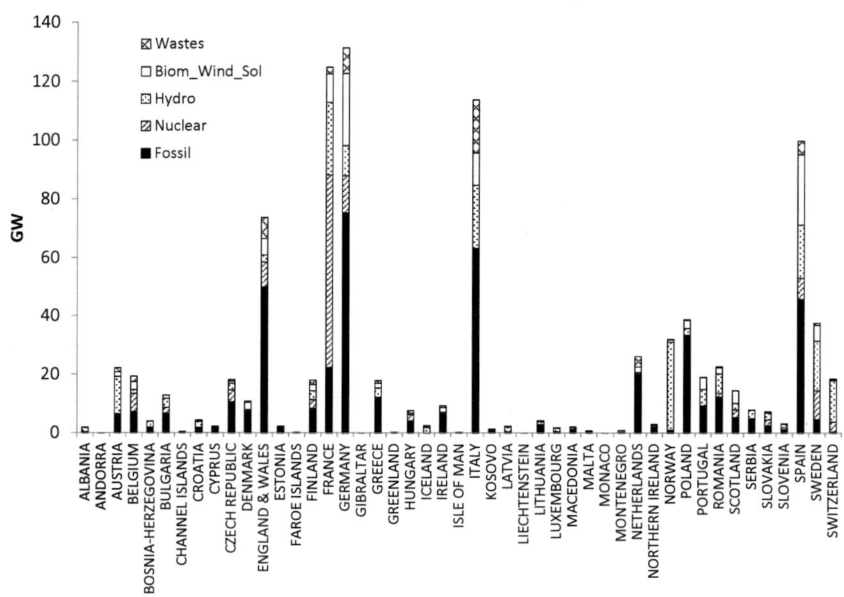


FIGURE 1 Power generation structure for European countries.

in the EU Roadmap to a Resource Efficient Europe, which states that water abstraction should remain below 20% of available renewable resources. The EU water blueprint advocates a series of tools and measures to increase water efficiency, from water pricing to leakage reduction. Southern Europe (Spain, France, Portugal) already faces serious issues related to water scarcity at some times of year.

The interplay between water and power for Europe will constitute the core of this chapter. We will illustrate how countries that are highly diverse in terms of socioeconomic development, energy sources, power mixes, and water access and availability must at the same time deal with power system development and water needs, and try to pinpoint the limits of this exercise. In Spain, for instance, the modernization of irrigation systems from 2002 to 2009 induced an increase in water-related electricity demand, with a possible rebound effect of increased water consumption (UN, 2014).

## 1.3 Complex Interplays Between Water and Electricity

Water is pervasive in energy and electric systems. Looking at the entire energy chain, water is required upstream for extraction, refining, and processing. It is a key factor in producing nonconventional oil and gas resources (eg, oil sands, shale gas, tight gas, etc.), the use of which is currently on the rise (Kleinberg et al., 2007; Perry et al., 2007; Stevens, 2010; Wu et al., 2009). The quantity of water varies with the method. For example, secondary oil recovery using water injection can require up to 600 $m^3$/TJ, whereas primary recovery requires only 5 $m^3$/TJ. The same applies for coal and uranium, for which the quantities of water used are highly dependent on the type of mine (Gleick, 1994; Mielke et al., 2010; DOE, 2006).

Further down the energy chain, in electricity production, five major water needs can be identified:

- Hydropower: beyond the clear link associating this technology with water, it is worth mentioning the evaporative water issue in the specific case of hydropower dams. Indeed, these evaporative water losses occur from the reservoir surfaces as the result of water being naturally evaporated and hence not available for electricity generation. Evaporative losses are climate zone and site dependent (Bakken et al., 2013).
- Cooling systems (Lemoine, 1986): the quantity of water used in cooling systems depends on the kind of technology. In wet closed-loop cooling systems, as water is recycled, water withdrawals are less significant than for open-loop technologies, which require 30–80 times more water. Nevertheless, water consumption is higher in a wet closed loop, mainly because of evaporation in the cooling tower. Dry systems, by definition, do not involve water withdrawals and consumption (Mielke et al., 2010; DOE, 2006; NETL, 2011).
- Heat transfer fluid: water withdrawal or consumption levels related to the main loop of the power plant are negligible compared to those related to the cooling system; we will therefore not deal with this water use (Nordmann, 2008).
- Emission control: new processes, like flue gas desulfurization or CCS, implemented to reduce the discharge of pollutants or greenhouse gases, turn out to be high water consumers. Desulfurization systems increase the volume of water consumed by a power plant using a closed-loop cooling system by 10%; carbon capture nearly doubles it (NETL, 2007, 2009; Shuster and Hoffmann, 2009).

- Gasification processes: as two-thirds of the electricity produced by a combined cycle power plant is issued from the gas turbine, integrated gasification combined cycle power plants require less water than classic coal-fired thermal power plants. The additional amount of water needed for the gasification process corresponds to 10% of the amount needed for the cooling systems (when considering a closed-loop cooling system) (NETL, 2007).

In terms of European countries' water needs for power generation, the three main technologies that raise concerns are: thermoelectric plants, which employ water in their cooling systems to discharge waste heat; hydropower, which is directly linked to water (dam requirements and evaporation); hydropower once again, when used to balance the widespread integration of intermittent renewable sources (solar and wind), because it provides a solution to reliability issues for the electrical system—through water storage it can be used to easily adjust load fluctuation and balance, and can be used to control flow variability. The levels of these needs by share of water per European country are shown in Fig. 2.

Consequently, the shift to a decarbonized power system in Europe will have to rely on options that will impact water. In what follows, we tackle this issue from two points of view, that is, from a hydropower development perspective and in terms of thermoelectric power plant cooling issues, and then cross the experiences of different European countries.

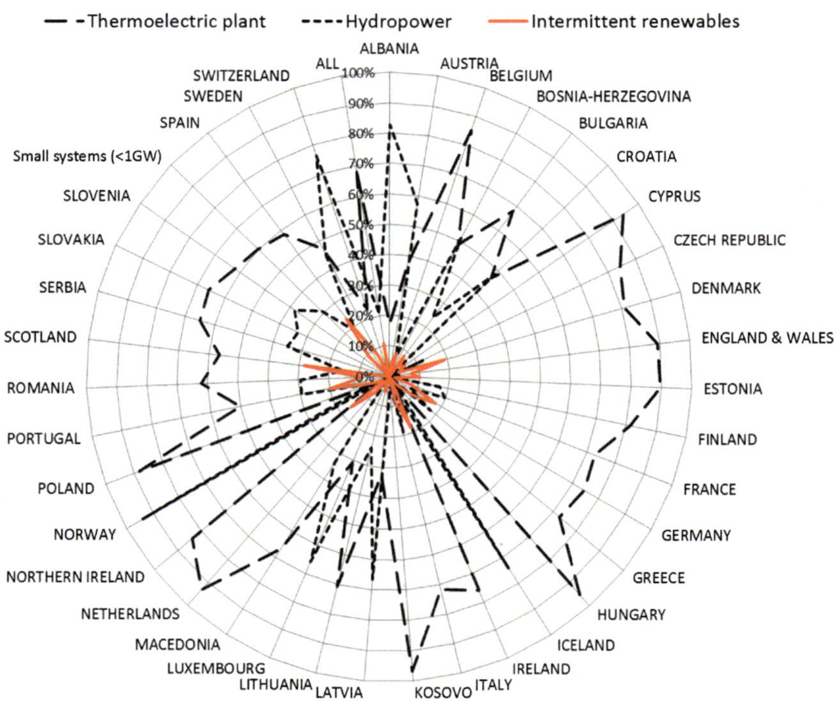


FIGURE 2 Water-related issues for power systems in European countries: shares of hydropower, thermoelectric and balancing capacity requirements.

We analyze each country's high water dependency in terms of its mix, the potential transition towards low-carbon technologies (intermittency vs. CCS), and its ability to move to more advanced cooling systems.

The analysis proposed in this chapter combines a synthesis of published studies with a dedicated analysis using the TIMES energy system modeling approach. TIMES belongs to a category of technological bottom-up models that enables the explicit representation of technologies in the energy sector for which demand is generally exogenous. This formulation allows for the creation of a virtual economy where different technologies compete with each other, as it leads to a technical optimum minimizing the discounted global system cost under constraints (technical, demand fulfillment, capacity, and activity bounds) (Loulou and Labriet, 2008).

## 2. WATER AND HYDROPOWER IN EUROPE'S ELECTRIC SYSTEMS

Hydropower is the largest source of renewable electricity in Europe and in the world. It harvests the mechanical energy of water from a reservoir or river and converts it into electricity. Its simple requirements are thus an appropriate hydrology (availability of water from rainfall or melting ice) and topology (sufficient height).

Hydropower was Europe's first source of electricity generation, and thousands of large dams were built in the last century and in particular during the postwar reconstruction period after 1950. Thanks to its flexibility, the maturity of its concepts, and a relatively good geographical distribution, hydropower already plays an important role in the current European electric system and is expected to be a central element of the transition to a low-carbon power mix (Eurelectric, 2011).

Europe can globally be defined as a "mature" region in terms of the evaluation and development of its hydropower potential, but significant differences persist among countries, both in terms of absolute hydropower potential and the level of effective development. As shown in Fig. 3, the four main hydropower regions in terms of volume are the European part of Russia, the mountains of Scandinavia, the Alps, and the Pyrenees. Fig. 3 also shows that Russia has considerable undeveloped potential, although this resource is located in the Asian part of Russia, far from urban consumption centers, and its development is not economically viable today.

The strategic focus naturally differs with the regional context. Where the potential has already been developed, or where social opposition to a new infrastructure is strong, the main issue is the seasonal management and sharing of the water resource between various economic activities. However, regions such as eastern and southeastern Europe experience high demand for new, small-scale hydropower plants, mainly run of river: these are hydropower plants that may have a limited amount of storage or no storage at all. As shown in Fig. 4, southeastern Europe is characterized by small electric systems that already contribute significant hydropower. However, a large share of the technical potential remains unexploited. To power economic development and the ensuing growing electricity demand, the incentive to construct new dams is high. Hundreds of small-scale hydropower projects are hence anticipated in the area, involving many large rivers in different neighboring countries (mainly Albania and Macedonia), such as the Sava, Bosna, Moraca, Vjona, Devoll, and Drim (Esch and Kabus, 2013). This so-called "Balkans dam boom" (Neslen, 2015) raises biodiversity concerns and several environmental associations have pointed out the considerable adverse effects on some of Europe's last preserved rivers.

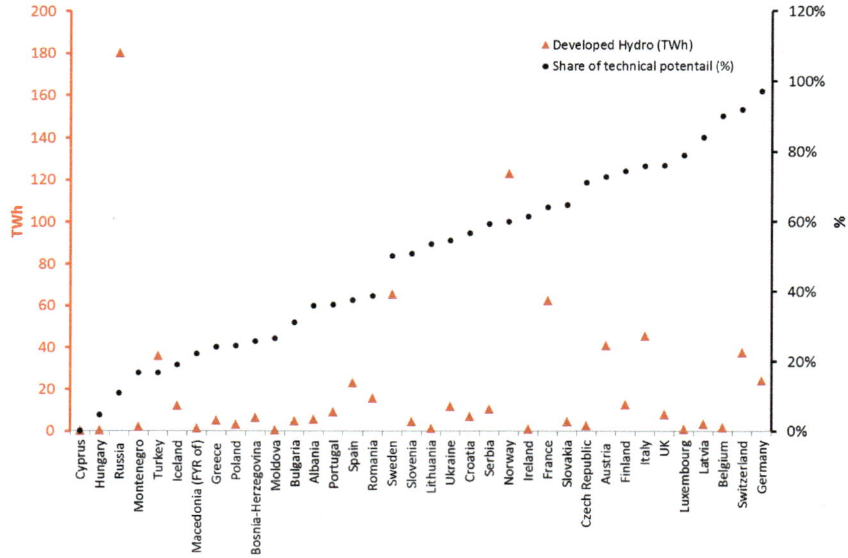


FIGURE 3 Hydropower production and share of technical potential developed in Europe.

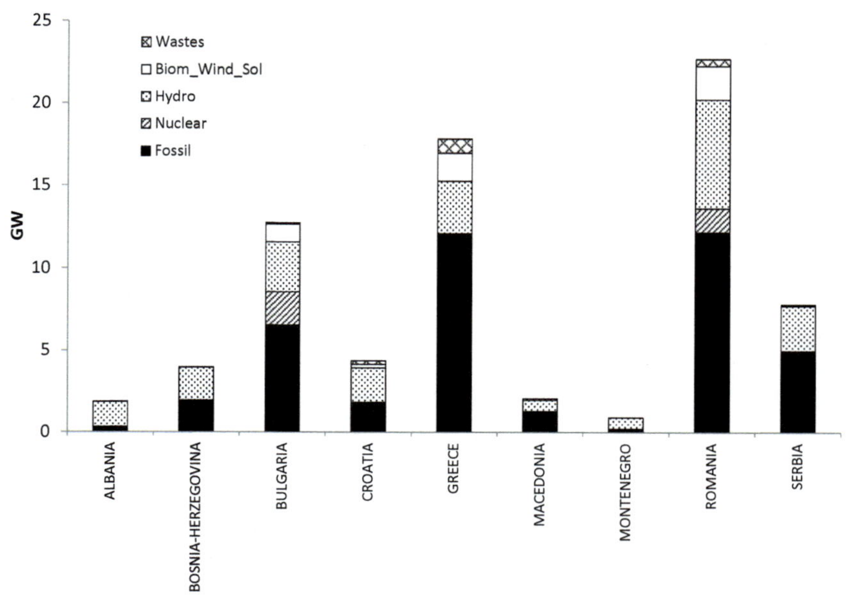


FIGURE 4 Power capacities in the Balkans. *Compiled data from Platts, 2013.*

Hydropower also offers operational flexibility, and the development of additional pumped hydrostorage capacities in Europe is the focus of increasing attention. This interest is essentially fostered by the EU's ambitious low-carbon power system objectives for 2050, which will require an increased share of intermittent renewable sources. In such a context, pumped hydropower plants represent a cost-effective way to store potentially large quantities of excess electricity and regulate supply—demand imbalances caused by the intermittency of solar and wind power. A report by the Joint Report Center of the EU (Gimeno-Gutiérrez et al., 2013) features an assessment of suitable pumped storage locations for different topographic conditions. The authors use a bottom-up description of candidate reservoirs to estimate a theoretical potential that could be as high as 10 times the installed pumped hydro capacity.

Yet the increased mean temperatures and more frequent extreme events expected as a result of climate change could threaten the availability of hydropower for the European electric system. The broad impacts for Europe are likely to be an increase in water availability in the Nordic countries and a decrease in southern Europe. In van Vliet et al. (2013) a 4—5% decrease in Europe's gross hydropower potential (without Russia) is estimated for 2031—2060 compared to the 1971—2000 climate. This combines regional variations ranging from a 15% increase in Norway to a 25% decrease in the Iberian region. In Gaudard et al. (2014) the effect of climate change in three catchment areas in the Alps is evaluated. The authors show that the effects are highly site specific, with significantly different results for geographically close sites. They also underline the difficulty of the task as, "the signal is below the noise induced by stochastic climatic variability."

This overview hence shows that hydropower is a key element in the operation of the European electricity system and has the ability to provide tailored technical solutions for the region's diverse energy system realities. The next two subsections focus on the roles of hydropower for two very different settings: a review of hydropower in Norway and model results to illustrate hydropower's versatility in a prospective 100% renewable scenario for France.

## 2.1 Norway: A Living Laboratory of Hydropower Systemic Interactions in an Interconnected Electric System

Norway is an iconic case for hydropower in Europe thanks to its atypical power generation mix that relies mainly on hydropower, and its multiple interconnections with countries that either feature a high number of thermal plants or offer considerable potential and ambitious objectives for wind power production. The following statistics demonstrate its singularity: with a population of only 5.1 million Norway produces approximately 20%[1] of Europe's hydropower, harbors 50% of Europe's conventional water storage (reservoir) capacity (Eurelectric, 2011), and according to Gimeno-Gutiérrez et al. (2013) represents a feasible potential of up to 13 TWh additional pumped storage.

### *2.1.1 Hydropower as the Basis for Electricity Generation*

Norway has a total installed capacity of 32.8 GW and produced 134 TWh (NMPE, 2015) of electricity in 2013. Hydropower plants accounted for 96% of this production, while the

1. Excluding Russia.

remaining 4% came from a small number of thermal power plants (3.3 TWh) and wind power plants (1.9 TWh). Looking at the distribution of currently operating Norwegian power plants per construction period, the most noticeable development pattern is the three decades of steady hydropower growth between 1960 and 1990 at an average level of 0.7 GW per year. Another clear feature is the fact that Norway's high wind power potential remains untapped.

The remaining potential for hydropower in Norway is still high, but around 50 TWh is situated in areas where hydropower development is prohibited under Norway's "Protection Plan for Watercourses." This illustrates the tradeoff between hydropower, other water usages, and environmental protection. It still leaves an additional production of 33.8 TWh, which is, for instance, comparable to the total hydropower production in Switzerland.

### *2.1.2 Hydropower at the Heart of an Interconnected System*

A second important feature of Norway's electric system is the high level of interconnection and power trade with its neighbors. Norway exported an average 15 TWh and imported 9 TWh per year between 2010 and 2014. As illustrated in Fig. 5, the country's net trade balance fluctuates strongly with the annual water inflow. 2010 was a cold year and the inflow to Norway's hydropower reservoirs was 18% below normal (NVE, 2010); Norway was a net importer from all countries. In subsequent years, and in 2012 in particular, a higher hydropower production level (NMPE, 2015) led to significant exports to all countries, making Norway a net exporter. The integrated Nordic electricity market thus takes advantage of the hydropower mix in Norway and in return reduces Norway's vulnerability to annual hydrological variations. Norway is always a net importer of electricity from Russia. Although this relates to small volumes, it is because of the specific conditions in northern Norway, which is poorly interconnected with the rest of the country, and highlights the existence of internal grid bottlenecks.

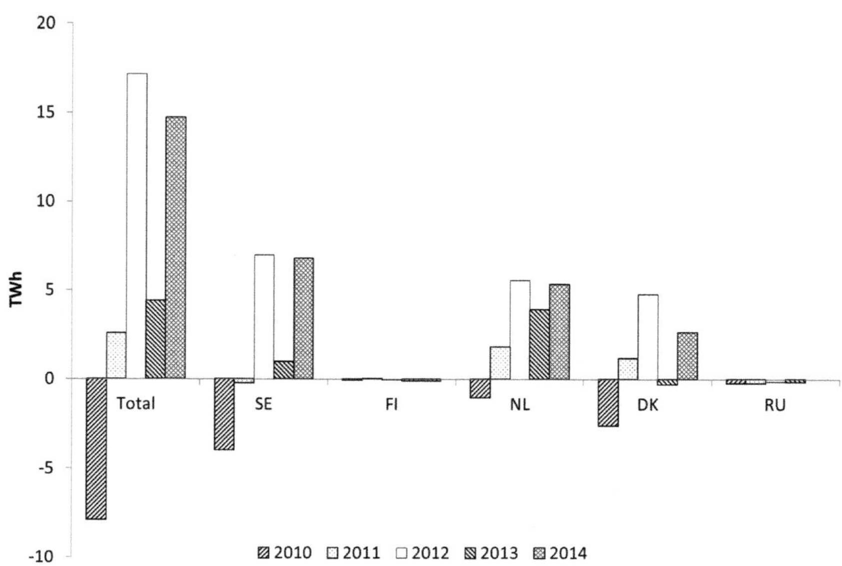


FIGURE 5 Net trade between Norway and neighboring European countries. *Compiled data from Platts, 2013.*

The new Skagerrak fourth interconnector of 700 MW (SK4 700 MW) with Denmark commissioned in 2014 increases Norway's trading capacity to 6.1 GW. Two planned subsea cables (ENTSOE, 2015), the 1400 MW NordLink project with Germany, and the North Sea Network interconnector of 1400 MW (1400 MW NSN) project with the United Kingdom, will increase this level and bring Norway's total gross trade capacity to 8.5 GW by around 2021. These new grid infrastructures will connect Norway to two large European electric systems and trade capacities will represent more than 25% of its current installed capacity.

### *2.1.3 European Demonstrator for Hydropower as a Balancing Option*

Although not economical under current market conditions, Norway's absolute wind power potential is higher than its hydropower potential. However, as mentioned previously, Norway is already connected, or is planning interconnection, to countries with a high share of fossil-based electricity and ambitious wind power development objectives: Denmark, the United Kingdom, and Germany. These developments illustrate hydropower's strategic pivotal value to regulate electric systems with increasing intermittency. The interaction between Norway's hydropower and wind power has been investigated in several publications that are now briefly summarized:

A dedicated International Energy Agency task force on wind and hydropower interaction (IEA, 2011b) proposed two case studies on wind power development in Norwegian regional subsystems with constrained power transfer capacity. In Rosenberg et al. (2013) the TIMES-Norway model is used to assess the future electricity supply and renewable portfolio for contrasted regional energy demand scenarios. Industrial and building sector electricity demands are estimated for seven regions. In Ely et al. (2013), simplified power sector and demand models are used, but the authors focus on the correlation between wind, hydropower, and weather-related demand in Norway and the United Kingdom to discuss the effect of increased interconnection by 2050. Although the focus and methods differ, lessons from these studies include:

- Significant quantities of wind power can technically be integrated, even in areas with low power transfer capacities, if wind power and hydropower plants are coordinated.
- However, at country level, the most economical long-term option is to use the additional hydropower potential. The contribution of domestic wind power in Norway is likely to remain low until 2030 and will only become significant by 2050 if industrial electricity demand remains high.
- Although based on simplified power models, the correlation between the weekly demand level, hydropower, and wind in the North Atlantic Oscillation analysis proposed for the United Kingdom raises similar concerns for the Germany–Norway interconnection. The weekly water availability at the end of the winter season could make this the critical balancing time. It can be expected that with a higher share of wind by 2030 or 2050, some thermal power plants will be used for balancing in addition to hydropower.

Finally Gullberg (2013) questions the political feasibility of using Norway as a freely available flexible source of hydropower to optimize the EU power system. The author confronts different future technical paradigms with the political decision process, which has its own pace. France future power strategy provides a significant example of this dynamical confrontation.

### 2.1.4 France: Interaction of Hydropower and Nuclear as $CO_2$-Free Electricity Generation Options

The French electricity generation system is unique. It relies on the highest share of nuclear energy in the world as shown in Fig. 6: in France (RTE, 2014), 52.3% of installed capacity and 73.5% of electricity supply comes from nuclear power plants. Hydropower is the second largest contributor to electricity generation, at 13.6%, split between dam, river, and storage (Fig. 6), which makes France the second largest hydropower producer in the European Union. Fossil power plants (half coal, half gas and oil) account for a mere 8% and are mainly used for peaks and system operation.

The nuclear power replacement strategy will be a major issue in the future, as we can see by looking at the lifespan of the residual capacities of the French power system from 2000 to 2050 (Fig. 7).

Thus, and as the power sector is clearly characterized by low emission levels, a future electricity generation mix combining nuclear and renewable intermittent energy sources constitutes a major issue. This future mix for electricity generation has to be assessed in a context involving numerous environmental constraints: for France, the long-term environment target, specified in an energy orientation law dated March 2005 and confirmed in an energy transition paper (LOI, 2015-992, 2015), is to quarter total GHG emissions by 2050 with respect to 2000 levels with an intermediate mitigation level of 40% in 2030 compared to 1990. We used the TIMES-FR-electricity model to investigate, at the level of France, the feasibility of a transition towards a 100% renewable electricity system by 2050.

This TIMES model for mainland France (Maïzi and Assoumou, 2014) and power exchanges with neighboring countries is constructed as follows. The base year is 2012 with load and production data calibrated on this year, and the end year is 2050. The time period

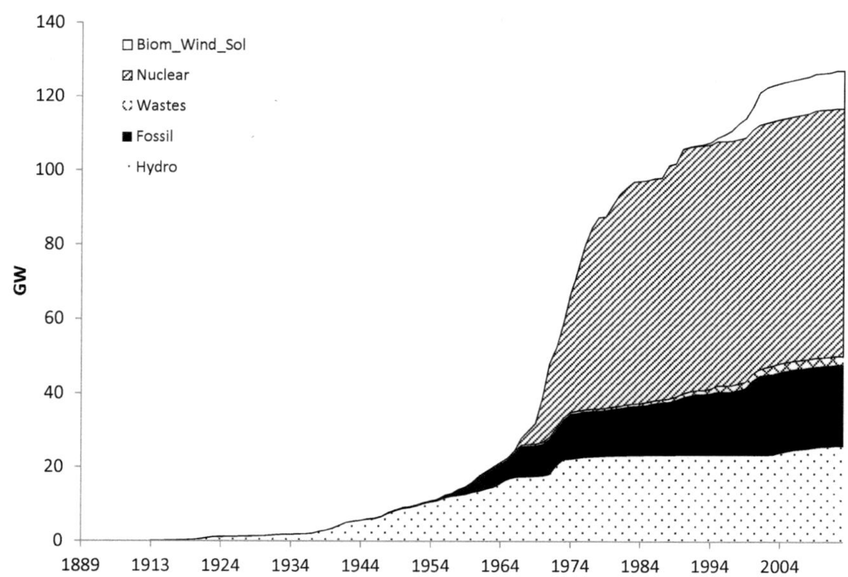


FIGURE 6 Breakdown of French electricity generation capacities. *Compiled data from Platts, 2013.*

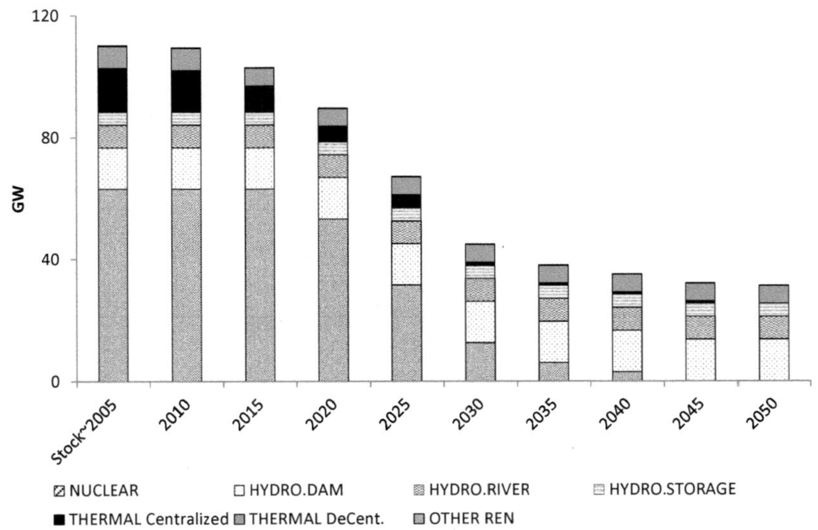


**FIGURE 7** Evolution of French residual capacities from 2005.

is divided into 13 periods of several years. Each year is further divided into seven seasonal periods, six monthly periods, and one that represents a winter week with low solar and wind production as well as restrictions on imports. Each seasonal period is then split into two typical days, one on a working day and one at the weekend. Finally, each typical day is divided into six hourly periods (two for the night, two for the morning, one for the afternoon, and one corresponding to peaking demand). This high time resolution better captures the seasonal variation of demand and intermittent renewables. We calibrated the production patterns for solar and wind power plants based on their 2012 output (Krakowski et al., 2015).

TIMES-FR results reveal what a transition towards 100% renewables would entail in terms of technological choices and their installed capacities. Indeed, the massive penetration of renewable energy requires transforming the electricity system and rethinking the way it is operated, as illustrated by the use of hydroelectric capacities, for which maximum load factors per season are also introduced.

The two scenarios assessed are based on the following assumptions:

*The 100%RES scenario (where RES stands for renewable energy sources) has to follow specific targets: 50% maximum nuclear energy in 2025, 27% minimum RES in the power generation mix in 2020, 50% minimum RES in 2030, 75% minimum RES in 2040, 100% in 2050.*

*The 100%RES_fewInterco scenario follows the same targets as the 100%RES scenario but places restrictions on the development of new interconnections, that is, no new interconnections except those under construction.*

The penetration of renewables in the French electricity system has several very clear impacts on the electricity production mix, as TIMES-FR results of the 100%RE scenario show in Fig. 8: steady exit of nuclear and fossil, but with a significant role for nuclear

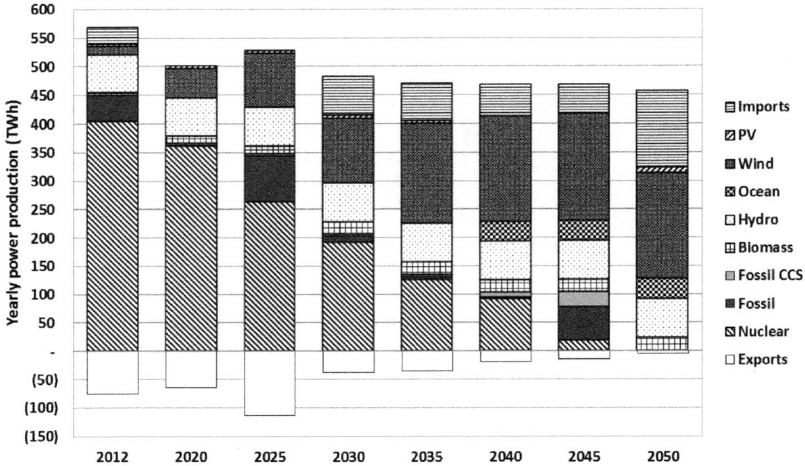


FIGURE 8 Evolution of power production between 2012 and 2050 in a 100% RES scenario.

power stations up until 2030 and 2040, as well as coal and gas power plants in 2025 and 2045, constant hydroelectric production over the period, a rapid rise of wind power, which represents about half of electricity production after 2035, moderate contributions from marine power and biomass, although higher than at present, a minor role for photovoltaic, and a surge in electricity imports in 2050.

In 2050, in this context, the first three means of production are wind, hydroelectric, and marine power, but imports make up almost 30% of French electricity consumption. Indeed, in the 100%RE scenario, imports are used massively to satisfy demand, without taking into account the specific constraints of electricity systems in neighboring countries or the availability of the electricity network. For this reason, we also evaluated a scenario with more limited import capacities: 100%RES_fewInterco. Of note in these results is the fact that a 100% renewable target in 2050 for France leads to the use of hydroelectric plants and biomass to satisfy the peak. In 2050, hydroelectric power makes up most of the base (domestic generation only), while peak and extreme peak demands are satisfied by biogas and biomass plants at a similar level to hydroelectric power, as shown in Fig. 9 (where in the 100%RES scenario the base is in reality mostly made up of imports).

In terms of managing the supply–demand balance, hydroelectric thus maintains its current role, acting as a base energy (run-of-the-river hydroelectricity), and intervening during peaks (dams and pumped-storage hydroelectricity), although with a greater role for peaks than currently. Biomass and biogas plants are also installed to supplement hydropower during peaks, replacing fossil plants. This use of hydroelectric stations and biomass requires increased flexibility of these means of production, which could prove difficult because of inherent constraints: the availability of hydroelectric depends on meteorological conditions, with seasonal and annual variations, and competes with the use of rivers for other purposes than electricity production. In addition, water availability can also be affected by climate change effects, such as the recent warm, dry summers. For biomass, the way the resource is managed is also important since it can only be considered as renewable if its upstream supply is well managed. In addition, water needed in the biomass process and for crop irrigation also raises questions of resource availability.

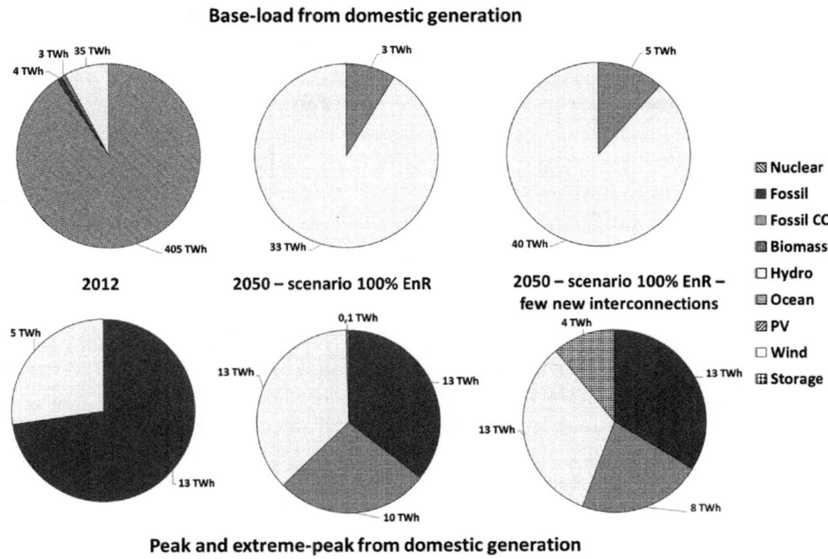


FIGURE 9 Composition of base (domestic generation only) energy (*top*) and peak and extreme peak demand during the same periods (*bottom*) in 2012, and in 2050 for the 100%RES and 100% RES_fewInterco scenarios.

## 3. WATER ISSUES FOR EUROPE'S THERMOELECTRIC POWER PLANTS

Thermoelectric plants require significant amounts of water. In the power sector, several past incidents corroborate the vulnerability associated with cooling systems: during the 2003 heatwave in Europe (as well as 2006 and 2009 events), power plants were forced to shut down or their efficiency was reduced, such as nuclear power plants in France subject to increased water-cooling needs while river levels were decreasing. Here we discuss this second dimension of water electricity interaction in Europe.

### 3.1 European Water and Thermoelectric Plant Challenges: Examples From Literature

Some pioneering research works have investigated the challenges posed by water availability for power plant cooling systems in Europe. In Rübbelke and Vögele (2011) and Van Vliet et al. (2013) this question is addressed with EU-wide coverage, while Byers et al. (2014), Hoffman et al. (2013), and Koch et al. (2014) provide country-level perspectives.

The strategic concern in the analysis proposed by Rübbelke and Vögele (2011) is the impact of cooling water availability on the direction and level of electricity exchanges between European countries and on a benchmark day. Their investigation focuses on nuclear power plants and a typical summer's day in two hypothetical settings (higher air temperature with or without a 10% reduction in water availability). The methodological approach is based on a detailed climate module with a 1 km$^2$ resolution and an equation describing the relationship between air temperature and water temperature. The resulting imbalances are then propagated in a representation of trade flows. A key insight is to show evidence (for the selected conditions) of possible indirect but systemic impacts in countries such as Italy that do not have constrained nuclear power plants. In UN (2014) and van Vliet et al. (2012) it is estimated that by 2030 half of the river basins in the European Union could be affected by an increase in temperature and water scarcity and stress. The methodology used in Rübbelke and Vögele (2011) is extended to include more types of thermoelectric plants, exogenous scenarios on both future installed capacities and alternative cooling options, and hydropower (as discussed in Section 2). The authors confirm a strong sensitivity to cooling water scarcity and give new insight into the adaptive capacity of the electric system via exogenous fuel and cooling system scenarios.

Country-scale analyses have been performed for Germany and the United Kingdom, which are respectively the first and third European countries in terms of installed capacity of thermal power plants. Together they represent 29% of the thermal power capacity. In Hoffman et al. (2013), the cooling water problem for 26 identified German power plants is studied. The authors do not focus on an electric system-wide approach, but use detailed dynamic models for each power plant, with feedback from the water temperature and availability on the efficiency of the plants. The model uses a daily resolution, computes the efficiency of the cooling process using thermodynamics, and considers legal thresholds. Hence the authors highlight a sensitivity to both output reduction and efficiency reduction. Like the previous studies, the authors simulate different cooling system adaptation scenarios and runoff reductions of up to 50%. They confirm the sensitivity to cooling water illustrated in existing studies, but point to an even greater output reduction when a daily time resolution is used. Koch et al. (2014) use a more detailed hydrological model to assess water temperature and thermal plant cooling in selected nuclear plants in Germany. They find that for some of the plants in their selection, the maximum threshold could lead to complete shutdown.

In Byers et al. (2014) a different perspective is chosen. The core question is a comparative analysis of the global amount of water use for cooling in thermal plants in the United Kingdom by 2050 in six different decarbonization paths. The authors recall that 35% of catchments in England and Wales have no availability for further licensing. The issue is hence not the climate-related temperature change, but water demand and competition. The methodology uses exogenous plant capacity, scenarios for distribution of cooling technologies, and simple water factor per type of generation. Interestingly in Rübbelke and Vögele (2011) and van Vliet et al. (2013) the United Kingdom was not found to be at risk with mean and summer temperature increases. A specific insight provided here is the strong impact of carbon capture and sequestration, in particular on future freshwater abstraction and consumption.

This overview demonstrates that the availability of sufficient water cooling is a multifaceted issue that is increasingly recognized as strategic for the electric system in Europe. Adapted methodologies have been developed with different strategic focuses, levels of detail of power plants and water availability models, geographical scopes, and time resolutions. This water dimension is of utmost importance for a European power system dominated by thermoelectric power plants. The analyses reviewed outline the following broad lessons:

- The effect could be considerable, ranging from output reduction to complete shutdown of large plants.
- A systemic link exists through exchange between countries, so that countries with no power plants directly at risk could be affected.

- Advanced cooling systems can provide adaptive responses.
- Apart from output reduction, the amount of cooling water needed could generate water competition; some low-carbon options, such as CCS or nuclear, can generate adverse water effects.

## 3.2 Water/Thermal Plant Interactions in Europe Using TIAM-FR

In what follows, our TIAM-FR model is used as a complementary tool to illustrate how water requirements for Europe's thermoelectric plants can remain a relevant issue, even in a global GHG mitigation analysis, to ascertain whether future energy mixes in Europe might vary with changing water availability conditions. TIAM-FR is a global TIMES model divided into 15 world regions with a dedicated water module that includes available cooling technologies (Bouckaert et al., 2014). It models water withdrawal and consumption by transformation processes in the energy chain. The structure of the model is shown in Fig. 10. Two types of water use are considered: water withdrawal (removed permanently or temporarily) and water consumption (no longer available for use). In the electricity sector, the model computes a cost optimal allocation between plant technologies and cooling systems that do not use water, such as photovoltaic, and the type of cooling system used, taking into account the power plant's additional electricity consumption.

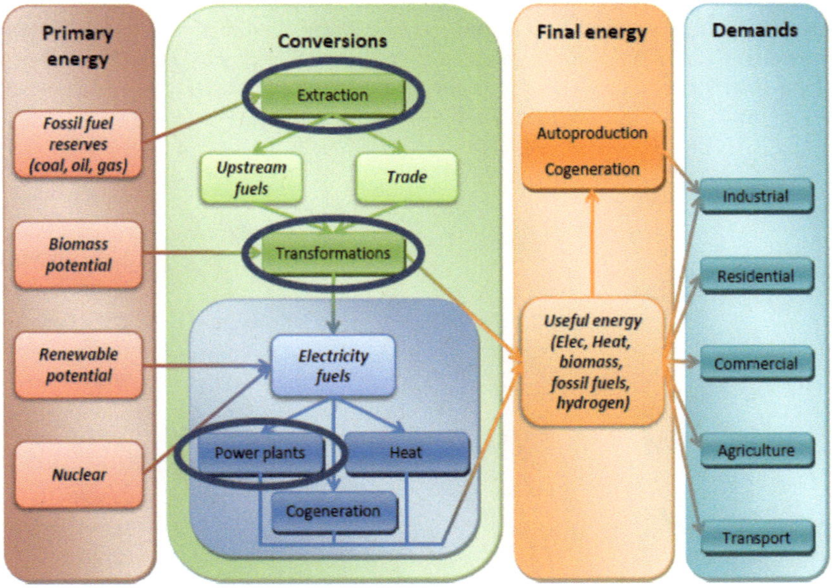


**FIGURE 10** TIAM-FR reference energy system. The ellipses identify sectors with enhanced water processes.

The model is run for four global scenarios including water conservation and greenhouse gas targets:

- REF is the reference scenario with no carbon or water conservation policy.
- CLIM50_80 is a global mitigation scenario where the world's energy system has to halve its $CO_2$ emissions compared to their 2005 levels with a specific target of 80% for Western Europe.
- CLIM50_80-Wconv is identical to CLIM50_80 in terms of mitigation objectives. In addition, water withdrawal and consumption for power plants in Western Europe are limited to their current value. The current share of cooling systems is maintained.
- CLIM50_80-Wflex also assumes mitigation objectives and a limitation of water withdrawal and consumption for power plants in Western Europe at their 2012 value. However, the choice of cooling systems is flexible and endogenous.

Although the demand can be adjusted through price elasticity, this capability was not included to keep energy services strictly comparable across scenarios and simplify the discussion. From the energy system-wide allocation problem, we focus here on the implications for the future electricity mix and choice of cooling system.

As depicted in Fig. 11, the European Union's ambitious climate objective implies a 20% increase in electricity generation to substitute fossil fuels in final end-uses, and a simultaneous deep decarbonization of the power mix. Decentralized and difficult-to-control emissions are substituted by more easily manageable ones. The preferred solutions combine various renewables, plants with CCS (fossils and biomass), and more nuclear. However, as more thermal plants with CCS are used, freshwater consumption and withdrawals are respectively multiplied by 3.4 and 3.9 (relatively to REF in 2050). Because closed systems reduce withdrawals but increase water consumption, a combined water conservation and climate policy leads to slightly lower electricity generation when the current share of cooling systems is maintained (Fig. 12). Two costly but water efficient

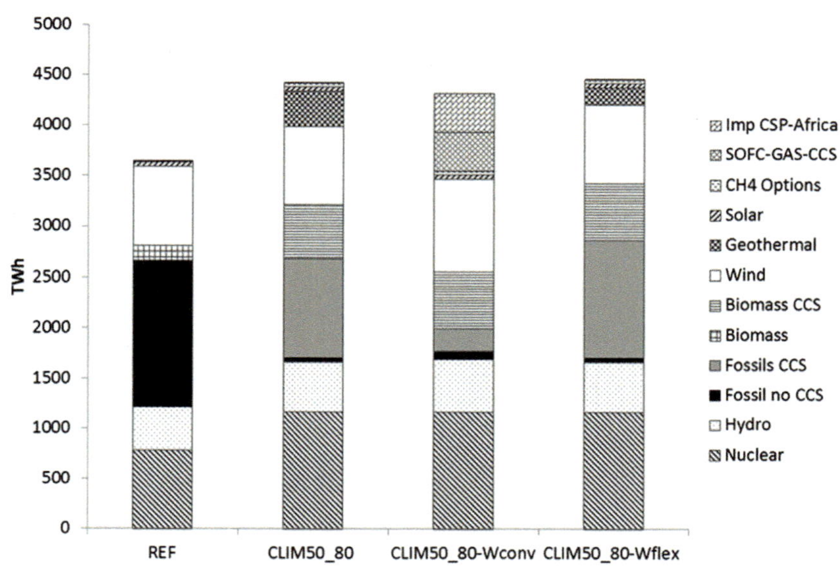


FIGURE 11 2050 Evolution of the power generation mix in Western Europe.

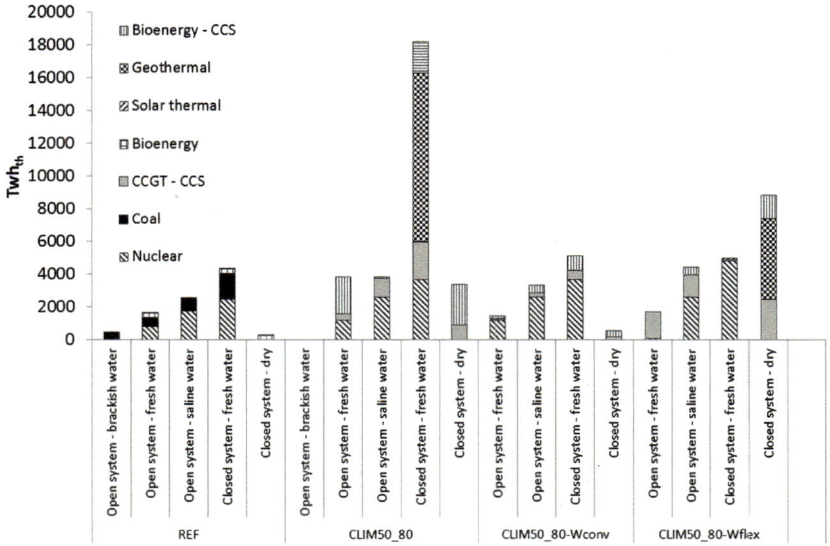


**FIGURE 12** 2050 Adaptive allocation of cooling system technologies.

options are used: natural gas-fueled solid oxide fuel cells (SOFC) with CCS and imported concentrated solar power (CSP) electricity from Africa. This last option, which is water efficient from a European point of view, illustrates the externalities of limiting water withdrawal and consumption for power plants in Western Europe: providing enough electricity for European consumption involves investing in Africa in technology based on converting thermal solar energy into electricity, which uses steam turbines and needs to be cooled. As a result, the problem is moved to a different area, which may be drier than the initial location, where water was already limited! This can be seen as a water issue "leakage." However, there is an alternative, as the CLIM50_80-Wflex scenario shows for the same freshwater budget as that allocated to the electricity sector, with the advantage of simultaneously adapted generation (less geothermal) and endogenous cooling systems (more dry systems).

These scenarios help us understand that water consumption and electricity generation must be assessed together to achieve a tradeoff within a global framework while respecting local constraints. This complex formulation is mandatory to establish a sustainable compromise!

## REFERENCES

Bakken, T.H., Killingtveit, Å., Engeland, K., Alfredsen, K., Harby, A., 2013. Water consumption from hydropower plants – review of published estimates and an assessment of the concept. Hydrology and Earth Systems Sciences 17, 3983–4000.

Bouckaert, S., Assoumou, E., Selosse, S., Maïzi, N., April 15, 2014. A prospective analysis of waste heat management at power plants and water conservation issues using a global TIMES model. Energy. ISSN: 0360-5442 68, 80–91. http://dx.doi.org/10.1016/j.energy.2014.02.008.

Byers, E.A., Hall, J.W., Amezaga, J.M., March 2014. Electricity generation and cooling water use: UK pathways to 2050. Global Environmental Change 25, 16–30. http://dx.doi.org/10.1016/j.gloenvcha.2014.01.005.

DOE, 2006. Energy Demands on Water Resources. The United States Department of Energy.

Ely, C.R., Brayshaw, D.J., Methven, J., Cox, J., Pearce, O., 2013. Implications of the North Atlantic Oscillation for a UK–Norway renewable power system. Energy Policy 62, 1420–1427. http://dx.doi.org/10.1016/j.enpol.2013.06.037, 2013.

ENTSOE, June 24, 2015. Regional Investment Plan North Sea Region. ENTSO-E. Draft for Consultation.

Esch, V., Kabus, J. (Eds.), 2013. A Future Security Architecture for Southeast Europe. The Aspen Institute.

EU, 2009. Parlement Européen et Conseil, « Journal officiel L 140/2009, Directive 2009/28/CE du Parlement Européen et du Conseil du 23 avril 2009 relative à la promotion de l'utilisation de l'énergie produite à partir de sources renouvelables et modifiant puis abrogeant les directives 2001/77/CE et 2003/30/CE ». http://eur-lex.europa.eu/legal-content/FR/TXT/HTML/?uri=OJ. L:2009:140:FULL&from=FR.

EU, 2014. Commission Européenne, « Communication de la commission au parlement européen, au conseil, au comité économique et social européen et au comité des régions – Un cadre d'action en matière de climat et d'énergie pour la période comprise entre 2020 et 2030 », EUR-Lex – 52014DC0015-EN. http://eur-lex.europa.eu/legal-content/FR/TXT/?uri=CELEX:52014DC0015.

Eurelectric, 2011. Hydro in Europe: Powering Renewables. Eurelectric report and annexes.

Gaudard, L., Romerio, F., Dalla Valle, F., Gorret, R., Maran, S., Ravazzani, G., et al., 2014. Climate change impacts on hydropower in the Swiss and Italian Alps. Science of the Total Environment 493, 1211–1221. http://dx.doi.org/10.1016/j.scitotenv.2013.10.012.

Gimeno-Gutiérrez, M., Lacal-Arántegui, R., 2013. Assessment of the European Potential for Pumped Hydropower Energy Storage: A GIS-Based Assessment of Pumped Hydropower Storage Potential. European Commission. Joint Research Center Scientific and Policy Reports.

Gleick, P.H., 1994. Water and energy. Annual Review of Energy and the Environment 19, 267–299.

Gullberg, A.T., June 2013. The political feasibility of Norway as the 'green battery' of Europe. Energy Policy. ISSN: 0301-4215 57, 615–623. http://dx.doi.org/10.1016/j.enpol.2013.02.037.

Hoffmann, B., Hofele, S., Karl, U., 2013. Analysis of performance losses of thermal power plants in Germany: a system dynamics model approach using data from regional climate modelling. Energy 49, 193–203.

IEA, 2011a. Climate & Electricity Annual: Data and Analysis 2011. International Energy Agency.

IEA, 2011b. IEA Wind Task 24 Integration of Wind and Hydropower Systems Volume 2: Participant Case Studies. Department of Energy Wind and Hydropower Program Technical Report NREL/TP-5000-50182.

IEA, 2013. $CO_2$ Emissions from Fuel Combustion, 2013. Highlights. International Energy Agency.

Kleinberg, R., Allison, E.C., Collett, T.S., Holditch, S.A., Howard, J.J., Sloan, E.D., 2007. Oil Shale. Tech. rep., Oil Shales and Hydrates Subgroup of the Technology Task Group of the NPC Committee on Global Oil and Gas.

Koch, H., Vögele, S., Hattermann, F., Huang, S., 2014. Hydro-climatic conditions and thermoelectric electricity generation – Part II: model application to 17 nuclear power plants in Germany. Energy 69, 700–707. http://dx.doi.org/10.1016/j.energy.2014.03.071.

Krakowski, V., Assoumou, E., Maïzi, N., Mazauric, V., 2015. A prospective analysis of contrasted renewable energy penetration targets in French power system. In: Proceedings of the 10th Conference on Sustainable Development of Energy, Water and Environment Systems, Dubrovnik, Croatia.

Lemoine, P., 1986. Refroidissement des eaux. Techniques de l'Ingénieur.

LOI n° 2015-992 du 17 août 2015 relative à la transition énergétique pour la croissance verte. Journal Officiel de la République Française 18, Août 2015.

Loulou, R., Labriet, M., 2008. ETSAP-TIAM: the TIMES integrated assessment model, part I: model Structure. CMS 5, 7–40.

Maïzi, N., Assoumou, E., December 31, 2014. Future prospects for nuclear power in France. Applied Energy. ISSN: 0306-2619 136, 849–859. http://dx.doi.org/10.1016/j.apenergy.2014.03.056.

Mielke, E., Anadon, L.D., Narayanamurti, V., 2010. Water Consumption of Energy Resource Extraction, Processing, and Conversion: A Review of the Literature for Estimates of Water Intensity of Energy-Resource Extraction, Processing to Fuels, and Conversion to Electricity. Energy Technology Innovation Policy Discussion Paper No. 2010–15. Belfer Center for Science and International Affairs, Harvard Kennedy School, Harvard University.

Neslen, A., February 20, 2015. Balkan Dam Boom Threatens Europe's Last Wild Waterways. The Guardian. http://www.theguardian.com/environment/2015/feb/20/balkan-dam-boom-threatens-europes-last-wild-waterways.

NETL, 2007. Power Plant Water Usage and Loss Study. The United States Department of Energy. National Energy technology Laboratory.

NETL, 2009. Water Requirements for Existing and Emerging Thermoelectric Plant Technologies. The United States Department of Energy. National Energy technology Laboratory. DOE/NETL-402/080108.

NETL, 2011. Estimating Freshwater Needs to Meet Future Thermoelectric Generation Requirements. The United States Department of Energy/National Energy technology Laboratory. DOE/NETL 2011/1523.

NMPE, 2015. Norwegian Ministry of Petroleum and Energy. Energy and Water Resources in Norway.

Nordmann, F., 2008. Gestion industrielle de l'eau pour la production d'énergie nucléaire. Techniques de l'Ingénieur.

NVE, 2010. Norwegian Water Resources and Energy Directorate. Annual Report 2010 The Norwegian Energy Regulator 2011.

Perry, K., Lee, J., Holditch, S.A., 2007. Unconvenional Gas Reservoirs – Tight Gas, Coal Seams, and Shales. Tech. rep. National Petroleum Council, Unconventional Gas Subgroup of the technology task group of the NPC Commitee on global oil and gas.

Platts World electric power plants data base, September 2013, ISBN: 1-56760-219-3.

Rosenberg, E., Lind, A., Espegren, K.A., 2013. The impact of future energy demand on renewable energy production – case of Norway. Energy 61, 419–431. http://dx.doi.org/10.1016/j.energy.2013.08.044.

RTE, 2014. Bilan prévisionnel de l'équilibre offre-demande d'électricité en France, Edition 2014. RTE (Réseau De Transport d'Electricité). http://www.rte-france.com/sites/default/files/bilan_complet_2014.pdf.

Rübbelke, D., Vögele, S., 2011. Impacts of climate change on European critical infrastructures: the case of the power sector. Environmental Science & Policy 14 (1), 53–63.

Shuster, E., Hoffmann, J., 2009. Water requirements for fossil-based electricity plants with and without carbon capture. In: GWPC Annual Forum (NETL).

Stevens, P., 2010. The 'Shale Gas Revolution': Hype and Reality. Tech. rep. A Chatham House Report.

UN, 2006. The United Nations Report, 2006. Water a Shared Responsibility.

UN, 2009. The United Nations Report, 2009. Water in a Changing World.

UN, 2014. The United Nations Report, 2014. Water and Energy.

van Vliet, M., Yearsley, J., Ludwig, F., Vögele, S., Lettenmaier, D.P., Kabat, P., 2012. Vulnerability of US and European electricity supply to climate change. Nature Climate Change 2, 676–681.

van Vliet, M.T.H., Vögele, S., Rübbelke, D., 2013. Water constraints on European power supply under climate change: impacts on electricity prices. Environmental Research Letters 8 (3), 035010. http://dx.doi.org/10.1088/1748-9326/8/3/035010.

Voinov, A., Cardwell, H., 2009. The energy-water Nexus: Why should We Care? Journal of Contemporary Water Research & Education 143, 17–29.

Wu, M., Mintz, M., Wang, M., Arora, S., 2009. Consumptive Water Use in the Production of Ethanol and Petroleum Gasoline. Tech. rep. Center for Transportation Research, Energy Systems Division, Argonne National Laboratory.

Chapter 2.3

# Water–Energy–Food Nexus—Commonalities and Differences in the United States and Europe

J.M. Peterson
*University of Minnesota, Saint Paul, MN, United States*

## 1. INTRODUCTION

At the most basic level, the food–energy–water (FEW) nexus is defined by resource tradeoffs. Water is an essential input to produce both food and energy while also serving many other uses. At least in the short run when technologies and the allocation of water for other uses are fixed, allocating more water to produce more food implies that less will be available for energy. Similarly, more "water for energy" leaves less "water for food," and obtaining more water for other uses takes some away from both food and energy. These tradeoffs comprise the so-called trilemma of FEW resources, which is only expected to intensify in the future because of long-term global trends including population growth, income growth, and climate change. This view of the nexus motivates the need for research and investment that can conserve resources through more efficient technologies.

The contributions in this book take a more behavioral view of the FEW nexus, examining it as a collection of human institutions. The preceding chapters illustrate how the allocation of water we observe is the product of various institutions–the combination of laws, programs, rules, markets, and private and public entities—that prevail in a given location. In both the United States and in Europe, water allocations naturally depend on institutions specific to water resources, but they also interact with the institutions of agriculture, energy, and other sectors. These interactions generate a number of feedback loops between human actions and resource conditions. The research agenda pursued in this book is driven by the need for better management of water resources through a better understanding of institutions and their function.

Institutional analysis is a useful framework for the current chapter, whose purpose is to synthesize the contributions of Part 1 and Part 2 of this volume and set the stage for Part 3. The next section summarizes some of the lessons about water management from the preceding chapters, identifying commonalities and differences in the United States and in Europe. A number of these chapters discussed the ways that institutions in the FEW nexus

**Competition for Water Resources. http://dx.doi.org/10.1016/B978-0-12-803237-4.00014-8**

function during extreme events such as droughts when competition for water resources is highest. These analyses revealed various interactions and feedbacks within and among institutions. While the details vary, a recurring theme is that existing institutions fail to absorb all the system stresses during extreme events. The following section then considers some of the potential causes for what may be termed a lack institutional resilience regarding water resources. It is argued that water has a special set of inherent characteristics, which set it apart from other natural resources and require management approaches that are situational, multifaceted, and adaptive. The concluding section looks ahead to consider the opportunities and barriers for institutional innovations.

## 2. THE NEXUS OF FOOD, ENERGY, AND WATER INSTITUTIONS

In both the United states and in Europe, a number of institutions have evolved with a specific focus on managing water resources. Hansen (2016, Chapter 1.1), Huffaker (2016, Chapter 1.3), and Schaible and Aillery (2016, Chapter 2.1.1) review various water institutions in the United States. The authority to allocate water is held by individual states, whose governments have tailored their allocation systems to local climate conditions. Most western states have adopted the prior appropriation system, which establishes water rights for specific withdrawal volumes based on historical uses, while the humid eastern states generally follow the riparian or reasonable use doctrine. However, the federal government maintains an influential role through financing of water infrastructure and water quality regulations authorized by the Clean Water Act, while tribal governments also have claims to the water resources within their territories. In the European Union, member states have jurisdiction over water within their boundaries but must harmonize their own laws to EU directives. The primary directives with purview over water resources include the wide-ranging Water Framework Directive of 2000, the Groundwater Directive, and the Drinking Water Directive (Villamayor-Tomas, 2016, Chapter 2.1.3; Drabik and Venus, 2016, Chapter 2.2.2; Reins, 2016, Chapter 2.2.5). On both sides of the Atlantic, local levels of government such as irrigation districts also play an important role and have legal authority over certain functions.

While institutions of the water sector have a primary role in managing water resources, institutions in agriculture, energy, and other sectors also are influential. In the United States, federal agricultural policies affect water use through conservation programs that subsidize certain irrigation practices and technologies, as well as through crop insurance programs that may change the incentives for growing different crops (Schaible and Aillery, 2016, Chapter 2.1.1). US energy policies such as the Renewable Fuels Standard, which established biofuels production targets, influenced water use and water quality indirectly by inducing an expansion of acreage planted to biofuels crops such as corn grain (Bergtold et al., 2016, Chapter 2.2.1). The EU also established biofuel production targets in the Renewable Energy Directive (Drabik and Venus, 2016, Chapter 2.2.2), which are partially met by imports and may have led to changes in water use and water quality both inside and outside the European Union. In both the United States and in Europe, electricity markets are regulated by several laws and agencies at various levels of government (Maïzi et al., 2016, Chapter 2.2.7; Webber, 2016, Chapter 2.2.6). Water plays an important role in electricity production, while irrigated agriculture is a large electricity consumer. In California (Schwabe et al., 2016, Chapter 2.1.2) and in Spain (Villamayor-Tomas, 2016, Chapter 2.1.3), water use for irrigation and irrigators' incomes were found to depend strongly on electricity prices.

Institutions in domains beyond agriculture and energy, including municipalities and ecological and wildlife protection agencies, also have come to have an influence on water use. In practice, water allocation is determined from the interactions in a cluster of institutions that span multiple domains and levels of government. The functioning of this institutional system is put to especially strong tests during heat waves and droughts, when sectors compete for water to avoid dire consequences. Events that play out during these events illustrate the linkages between food, energy, and water created by institutional rules. During droughts, farmers face the possibility of crop failure, which may impact incomes over a multiple year period in the case of perennial or orchard crops. Reduced streamflow lowers hydroelectric production and puts thermoelectric plants at risk of losing sufficient access to cooling water (Webber, 2016, Chapter 2.2.6), and low streamflow often coincides with peak electricity demands that are driven in part by increased pumping for irrigation. Meanwhile, reduced streamflow impairs various instream uses including ecological services, recreation, and navigation. These impairments can lead to yet another wave of impacts such as fuel shortages for electricity plants relying on coal delivered by river barge (Webber, 2016). Any ensuing electricity shortages or increased rates would have further negative impacts on irrigators.

A number of episodes have illustrated how existing institutions fail to absorb the full set of stresses during extreme events, often precipitating short-term emergency actions and sometimes becoming a catalyst for institutional reform. When drought struck the Iberian Peninsula in 2005, the Spanish government immediately enacted Royal Decree 10/2005 providing financial assistance to drought-stricken farmers (Ramos et al., 2005) but also began discussions that to led to later reforms (Villamayor-Tomas, 2016). During drought conditions in 2011, the State of Kansas enacted an emergency measure allowing farmers to exceed their annual groundwater pumping limits, and later forgave overpumping sanctions for farmers who opted into a new 5-year allocation system (Peterson and Hendricks, 2016). In California, several state and local measures have been enacted in response to the multiyear drought that began in 2010, including financial assistance to affected sectors and longer-term reforms such as groundwater management plans (Harter, 2015). Webber (2016, Chapter 2.2.6) notes examples in France and along the Upper Mississippi River in the United States, in which electric power plants were given special exemptions from their thermal loads on water bodies to continue producing electricity during heat waves. As Maïzi et al. (2016, Chapter 2.2.7) note, planning efforts for future institutional reforms in Europe include simulations of possible drought events, accounting for localized water supply, electricity demands, and patterns of power import and export.

## 3. THE SPECIAL CHALLENGE OF MANAGING WATER

One explanation for the observed institutional breakdowns is that water is a particularly complex resource to manage. This complexity is partly from the special characteristics of water that distinguish it from other natural resources (Peterson and Hendricks, 2016; Young and Loomis, 2014). This section reviews these characteristics and why they pose particular challenges for water institutions.

### 3.1 Multiple Uses

A basic complication for water institutions is that water resources provide a multitude of uses. Broadly, these can be classified as consumptive uses, nonconsumptive uses, and passive uses. Consumptive uses, also known as out-of-stream uses, are those where water is diverted from its natural flow patterns and becomes an input into some human-driven process, such as agricultural irrigation or energy production. Nonconsumptive, or instream, uses are activities such as recreation and navigation that rely on water in its natural flow channels but do not consume water. Many consumptive uses, such as thermoelectric cooling, return much of their diversions to the watercourse; in these cases the consumptive use is the difference between diversions and return flows. Additionally, water resources provide the function of a waste receptor, both from diffuse overland flows (nonpoint sources) as well as from discrete discharge locations (point sources). Both consumptive and nonconsumptive uses often have water quality impacts including thermal pollution.

Passive uses of water resources require neither consumption nor even direct contact with those resources. Economic value of passive uses (also called nonuse values) arise from humans' willingness to pay to preserve water resources to protect the option for using them in the future (option value), to keep them available to future generations to use (bequest value), or because of their intrinsic worth independent of any human use (existence value). The latter may be derived from spiritual or cultural values of water resources, which have been important throughout human history and may be traced to the understanding of humans in the earliest civilizations that water is the basis of life.

## 3.2 Transport Constraints and Situational Value

Water is a naturally mobile resource but is also classified as bulky—meaning a high transport cost per unit of economic value—when diverted from its natural flow channels. The inexorable mobility of water has important economic implications. Activities in one location may have dramatic impacts on both the availability and quality of water resources at many distant locations. These connections create a complex web of externalities. So intricate is this web, and so difficult is measurement at all its nodes, that it becomes prohibitive to attribute an externality observed at some point (eg, impaired water quality) to all its upstream sources. The bulkiness of water implies that consumptive uses of water necessarily occur either near source water bodies and/or downstream of them. Consequently, users at a given location have few, if any, substitutes for their primary source if its supplies become limited or its quality becomes impaired.

The transport limitations of water resources also contribute to the highly situational economic benefits of different uses and the costs of water impairments. A concentration of some contaminant may render water unsuitable for some uses while creating no impairment for other uses. High concentrations of nutrients, for instance, create health hazards for drinking or swimming and may induce toxic algal growth that is harmful to ecosystem as well as human health. At the same time, high nutrient concentrations are beneficial for irrigating plants. The value of additional water during wet periods may be zero or even negative in the case of flood conditions, but the same user at the same location may place a high value on the incremental unit during dry, peak demand periods. Water can be distinguished from many other resources by the dramatic difference in its economic value across the dimensions of use, time, and space.

## 3.3 Situational Property Rights

A more fundamental challenge is that even the ability to define property rights for the different uses of water depends on the situation. In general, property rights for any good or service fall into four categories depending on whether it is excludable and whether it is subtractive (Easter et al., 1997). A good is excludable if the owner is able to exclude others from using it. A good is subtractive if the amount the owner uses reduces the total amount available to others.

Subtractive, excludable goods are *private goods*, for which property rights can be enforced on physically separated packets of that good such that economic transactions can occur on the basis of unit prices. An example of a private good derived from water resources is piped municipal tap water sourced from an aquifer. The piped water one customer receives subtracts from the amount available to other customers, but after the water is received the customer in question can easily exclude others from using it. Hence prices for quantities of water consumed can be charged to individual users.

However, the excludability property occurred in the piped water example only after the technological interventions of piping and metering as well as institutional rules that make "water poaching" unlawful. In the absence of these interventions, for example, where individual property owners drill their own wells to extract groundwater, the aquifer would be subtractive but nonexcludable. Such *common-pool resources* are vulnerable to overexploitation because of the "tragedy of the commons": each user only pays for the cost of extracting the resource but not for the value of the diminished stocks that could have been used by others. In still other circumstances, water resources are nonsubtractive but excludable; this is the case of *club goods* in which those accessing the resource do not diminish its total supply but can exclude outsiders. An example of a water-related club good is a lake that has no public access point and is completely surrounded by privately owned homes. The lake provides nonconsumptive use values such as aesthetic and recreational benefits to only those homeowners. Finally, water resources are also sometimes *public goods*, which are nonsubtractive as well as nonexcludable and are usually provided through public financing. The various nonuse values of water resources are public goods.

With all of these complicating factors, designing institutions to manage water resources is an exceedingly complex challenge. Institutions must account for the fundamental property rights of water, which differ across a multiplicity of uses that a given water body might deliver at any given time. For example, with supporting technologies and infrastructure, it is possible to define private property rights for competing consumptive users of water (eg, agriculture, energy, and municipal/industrial), so that allocation of limited supplies to these sectors could be achieved by pricing or market trading. However, institutions such as pricing and trading are not automatically suited to allocate the nonprivate goods (nonconsumptive uses and nonuse values) obtained from the same water supplies, or to balance consumptive uses against other uses. In cases where several consumptive users withdraw from the same water body, institutions must ensure that the scarcity value of the resource is properly incorporated in allocation choices. Finally, there must be some mechanism to address the externalities that diffuse through the network of resource flows.

## 4. PROSPECTS FOR INSTITUTIONAL INNOVATION

Sustainable water resource management requires institutions that can adapt to evolving conditions. Such institutions would grant individual actors the flexibility to adjust to shocks and situational needs, but also would include constraints to ensure the resource has enough capacity for system resilience. The inherent properties of water make this a complex challenge. Where legacy institutions prevail, the challenge is often compounded by historical rigidities. The prior-appropriation system in the western

United States, for example, relies entirely on two-party water right transfers to adjust water allocations across uses as water demands change. However, transfers via negotiated transactions are quite rare because of high transaction costs and the paucity of willing sellers (Hansen, 2016, Chapter 1.1). Reforming the legal framework of water rights could be seen as mandating involuntary transfers, which would likely be challenged as an unconstitutional taking of property (Peck, 2015). Yet another barrier is that the institutions of the FEW nexus span not only different sectors but multiple layers of government. Institutional reform is hindered by the fact that changes at any one level may be inconsistent with prevailing rules at another. The transposition of the EU Water Framework Directive into the Spanish system, for example, was a difficult and lengthy process (Villamayor-Tomas, 2016, Chapter 2.1.3).

Nevertheless, institutional innovation is happening. One testbed for new institutional structures is in more water plentiful regions where increasing amounts of water are being allocated for supplemental irrigation. Within the 48 contiguous states in the United States, most of the irrigated cropland is in the 17 westernmost states, but the 31 eastern states are responsible for much of the recent growth (Schaible and Aillery, 2016, Chapter 2.1.1). Meanwhile, growing municipal water use and emerging water uses such as unconventional natural gas extraction add to the competition for water resources in humid regions. Here, institutions to allocate limited water supplies are not as well developed, but expanding water demands are an impetus for mechanisms to protect aquifers, ecosystems, and other instream uses. In many cases, the focus in these regions is squarely on extreme event management. For example, in the Susquehanna River Basin, new institutions have been created to permit withdrawals from rivers and streams for unconventional development of natural gas as those needs evolve, but withdrawal rights were defined to be interruptible and subject to passby flow requirements for resource protection during extreme events (Shank et al., 2016, Chapter 2.2.4).

New innovations are emerging in arid regions too. Huffaker (2016, Chapter 1.3) and Hansen (2016, Chapter 1.1) both discuss water markets in these regions as a way to introduce more adaptive flexibility and discuss institutional impediments to developing them. Hansen discusses a number of market innovations that allow for individuals to adjust during extreme events such as interruptible leases and water banks. Part 3 of this book provides a number of additional case studies in the United States and Europe revealing past experience and emerging innovations for water management.

## REFERENCES

Bergtold, J.S., Sant'Anna, A.C., Miller, N., Ramsey, S., Fewell, J.E., 2016. Water scarcity and conservation along the biofuel supply chain in the United States: from farm to refinery. In: Ziolkowska, J.R., Peterson, J.M. (Eds.), Competition for Water Resources: Experiences and Management Approaches in the US and Europe. Elsevier.

Drabik, D., Venus, T.J., 2016. Water use for biofuels in Europe. In: Ziolkowska, J.R., Peterson, J.M. (Eds.), Competition for Water Resources: Experiences and Management Approaches in the US and Europe. Elsevier.

Easter, K.W., Becker, N., Tsur, Y., 1997. Economic mechanisms for managing water resources: pricing, permits, and markets. In: Biswas, A. (Ed.), Water Resources: Environmental Planning, Management, and Development. McGraw-Hill, New York.

Hansen, K., 2016. Meeting the challenge of water scarcity in the western U.S. In: Ziolkowska, J.R., Peterson, J.M. (Eds.), Competition for Water Resources: Experiences and Management Approaches in the US and Europe. Elsevier.

Harter, T., 2015. California's agricultural regions gear up to actively manage groundwater use and protection. California Agriculture 69 (3), 193–201.

Huffaker, R., 2016. Institutional aspects and policy background of water scarcity problems in the U.S. In: Ziolkowska, J.R., Peterson, J.M. (Eds.), Competition for Water Resources: Experiences and Management Approaches in the US and Europe. Elsevier.

Maïzi, N., Bouckaert, S., Assoumou, E., 2016. Long term water and energy issues in European power systems. In: Ziolkowska, J.R., Peterson, J.M. (Eds.), Competition for Water Resources: Experiences and Management Approaches in the US and Europe. Elsevier.

Peck, J.C., 2015. Legal challenges in government imposition of water conservation: the Kansas example. Agronomy Journal 107 (4), 1–6.

Peterson, J.M., Hendricks, N., 2016. Economics of Water. Oxford Handbook of Water Politics and Policy. Oxford University Press (forthcoming).

Ramos, L., Munoz, P., McGrath, C., July 2005. The Iberian Peninsula Drought: Agricultural Situation. USDA Foreign Agricultural Service, Global Agriculture Information Network Report PO5018. http://apps.fas.usda.gov/gainfiles/200508/146130518.pdf.

Reins, L., 2016. Water use for unconventional gas production in the European Union. In: Ziolkowska, J.R., Peterson, J.M. (Eds.), Competition for Water Resources: Experiences and Management Approaches in the US and Europe. Elsevier.

Schaible, G., Aillery, M., 2016. Challenges for U.S. irrigated agriculture in the face of emerging demands and climate change. In: Ziolkowska, J.R., Peterson, J.M. (Eds.), Competition for Water Resources: Experiences and Management Approaches in the US and Europe. Elsevier.

Schwabe, K., Knapp, K., Luviano, I., 2016. The water-energy nexus and irrigated agriculture in the US: trends and analyses. In: Ziolkowska, J.R., Peterson, J.M. (Eds.), Competition for Water Resources: Experiences and Management Approaches in the US and Europe. Elsevier.

Shank, M.K., Balay, J.W., Richenderfer, J., 2016. Water use for unconventional natural gas development within the Susquehanna River Basin. In: Ziolkowska, J.R., Peterson, J.M. (Eds.), Competition for Water Resources: Experiences and Management Approaches in the US and Europe. Elsevier.

Villamayor-Tomas, S., 2016. The water energy nexus in Europe and Spain: an institutional analysis from the perspective of the Spanish irrigation sector. In: Ziolkowska, J.R., Peterson, J.M. (Eds.), Competition for Water Resources: Experiences and Management Approaches in the US and Europe. Elsevier.

Webber, M.E., 2016. Water for electricity generation in the United States. In: Ziolkowska, J.R., Peterson, J.M. (Eds.), Competition for Water Resources: Experiences and Management Approaches in the US and Europe. Elsevier.

Young, R.A., Loomis, J.B., 2014. Determining the Economic Value of Water: Concepts and Methods. Routledge.

# PART 3

# Water Management Approaches to Mitigate/Adapt to Scarce Water Resources - Case Studies From the United States and Europe

SUBPART 3.1

# Management Approaches in the United States

Chapter 3.1.1

# Willingness to Pay for Reclaimed Water: A Case Study for Oklahoma

**T.A. Boyer, M. Hopkins, J.Q. Moss**
*Oklahoma State University, Stillwater, OK, United States*

## 1. INTRODUCTION

Increased human consumption of water and intensified drought conditions in certain regions in the United States have increased demand for water while also decreasing surface water supplies and increasing the depletion of groundwater aquifers. During the drought of 2010–2014, Oklahoma endured precipitation deposition patterns less than those of the Dust Bowl in the 1930s (Parker, 2014). In January 2013 the US Department of Agriculture designated 76 of all 77 counties in Oklahoma as disaster areas because of drought and heat (USDA, 2013). In early January 2013 one of the primary drinking water sources for the Oklahoma City metro area, Lake Hefner, was at 17 ft below its maximum capacity, the lowest in the lake's 66-year history. Because of dangerously low lake levels, Oklahoma City released water from Canton Lake, to which it owned the water rights, to raise Lake Hefner (Layden, 2013). To combat future water supply shortages, Oklahoma City, like many other communities in Oklahoma, is considering utilizing alternative water sources, such as reclaimed water and additional pipelines for water transfers. This study estimates Oklahomans' willingness to pay (WTP) for reclaimed water to augment future municipal water supplies.

Reclaimed water use may be a sound strategy to offset shortfalls in fresh water supply. Reclaimed water, also referred to as recycled water, can provide economic and environmental benefits to communities through various applications that replace potable water, including groundwater recharge, landscape irrigation, agricultural irrigation, and potable water supply augmentation (USEPA, 2013). Municipalities can often utilize reclaimed water projects to meet demand at a lower cost, rather than adding additional capacity to the supply system, such as pipelines for long-distance water transfers. The economic feasibility of reclaimed water can be assessed with the benefit–cost analysis method. A project is feasible when the benefits of reclaimed water implementation (avoided cost of adding a pipeline and/or acquiring additional water rights) are greater than the costs of implementing a reclaimed water system. Reclaimed water use projects are often met with public

**Competition for Water Resources. http://dx.doi.org/10.1016/B978-0-12-803237-4.00015-X**

opposition because of the connection of reclaimed water to sewage, creating a negative public health perception (Hartley, 2006). Communities, such as San Diego, California, Tampa, Florida, and Queensland, Australia, had to shut down reclaimed water use projects because of failed community outreach and heavy pushback from citizens regarding the safety of public health (DeSena, 1999; Hurlimann and Dolnicar, 2010). This chapter will provide (1) an overview of reclaimed water use for municipal water supplies, (2) an overview of literature about public acceptance of reclaimed water, and (3) a case study on the WTP for reclaimed water use in Oklahoma.

## 2. DEFINITION OF RECLAIMED WATER

According to the Oklahoma Department of Environmental Quality (ODEQ), reclaimed water is wastewater that has gone through various treatment processes to meet specific water quality criteria with the intent of being utilized in a beneficial manner (ODEQ, 2012). Reclaimed water can be used both for nonpotable and potable purposes. Potable reclaimed water use projects can be further categorized as direct potable and indirect potable reuse. Direct potable reuse projects introduce reclaimed water directly into potable water distribution systems without prior storage. Indirect potable reuse projects place treated reclaimed water into an environmental buffer, such as a lake or reservoir, before entering potable water distribution systems. De facto (indirect) potable reuse has been in existence for centuries, because drinking water is often drawn from river systems where treated wastewater has been discharged from wastewater sewage treatment plants from cities upstream.

Treatment processes for reclaimed water vary according to state regulations and the intended end use, but the primary goal is to disinfect wastewater to ensure the protection of the public's health and the environment. Reclaimed water undergoes primary and secondary treatment, just as traditional wastewater, but must also undergo tertiary and disinfection treatment processes prior to use. Tertiary, or advanced, treatment technology and processes are constantly evolving and are used to remove additional organic, chemical, and biological contaminants from wastewater leftover after conventional primary and secondary treatments (Asano et al., 2007).

## 3. THE ROLE OF RECLAIMED WATER USE IN THE UNITED STATES

The increased global demand for water because of population growth has led many countries across the world to implement reclaimed water projects as an authentic management solution. The use of reclaimed water has become a prevalent water conservation strategy because it is considered a constant and reliable source. Reclaimed water projects have typically utilized this water for industrial purposes, landscaping, agriculture, environmental restoration, and augmentation of public water supply.

In the United States, agricultural irrigation represents the largest application of reclaimed water (Jimenez and Asano, 2008). If managed properly, reclaimed water can provide beneficial nutrients to crops and can cost less than potable water for irrigation. Farms in California, Florida, and Texas have used reclaimed water, referred to as effluent, for agricultural irrigation on crops such as cotton, wheat, grain sorghum, and corn, dating back to the early 1900s (National Research Council, 2012). The constant technological improvement of advanced treatment technologies has increased the use of reclaimed water for potable reuse projects in the United States within the last 50 years (National Research

Council, 2012). In 2010, about 0.1% of the municipal treated wastewater, 355 millions of gallons per day, was utilized for potable reuse projects in the United States (National Research Council, 2012). This small percentage may continue to grow as more communities face potable supply constraints caused by changing climate conditions and increased demand from population growth.

Reclaimed water use projects are not common in Oklahoma and are primarily concentrated in the Oklahoma City metro area for golf course irrigation and industrial cooling tower processes. Taghvaeian (2013) reported that in 2013 reclaimed water was used on only 18 farms in Oklahoma. He stated: "the total irrigated area in these farms was 2,205 acres, accounting for about half a percent of all irrigated acres in the state. This shows a decrease compared to 3775 acres that reported use of reclaimed water in 2008." In the 2012 Oklahoma Comprehensive Water Plan, the Oklahoma Legislature passed a conservation strategy known as the "Water for 2060 Act" establishing a "statewide goal of consuming no more fresh water in 2060 than consumed today" (OWRB, 2014). As part of this strategy, reclaimed water is expected to be utilized as an alternative water source to augment municipal water supplies. The ODEQ is currently developing regulations for direct and indirect potable water reuse, while nonpotable reclaimed water use regulations were established in 2012 (ODEQ, 2012). The establishment of official potable water reuse regulations will allow municipalities to integrate water reuse projects into their water management plans.

## 4. PUBLIC ACCEPTANCE AND WILLINGNESS TO PAY FOR RECLAIMED WATER

A substantial amount of research has emerged regarding public acceptance and WTP for reclaimed water use since the early 1970s (Bruvold and Ward, 1970; Bruvold, 1972; Sims and Baumann, 1974; Kasperson et al., 1974). One finding remains consistent across previous reclaimed water research: public acceptance of alternative water resources is greater for purposes that involve low human contact than for purposes of a closer personal contact, such as bathing (Dolnicar and Hurlimann, 2010). The public's support for environmental stewardship related to reclaimed water use reaches a tipping point when the end use becomes too personal, thus support decreases (Hurlimann and McKay, 2007).

The potential threat of disease and harmful bacteria transmission to humans and pets by contact with or ingestion of reclaimed water has been the most common public concern. However, research has reported no confirmed cases of human illness related to the use of reclaimed water systems and the potential risks are low (Rock et al., 2012). Advanced treatment technology can deliver reclaimed water that meets and exceeds national drinking water standards (Ormerod and Scott, 2012). Municipalities often make decisions regarding reclaimed water use projects based on public acceptance rather than scientific research because of negative opposition campaigns and public opposition. The general public may also struggle to rely on information given by government or academic institutions, since studies have shown that people are more apt to trust their own intuition than peer-reviewed scientific research (a concept also known as cognitive dissonance) (Rock et al., 2012). Without public education and communication outreach by municipalities to combat the negative public perception associated with reclaimed water use, such projects are unlikely to succeed.

Concerns regarding the health risks associated with reclaimed water stem from its association with "dirty" water or sewage. The majority of the public is unaware that in parts of the United States, drinking water contains a percentage of treated wastewater that was discharged from another municipality's treatment plant upstream and blended into surface water systems (Asano et al., 2007). The general public appears to have a strong cultural connection to water purity and a lack of education about the urban hydrologic cycle, directly relating to a negative perception of reclaimed water. The terms "Toilet to Tap" and "Sewage Beverage" began circulating in the mass media in the 1990s, during a time when a number of indirect potable reuse projects were proposed (Hartley, 2006). The negative media attention regarding reclaimed water projects further spread an "ick" or "yuck" factor associated with reclaimed water use, increasing public opposition to these projects.

The incorporation of public education, social media, and outreach campaigns about how the safety and necessity of a reclaimed water project for ensuring municipal water supply is important for municipal water utilities to successfully implement a new reclaimed water project. Educating the public about how reclaimed water systems work can lead to positive perceptions and acceptance of these projects (Dolnicar et al., 2011). Allowing the public to provide input and opinions can increase legitimacy of the project while decreasing potential opposition. Public involvement should be integrated from the early stages through the completion of reclaimed water use planning for every reclaimed water use project (Asano et al., 2007). The public should participate in the planning of reclaimed water use projects because they are directly affected as customers and because utilities are highly regulated as natural monopolies (Asano et al., 2007). In 1993 the City of San Diego, California, attempted to implement a potable reclaimed water use project into their municipal water system (DeSena, 1999). The City of San Diego failed to implement a public education and outreach program about the reclaimed water use project, and the project was eventually eradicated because of the public backlash and negative media attention, although interest has been renewed in 2015 because of historic drought conditions in California (Morin, 2015).

These cases exhibit that lack of public acceptance led to the failure of reclaimed water projects. However, there is evidence that the public is willing to pay for reclaimed water when supplies are constrained within the United States and worldwide. Little or no qualitative research on reclaimed water acceptability exists in the semiarid Midwest or in Oklahoma. Therefore this case study provides valuable insight regarding public WTP for reclaimed water in Oklahoma. Our results indicate that the public's WTP for reclaimed water varies by respondents' demographic and attitudinal characteristics. Furthermore, WTP differs when participants are presented with randomly assigned information regarding the bacterial and fecal coliform safety standards for treated reclaimed water versus untreated commonly contacted recreational waters from which potable sources are commonly drawn. These findings may be generally applicable for targeting groups for education about reclaimed water projects in other semiarid states in the United States and the country as a whole.

## 5. CONTINGENT VALUATION METHOD AND WILLINGNESS TO PAY

The contingent valuation method (CVM) estimates WTP for nonmarketed or hard to value services such as ecosystems. The CVM is categorized as a stated preference method (Grafton et al., 2004; Bakopoulou et al., 2010), and has been a prominent analytical tool for measuring use and nonuse values of water. Different formats can be used in a contingent valuation study, for example, an open-ended format or single bound format. The open-ended format asks survey respondents to state the maximum amount of money they would be willing to pay, while the single bound format asks respondents if they would be willing to pay a specific price (bid) (Genius et al., 2008). The single bound format more accurately imitates the market situations in which consumers pay a specific price for a commodity, and it is widely used in comparison to the open-ended format (Genius et al., 2008). For water resources the CVM has been valuable for analyzing use and nonuse values of water. Numerous studies have been conducted using the CVM to investigate the WTP for reclaimed water use (Chiueh et al., 2011; Bakopoulou et al., 2010; Tziakis et al., 2009; Genius et al., 2008). These studies primarily focus on using the CVM to evaluate the WTP for reclaimed water for agricultural purposes and the WTP for reclaimed water to augment public water supply for municipal and industrial uses. The overall message from this research indicates that WTP for reclaimed water use is higher during times of drought, when priced lower than potable water, and when utilized as an alternative management strategy to augment current water supplies.

The objective of this study was to investigate Oklahoman's hypothetical WTP for reclaimed water as a hedge against drought-driven shortages. Implementation of reclaimed water projects is costly, requires cost recovery, and involves establishing new pipelines or renovating neglected infrastructure. Therefore this study assumes these projects will increase costs to customers and thus uses a payment instrument that is stated as an additional cost to the respondents per 1000-gallon water fee charged by the respondent's water utilities. This study also assumes that a volumetric or block-rate pricing scheme is charged in commonly used 1000-gallon increments in Oklahoma. When the CVM method is used, it is important to explain clearly what is being valued and provide respondents with realistic price choices. The CVM method can be beneficial in providing hypothetical prices for reclaimed water use and projects as determined by survey respondents who may otherwise be unknown to water managers and municipalities.

## 6. SURVEY DESIGN

Our 33-question, qualitative, internet survey was designed with Qualtrics software and sent during October 25 to November 1, 2014 to approximately 486 Oklahomans recruited by Survey Sampling International. Human subject research approval was obtained on October 15, 2014 (OSU IRB # AG-14-43). Two versions of the WTP question within the survey were created to test the hypothesis that the inclusion of water quality data affected WTP for reclaimed water use. The survey began with definitions of "reclaimed water use system" and "reclaimed water" sources from the ODEQ.

One version of the survey included a table with precise water quality data about the standards for acceptable maximum common fecal coliform and *Escherichia coli* contaminant levels for recreational surface waters (from which municipal water is usually obtained and treated in Oklahoma when groundwater is not present) and reclaimed water categories in Oklahoma. In the scientific water quality treatment question, the data showed a table that compared the bacterial standards for fecal coliform and *E. coli* in recreational surface waters (126 cfu/100 mL) and posttreatment Category 2 reclaimed water (23 cfu/100 mL) (USEPA, 2003; ODEQ, 2012). Thus Category 2 reclaimed water would be returned after treatment to surface water cleaner than the standard required in a recreational water body. The second

EPA (2008a) reported that there were 549 reported impairments because of *E. coli*, enterococci, or fecal coliform for water bodies in Oklahoma in 2006. Only 10 of these impairments occurred in the IRW (one water body was considered impaired for both enterococci and fecal coliform). Oklahoma's 2008 report was approved by EPA on October 22, 2008 (ODEQ 2008a). The 2008 report lists twelve water bodies in the IRW impaired for PBCR as the result of bacteria. Of those twelve, a source of the bacteria for three water bodies is indicated as Confined Animal Feeding Operations but not specifying poultry (ODEQ 2008a). Possible other sources for all three water bodies include on-site treatment systems (septic systems), grazing in riparian zones, rangeland grazing, wildlife other than waterfowl, and "unknown" (State of Oklahoma VS. Tyson Foods Inc., Case No. 4:05-CV-329-GKF (SAJ)). One of the citizen concerns raised by reuse water is for water quality and public health.

For background, we explain Oklahoma regulations for common bacterial contaminants. The presence of coliform bacteria, including *E. coli* and fecal coliform in lakes, rivers, and streams is a signal of water contamination by human or animal feces. The Environmental Protection Agency maximum for E. coli in recreational water samples is exceeded when above 126 colony-forming units (cfu) per 100 mL of water This means an approximate risk of 8 per 1000 people contracting stomach illness, vomiting or diarrhea as result of contact. In 2012, 77 Oklahoma recreational water bodies went over the *E. coli* bacteria limit By contrast, for water to be considered a Category 2 for reclaimed water in Oklahoma, the Department of Environment Quality states there can be no detectable fecal coliform organisms in four of seven samples, with a maximum positive fecal coliform organism sample of 23 cfu/100mL in three of the samples. Reclaimed water at this level may be used for outdoor irrigation purposes. Water systems following the regulated standards will typically not have problems with treated reuse water.

A summary table of the standard is provided below:

| Oklahoma Standards | Recreational Water | Non-Potable Reclaimed Water, Category 2 |
|---|---|---|
| | Max 126 cfu/100ml | Max 23 cfu/100 ml |

**FIGURE 1** Water quality data for surface versus reclaimed waters in Oklahoma.

version of the survey included the contingent valuation question without the comparison table on relative water quality. This randomly assigned treatment was designed to test the hypothesis that respondents who viewed the water quality data would be more likely to support more widespread reclaimed water use versus those respondents who did not view the water quality data because the bacterial standard of fecal coliform for reclaimed water is much more stringent than allowed in surface waters (Fig. 1).

The WTP portions of our survey asked respondents if they would pay a fee per 1000 gallons for reclaimed water use in addition to what they currently pay per 1000 gallons for municipal treated surface water that did not originate from reclaimed sources (Fig. 2). The bid amounts in the WTP question varied and were randomly assigned in each questionnaire at six values between \$0.35 and \$3.35 (\$0.35, \$0.85, \$1.35, \$1.85, \$2.35, \$2.85, and \$3.35) as an additional charge per 1000 gallons delivered. The survey also included attitudinal questions pertaining to reclaimed water use. For example, the survey asked respondents to identify whether they believed reclaimed water use was hazardous using a 5-level Likert scale. In addition, two questions in the survey were related to reclaimed water policy in Oklahoma—asking respondents whether they would support policy that encouraged more widespread reclaimed water use and more local reclaimed water use in their community. These questions were included in the survey to assess respondents' acceptance of reclaimed water use when monetary costs were not assigned to the question. We also included a survey question asking respondents to identify whether they believed drought conditions in their community would increase over the next 25 years, using a 5-level Likert scale. This question was used to assess respondents' belief that future drought conditions would worsen and persist.

In Oklahoma, water prices are increasing due to old infrastructure, urban growth, and the need for water conservation. New pricing rates are often based on usage. Currently on average water costs $4.90 per 1000 gallons across Oklahoma ($2.73 per 1000 gallons in OKC (Oct. 9), $3.18 per 1000 gallons in Tulsa). The average household uses 7000 gallons per month in the summer.

Would you be willing to pay an extra charge of $X per 1000 gallons to increase reclaimed water use, and thus maintain a sustainable water supply?

| | |
|---|---|
| O | Yes |
| O | No |

FIGURE 2 Willingness to pay for reclaimed water use in Oklahoma question.

Several sociodemographic and behavioral variables were also included in our survey, such as: gender, education, household income, employment status, whether the respondent rents or owns his or her home, and whether the respondent recycles consumer waste at home. Summary statistics and variable descriptions for the models are presented in Table 1. Based on the literature regarding WTP for reclaimed water (Tsagarakis et al., 2007; Dolnicar and Shafer, 2009; Burfurd et al., 2012; Rock et al., 2012), we hypothesized that respondents who were males, owned their homes, had an advanced degree, were employed, had an annual income over $80,000, and supported reclaimed water use policy in Oklahoma would choose to pay an additional fee per 1000 gallons of water for reclaimed water.

## 7. PROBIT MODELS

We utilized a probit model maximum likelihood estimation to obtain the mean WTP for reclaimed water in Statistical Analysis System (SAS) 9.3. The probit model uses the Maximum Likelihood Estimate (MLE) to provide a set of values for the model's parameters that maximize the likelihood/probability function. The survey data was modeled using MLE, with the likelihood of accepting the bid for paying an additional fee for reclaimed water estimated as a binary dichotomous choice (1 if accepted and 0 if not). We assumed a type I extreme value distribution for the error terms, where the following expression results in the probability of a respondent responding "yes" (Grafton et al., 2004):

$$\Pr(\text{Yes}) = 1 - \frac{e^{(-\alpha+\beta B)}}{1 + e^{(-\alpha+\beta B)}} \tag{1}$$

If we assume a normal distribution, the probability that a respondent says yes to the reclaimed water bid question becomes:

$$\Pr(\text{Yes}) = 1 - \Phi(-\alpha + \beta B) \tag{2}$$

**TABLE 1 Reclaimed Water Survey—Descriptive Statistics, Variable Descriptions, and Expected Effect on Willingness to Pay (WTP)**

| Variable | Description[a] | Measure | Expected Effect on WTP | Mean | Std. Dev. | Min. | Max. | Frequency |
|---|---|---|---|---|---|---|---|---|
| *Quality* | Water quality data for Oklahoma surface versus RH2O | 1 if provided data, 0 otherwise | + | 0.45 | 0.50 | 0 | 1 | 581 |
| *Acceptr* | WTP for RH2O | 1 if accept bid, 0 otherwise | + | 0.58 | 0.49 | 0 | 1 | 491 |
| *Hazard* | Considered RH2O hazardous to humans and animals | 1 definitely yes to 5 definitely no | − | 3 | 1.05 | 1 | 5 | 503 |
| *BidH2O* | Bid amount for WTP for RH2O | | −/+ | 1.88 | 0.98 | 0.35 | 3.35 | 581 |
| *Rent* | Rent or own their home | 1 if renting, 0 if otherwise | − | 0.44 | 0.5 | 0 | 1 | 494 |
| *Drought* | Drought will increase in region over the next 25 years | 1 if definitely yes to 5 if definitely no | + | 2.66 | 0.98 | 1 | 5 | 491 |
| *Recuse* | Support RH2O use in their local municipal water system | 1 if support more RH2O use, 0 otherwise | + | 0.79 | 0.41 | 0 | 1 | 487 |
| *RegPol* | Support regulations to promote RH2O use in Oklahoma | 1 if support regulations, 0 otherwise | + | 0.8 | 0.4 | 0 | 1 | 479 |

| *Age* | Age in years | | −/+ | 41.91 | 16.07 | 18 | 99 | 488 |
|---|---|---|---|---|---|---|---|---|
| *Female* | Gender | 1 if female, 0 otherwise | − | 0.71 | 0.45 | 0 | 1 | 487 |
| *Apt* | Live in an apartment | 1 if live in apartment, 0 otherwise | − | 0.16 | 0.37 | 0 | 1 | 486 |
| *Twenty* | Household annual income of $21,000–40,000 | 1 if annual income is $21,000–40,000, 0 otherwise | − | 0.3 | 0.46 | 0 | 1 | 481 |
| *Forty* | Household annual income of $41,000–60,000 | 1 if annual income is $41,000–60,000, 0 otherwise | − | 0.2 | 0.4 | 0 | 1 | 481 |
| *Sixty* | Household annual income of $61,000–80,000 | 1 if annual income is $61,000–80,000, 0 otherwise | + | 0.1 | 0.30 | 0 | 1 | 481 |
| *Eighty* | Household annual income of $81,000–100,000 | 1 if annual income is $81,000–100,000, 0 otherwise | + | 0.06 | 0.23 | 0 | 1 | 481 |
| *Hundred* | Household annual income of over $100,000 | 1 if annual income is $100,000+, 0 otherwise | + | 0.11 | 0.31 | 0 | 1 | 481 |
| *HS* | Have high school degree | 1 if have high school degree, 0 otherwise | − | 0.23 | 0.42 | 0 | 1 | 486 |
| *BS* | Have a bachelor's degree or higher | 1 if have bachelor's degree or higher, 0 otherwise | + | 0.32 | 0.47 | 0 | 1 | 486 |
| *Unemploy* | Unemployed | 1 if unemployed, 0 otherwise | − | 0.10 | 0.3 | 0 | 1 | 481 |
| *Employed* | Employed | 1 is employed, 0 otherwise | + | 0.50 | 0.5 | 0 | 1 | 481 |

*Continued*

**TABLE 1** Reclaimed Water Survey—Descriptive Statistics, Variable Descriptions, and Expected Effect on Willingness to Pay (WTP)—cont'd

| Variable | Description[a] | Measure | Expected Effect on WTP | Mean | Std. Dev. | Min. | Max. | Frequency |
|---|---|---|---|---|---|---|---|---|
| *Home2* | Live in a house | 1 is live in a house, 0 otherwise | + | 0.7 | 0.46 | 0 | 1 | 486 |
| *Recycl2* | Recycle at home | 1 if recycle, 0 otherwise | + | 0.57 | 0.50 | 0 | 1 | 483 |

*[a]Variables measured on a Likert scale where 1, Definitely Yes; 5, Definitely No; RH2O, reclaimed water.*

The expected value for compensating variation is:

$$C = \int_{-\infty}^{0} F(B)dB + \int_{0}^{\infty} (1 - F(B))dB \tag{3}$$

The following probit model was estimated to empirically model a respondent's WTP for reclaimed water use in Oklahoma:

$$\begin{aligned} WTP_i = {} & \beta_0 + \beta_1 H20Bid_i + \beta_2 Quality_i + \beta_3 Gender_i + \beta_4 Twenty_i + \beta_5 Forty_i \\ & + B_6 Sixty_i + \beta_7 Eighty_i + \beta_8 Hundred_i + \beta_9 Rent_i + \beta_{10} Recycle_i \\ & + \beta_{11} RegPol_i + \beta_{12} Drought_i + \beta_{13} Hazard_i + \varepsilon_i \end{aligned} \tag{4}$$

Here, $Bid_i$ represents the amount that the respondent was asked to pay. The following variables are represented in our model as binaries: $Quality_i$ is 1 if the respondent received the water quality data, or 0 if not; $Gender_i$ represents whether the respondent identified as being a female (1) or male (0); $Twenty_i$ represents whether the respondent indicated having an annual income of \$20,000−40,000 (1) or not (0); $Forty_i$ represents whether the respondent indicated having an annual income of \$40,001−60,000 (1) or not (0); $Sixty_i$ represents whether the respondent indicated having an annual income of \$60,001−80,000 (1) or not (0); $Eighty_i$ represents whether the respondent indicated having an annual income of \$80,001−100,000 (1) or not (0); $Hundred_i$ represents whether the respondent indicated having an annual income of more than \$100,000 (1) or not (0); $Rent_i$ represents whether the respondent indicated that he or she rents his or her home (1) or not (0); $RegPol_i$ represents whether the respondent supports reclaimed water use policy and regulation in Oklahoma (1) or not (0); and $Hazard_i$ represents whether the respondent believes reclaimed water use is hazardous (1) or not (0). $Drought_i$ represents the respondent's perception of drought increase over the next 25 years on a 5-level Likert scale with 1 as "definitely yes" to 5 as "definitely no." Finally, $\varepsilon_i$ is the error term.

The mean WTP estimates were calculated using a "grand constant" (Giraud et al., 1999), which is determined by multiplying the variable coefficients by their respective mean then summing over all coefficients (without the bid) and dividing by the bid term (Loureiro and Umberger, 2003).

$$\text{Mean } \widehat{\text{WTP}} = \frac{\widehat{\beta_0} + \sum_{j=2}^{12} \left( \widehat{\beta_j} \overline{x_i} \right)}{\widehat{\beta_1}} \tag{5}$$

## 8. RESULTS AND DISCUSSION

The coefficients for the contingent valuation probit models with and without attitudinal variables are located in Table 2. The models are estimated using the maximum likelihood and therefore the variable coefficients contribute to the likelihood of a "yes" response. In both the baseline and attitudinal variables models, the reclaimed water bid coefficients were significant at the 1% and 0.5% levels, respectively, and were also negative, indicating that the probability of a "yes" response for WTP decreases as the bid amount increases. According to demand theory, the higher the amount requested of the

**TABLE 2 Probit Model Results of WTP for Reclaimed Water in Oklahoma**

| | Baseline Model | | | Attitudinal Variable Model | | |
|---|---|---|---|---|---|---|
| **Variable** | **Coefficient** | **Std. Error** | **Pr > ChiSq** | **Coefficient** | **Std. Error** | **Pr > ChiSq** |
| *Intercept* | 0.1085 | 0.3795 | 0.7749 | −0.8263 | 0.4965 | 0.0961[a] |
| *BidH2O* | −0.1132 | 0.0624 | 0.0700[a] | −0.2059 | 0.0690 | 0.0028[b] |
| *Quality* | −0.2132 | 0.1257 | 0.0898[a] | −0.2205 | 0.1380 | 0.1100 |
| *Age* | −0.0019 | 0.0045 | 0.6651 | −0.0047 | 0.0049 | 0.3390 |
| *Female* | −0.3040 | 0.1409 | 0.0310[b] | −0.3351 | 0.1565 | 0.0323[b] |
| *Employed* | −0.1364 | 0.1458 | 0.3494 | −0.0420 | 0.1569 | 0.7887 |
| *Unemploy* | −0.208 | 0.2288 | 0.3633 | −0.0935 | 0.2517 | 0.7102 |
| *Home2* | 0.2329 | 0.1906 | 0.2218 | 0.1150 | 0.2110 | 0.5857 |
| *Apt* | 0.2334 | 0.2321 | 0.3146 | 0.2422 | 0.2572 | 0.3464 |
| *Twenty* | 0.4018 | 0.1749 | 0.0216[b] | 0.3114 | 0.1917 | 0.1042 |
| *Forty* | 0.7465 | 0.2054 | 0.0003[c] | 0.8168 | 0.2251 | 0.0003[c] |
| *Sixty* | 0.7324 | 0.248 | 0.0031[b] | 0.6033 | 0.2697 | 0.0253[b] |
| *Eighty* | 1.2352 | 0.3268 | 0.0002[c] | 1.1389 | 0.3583 | 0.0015[b] |
| *Hundred* | 1.0589 | 0.2708 | <0.0001[c] | 0.9118 | 0.2923 | 0.0018[b] |
| *HS* | −0.1352 | 0.1588 | 0.3946 | −0.0824 | 0.1743 | 0.6363 |
| *BS* | −0.1548 | 0.1580 | 0.3270 | −0.2756 | 0.1736 | 0.1124 |
| *Rent* | 0.3518 | 0.1598 | 0.0277[b] | 0.3428 | 0.1736 | 0.0483[b] |
| *Recycl2* | | | | 0.3697 | 0.1354 | 0.0063[b] |
| *RegPol* | | | | 0.8375 | 0.2299 | 0.0003[c] |
| *Recuse* | | | | 0.1520 | 0.2409 | 0.5281 |
| *Drought* | | | | −0.1282 | 0.0704 | 0.0685[a] |
| *Hazard* | | | | 0.2353 | 0.0733 | 0.0013[b] |
| *Pseudo* $R^2$ | 629.652 | | | 604.576 | | |
| *Log Likelihood* | 44.5149 | | | 113.5604 | | |
| *Sample Size* | 463 | | | 446 | | |

[a]*Represents significance at 10%.*
[b]*Represents significance at 5%.*
[c]*Represents significance at 1%.*

respondent to pay, the lower the probability that the respondent would be willing to pay the amount. The water quality information coefficient was statistically significant at the 0.01% level and negative in the baseline model, indicating that as respondents saw the water quality information at the beginning of the survey, the probability of a "yes" response decreased. This finding could be related to the idea of cognitive dissonance, in which the respondent chose not to believe the scientific data, but rather rely on his or her own judgment regarding the quality and safety of reclaimed water. An alternative theory is that the respondent chose not to support reclaimed water use because the water quality information may have been too complicated to comprehend. Finally, the female coefficient was statistically significant and negative in the baseline and attitudinal variables models, meaning that the probability of a "yes" response declines if the respondent is female. This finding supports previous research that males are more likely to pay for reclaimed water use.

The coefficients for the five income levels included in both models were statistically significant and positive, which negates our hypothesis that income levels of $80,000 and over will contribute to a higher probability of "yes" responses. This variable's significance also suggests that Oklahomans from all income brackets are in favor of reclaimed water use compared to the lowest income bracket of less than an annual income of $20,000 (not just those with a higher amount of discretionary funds). Municipalities that are considering implementing reclaimed water use in their communities, but are concerned about the financial impact, may find these income data encouraging.

The coefficients for the rent variable in the baseline and attitudinal variables models were both significant and positive at the 0.05 level, indicating a higher probability of a "yes" response from renters to the contingent bid. Homeowners were hypothesized to be more willing to pay for reclaimed water use in Oklahoma to ensure water supply to the area, but our findings did not confirm this hypothesis. In fact, we found that renters had a higher WTP for reclaimed water, which suggests that reclaimed water use can appeal to a larger market of citizens than we previously hypothesized. Also we theorize that homeowners may be concerned that reclaimed water use might lower their residential property value, while renters may not share this concern. As for dwelling type, coefficients for neither apartments nor homes were significant in either model. This suggests that WTP does not differ by dwelling type, only by ownership. In addition, age did not prove to be statistically significant in either model. Education and employment status also proved to be statistically insignificant in either of the model's results, contrary to our hypothesis. These findings suggest that employed and more educated citizens do not place a higher value on reclaimed water than those with less than a high school education. Again this bolsters the need for public education and community outreach by municipal utility managers in an effort to gain support for reclaimed water use in their communities.

Four behavioral and attitudinal variable coefficients were statistically significant in the attitudinal variables model, including: recycle, reclaimed water policy, drought increase, and safety hazard of reclaimed water. The coefficient for recycling at home was significant and positive, indicating that the probability of a "yes" response increases if the respondent recycles consumer waste such as paper and bottles at home. The reclaimed water policy coefficient was positive and significant at the 0.001% level, meaning that the probability of a "yes" response increases if the respondent supports reclaimed water policy and regulation in Oklahoma. This finding is notable because citizen's support for reclaimed water policy and regulation is an easily measurable

**TABLE 3** Reclaimed Water Survey Mean Willingness to Pay Estimates (2014, $USD)

| Model | Mean Willingness to Pay Estimate (In Addition to Current Price per 1000 gallons) |
|---|---|
| Baseline | $4.20 |
| Attitudinal variables | $3.47 |

variable for municipalities to target for education and to assess in surveys. In regards to the drought coefficient the sign was negative and significant, indicating that if respondents did not believe drought would increase in the next 25 years in their region, the probability of a "yes" response decreased. The coefficient for the hazard variable was positive and significant. This result indicated that if respondents believed that reclaimed water use is not hazardous to humans or animals, the probability of a "yes" response increased. This finding is notable because if customers can be educated to understand the low risk of hazards using reclaimed water, they will be more likely to support its use and pay for its provision. The difficulty, given the results for the water quality standard treatment, is finding a way to accurately and scientifically communicate the low risk of using reclaimed water to customers.

The mean WTP estimates for the baseline and attitudinal variables models are displayed in Table 3. The baseline model mean WTP estimate per 1000 gallons reclaimed water in addition to the current fee is $4.20 (US$, 2014). The attitudinal variables model mean WTP estimate is $3.47 per 1000 gallons. These WTP estimates are higher than the bid amounts provided in both treatments of our survey. This indicates that many respondents were willing to pay for reclaimed water use even with bid amounts as high as $3.47 in addition to the rate charged per 1000 gallons of water usage. In the case for Oklahoma City, average household water consumption in the summer of 2012 was 10,100 gallons (Boyer et al., 2015). This amount multiplied by the average price of water per 1000 gallons in Oklahoma ($4.90) equates to a $49.49 increase in an average monthly household bill. Adding the mean WTP estimate of $4.20 and $3.47 per 1000 gallons for reclaimed water use to the average monthly household water bill brings the monthly total to $91.91 and $84.54 per household. However, because the CVM is a hypothetical survey method, the actual WTP amount may be inflated. Future research should examine the public's preferences regarding various water sources (eg, selection from a portfolio of water choices). While price elasticity of demand for water in Oklahoma is low (ie, price does not significantly affect demand), higher prices for reclaimed water may encourage water conservation efforts. In fact, Oklahoma City Water Utilities Trust adopted an inclining block two-tier rate structure. This rate structure applies a second-tier rate for customers that exceed the first 10,000 gallons in a billing month, charging a higher rate as more water is used. This pricing structure was implemented to encourage water conservation, to meet increasing demand, and to help pay for a new pipeline project for transportation of water from Southeast Oklahoma to Oklahoma City reservoirs (City of Oklahoma City, 2014).

## 9. CONCLUSIONS

This case study estimated Oklahoman's WTP for reclaimed water use given various behavioral and attitudinal variables, as well as the inclusion of water quality information for surface and reclaimed waters. Our survey was conducted via the internet with respondents from communities throughout Oklahoma. These results indicate that respondents are generally supportive of reclaimed water use and are willing to pay an additional fee per 1000 gallons for more widespread use in Oklahoma to ensure future supplies given cyclical drought.

In particular, the results suggest that Oklahomans who are males, have incomes of \$20,000–100,000+, rent their homes, support reclaimed water use policy, recycle at home, and believe reclaimed water is not hazardous are more likely to support reclaimed water use in Oklahoma. Respondents' perceptions on drought indicated that when they believe the potential for drought over the next 25 years is decreased, they were less willing to pay for reclaimed water. Surprisingly, education, employment, and age did not significantly affect WTP for reclaimed water. However, the findings do indicate that homeowners may be more concerned about a long-term stigma on house value caused by reclaimed water use. Overall, our findings contribute to the wealth of literature on reclaimed water use by illustrating how the combination of educational information and water quality data with demographic, attitudinal, and behavioral variables impact WTP for reclaimed water.

WTP is an integral part of reclaimed water use projects, as the provision of infrastructure to implement reuse may be substantial, albeit potentially more cost effective than securing additional surface or groundwater supplies or desalinization/treatment of nonpotable sources. Opposition to reclaimed water projects often stems from a lack of public education and misconceptions regarding public health. For this reason we suggest municipalities incorporate the public when planning reclaimed water use projects. However, the inclusion of scientific water quality standards in our survey shows that respondents were less willing to pay for reclaimed water compared to those that did not get this information. We theorize that this information may have induced more participants to reject the bid by reminding them of the "ick" factor or *E. coli* risk, however low, or that the data presentation was simply confusing. Therefore we recommend that public education and information regarding the quality and safety of reclaimed water use be communicated simply and clearly to the public. Future research should test the mode of delivery and format of educational materials to gauge which messages best decrease misconceptions regarding reclaimed water use and the cyclical nature of droughts in Oklahoma. Our results, and future research in this realm, may aid water managers and city officials in Oklahoma and elsewhere in targeting groups and identifying educational gaps on the part of utility customers. As the current water crisis over lead contamination in water in Flint, Michigan's, supply systems shows, public trust may be difficult or impossible to regain after utility customers are not given full information and consideration in the management of their water supplies, and costs to fix problems may be prohibitive (Bosman et al., 2016).

## ACKNOWLEDGMENTS

This work was supported by the USDA National Institute of Food and Agriculture, Hatch project # OKL028252.

## REFERENCES


Asano, T., Burton, F.L., Leverenz, H.L., Tsuchihashi, R., Tchohanoglous, G., 2007. Reclaimed Water Use: Issues, Technologies, and Applications. Metcalf & Eddy, AECOM. McGraw-Hill, New York.

Bakopoulou, S., Polyzos, S., Kungolos, A., 2010. Investigation of farmers' willingness to pay for using recycled water for irrigation in Thessaly region, Greece. Desalination 250, 329–334.

Bosman, J., Davey, M., Smith, M., January 20, 2016. As Water Problems Grew, Officials Belittled Complaints from Flint. New York Times.

Boyer, T., Kanza, P., Ghimire, M., Moss, J., 2015. Household adoption of water conservation and resilience under drought: the case of Oklahoma city. Water Economics and Policy 1, 1–21.

Bruvold, W.H., Ward, P.C., 1970. Public attitudes toward uses of reclaimed wastewater. Water & Sewage Water 117, 120–122.

Bruvold, W.H., 1972. Public Attitudes toward Reuse of Reclaimed Water. Water Resources Center, University of California.

Burfurd, I., Gangadharan, L., Nemes, V., 2012. Stars and standards: energy efficiency in rental markets. Journal of Environmental Economics and Management 64, 153–168.

Chiueh, Y., Chen, H.H., Deng, C.F., 2011. The willingness to pay of industrial water users for reclaimed water in Taiwan. In: Uhlig, U. (Ed.), Current Issues of Water Management. InTech, Shanghai, China, pp. 261–270.

City of Oklahoma City, 2014. Council Agenda Item No. VIII. H. The City of Oklahoma City.

DeSena, M., 1999. Public opposition sidelines indirect potable reuse projects. Water & Environment Technology 11, 16–18.

Dolnicar, S., Hurlimann, A., 2010. Water alternatives-who and what influences public acceptance. Journal of Public Affairs 11, 49–59.

Dolnicar, S., Schafer, A.I., 2009. Desalinated versus recycled water – public perceptions and profiles of the acceptors. Journal of Environmental Management 90, 888–900.

Dolnicar, S., Hurlimann, A., Grun, B., 2011. What affects public acceptance of recycled and desalinated water? International Water Association. Water Research Journal 45, 933–943.

Genuis, M., Hatzaki, E., Kouromichelaki, M., Kouvakis, G., Nikiforaki, S., Tsagarakis, K.P., 2008. Evaluating consumers' willingness to pay for improved potable water quality and quantity. Water Resources Management 22, 1825–1834.

Giraud, K.L., Loomis, J.B., Johnson, R.L., 1999. Internal and external scope in willingness-to-pay estimates for threatened and endangered wildlife. Journal of Environmental Management 56, 221–229.

Grafton, R.Q., Adamowicz, W., Dupont, D., Nelson, H., Hill, R.J., Renzetti, S., 2004. The Economics of the Environment and Natural Resources. Blackwell Publishing Ltd., Malden, MA.

Hartley, T.W., 2006. Public perception and participation in reclaimed water use. Desalination 187, 115–126.

Hurlimann, A., Dolnicar, S., 2010. When public opposition defeats alternative water projects – the case of Toowoomba, Australia. Water Research 44, 287–297.

Hurlimann, A., McKay, J., 2007. Urban Australians using recycled water for domestic use – an evaluations of the attributes of price, saltiness, colour, and odour using conjoint analysis. Journal Environmental Management 83, 93–104.

Jimenez, B., Asano, T., 2008. Water reclamation and reuse around the world. In: Jimenez, B., Asano, T. (Eds.), Water Reuse: An International Survey of Current Practice, Issues, and Needs. IWA Publishing, London, pp. 3–26.

Kasperson, R.E., Baumann, B., Dworkin, D., McCauley, D., Reynolds, J., Sims, J., 1974. Community Adoption of Reclaimed Water Use System in the United States. US Department of the Interior. Office of Water Resources Research, Washington, DC.

Layden, L., 2013. Improving Drought Conditions Haven't Helped Canton Lake Recover from OKC Withdrawal. StateImpact Oklahoma. Oklahoma Public Media Exchange.

Loureiro, M.L., Umberger, W.J., 2003. Estimating consumer willingness-to-pay for country-of-origin labels. Journal of Agricultural Resource Economics 28, 287–301.

Morin, M., 2015. Turning Sewage into Drinking Water Gains Appeal as Drought Lingers. Los Angeles Times.

National Research Council, 2012. Water Reuse: Potential for Expanding the Nation's Water Supply through Reuse of Municipal Wastewater. The National Academies Press, Washington DC.

ODEQ, 2012. Title 252. Chapter 627: Operation and Maintenance of Reclaimed Water Use Systems. ODEQ, Oklahoma City, OK.

Ormerod, K.J., Scott, C.A., 2012. Drinking wastewater: public trust in potable reuse. Science, Technology, & Human Values 38, 351–373.

OWRB, 2014. Water for 2060 Status Report. OWRB, Oklahoma City, OK.

Parker, L., 2014. Parched: A New Dust Bowl Forms in the Heartland. National Geographic News.

Rock, C., Solop, F.I., Gerrity, D., 2012. Survey of statewide public perceptions regarding reclaimed water use in Arizona. Journal of Water Supply: Research and Technology-AQUA 61, 506–517.

Sims, J.H., Baumann, D., 1974. Renovated waste water: the questions of public acceptance. Water Resources Research 10, 659–665.

Taghevaeian, S., 2013. Irrigated Agriculture in Oklahoma. Oklahoma Cooperative Extenstion Service Factsheet. BAE-1530. Oklahoma State University, Stillwater, OK.

Tsagarakis, K.P., Mellon, R.C., Stamaki, E., Kounalaki, E., 2007. Identification of recycled water with an empirically derived symbol increases its probability of use. Environmental Science & Technology 41, 6901–6908.

Tziakis, I., Pachiadakis, I., Moraitakis, M., Xideas, K., Theologis, G., Tsagarakis, K.P., 2009. Valuing benefits from wastewater treatment and reuse using contingent valuation methodology. Desalination 237, 117–125.

USDA, 2013. USDA Designates 597 Counties in 2013 as Disaster Areas Due to Drought. USDA, Washington, DC. Office of Communications. Release No. 0002.13.

USEPA, 2003. Bacterial Water Quality Standards for Recreational Waters. USEPA, Washington DC. Office of Water (4305T).

USEPA, 2013. Water Recycling and Reuse: The Environmental Benefits. USEPA, Washington, DC. Water Division Region IX-USEPA-909-F-98-001.

Chapter 3.1.2

# Conjunctive Water Management in Hydraulically Connected Regions in the Western United States

K.M. Cobourn[1], L. Elbakidze[2], S. Ghosh[3]

[1]*Virginia Polytechnic Institute and State University, Blacksburg, VA, United States;* [2]*West Virginia University, Morgantown, WV, United States;* [3]*Northeastern State University, Tahlequah, OK, United States*

## 1. INTRODUCTION

In the western United States, surface water and groundwater were for a long time, and often still are, managed as separate resources. In some cases, surface water is managed based on private property principles, while groundwater is managed based on common property principles (Griffin, 2006). However, in many areas, surface water and groundwater are hydraulically connected, which implies that the use and management of one water resource influences the availability of the other. As water has become increasingly scarce over the past several decades and conflict between surface water and groundwater users has escalated, policymakers have turned their attention toward developing conjunctive management policies for surface and groundwater. Conjunctive management policies differ widely across states, but they share a common goal—to jointly manage surface and groundwater to maximize the availability and reliability of water supplies for multiple uses.

A diverse set of policies may be included under the umbrella of conjunctive management. For example, one important conjunctive management practice in the western United States is the allocation of surface water to aquifer recharge.[1] This creates a mechanism to store surface water in years with abundant flows. That water may later be extracted to augment scant surface water flows in relatively dry years. Groundwater storage thus smooths water use across years, facilitating more consistent access to water and providing value as a buffer against natural variability in surface water flows (Gemma and Tsur, 2007; Knapp and Olson, 1995; Tsur and Graham-Tomasi, 1991; Tsur, 1990; Burt, 1964).

1. This is often referred to as artificial recharge. Some of the states that have used artificial recharge to date are Arizona, California, Colorado, Idaho, Texas, New Mexico, Oregon, Utah, and Washington.

**Competition for Water Resources. http://dx.doi.org/10.1016/B978-0-12-803237-4.00016-1**

Another important component of conjunctive management involves specifying how to administer property rights for groundwater vis-à-vis property rights for surface water. The joint administration of water rights may be distinguished from other conjunctive management activities by referring to it as "conjunctive administration" (Tuthill et al., 2013). It is often necessary to establish new policies for conjunctive administration because property rights for water were developed independently by water source. Some of the oldest property rights were assigned for surface water in the mid- to late 19th century; property rights for groundwater generally were not established until the mid-20th century.[2] As a result of this difference in timing, as well as a relative lack of awareness about the extent of connectivity between surface water and groundwater in many regions at the time of surface water rights development, property rights rarely reflect the hydraulic relationships between surface water and groundwater. Poor integration of surface and groundwater rights administration in areas where the resources are hydraulically connected can compromise the economic efficiency of water use and distribution (Griffin, 2006).

In the late 1900s and into the early 2000s, protracted legal conflicts between surface water and groundwater users, along with an improved scientific understanding of hydrology, revealed the inadequacies of these preexisting, separate property rights systems for surface water and groundwater. For example, Hathaway (2011) discussed numerous instances of cross-boundary water conflicts between surface water and groundwater users in adjacent states. In many of these cases, resolution was sought by regulating groundwater pumping to ensure that surface water flows are sufficient to satisfy interstate compacts. Examples include the Pecos River in New Mexico and Texas, the Arkansas River in Colorado and Kansas, and the Republican River in Nebraska and Kansas (Hathaway, 2011; Burt et al., 2002).

In addition to these cross-boundary conflicts, clashes between surface water and groundwater users commonly occur within states. For example, Idaho has a history of legal conflict between surface and groundwater users dating to 1984. In the Swan Falls case, Idaho Power Company successfully argued that groundwater pumping by agricultural irrigators reduced river flows, precluding the company's ability to fulfill their surface water rights. This conflict played a pivotal role in highlighting the relationships between groundwater and surface water and ultimately led the state to establish policies for conjunctive administration. Even so, the practical implementation of conjunctive administration is an ongoing challenge. Conflict in the state has continued and noticeably increased in dry years as surface water users have sought to reduce groundwater pumping in an effort to augment declining surface water flows (Brockman, 2009; Fereday, 2008).[3]

2. Property rights for surface water were developed in response to a need to divert water from rivers and streams for use in mining (Coman, 1911). Property rights for groundwater developed following the advent of high-lift pumping technology, which allowed widespread access to groundwater stores. Many states rely on different laws to administer property rights for the two water sources, for example, California, Nebraska, Oklahoma, and Texas.

3. The mechanism by which surface water users can reduce groundwater pumping is by filing a curtailment call, defined as: "The exercise of a senior water right holder to obtain water. A 'call' requires junior upstream water rights to be passed down stream to the senior diversion making the 'call.' A junior right may continue to divert if proper augmentation is provided so as to meet the downstream senior right need" (Western Rivers Institute, 2015).

The precedent for reducing groundwater use to enhance surface water flows is rooted in the longstanding use of prior appropriation doctrine to administer surface water rights in the western United States. Prior appropriation, often synopsized as "first in time, first in right," allocates water to users based on the relative seniority of their water right. Each water right possesses a priority date, which is the first date on which the water was claimed for diversion in a prespecified beneficial use, for example, domestic, industrial, irrigation. Those water rights with the earliest priority dates are relatively senior; those with the latest priority dates are relatively junior. According to prior appropriation doctrine, when there are insufficient flows to satisfy all water rights, junior rights may be satisfied only if there is enough water to fulfill senior water rights.

Because groundwater rights were in large part established after surface water rights, a conjunctive administration approach that relies on prior appropriation doctrine in general treats groundwater rights as junior to surface water rights. It follows that this approach may involve reducing pumping by those with junior groundwater rights if doing so increases the surface water flows required to fulfill senior surface water rights. Since the Idaho Supreme Court awarded precedence to surface water rights in the Swan Falls case, this approach has been used by the Idaho Department of Water Resources to administer surface and groundwater rights conjunctively in hydraulically connected areas.

Reducing groundwater pumping to increase surface water flows is not an approach unique to Idaho. Colorado, Kansas, Nebraska, and Oregon have also given their state water agencies the power to regulate groundwater extraction to protect surface water rights (Hazard and Shively, 2011; Oklahoma Water Resources Board, 2010). However, most states in the western United States use different approaches to define when surface water and groundwater are hydraulically connected and water rights should be conjunctively administered. These include regional-scale groundwater simulation models (Idaho), expert judgment (New Mexico, Nevada, Wyoming, Utah), and predefined rules defining the extent of surface water−groundwater connectivity (Colorado, Oregon, Montana, Washington).

Although conjunctive management is a highly relevant issue to policymakers and water managers, little is known about how to design efficient conjunctive administration policies that reflect the hydraulic and economic relationships between surface water and groundwater resources. The objective of this chapter is to highlight key principles to consider in the conjunctive administration of surface water and groundwater rights. First, we present an overview of surface water−groundwater hydrology, demonstrating how these two resources interact in hydraulically connected regions and outlining the physical factors that govern the interaction. The second section builds on these hydrologic concepts to develop a stylized hydroeconomic model that we use to evaluate conjunctive administration policies. In presenting a simplified hydroeconomic model, we seek to develop intuition into how physical and economic factors affect economic outcomes under alternative conjunctive administration policies. We also seek to provide a generalizable modeling framework that can be applied to analyze diverse systems. We conclude by discussing how the results of our hydroeconomic model are borne out in studies of regional surface water−groundwater systems, and then summarize some of the practical challenges facing policymakers and water managers when implementing conjunctive administration policies.

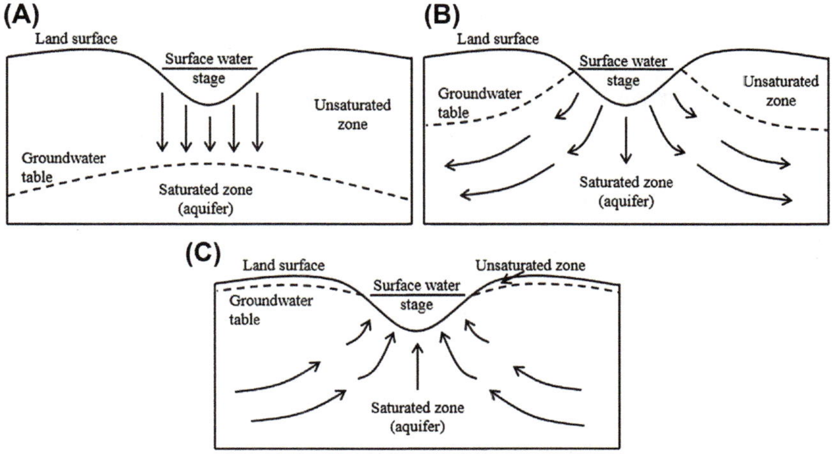


FIGURE 1 **Surface water–groundwater interactions.** Notes: Panel (A) illustrates a hydraulically disconnected surface water–groundwater system. Panels (B) and (C) illustrate a hydraulically connected surface water–groundwater system in which the surface waterway may be losing to the aquifer (connected-losing, panel B) or gaining from the aquifer (connected-gaining, panel C). *Figure adapted from Winter, T.C., Harvey, J.W., Franke, O.L., Alley, W.M., 1998. Groundwater and Surface Water: A Single Resource. U.S. Geological Survey, Reston. Circular 1139.*

## 2. SURFACE WATER–GROUNDWATER HYDROLOGY

When an unconfined aquifer underlies a surface waterway, such as a stream, river, or lake, surface water and groundwater may be hydraulically disconnected or connected.[4] When a system is disconnected, as illustrated in panel (A) of Fig. 1, an unsaturated zone separates the aquifer (the saturated zone) from the surface waterway. When disconnected, the surface waterway provides recharge to the aquifer at a rate that depends on the system's physical characteristics, such as soil properties. When a system is hydraulically connected, as illustrated in panels (B) and (C) of Fig. 1, the aquifer and the surface waterway are not separated by an unsaturated zone and may interact with one another. In this case, water can move back and forth between the aquifer and the surface waterway.

When a system is hydraulically connected, the direction and rate of water exchange depends on the relative heights of the groundwater table (the upper limit of the saturated zone) and the surface water stage (the uppermost water level in the surface waterway). If the groundwater table lies below the surface water stage, water moves from the surface waterway into the aquifer, providing recharge, as in panel (B) of Fig. 1. This is referred to as a connected-losing system. The direction of movement is the same in panel (B) as in

4. Confined aquifers are separated from the surface by an impermeable layer, and by definition cannot interact with surface water bodies. The classification of a system as connected or disconnected applies at a single point in space and time. Connectivity may vary spatially and temporally with storm events, diurnal differences in plant transpiration, or differences in topography, soil porosity, and vegetation. Some areas may be perennially connected or disconnected, in which case the classification of the system does not vary over time.

panel (A). However, the rate of recharge in a connected-losing system is influenced by the height of the groundwater table, whereas the groundwater table has no effect on the rate of recharge in a disconnected system. If the water table lies above the surface water stage, as in panel (C) of Fig. 1, water moves from the aquifer into the surface waterway, providing discharge (also called baseflow) to the surface waterway. This is referred to as a connected-gaining system. As in a connected-losing system, the height of the groundwater table affects the rate of water exchange in a connected-gaining system.

The key difference between a hydraulically disconnected system and a connected system, whether connected-losing or connected-gaining, is that groundwater pumping affects surface water flows in a connected system but not in a disconnected system. In a disconnected system, pumping causes the groundwater table to decline, but pumping cannot affect the flow of water through the surface waterway because the aquifer and the surface waterway are separated by an unsaturated zone. In contrast, in a hydraulically connected system, groundwater pumping lowers the groundwater table and alters the exchange of water between the aquifer and the surface waterway. As pumping lowers the groundwater table relative to the surface water stage, water is induced into the aquifer as the rate of recharge increases in a connected-losing system or as the rate of discharge decreases in a connected-gaining system.[5]

The timing of the effect of pumping on surface water flows is characterized by lags that depend on the properties of the aquifer and the distance between a groundwater well and a surface waterway. When groundwater is pumped, a cone of depression is formed and propagates radially over time until it reaches a hydraulically connected surface waterway. At that point, the drawdown in the groundwater table induces an increase in the flow of water out of the surface waterway and into the aquifer, or a decrease in discharge from the aquifer into the surface waterway. Over time, an amount of water equal to that pumped from the aquifer will be drawn out of the connected surface waterway.

The dynamic effect of groundwater pumping on surface water flows is often summarized using a response function (Kuwayama and Brozovic, 2013; Elbakidze et al., 2012; Harou and Lund, 2007; Cosgrove and Johnson, 2004, 2005; Miller et al., 2003). A response function expresses the proportion of the groundwater pumped that is drawn out of a surface waterway in each time period following pumping. Response functions vary in complexity and can be derived analytically, using numerical simulation, or they may be measured directly.

A straightforward analytical response function that is sufficient for our example is presented by Glover and Balmer (1954) based on the work of Theis (1941). Pumping at a constant rate from a single well in an initial year results in surface water depletion in year $i$ in the quantity:

$$q_i = Q_0 \cdot \operatorname{erfc}\left(\sqrt{\frac{d^2 S}{4 i \tau}}\right) \tag{1}$$

where $q_i$ is the surface water depletion rate in year $i$, $Q_0$ is the pumping rate in the initial year, $d$ is the distance of the well from the surface waterway, $S$ is aquifer storativity, $\tau$ is the

5. If pumping draws the groundwater table below the surface water stage, the system switches from connected-gaining to connected-losing. If pumping draws the groundwater table below the minimum level of the surface waterway, the system shifts from hydraulically connected to disconnected.

transmissivity of the aquifer, erfc is the complementary error function, and $i = 0, \ldots, \infty$.[6] The storativity coefficient describes the ability of the aquifer to release groundwater; transmissivity is defined as the rate of lateral flow of groundwater. The predicted proportion of surface water depletion in each year following pumping, per Eq. (1), defines the response function.

Eq. (1) can be modified to reflect that groundwater pumping occurs only over a portion of a year and to capture the cumulative effects of pumping across multiple wells and over multiple years, as in Kuwayama and Brozovic (2013).[7] The formula can also be applied directly to quantify the additional surface water flows obtained when groundwater pumping is reduced or when artificial recharge takes place. The advantages of this analytical specification are that it is straightforward to calculate and it captures the way in which the response function varies with aquifer characteristics and the distance of the well from the surface waterway. However, a disadvantage is that strong assumptions are required to formulate an analytical expression for stream depletion and hence the response function.[8]

# 3. CONJUNCTIVE MANAGEMENT POLICY ANALYSIS

This section presents an intuitive and stylized example of conjunctive administration in a hydraulically connected surface water–groundwater system. This example integrates the hydrologic response function from the previous section with an economic model of decision making by agricultural water users. We use this model to quantify the effects of conjunctive administration policies on surface water flows and economic outcomes. We focus on agriculture in particular because of its importance as a water user: irrigated agriculture accounts for 80–90% of the water consumed in the United States, with 74% of the irrigated acreage in the United States located in the 17 contiguous western states (Schaible and Aillery, 2012). Though our focus here is on agriculture, our model may be generalized to consider a variety of water users.

We assume that the reader is familiar with the fundamental economic concepts of discounting, net present value (NPV), marginality, and constrained profit maximization. For additional discussion of the economic concepts needed to develop a hydroeconomic

---

6. The complementary error function is defined as: $\mathrm{erfc}(z) = \frac{2}{\sqrt{\pi}} \int_z^\infty e^{-s^2} ds$.
7. If pumping occurs over proportion $b$ of the initial year, stream depletion in year $i$ is: $q_i = Q_0 \cdot \left[\mathrm{erfc}\left(\sqrt{\frac{d^2S}{4i\tau}}\right) - \mathrm{erfc}\left(\sqrt{\frac{d^2S}{4(i-b)\tau}}\right)\right]$. If pumping occurs at multiple wells indexed by $j = 1, \ldots J$, stream depletion in year $i$ from pumping at all wells in the initial year is: $q_i = \sum_{j=1}^{J} Q_{0j} \cdot \left[\mathrm{erfc}\left(\sqrt{\frac{d_j^2 S_j}{4i\tau_j}}\right) - \mathrm{erfc}\left(\sqrt{\frac{d_j^2 S_j}{4(i-b)\tau_j}}\right)\right]$, where the pumping rate, distance from the surface waterway, and aquifer characteristics may differ across wells. Cumulative stream depletion from pumping in multiple years may similarly be obtained by summing over years prior to year $i$. The stream depletion function presented in this example describes stream depletion at a single point along a surface waterway. However, pumping at one well can generate a stream depletion effect that differs along reaches of a river or stream, as described in Elbakidze et al. (2012).
8. The response function of Glover and Balmer (1954) requires that the aquifer is homogeneous, of infinite extent, and isotropic; the course of the river is a straight line that continues past the well; and groundwater can move freely into and out of the surface waterway.

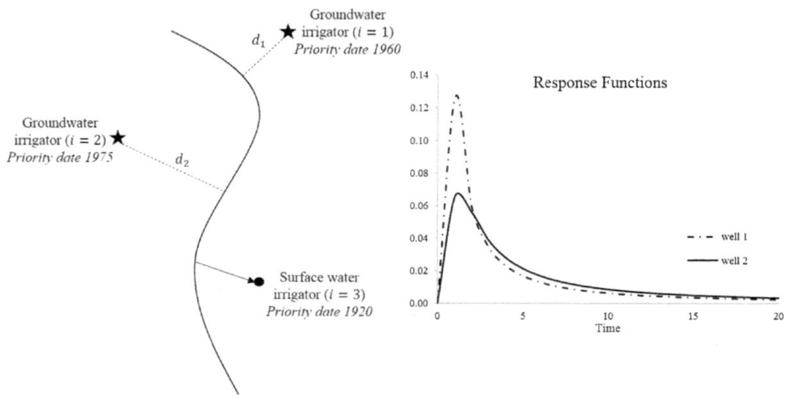


FIGURE 2 **Surface water–groundwater system and response functions by well.** Notes: Response functions describe the proportional decrease in streamflow that results from pumping in the initial year. Response functions are formulated using the analytical stream depletion formula from Glover and Balmer (1954).

model of the type presented here, we refer the reader to Elbakidze and Cobourn (2013), Booker et al. (2012), and Harou et al. (2009).

## 3.1 Surface Water–Groundwater System

Consider the simplified system depicted in Fig. 2. The system includes three irrigators indexed by $i = 1, 2, 3$. Two of the irrigators rely on groundwater ($i = 1, 2$) and one relies on surface water from a stream ($i = 3$).[9] Each groundwater irrigator has one well. The well for irrigator 1 is located closer to the stream than the well for irrigator 2, where $d_1$ and $d_2$ denote the distance of each well from the stream. Let $d_1 = 2.4$ km and $d_2 = 3.6$ km. Assume that the aquifer is homogeneous, with a storativity coefficient of 0.15 and transmissivity of 929 $m^2$/day.

Groundwater pumping from each well in year $t$ affects streamflow in year $t$ and in all subsequent years. These effects can be described using the analytical formula for stream depletion in Eq. (1) with a modification to reflect that pumping occurs over a portion of a calendar year. The response functions are formulated assuming that pumping occurs for 160 days per year, which is consistent with the length of an irrigation season in most of the western United States. The dynamic response of the stream to 1 year of pumping at each well is illustrated in Fig. 2. Because the aquifer is uniform, differences in the response functions for each well arise solely because of differences in the distance of each well from the stream. The effects of pumping at well 1 on streamflow peak in the first year and decline rapidly thereafter; the effects of pumping at well 2 on streamflow also peak in the first year, but the peak effect is smaller and the effects are more sustained over time than those at well 1.

Each irrigator owns a single water right with a unique priority date and a diversion constraint of 100 units. The surface water irrigator owns the most senior right in the

9. For illustrative purposes and without loss of generality we exclude the possibility that irrigators possess or can acquire rights to access both groundwater and surface water.

system. Groundwater irrigator 1 owns the next most senior right, and groundwater irrigator 2 owns the most junior right. For simplicity, we assume that each irrigator produces a single crop. The profit function for each irrigator can be written as: $\pi_i(w_i) = p \cdot f(w_i) - c_i(w_i) - F_i$, where $p$ is the market price of the crop, $w_i$ is the amount of water diverted by irrigator $i$, $f(w_i)$ is a concave production function ($f'(w_i) > 0$; $f''(w_i) < 0$) that defines the relationship between water diversions and crop yield, $c(w_i)$ is the variable cost of diverting water, and $F_i$ are fixed costs of diverting water.[10]

In this example, all irrigators face the same output price and production functions, but the costs of diverting water differ by source. Groundwater users must pay a per-unit cost to divert water, which depends on the depth to groundwater (or lift). Surface water irrigators face negligible per-unit costs of diverting water, but often pay fixed costs to divert water, usually in the form of an annual assessment to an irrigation district. The following parameter values and functions define the profit function for each irrigator: $p = 1$; $f(w_i) = 260w_i - w_i^2$; $c_i(w_i) = 20w_i$ for $i = 1, 2$; $c_i(w_i) = 0$ for $i = 3$; $F_i = 0$ for $i = 1, 2$; and $F_i = 2000$ for $i = 3$.

Each irrigator chooses diversions to maximize profit subject to the diversion constraint specified by their water right. Each irrigator's constrained optimization problem is:

$$\begin{aligned} \pi_i(w_i) &= p \cdot f(w_i) - c_i(w_i) - F_i \\ s.t.\ w_i &\le W_i \end{aligned} \tag{2}$$

where $W_i$ is the diversion limit associated with irrigator $i$'s water right. If diversions are unconstrained, each groundwater irrigator maximizes profit by diverting 120 units of water per period.[11] The surface water irrigator maximizes profit by diverting 130 units of water. The diversion constraint of 100 units for each irrigator is therefore binding, as is often the case in appropriative systems (Cobourn, 2015; Pfeiffer and Lin, 2012). In the constrained maximization problem, each irrigator will divert their maximum allowable 100 units per year.

Suppose that past groundwater extraction at a rate of 100 units per year from each well has resulted in stable streamflow of 90 units per year. Streamflow of 90 units per year is

10. In this example, we do not distinguish between water diversions and consumptive water use. Consumptive water use, or evapotranspiration, is the amount of water that is used by a crop for growth. Depending on the irrigation technology used, diversions and consumptive use may differ substantially. For example, consumptive use under gravity irrigation is approximately 40–65% of total diversions on average. Water that is applied but unconsumed by the crop may be captured as tailwater and returned to the stream, or it may percolate into the aquifer, providing incidental recharge (Cobourn, 2015; Chakravorty and Umetsu, 2003; Huffaker and Whittlesey, 2000). Under high-pressure sprinkler technology or low-pressure (microdrip) technologies, consumptive use is approximately 75–90% of the amount of water diverted. Most of the unconsumed water percolates back into the aquifer.

11. All prices and costs are in 1000 USD; diversion and streamflow units are unspecified here but are most often measured in cfs (cubic feet per second) in the western United States. The unconstrained profit-maximizing level of diversions for each irrigator occurs where the additional revenue earned for the last unit of water diverted (marginal value product) equals the additional cost of diverting that unit (marginal cost). This level can be obtained by taking the first-order necessary condition for a maximum of the irrigator's profit function. The concavity of the production function ensures that the first-order condition is also sufficient.

insufficient to satisfy the surface water irrigator's right to divert 100 units. As a result of this shortfall in streamflow, the surface water irrigator loses $2700 in profit each year. Over a 20-year time horizon and using a discount rate of 4%, the surface water irrigator earns an NPV of profit equal to 187.98 (in 1000 USD), as compared to 226.14 if 100 units of water are available for diversion. Over the same time period, each groundwater irrigator diverts 100 units per period and earns an NPV of profit equal to 197.88. In Table 1, this outcome is reported as the baseline scenario.

## 3.2 Conjunctive Management Policies

Groundwater pumping by irrigators who own junior water rights reduce streamflow and the profit earned by the surface water irrigator. If prior appropriation doctrine is applied across surface and groundwater rights, the senior surface water irrigator could file a call to reduce pumping by one or both of the relatively junior groundwater pumpers. Reducing or eliminating pumping by the groundwater irrigators may augment streamflow so that the surface water user can divert the full 100 units of water permitted under the surface water right.

To explore how decisions about diversions affect streamflow and economic outcomes over time, we combine the response functions from Fig. 2 and the profit maximization model described in Eq. (2) to develop a hydroeconomic model. Using this model, we compare outcomes across three policy scenarios: (a) prior appropriation-based water rights administration; (b) economically efficient groundwater pumping reductions within a prior appropriation framework; and (c) the economically efficient water allocation. In scenarios (a) and (b), groundwater pumping must be reduced to augment surface water flows to fulfill the senior surface water right. Both policy scenarios (a) and (b) honor the precedence of the senior surface water right, consistent with prior appropriation doctrine. In scenario (a), reductions in pumping are distributed across groundwater irrigators according to the seniority of the groundwater rights. In scenario (b), pumping reductions are distributed across the two groundwater irrigators in an economically efficient way, such that the sum of profit across the two groundwater irrigators is maximized subject to a constraint on streamflow availability for the surface water user. This scenario does not consider water right seniority across groundwater irrigators when choosing pumping reductions. In scenario (c), we solve for the economically efficient allocation of water across all three irrigators, while taking into account the hydraulic relationship between the aquifer and the stream. The economically efficient allocation of water maximizes the sum of profit across all three irrigators. This third scenario does not necessarily honor water right seniority for any of the irrigators.

Table 1 presents the results for each scenario. For policy scenarios (a) and (b), we consider scenario variants in which streamflow must be sufficient to satisfy the surface water right for 1 or multiple years.[12] We present results for variants in which 100 units of streamflow must be sustained for 1, 5, 10, 15, and 20 years. For policy scenario (c), we

12. Generally, prior appropriation doctrine provides a legal precedent for reducing water use by junior rights to fulfill senior rights only during a single year. However, given the time lags associated with changes in groundwater pumping, it is informative to consider how those lags affect the pumping reductions required to augment streamflow in future years. Furthermore, the possibility of repeated water calls in consecutive years justifies examination of multiyear scenarios.

**TABLE 1** Net Present Value (NPV) of Profit by Irrigator and Policy Scenario Over 20 Years

| | | | 20-Year NPV Profit (1000 USD) | | | |
|---|---|---|---|---|---|---|
| **Policy Scenario** | **Total Pumping Reduction** | **Total Streamflow** | **GW 1 Profit** | **GW 2 Profit** | **SW Profit** | **Total Profit** |
| *Baseline* | – | 1800.00 | 197.88 | 197.88 | 187.98 | 583.73 |
| **Policy (a): Prior Appropriation-Based Groundwater Pumping Reductions** | | | | | | |
| 1-year | 127.29 | 1840.80 | 196.04 | 183.88 | 190.62 | 570.53 |
| 5-year | 289.89 | 1890.84 | 196.04 | 171.90 | 193.39 | 561.33 |
| 10-year | 458.76 | 1939.34 | 196.04 | 162.38 | 196.65 | 554.07 |
| 15-year | 614.17 | 1977.92 | 196.04 | 155.46 | 197.15 | 548.65 |
| 20-year | 762.77 | 2000.00 | 196.04 | 150.13 | 197.88 | 544.05 |
| **Policy (b): Economically Efficient Groundwater Pumping Reductions** | | | | | | |
| 1-year | 92.66 | 1830.63 | 191.13 | 195.96 | 190.01 | 577.10 |
| 5-year | 248.62 | 1880.92 | 186.06 | 192.40 | 192.89 | 571.36 |
| 10-year | 412.25 | 1930.88 | 182.00 | 189.35 | 195.29 | 566.64 |
| 15-year | 564.16 | 1972.50 | 179.00 | 187.03 | 196.96 | 562.98 |
| 20-year | 709.26 | 2000.00 | 176.64 | 185.23 | 197.88 | 559.75 |
| **Policy (c): Economically Efficient Water Allocation** | | | | | | |
| Constrained | – | 1800.70 | 197.40 | 198.33 | 188.02 | 583.75 |
| Unconstrained | −360.85 | 1709.33 | 201.38 | 201.93 | 182.94 | 586.25 |

Notes: Total pumping reduction and total streamflow are summed over 20 years. Units are unspecified. Streamflow and pumping are most often measured in cfs (cubic feet per second) in the western United States. *GW 1* and *GW 2*, groundwater irrigators 1 and 2, respectively; *SW*, surface water irrigator.

consider two variants: in the first, groundwater pumping is limited to the combined total allowed under the two groundwater rights (200 units); in the second, groundwater pumping is unconstrained. In both variants of scenario (c), the optimal choice of groundwater and surface water diversions takes into account the dynamic effects of groundwater pumping on surface water flows.

### 3.2.1 Policy Scenario (a): Prior Appropriation-Based Groundwater Pumping Reductions

In scenario (a), applying prior appropriation across groundwater irrigators implies that pumping at well 2, the junior well, will be reduced first in an effort to increase streamflow. If reducing pumping at well 2 is not sufficient to increase streamflow to the targeted level, pumping at well 1 may also be reduced. To meet a 1-year streamflow constraint, pumping at well 2 must be reduced to zero in year 1. The response function for well 2 implies that this decrease in pumping increases streamflow by 6.55 units in the same year (all changes are relative to the baseline scenario). To fulfill the remaining shortfall in streamflow of 3.45 units, pumping at well 1 must also be reduced by 27.29 units. These reductions in pumping reduce profit for each groundwater irrigator in year 1. Over a 20-year time horizon, these losses in year 1 decrease the NPV of profit for groundwater irrigators 2 and 1 by 7.08% and 0.93%, respectively.

Although pumping is reduced only in year 1, the pumping reduction continues to provide increased streamflow in future years because of the lagged hydrologic effects illustrated in Fig. 2. In addition to the effect in year 1, eliminating pumping from well 2 in year 1 increases streamflow by 5.56 units in year 2, 3.82 units in year 3, and 2.76 units in year 4. The streamflow increases provided by the year 1 pumping reduction decline each year but continue, reaching 0.31 units in year 20. Over the course of 20 years, the 1-year reduction in pumping at well 2 provides an additional 40.80 units of streamflow. The economic benefit of this dynamic increase in streamflow accrues to the surface water irrigator, increasing the NPV of profit for irrigator 3 by 1.4% relative to the baseline. Despite these sustained benefits to the surface water irrigator, the loss in profit for the two groundwater irrigators in year 1 outweighs the discounted gains to the surface water irrigator. As a result, the three irrigators suffer a loss in aggregate discounted profit of 2.26% relative to the baseline.

This decrease in aggregate discounted profit in scenario (a) is driven by two factors. First, the loss in groundwater from the pumping reduction in year 1 exceeds the increase in streamflow over the next 20 years (127.29 units of groundwater lost vs. 40.80 units of surface water gained). The second factor, which compounds this effect, is discounting. The decrease in profit to the groundwater irrigators is experienced immediately, whereas the increase in profit to the surface water irrigator is distributed over time. Discounting implies that the benefits to the surface water irrigator in future periods are weighed less heavily than the current losses to the groundwater irrigators.

In the multiyear variants of policy scenario (a), the lagged effects of the pumping reduction in year 1 imply that less severe pumping reductions are required to increase streamflow in subsequent years. For example, increasing streamflow in year 2 requires that pumping at well 2 is reduced by only 43 units in year 2. By year 5, only a 37 unit reduction in pumping at well 2 is required to meet the streamflow constraint. No additional pumping reductions by well 1 are required to meet the multiyear streamflow constraints. In all multiyear variants of policy scenario (a), the losses to the groundwater irrigators from early pumping reductions outweigh the later, more heavily discounted gains to the surface water irrigator. An interesting result is that the marginal cost of extending the duration of the streamflow constraint diminishes. This follows from the lagged hydrologic effects provided by pumping reductions in earlier years. For a 1-year streamflow constraint, the aggregate loss in discounted profit is 2.26%. Extending the streamflow constraint to 5 years increases aggregate losses by another 1.58%. The marginal increases in aggregate losses associated with extending the constraint to 10, 15, and 20 years are 1.24%, 0.93%, and 0.79%, respectively.

It should perhaps be intuitively apparent that reducing pumping at a well located closer to the stream, rather than at a well located further from the stream, is likely to provide the desired increase in streamflow for a smaller reduction in pumping. Consider, for example, reducing pumping at well 1 first, rather than well 2. Meeting the 1-year streamflow constraint requires a reduction in pumping at well 1 of only 79.11 units. No pumping reduction at well 2 is necessary. By shifting the burden of the pumping reduction to the well located closer to the stream, it is possible to increase aggregate discounted profit by 0.98% in the scenario variant with a 1-year streamflow constraint.

### 3.2.2 Policy Scenario (b): Economically Efficient Groundwater Pumping Reductions

Reducing pumping at well 2 prior to well 1, as in policy scenario (a), is consistent with the straightforward application of prior appropriation doctrine across groundwater rights. However, this may not be the most efficient means of augmenting surface water flows because the streamflow constraint(s) from scenario (a) can be met with a smaller pumping reduction by shifting some of the burden of the reduction to the well located closest to the stream. Rather than enforce prior appropriation across groundwater rights, a water manager could set a target level of streamflow and allow groundwater irrigators to choose the reductions in pumping at each well that maximize their joint profit. Scenario (b) captures this possibility. For each scenario variant considered, streamflow must be at least 100 units per year as in scenario (a), but the reduction in pumping required to increase streamflow is distributed across groundwater irrigators in an economically efficient way (and does not necessarily honor the relative seniority of groundwater rights).

The results in Table 1 indicate that aggregate discounted profit increases in scenario (b) relative to scenario (a) for all scenario variants. For example, with a 1-year streamflow constraint, aggregate discounted profit increases from 570.53 in scenario (a) to 577.10 in scenario (b). The same applies for each of the multiyear scenario variants. In fact, the relative gain under scenario (b) as compared to scenario (a) increases with the duration of the streamflow constraint. For a 1-year streamflow constraint, scenario (b) increases aggregate profit by 1.15%; for a 20-year streamflow constraint, the gain in aggregate profit increases to a maximum of 2.89%.

In each variant of scenario (b), as compared to scenario (a), the burden of pumping reductions is redistributed toward well 1. As a result, irrigator 1 suffers a greater loss in profit than in scenario (a), irrigator 2 sees an increase in profit relative to scenario (a), and irrigator 1 earns less profit than irrigator 2. Although the distribution of pumping reductions is to the disadvantage of irrigator 1, scenario (b) results in a water allocation that is a potential Pareto improvement over scenario (a). Specifically, the gains for irrigator 2 exceed the losses for irrigator 1, which implies that irrigator 2 could potentially compensate irrigator 1 for their pumping reductions and both irrigators would be better off than they are in scenario (a) (Griffin, 1995).

### 3.2.3 Policy Scenario (c): Economically Efficient Water Allocation

In scenario (c), water may be reallocated freely between any of the three irrigators in an economically optimal way. The economically optimal distribution of water in this scenario differs from privately optimal water use because this scenario takes into account how groundwater pumping affects streamflow. In economic parlance, this scenario internalizes the streamflow externality associated with groundwater pumping. We consider two variants in this policy scenario. In the first, we assume that total groundwater pumping cannot exceed the sum of the diversion limits for the two groundwater rights (200 units per period). The first variant represents a constrained maximum that honors existing limits to groundwater pumping. In the second variant, we allow for unlimited groundwater pumping. Though unlimited pumping may not be realistic in many systems, the outcome under this scenario forms a useful benchmark because it bounds the upper level of aggregate discounted profit. In both scenario variants, surface water diversions are limited only by streamflow.

The results in Table 1 suggest that when groundwater pumping is constrained, it is optimal to reduce pumping at well 1 to increase streamflow and surface water diversions. This offsets a small increase in pumping at well 2, which reduces streamflow. The net effect is a minor increase in streamflow of 0.70 units over 20 years. This scenario generates an increase in total profit, though the magnitude of the gain is small. In general, the magnitude of the gain depends on the relative value of a unit of water in production by groundwater and surface water irrigators. If a unit of water generates greater value when used by the surface water irrigator, it becomes optimal to reduce groundwater pumping more to increase surface water flows. The converse also holds: if a unit of water generates greater value when used by the groundwater irrigators, it is optimal to increase groundwater pumping, if possible. The second variant of scenario (c) explores whether it is economically optimal to increase total groundwater pumping over 200 units per period and accept the concomitant decrease in streamflow. The results in Table 1 indicate that for our example it is. In this scenario, groundwater pumping at both wells increases by 360.85 units over the course of 20 years, relative to the baseline. The increase in pumping is slightly greater for well 2 given that it generates smaller losses in streamflow than does pumping at well 1. The increase in pumping at both wells reduces surface water flows by 148 units over 20 years.

## 3.3 Accounting for Differences in Productivity

Up to this point, we have maintained the assumption that the production functions are the same for each of the irrigators in the surface water−groundwater system. This implies that the marginal product of water, defined as the change in yield from a one-unit change in diversions, is the same for each irrigator. However, the marginal product of water is likely to differ from farm to farm because of, for example, differences in physical factors such as soil quality. A logical question to ask in the context of our analysis is: how do differences in the marginal product of water across irrigators affect the economic performance of conjunctive administration policies?

To answer this question, we consider several productivity scenarios that introduce differences in the marginal product of water across the three irrigators. In these scenarios, we allow one irrigator to possess a relative productivity advantage, such that the marginal product of water for that irrigator is twice that for the other two irrigators. For example, if irrigator 1 has a relative productivity advantage, setting $f(w_1) = 2f(w_2) = 2f(w_3)$ in Eq. (2) implies that the marginal product of water for irrigator 1 will be twice that for irrigators 2 and 3. Table 2 summarizes outcomes for each policy and productivity scenario, including

**TABLE 2** Change Relative to Baseline by Productivity and Policy Scenario

| Productivity and Policy Scenarios | GW Pumping | Streamflow | 20-Year NPV Profit (1000 USD) | | | |
|---|---|---|---|---|---|---|
| | | | GW 1 | GW 2 | SW | Total |
| **Equal Productivity Across Irrigators** | | | | | | |
| Policy Scenario (a), 1-year | 127.29 | 40.80 | −1.84 | −14.00 | 2.64 | −13.20 |
| Policy Scenario (b), 1-year | 92.66 | 30.63 | −6.75 | −1.91 | 2.03 | −6.64 |
| Policy Scenario (c), constrained | – | 0.70 | −0.47 | 0.45 | 0.04 | 0.02 |
| Policy Scenario (c), unconstrained | −360.85 | −90.67 | 3.51 | 4.06 | −5.05 | 2.52 |
| **Productivity Advantage, Groundwater Irrigator 1** | | | | | | |
| Policy Scenario (a), 1-year | 127.29 | 40.80 | −4.22 | −14.00 | 2.64 | −15.58 |
| Policy Scenario (b), 1-year | 108.00 | 35.14 | −9.42 | −5.99 | 2.30 | −13.12 |
| Policy Scenario (c), constrained | – | −7.08 | 10.83 | −6.61 | −0.44 | 3.79 |
| Policy Scenario (c), unconstrained | −558.77 | −148.46 | 16.45 | 3.84 | −8.68 | 11.61 |
| **Productivity Advantage, Groundwater Irrigator 2** | | | | | | |
| Policy Scenario (a), 1-year | 127.29 | 40.80 | −1.84 | −30.00 | 2.64 | −29.20 |
| Policy Scenario (b), 1-year | 80.32 | 27.02 | −9.17 | −0.26 | 1.81 | −7.63 |
| Policy Scenario (c), constrained | – | 8.02 | −7.53 | 11.75 | 0.48 | 4.71 |
| Policy Scenario (c), unconstrained | −542.91 | −136.73 | 3.28 | 16.78 | 7.88 | 12.18 |

*Continued*

**TABLE 2** Change Relative to Baseline by Productivity and Policy Scenario—cont'd

| Productivity and Policy Scenarios | GW Pumping | Streamflow | 20-Year NPV Profit (1000 USD) | | | |
|---|---|---|---|---|---|---|
| | | | GW 1 | GW 2 | SW | Total |
| **Productivity Advantage, Surface Water Irrigator** | | | | | | |
| Policy Scenario (a), 1-year | 127.29 | 40.80 | −1.84 | −14.00 | 5.27 | −10.57 |
| Policy Scenario (b), 1-year | 94.28 | 31.11 | −6.46 | −2.25 | 4.11 | −4.60 |
| Policy Scenario (c), constrained | 38.93 | 13.54 | −1.69 | 0.21 | 1.63 | 0.16 |
| Policy Scenario (c), unconstrained | −18.73 | 5.03 | −1.16 | 0.60 | 0.95 | 0.40 |

Notes: Policy Scenario (a) is prior appropriation-based groundwater pumping reductions; Policy Scenario (b) is economically efficient groundwater pumping reductions; Policy Scenario (c) is the economically efficient water allocation. All changes are relative to the baseline scenario in which groundwater irrigators pump 100 units per period and 90 units of surface water flow are available. All changes are summed over 20 years. Units are unspecified. Streamflow and pumping are most often measured in cfs (cubic feet per second) in the western United States. *GW 1 and GW 2*, groundwater irrigators 1 and 2, respectively; *NPV*, net present value; *SW*, surface water irrigator.

the case in which the marginal product of water is equal across irrigators. The outcomes in Table 2 are expressed as changes relative to the baseline policy scenario, in which the groundwater irrigators extract 100 units per period and 90 units of streamflow are available per period for diversion by the surface water irrigator.

For policy scenario (a), altering the relative productivity of each irrigator has no effect on pumping reductions, which are determined based solely on the application of prior appropriation across groundwater rights. However, changes in the relative productivity of irrigators affect the magnitude of the aggregate profit losses in scenario (a) relative to the baseline. For example, when irrigator 1 has a relative productivity advantage, the total loss in the NPV of profit for all irrigators is 15.58. When the productivity advantage is shifted to groundwater irrigator 2, the aggregate loss in profit nearly doubles to 29.20. This substantial increase in losses arises because the cost of reducing pumping is highest for irrigator 2, yet irrigator 2 has the lowest priority water right and bears the majority of the pumping reductions needed to augment streamflow.

If the burden of pumping reductions is shifted to the lower-cost, or lower-productivity irrigator, the same increase in streamflow can be obtained at lower total cost. This is evident when comparing policy scenarios (b) and (a) across productivity scenarios. When irrigator 2 has a relative productivity advantage, scenario (b) reduces aggregate profit losses by 21.67 (from 29.20 to 7.63). When irrigator 1 has a productivity advantage, scenario (b) reduces aggregate profit losses by only 2.46 (from 15.58 to 13.12). Scenario (b) thus yields a greater improvement in economic outcomes when irrigator 2 has the productivity advantage because this scenario allows pumping reductions to be shifted towards the relatively low-productivity irrigator, for whom the cost of reducing pumping is lowest. When irrigator 2 has the productivity advantage, the increase in profit in scenario (b) over scenario (a) is 2.28% as compared to 1.15% when the two groundwater irrigators are equally productive.

We also consider the case in which the surface water irrigator has a relative productivity advantage. The losses under scenario (a) are at their lowest when the surface water irrigator has a productivity advantage (10.57 as compared to 13.20, 15.18, and 29.20 in the other three productivity scenarios). As the productivity of the surface water irrigator increases relative to the groundwater irrigators, it is economically optimal to sacrifice more groundwater pumping to augment streamflow. However, the productivity advantage held by the surface water irrigator would have to be substantial to fully offset the costs of pumping reductions in policy scenario (a): in this example, the productivity of the surface water irrigator would have to be on the order of seven times that of the groundwater irrigators to yield an increase in total discounted profit.

## 4. DISCUSSION AND CONCLUSIONS

The analysis in the preceding section is based on a simplified and stylized system, but it highlights several important issues to consider when evaluating and implementing policies for conjunctive administration. The first is that spatial variability in response functions implies that pumping at each groundwater well contributes to different quantities of surface water depletion over time. In general, policies for conjunctive administration that reflect these differences will outperform policies that do not by increasing surface water flows at least total cost to groundwater users. In our example, scenario (b) improves economic outcomes relative to scenario (a) because it reflects the fact that increases in streamflow can be obtained with smaller pumping reductions from the well located closest to the surface waterway.[13] This result has been demonstrated for the Republican River Basin of Nebraska and the Eastern Snake River Plain of Idaho by Kuwayama and Brozovic (2013) and Elbakidze et al. (2012), respectively. Both of these studies find that at a regional level, distributing pumping reductions across wells in a way that reflects differences in hydrologic response functions yields economic gains over a policy that is spatially uniform or based on the application of prior appropriation across groundwater rights.

These two studies also highlight an important policy consideration that is explored in this chapter: the cost of reducing pumping differs across space because of differences in the marginal productivity of water across groundwater irrigators. These differences interact with differences in hydrologic response functions and/or water rights ownership to affect the economically optimal allocation of pumping reductions across wells. The fundamental result can be summarized as follows: a conjunctive administration policy that applies prior appropriation across all irrigators will achieve an economically efficient outcome if those irrigators who can reduce pumping at lowest cost provide the greatest surface water increase for a unit of pumping reductions *and* possess the least senior water rights. If these three factors do not align in this way, then policy instruments that allow groundwater pumping reductions to be distributed flexibly across wells will outperform a prior appropriation-based approach to conjunctive administration in which the reallocation of water use is not an option. However, the magnitude of the gain from a more flexible policy instrument will depend on the spatial arrangement of these factors within a specific region.

Three studies in the literature touch on how these factors combine to affect the efficiency of conjunctive administration policies. Kuwayama and Brozovic (2013) find that spatial variation in the costs of reducing pumping are correlated with the hydrologic response functions for individual wells in the Republican River Basin. The consequence of this correlation is that it reduces the gains to a spatially differentiated permit trading system over a spatially homogeneous approach. Ghosh et al. (2014) find that farms in the Eastern Snake River Plain with higher costs of pumping reductions tend to have the least senior water rights. This negative correlation enhances the gains to reallocating groundwater pumping reductions through a water banking system, relative to the allocation implied by prior appropriation doctrine. Elbakidze et al. (2012) allow for the interaction of all three factors—productivity, response functions, and water rights—when estimating the gains to reallocating groundwater pumping reductions in the Eastern Snake River Plain. Taking these factors in combination, they estimate that redistributing groundwater pumping reductions increases discounted irrigator profits relative to the strict application of prior-appropriation doctrine across groundwater irrigators.

The interaction of productivity, response functions, and water rights also affects whether it is economically optimal to sacrifice groundwater pumping to increase surface water flows. The example presented in this chapter suggests that this is unlikely unless the marginal product of water is much greater for surface water users than for groundwater users. Whether this is the case will differ by study region. However, surface water users will not likely hold a productivity advantage if the need to locate close to surface

13. In this chapter, we model differences in response functions due solely to distance between a well and a surface waterway, but aquifer characteristics, such as storativity and transmissivity are also variable across space and affect response functions. It is not merely *absolute* distance between a well and a surface waterway that matters, but *hydrologic* distance, which captures all of these factors, that determines how pumping reductions affect streamflow.

waterways required the first surface water irrigators to settle less productive lands than later groundwater irrigators. Unfortunately, there are relatively few studies that examine this question. The few that do address the question in the context of the Eastern Snake River Plain. Snyder and Coupal (2005) find that the losses from applying prior appropriation-based groundwater pumping reductions exceed the gains to surface water users by up to an order of magnitude. Similarly, Ghosh et al. (2014) demonstrate that it is optimal to reduce surface water diversions during a drought, rather than placing the sole burden of water use reductions on groundwater irrigators. Under severe drought conditions, they show that redistributing water from surface water to groundwater irrigators may increase aggregate economic welfare by nearly 50%.

Although this chapter illustrates and discusses some important factors to consider when implementing conjunctive administration, there are several policy relevant issues that are not incorporated into this analysis. Though certainly not exhaustive, the basic hydro-economic modeling framework presented can provide a foundation that can be extended to consider other water policy problems and additional hydrologic complexities in surface water–groundwater systems. For example, preserving streamflow is often motivated by ecological concerns, which are not incorporated into our analysis. It is possible to incorporate the nonmarket ecological benefits of streamflow into the model so that these benefits are reflected in the optimal allocation of water. Another complication identified in the literature is that hydrologic response functions like those used in this chapter may change over time as groundwater levels decline (Cobourn, 2015). This is a particularly relevant concern when considering the long-run effects of conjunctive administration policies. Cobourn (2015) and Taylor et al. (2014) demonstrate that groundwater pumping is not the only factor that contributes to declining groundwater levels. Changes in irrigation technology, such as lining canals to prevent seepage or switching from gravity to sprinkler irrigation, affect aquifer recharge and groundwater levels. These studies suggest that the hydraulic response functions included in this analysis are not fixed, but depend on the ways in which water use decisions by both surface water and groundwater users affect groundwater levels. To consider these changes, the model must account for the decisions irrigators make about technology adoption, and it must differentiate between the proportion of water consumed by crops and the proportion returned to surface waterways and the aquifer (Cobourn, 2015; Taylor et al., 2014; Chakravorty and Umetsu, 2003; Huffaker and Whittlesey, 2000).

Conjunctive management of surface water and groundwater in practice has posed a long-running challenge throughout the western United States. Establishing the hydraulic relationships between surface water and groundwater resources is itself a daunting task for scientists and policymakers; the preexistence of property rights systems that often fail to reflect hydraulic connectivity adds another layer of complexity to developing efficient conjunctive management policies. In most cases, the problem of inefficient and/or unsustainable water use in linked groundwater–surface water systems stems from deficiencies in groundwater law relative to surface water law (Griffin, 2006). This chapter demonstrates that extending current institutions, such as prior appropriation, to administer groundwater rights may not lead to economically efficient outcomes. As groundwater and conjunctive administration policies evolve, a greater understanding of the complex spatial and dynamic interactions between productivity, hydrologic systems, and water rights is necessary. This is particularly true given that increasing water scarcity is likely to intensify debate and conflict over water policies in the western United States and worldwide into the future.

## REFERENCES


Booker, J.F., Howitt, R.E., Michelsen, A.M., Young, R.A., 2012. Economics and modeling of water resources and policies. Natural Resource Modeling 25 (1), 168–218.

Brockmann, J.C., 2009. Overview of New Mexico's groundwater code. In: Paper Presented at the 27th Biennial Groundwater Conference and 18th Annual Meeting of the Groundwater Resources Association of California, Sacramento, CA.

Burt, O.R., Baker, M., Helmers, G.A., 2002. Statistical estimation of streamflow depletion from irrigation wells. Water Resources Research 38 (12), pp. 32-1–32-13.

Burt, O.R., 1964. The economics of conjunctive use of ground and surface water. Hilgardia 36 (2), 31–111.

Chakravorty, U., Umetsu, C., 2003. Basinwide water management: a spatial model. Journal of Environmental Economics and Management 45 (1), 1–23.

Cobourn, K.M., 2015. Externalities and simultaneity in surface water-groundwater systems: challenges for water rights institutions. American Journal of Agricultural Economics 97 (3), 786–808.

Coman, K., 1911. Some unsettled problems in irrigation. American Economic Review 1 (1), 1–19.

Cosgrove, D.M., Johnson, G.S., 2004. Transient response functions for conjunctive water management in the Snake River Plain, Idaho. Journal of the American Water Resources Association 40 (6), 1469–1482.

Cosgrove, D.M., Johnson, G.S., 2005. Aquifer management zones based on simulated surface-water response functions. Journal of Water Resources Planning and Management 131 (2), 89–100.

Elbakidze, L., Cobourn, K.M., 2013. Economic foundations for the interdisciplinary modeling of water resources under climate change. Journal of Contemporary Water Research and Education 152 (1), 32–41.

Elbakidze, L., Shen, X., Taylor, G., Mooney, S., 2012. Spatiotemporal analysis of prior appropriations water calls. Water Resources Research 48 (6). http://dx.doi.org/10.1029/2011WR010609.

Fereday, J.C., 2008. Idaho decisions address groundwater–surface water conflicts in Snake River Basin. Water Law Newsletter XLI (2), 1–8.

Gemma, M., Tsur, Y., 2007. The stabilization value of groundwater and conjunctive water management under uncertainty. Review of Agricultural Economics 29 (3), 540–548.

Ghosh, S., Cobourn, K.M., Elbakidze, L., 2014. Water banking, conjunctive administration, and drought: the interaction of water markets and prior appropriation in Southeastern Idaho. Water Resources Research 50 (8), 6927–6949.

Glover, R.E., Balmer, G.G., 1954. River depletion resulting from pumping a well near a river. Transactions, American Geophysical Union 35 (3), 468–470.

Griffin, R.C., 1995. On the meaning of economic efficiency in policy analysis. Land Economics 71 (1), 1–15.

Griffin, R.C., 2006. Water Resource Economics: The Analysis of Scarcity, Policies, and Projects. MIT Press, Cambridge.

Harou, J.J., Lund, J.R., 2007. Hydroeconomic conjunctive use model formulation for California's Central Valley using response functions. In: Kabbes, K.C. (Ed.), World Environmental and Water Resources Congress 2007. ASCE, Tampa, pp. 1–13.

Harou, J.J., Pulido-Velazquez, M., Rosenberg, D.E., Medellin-Azuara, J., Lund, J.R., Howitt, R.E., 2009. Hydro-economic models: concepts, design, applications, and future prospects. Journal of Hydrology 375 (3–4), 627–643.

Hathaway, D.L., 2011. Transboundary groundwater policy: developing approaches in the western and southwestern United States. Journal of the American Water Resources Association 47 (1), 103–113.

Hazard, J., Shively, D., 2011. Conjunctive Management of Surface and Ground Water Resources in the Western United States. Montana Department of Natural Resources. Available from: dnrc.mt.gov/divisions/water/management/docs/clark-forktask-force/policy-administrative_tools-shively (30 December 2014).

Huffaker, R., Whittlesey, N., 2000. The allocative efficiency and conservation potential of water laws encouraging investments in on-farm irrigation technology. Agricultural Economics 24 (1), 47–60.

Knapp, K.C., Olson, L.J., 1995. The economics of conjunctive groundwater management with stochastic surface supplies. Journal of Environmental Economics and Management 28 (3), 340–356.

Kuwayama, Y., Brozović, N., 2013. The regulation of a spatially heterogeneous externality: tradable groundwater permits to protect streams. Journal of Environmental Economics and Management 66 (2), 364–382.

Miller, S.A., Johnson, G.S., Cosgrove, D.M., Larson, R., 2003. Regional scale modeling of surface and ground water interaction in the Snake River Basin. Journal of the American Water Resources Association 39 (3), 517–528.

Oklahoma Water Resources Board, 2010. Conjunctive Water Management in Oklahoma and Other States, State of Oklahoma. Available from: http://www.owrb.ok.gov/supply/ocwp/pdf_ocwp/WaterPlanUpdate/ConjunctiveWaterManagement.pdf (4 August 2015).

Pfeiffer, L., Lin, C.Y.C., 2012. Groundwater pumping and spatial externalities in agriculture. Journal of Environmental Economics and Management 64 (1), 16–30.

Schaible, G., Aillery, M., 2012. Water Conservation in Irrigated Agriculture: Trends and Challenges in the Face of Emerging Demands. U.S. Department of Agriculture. Available from: http://www.ers.usda.gov/publications/eib-economic-information-bulletin/eib99.aspx (4 July 2015).

Snyder, D.L., Coupal, R.H., 2005. Assessment of Relative Economic Consequences of Curtailment of Eastern Snake Plain Aquifer Groundwater Irrigation Rights, State of Idaho. Available from: http://www.researchgate.net/profile/Roger_Coupal/publication/266337939_Assessment_of_Relative_Economic_Consequences_of_Curtailment_of_Eastern_Snake_Plain_Aquifer_Ground_Water_Irrigation_Rights/links/54bcf7520cf253b50e2da383.pdf (4 August 2015).

Taylor, R.G., Schmidt, R.D., Stodick, L., Contor, B.A., 2014. Modeling conjunctive water use as a reciprocal externality. American Journal of Agricultural Economics 96 (3), 753–768.

Theis, C.V., 1941. The effect of a well on the flow of a nearby stream. Transactions, American Geophysical Union 22 (3), 734–738.

Tsur, Y., Graham-Tomasi, T., 1991. The buffer value of groundwater with stochastic surface water supplies. Journal of Environmental Economics and Management 21 (3), 201–224.

Tsur, Y., 1990. The stabilization role of groundwater when surface water supplies are uncertain: the implications for groundwater development. Water Resources Research 26 (5), 811–818.

Tuthill Jr., D.R., Rassier, P.J., Anderson, H.N., 2013. Conjunctive management in Idaho: public-private partnerships and conjunctive management of surface and ground water. In: The Water Report, vol. 108, pp. 1–11.

Western Rivers Institute, 2015. Water Terms and Concepts. Available from: http://westernriversinstitute.org/?page_id=186 (4 August 2015).

Winter, T.C., Harvey, J.W., Franke, O.L., Alley, W.M., 1998. Groundwater and Surface Water: A Single Resource. U.S. Geological Survey, Reston. Circular 1139.

Chapter 3.1.3

# Prospects for Desalination in the United States—Experiences From California, Florida, and Texas

J.R. Ziolkowska, R. Reyes
*The University of Oklahoma, Norman, OK, United States*

## 1. INTRODUCTION

Desalination (commonly called "desal") is a process of removing salts and minerals from seawater or brackish groundwater. In the past half century, desalination has emerged in the United States as a prospective technology for mitigating water scarcity and providing an alternative water source for municipal use as well as industrial and power generation processes. This development was triggered by two opposite trends occurring simultaneously: (1) an exponential growth in water demand and (2) a steady decline in available groundwater and surface water resources. In addition, recurring extreme and exceptional droughts have exposed many regions in the United States to a dire tradeoff situation of allocating water between competing economic sectors.

Brackish groundwater desalination in the United States initially started in 1952, when the Saline Water Act passed by Congress provided federal support for desalination. The first seawater desalination plant was launched in the 1960s at Guantanamo Bay Naval Base, Cuba, as a result of military and political changes occurring in that region[1] (West Basin Municipal Water District, 2014). The rapid increase in desalination (especially since the 1990s) was a result of a growing water demand, improving desalination technology, and state-funded subsidies that encouraged investments in the desalination sector.

As of 2013, 68% of the installed desalination plant capacity in the United States was produced from brackish or inland water, 23% from river water, and only 4% from seawater [authors' calculations based on GWI (2013)]. Water source (feed water) is one

1. When water supplies to the naval base were cut off in retaliation for the Cuban Missile Crisis, the base became self-sufficient, desalinating 3.4 million gallons of water every day.

**Competition for Water Resources. http://dx.doi.org/10.1016/B978-0-12-803237-4.00017-3**

of the crucial determinants of the final desalination costs. Increased feed water salinity (measured in TDS—total dissolved solids[2]) also raises the final price for desalinated water.

In 2013, ~73% of desalinated water in the United States was supplied to municipalities as drinking water, ~22% to the industry sector, 5% was used in power stations, 2% in tourist facilities, and 1% in the agricultural sector for irrigation and for military purposes, respectively. Among the commercialized desalination technologies, reverse osmosis (RO)[3] provides 88% of desalinated water in the country, electrodialysis constitutes 8%, nanofiltration (also used as a pretreatment stage of RO) accounts for 3%, and electrodeionization makes only 1% of the total desalinated water in the country. The RO technology using a range of membranes (filters) for water purification has proven to be most efficient. Also distillation processes such as multistage flash distillation (MSF)[4] and multieffect distillation (MED) are still in use at a low scale with 18 MGD (million gallons per day) and 66.6 MGD of desalinated water in 2013, respectively [authors' calculations based on GWI (2013)].

At the global scale, the United States is among the leaders in the desalination sector with its daily production capacity of 2 BGD (billion gallons per day) and around 1336 operating plants in 2013. With its current 13% share in the global sector (measured with the desalination production in MGD) (GWI, 2013), the United States competes for the top desalination position with Saudi Arabia and United Arab Emirates. Whereas the global position of the United States in the desalination sector is important, this chapter aims to present a regional perspective of desalination within the United States, based on the idea that solutions for water problems should be designed at the regional level, according to the phrase: "Think globally, act locally."

To depict a broad picture of developments in the desalination sector as well as prospects for this emerging technology, this chapter will conceptualize a SWOT (strengths/weaknesses/opportunities/threats) analysis for desalination in the United States. Further, it will also present experiences from and examples of different desalination projects in California, Florida, and Texas—the states characterized by the highest desalination capacity and the largest number of desalination plants in the country.

---

2. Total dissolved solids (TDS) describe the total amount of mobile charged ions (including minerals, salts or metals) dissolved in a given volume of water, expressed in units of mg per unit volume of water (mg/L), also referred to as parts per million (ppm).
3. In the RO process water molecules are moved from a concentrated solution to a dilute solution by applying pressure on the concentrate. It occurs by means of a set of filters serving as semipermeable membranes. The process is opposite to osmosis occurring naturally where solvent moves through a membrane from an area of low concentration to an area of high concentration.
4. Both the MSF and MED processes are based on distillation that takes place in a series of chambers (or effects) operating at progressively lower pressures. The main difference between them regards the method of evaporation and heat transfer. In a MED plant, evaporation occurs through the contact of the seawater film with the heat transfer source, whereas in a MSF plant heating of seawater occurs within the tubes and evaporation is a result of the brine flow (flashing) in each stage to produce vapor.

# 2. PROSPECTS FOR DESALINATION IN THE UNITED STATES—SWOT ANALYSIS

The desalination sector in the United States is directly affected by variability in the US economy and water demands, and thus assessments about the future of desalination vary among scientists. Several research studies have claimed that desalination will become a reliable technology (Voutchkov, 2010) providing sustainable water supply and lessening pressure on traditional water sources. Other studies anticipate that high desalination costs and demanding environmental standards will hinder desalination developments in the years to come (Ghaffour et al., 2013).

To understand the underlying issues that might affect the long-term developments in the sector in the United States, a SWOT analysis has been developed in this chapter. The SWOT methodology originated from work of Albert S. Humphrey in the 1960s. This analysis has been used for many years in business settings to help companies and businesses find a sustainable niche in their market place. Even though the method is normally applied as a preevaluation of feasibility studies, it can be used for both ex-ante and ex-post assessments at the enterprise, sector, household, and/or personal assessment levels.

By analyzing strengths the SWOT analysis allows for unveiling unrevealed opportunities, while by evaluating weaknesses, potential threats can be mitigated ahead of time. In the context of desalination assessment, we define the "threats" as risks, uncertainties, and challenges. The presented analysis defines and embraces the most important aspects of desalination, while the list could be extended in the face of constant changes occurring in the sector. Moreover, it needs to be emphasized that some weaknesses and challenges create direct opportunities for the sector's development. At the same time, the strengths and opportunities generated by the sector provide benefits for the entire economy. Those benefits are not included directly in the SWOT analysis, which, in this specific case, is focused only on the desalination technology and sector (when evaluating the strengths and weaknesses), and on the external environment (when evaluating opportunities and threats/challenges). The SWOT analysis for the desalination sector encompasses four elements (Table 1):

1. **S**trengths of the sector and technology fostering success
2. **W**eaknesses of the sector and technology hindering success
3. **O**pportunities created by the environment fostering success
4. **T**hreats/challenges created by the environment hindering success

A detailed deliberation on the most important SWOT factors is presented in Sections 2.1–2.4.

## 2.1 Strengths of the Desalination Sector

Desalination has been used for many purposes: municipal, industrial, power generation, in tourist facilities, and for military services, with the most pronounced benefits being drinking water provision in times of extreme and unexpected weather events like drought.

Because of growing investments and competition on the desalination and material supplier markets, the desalination sector has been fast and flexible when adjusting to new and more efficient technological innovations over time. It has been proven that more precise desalination membranes increase efficiency of desalination and reduce the total desalination costs, and thus the price for desalinated water for the final consumer (Ghaffour

**TABLE 1** **SWOT Analysis for the Desalination Sector in the United States**

| **Strengths** | **Weaknesses** |
|---|---|
| Alternative water source in drought-affected or high-water-demand areas | Relatively high desalination costs |
| High yield of produced water | High energy requirements |
| Alleviates stress and pressure on traditional water sources (protecting habitats and environmental flows) | Maintenance costs even if plants are temporarily offline (mothballed) |
| Independent from weather events | Environmental effects of brine disposal |
| Fast and flexible to adjust to new and efficient technology innovations | Dependency on proximity to water (brackish groundwater, river water, or seawater) |
| Flexible in increasing water production in emergency cases (up to the total capacity) | Operating at a large scale[a] (portable units—desal skids are too expensive) |
| **Opportunities** | **Threats (Risks/Challenges)** |
| Brackish groundwater resources potentially abundant (USGS, 2014) | Mothballing risk in situations of low water demand |
| Coast line and easy accessibility to seawater | Potential sustainability regulations might increase desalination costs |
| Availability of state subsidies for desalination projects | Volatility of oil prices and energy production might impact desalination costs |
| Well-developed R&D infrastructure in the country (institutional collaborations: USGS, State Water Resources Boards, regional governments) | Administrative limitations and long waiting times for permit issuance |
| Market strength—growing investments in the desalination sector | Potential competing technologies |
| Recurring droughts and growing population create elevated demand for water | No recent studies on brackish water availability |
| Debates about value of water, water markets, and pricing water as a natural resource (comparable with oil/gas prices) | Low social awareness about the need for additional water supply |
| Renewable energies can help lower desalination costs in the long term | Potential negative environmental impacts |

*USGS*, US Geological Survey.
[a]*Large-scale operation refers to any industrial operation in a traditional desalination plant, even with very low production capacity. In contrast, small-scale operations refer to deployable mobile units (desal skids) that do not need the industrial infrastructure, but can be transported and used in any place at any time.*
Authors' analysis and presentation.

et al., 2013; Nair and Kumar, 2013; Penate and Garcia, 2012; Zhou and Tol, 2004; Van der Bruggen, 2003).

In the face of exponentially growing population and recurring droughts in many US regions, desalination has been acknowledged as a technology that can help alleviate the identified stress on water resources. In this way, also natural habitats and environmental flows can be protected. In addition, desalination is highly effective as large plants have desalination capacities of 25−30 MGD, while they are ready for emergency situations and urgent spikes in water demand. Because of relatively high prices of desalinated water compared with water from traditional sources, a significant increase of desalination in municipal water portfolios could result in growing water rates paid by the final customers. However, in situations of long-term water scarcity, desalination can be life-saving. Also, using the full production capacity of desalination plants could generate economies of size, and thus allow for lower water costs.

## 2.2 Weaknesses of the Desalination Sector

The main factor hindering the fast development of desalination is high production costs of desalinated water and the resulting high water rates. The prices for desalinated water are highly variable and on average two or three times higher than prices for water from traditional water sources (Afgan et al., 1999; Daniels and Daniels, 2003). In 2010, they ranged between \$0.2−1.2/m$^3$ (\$0.8−4.5/kgal) for desalinated brackish groundwater and \$0.3−3.2/m$^3$ (\$1.1−12.1/kgal) for desalinated seawater (with technologies using conventional energy sources) (Gude et al., 2010; Karagiannis and Soldatos, 2008). Thus, in normal wet years, desalination might be economically infeasible. This weakness of the technology can be seen as a challenge at the same time. New more efficient membranes are being developed and solutions are explored that would help decrease desalination production costs and thus make the technology a complementary option in the regional water plans and portfolios. It needs to be emphasized that cost feasibility of desalination is not necessarily a fair comparison because the current prices paid for conventionally treated fresh water usually do not reflect the true value of water resources. Moreover, the prices for water from conventional sources do not include the water scarcity value that would otherwise elevate the water rates and theoretically make desalination a very affordable option.

Desalination costs can be divided into: (1) capital costs (CAPEX) occurring as a one-time investment in the construction phase of the desalination plant and (2) operational and maintenance (O&M) costs (OPEX) occurring continuously during the operation period of the plant. While capital costs are amortized for every single year of operating the plant, operational costs can make the final price of desalinated water unaffordable, compared to cheaper and easier accessible, traditional (even though depleted) water sources. OPEX costs include: energy costs, costs of chemicals, labor and management costs, maintenance and membrane exchange costs, disposal costs of brine (a highly concentrated saline by-product in the desalination process), and institutional charges (compliance and regulatory costs, access charges, etc.).

The main cost components driving the final price of desalinated water are energy costs and brine disposal costs; whereas the other cost determinants include the applied desalination technology, capital and other operational costs, production capacity, and water salinity.

According to the Australian Government (2008), energy costs account for 60% of the total O&M costs of operating seawater desalination plants. A study of Mabrouk et al. (2010) found that energy costs make between 46% and 73% of the total desalinated water cost, depending on the salinity levels. Also, a study by Ziolkowska (2015) confirmed a linear relationship and a very strong correlation between the energy pieces and the final water prices. Because of the complexity of desalination processes, as well as many factors determining the final price, both energy requirements and final prices of desalinated water can vary considerably from region to region and from plant to plant. Table 2 presents energy requirements subject to the applied desalination technology (membrane vs. distillation processes). It can be clearly stated that distillation processes (MSF and MED) are more energy intensive, which directly impacts the final price of desalinated water.

Another major cost component is disposal costs of brine. The most common brine disposal approaches include: (1) surface water discharge (direct ocean outfall, shore outfall, colocated outfall, discharge to rivers, canals, lakes), (2) disposal to sewer (sewer line, direct line to wastewater treatment plant, tracking concentrate to wastewater treatment plant), (3) subsurface injection (deep or shallow well injection), and (4) evaporation ponds. Because of additional capital costs, evaporation ponds and Zero Liquid Discharge (ZLD) treatment processes are most expensive (Table 3). Brine disposal costs can be kept low or completely eliminated if brine is disposed back to the ocean (in the case of seawater desalination or brackish water plants located at the shore, like, for instance, the Southmost desalination plant in Brownsville, Texas). In the case of seawater desalination, it needs to be mentioned that the favorable location of seawater desalination plants (eliminating brine disposal costs) cannot completely leverage the energy requirements that are very high because of high water salinity. Thus, brackish desalination plants are more desirable and affordable. Moreover, taking into consideration that because of missing legal regulations, many brackish desalination plants dispose of brine directly into sewers, the disposal costs in those plants are kept at a minimum in many cases, despite potential environmental harm created by those practices.

Another relevant weakness is the potential environmental effects of brine disposal. Environmentalists are concerned that brine from seawater RO plants accumulates on the sea floor in shallow coastal waters and negatively affects benthic communities. In the case of brackish groundwater desalination, disposal to the sewer can negatively impact river ecosystems and increase wastewater treatment costs. Those concerns are not always confirmed in practice, as depicted with the example from Florida (Section 3). However, little solid evidence exists about the extent of environmental impacts and the associated external costs.

Currently, desalination plants operate at a high capacity and require a demand market to be efficient. In some situations, smaller units (desalination skids) would be enough to satisfy local water needs. Even though mobile units are available from water treatment companies, such as SevenSeasWater, Osmoflo, Aquamove, Aquatech, and others, the lease price ($700,000–3,000,000) for production capacity between 200,000 GPD (gallons per day) and 1 million GPD (BenJemaa, 2009) constitute a considerable financial limitation for a broad application of portable appliances.

Because of the currently high costs, municipalities and policymakers are hesitant about incorporating desalination as a substitute for water from traditional sources. However, in times of water scarcity, desalination could turn out to be the most reliable water supply option.

**TABLE 2** Energy Consumption and Average Water Cost of Large-Scale Commercial Desalination Processes

| Process | Thermal Energy (kWh/m$^3$) | Electrical Energy (kWh/m$^3$) | Total Energy (kWh/m$^3$) | Investment Cost (\$/m$^3$/day) | Total Water Cost (\$/m$^3$) |
|---|---|---|---|---|---|
| **MSF** (multistage flash distillation) | 7.5–12 | 2.5–4 | 10–16 | 1200–2500 | 0.8–1.5[a] |
| **MED** (multieffect distillation) | 4–7 | 1.5–2 | 5.5–9 | 900–2000 | 0.7–1.2 |
| **SWRO** (seawater reverse osmosis) | – | 3–4[b] | 3–4 | 900–2500 | 0.5–1.2 |
| **BWRO** (brackish water reverse osmosis) | – | 0.5–2.5 | 0.5–2.5 | 300–1200 | 0.2–0.4 |

[a]*Includes subsidies (price of fuel).*
[b]*Includes energy recovery system.*
Authors' presentation based on data from GWI (2013), Quteishat (2009), Reddy and Ghaffour (2007), Pankratz (2008), Borsani and Rebagliati (2005), Sommariva et al. (2003), Darwish and Al-Najem (1987), Maurel (2006), Australian Government (2008).

**TABLE 3 Cost Comparison of Brine Disposal Options**

| Disposal Option | Cost ($/m$^3$) | Critical Factors |
|---|---|---|
| Surface water | 0.03–0.30 | Piping, pumping, and outfall construction |
| Evaporation pond | 1.18–10.04 | Pond size and depth, salt concentration, evaporation rate, pond liner cost |
| Deep well injection | 0.33–2.64 | Tubing diameter and depth, injection rate, chemical costs |
| Sewer | 0.30–0.66 | Disposal rate, salinity, sewer capacity, fees |
| Brine concentrator (ZLD) | 0.66–26.41 | Disposal rate, energy costs, salinity |

Note: Costs for concentrate disposal include both capital and operational and maintenance costs.
*ZLD*, Zero liquid discharge.
Authors' presentation based on Greenlee et al. (2009), Koyuncu et al. (2001), Mickley (2001), Miller (2003), Sethi (2007).

## 2.3 Opportunities for Desalination

Statistics show that the United States has been affected by drought every single year for many decades and centuries, with regional variations and recurring droughts in the West, Midwest or the South of the country. Since mid-2010, the United States has been experiencing extreme and exceptional drought[5] (Drought Monitor, 2015) [with the Standardized Precipitation Index[6] (SPI) equal to −2.0 or less] (NOAA, 2013).

As of March 2015, more than 50% of the contiguous US area has been affected by drought, with about 8% constituting extreme or exceptional drought in California, Texas, and Oklahoma (Drought Monitor, 2015).

While water conservation measures in the affected regions are most effective in the short term, they do not provide an additional water supply needed to mitigate the effects of drought and water demands resulting from a growing population. In many states, aquifers have been largely depleted or are close to depletion because of anthropogenic activities, while streamflow conditions have been considerably worsening, especially during drought periods. Desalination could constitute a viable technology to meet urgent water demand resulting from drought as well as to provide a long-term solution by incorporating it into municipal water portfolios and leveraging pressure on traditional water sources (groundwater and surface water).

Another significant opportunity for an effective long-term development of desalination is the abundant resource availability of seawater and brackish water in the United States.

5. The USDA and NOAA use the following classification for drought severity: D0—Abnormally dry, D1—Moderate drought, D2—Severe drought, D3—Extreme drought, and D5—Exceptional drought.
6. The Standardized Precipitation Index (SPI) is an index based on the probability of recording a given amount of precipitation. The index is negative for drought and positive for wet conditions. As the dry or wet conditions become more severe, the index becomes more negative or positive.

According to USGS (2014), mineralized brackish groundwater underlies most of the country. Despite many definitions of brackish water, we define it as water with a greater dissolved solids content (1000 and 10,000 mg/L TDS) than present in fresh water, but lower than the one observed in seawater (35,000 mg/L TDS). Brackish water is also classified as "saline" water characterized with a TDS greater than 1000 mg/L. Although brackish groundwater is used for different purposes such as cooling water for power generation, aquaculture, and in the oil and gas industry (drilling, enhancing recovery, and hydraulic fracturing), this contribution is mainly focused on brackish water for drinking purposes (municipal use), industry purposes, and power generation. With 95,471 miles of shore line, the United States has a considerable advantage for developing seawater desalination. Currently, most desalination plants are located in the coastal states (both on the East and West Coast) (Ziolkowska and Reyes, 2016).

Governmental and regional support for desalination in the form of subsidies presents one of the opportunities for a fast development of desalination. Examples from Tampa Bay, Florida, and five projects in Southern California subsidized by the Metropolitan Water District of Southern California confirm that the majority of desalination projects in the United States are subsidized to some degree. As subsidies are often hidden, the estimation of the reported and actual desalination costs might be nonrepresentative and problematic. According to Cooley et al. (2006), public subsidies for desalination plants should be granted only when explicit public benefits are guaranteed, for instance restoration of ecosystem flows. Nowadays, subsidies for establishing new desalination plants create a boost to the development of the sector and attract private investors. As with any developing sector and market, subsidies are meant to be a short-term stimulus crucial for their uptake, similarly as experienced with biofuels production. The strong position of the desalination sector in the market place substantiates its readiness to stand for itself and remain resistant to market fluctuations in the intermediate and long term. Global investments in desalination have been continuously growing since the 1980s and reached the peak of $9 billion in capital expenditures in 2007, with a compound annual growth rate (based on a constant rate of return) anticipated to reach 8.3% in the 2010–2018 time period (Gasson, 2013). The well-developed R&D infrastructure in the country as well as institutional collaborations among many water entities, such as the US Geological Survey (USGS), State Water Resources Boards, and regional governments can foster successful long-term development of the desalination sector.

Another relevant opportunity in this context (and a positive spillover effect) is a chance for initiating regional and national debates about the real value of water as a resource. For centuries, water has been underpriced, while the current water rates represent only the costs of extracting water from aquifers, water treatment costs, delivery costs to the final consumer, and administrative costs of water utility companies. Although water rates differ in different regions of the country, depending on the bed rock, water accessibility, water pollution, etc. (eg, $2.65/kgal in Oklahoma City in 2015), they do not represent the actual value of water. For years, water from traditional sources (groundwater and surface water) has been available as a free resource, thus making high costs of desalinated water an unacceptable option. However, in the face of extreme droughts in the past several years policymakers and researchers often debate the establishment of water markets. This would allow for a reliable market mechanism of selling and buying water rights and determining the price for water resources based on its real value as a natural resource. The debates could induce training and education workshops for citizens, which in turn could have positive implications on water conservation measures. Although water markets already exist in some states, for example, Arizona, California, Colorado, Nevada, they are said to be insufficient, imperfect, and serving as water banks rather than real water markets per se. Concrete measures are needed to establish social awareness about the value of water, and the desalination issue could create a platform for a broader debate and implementation of appropriate programs and policies.

## 2.4 Challenges for Desalination

An important group of challenges for desalination constitute environmental concerns, while seawater desalination raises more issues at hand. With seawater intake, the risk of loss of aquatic organisms through impingement (organisms collide with intake screens) or entrainment (organisms are drawn into the plant with the source water) is accentuated in environmental debates. Also construction of the intake infrastructure and piping could potentially disturb the seabed and cause resuspension of sediments, nutrients, or pollutants into the water column. Moreover, desalination plants could increase ambient seawater salinity in the ocean and contribute to seawater pollution through chemical additives in the desalination process, for example: sodium hypochlorite, ferric chloride, or aluminum chloride, antiscale additives (sodium hexametaphosphate), and acids, for example, sulfuric acid or hydrochloric acid, among others. Another potential impact on marine life regards changes in seawater temperature caused by elevated outlet water temperature from cooling processes in situations when desalination plants operate in conjunction with power plants. While most organisms can adapt to minor temperature deviations and salinity level changes, continuous exposure could cause long-lasting change in species composition. Other concerns regard dissolved oxygen, chlorine concentration, heavy metals, and un-ionized ammonia removed in the desalination process (Romeil, 1977; Lee et al., 2008; Abdul-Wahab and Jupp, 2009), though recent comprehensive studies addressing those issues are still limited. A solution to those concerns provides dilution of brine water with seawater or cooling water before discharge to the sea or else brine harvesting.

Because of high energy demands (and energy production based on fossil fuels), desalination has been discussed in terms of greenhouse gas (GHG) emissions. A study by Dawoud and Al Mulla (2012) for Gulf countries has found that in 1996–2010, $CO_2$ emissions from desalination doubled in Bahrain, Saudi Arabia, Qatar, and Oman, while they increased by 50% in United Arab Emirates and by 87% in Kuwait. According to Rauly et al. (2005), depending on the energy source, RO desalination plants generate between 1.75 and 2.79 kg (3.8–6.1 lb) of $CO_2$ emissions per cubic meter ($m^3$) (264.2 US gal) of produced water. This translates to 14.7–23.3 lb $CO_2$ per 1000 gal of desalinated water. $CO_2$ emissions from distillation processes (MSF and MED) range from 1.19 to 23.41 kg/$m^3$ (9.9–195.3 lb/kgal).

Currently, no regulatory standards exist regarding energy use and GHG emissions generated by desalination in the United States. However, according to Pankratz (2012) and Cooley and Heberger (2013) there are a variety of state programs, policies, and agencies that must be considered when developing a desalination project. In California, for example, these include environmental review requirements, for example, issuance of permits by the Coastal Commission, the Integrated Regional Water Management Planning process, and policies of other state agencies, such as the State Lands Commission, the State Water Resources Control Board, and regional standards like the California Environmental Quality Act. Using natural gas and renewable energy sources could alleviate the problem of GHG emissions. From the economic perspective, however, and given current high prices for renewable energies, using conventional energy for desalination is currently most cost efficient (Gude et al., 2010; Karagiannis and Soldatos, 2008).

Another challenge is a potential risk related to various factors including design and technology, financing sources, permits, construction, operation/performance, financing, markets, and policy regulations, as well as ways of mitigating those problems (Cooley and Ajami, 2012). One of the major risks and uncertainties refers to mothballing desalination plants if desalinated water is not needed and/or it is not competitive with current rates for water from traditional sources. Experiences from Australia, when it was plagued by the Millennium Drought (1995–2009), show a number of plants that were mothballed after the end of the drought, when replenished aquifers and rivers provided a cheaper water supply. Those plants, however, were built without a long-term business plan, but rather specifically for the purpose of mitigating drought effects in the short run; leaving the municipalities locked into long-term (usually 30-year) contracts and owing high debts without generating the expected revenue from desalinated water sales. Only a few plants have been mothballed in the United States. Examples include the Santa Barbara desalination plant, which was built in 1992 and mothballed after 2 months of operation, and the Yuma plant in Arizona mothballed 1 year after the start of its operation (Bureau of Reclamation, 2012). Several plants have been decommissioned because of their long-time operation, but in most cases they were replaced with new plants at the same sites. However, most desalination plants have been successfully operating continuously for years. Although unsuccessful cases are isolated considering the large scope of desalination in the United States, mothballing desalination plants has raised questions about the feasibility of desalination projects.

At the current high costs for desalinated water, one of the main challenges is to lower final prices or at least keep them at a stable level. In the long term, however, potential sustainability and environmental regulations could theoretically increase costs of desalination making it even less affordable. There are currently no legislative plans indicating or anticipating such standards at the national level. Nevertheless, strong environmental concerns could create a trigger for legal changes in the future. Ensuring low prices for desalinated water is challenging also because of high volatility of oil and gas prices. Many desalination plants use oil, gas, coal, or nuclear power as an energy source. Also for plants using electricity, volatility of oil and gas prices could affect the final price of desalinated water. As desalination requires high energy inputs, even a slight increase in oil/gas prices could significantly affect desalination costs and negatively impact the development of the sector. The currently low oil prices (first quarter of 2015) have a positive impact on investments in desalination and can contribute to a significant decrease in desalination costs.

## 3. EXPERIENCES WITH DESALINATION FROM CALIFORNIA, FLORIDA, AND TEXAS

California, Florida, and Texas are the leading states in the desalination sector in the United States, both in terms of the installed cumulative capacity and the total number of operating plants. In all those states, brackish groundwater, river water, and seawater desalination are predominant. This chapter provides a comparative analysis, as well as experiences with desalination in those states.

California, Florida, and Texas are worth considering also because of their differing climatic conditions. California has been suffering from drought since early 2012, with 100% of the state's area affected, and almost 80% of the area plagued by extreme and exceptional drought since 2014. The extreme and exceptional drought in Texas in 2011 and

2012 affected 98% of its land area. As of March 2015, 60% of the area in Texas was affected, while only 15% can be attributed to extreme and exceptional drought. Florida was impacted by drought mainly in 2011–2013; however, extreme drought affected only 40% of the state in the peak times during that period (Drought Monitor, 2015). As California and Florida are among the largest producers of agricultural produce in the United States, water availability for agriculture is crucial and a relevant topic of policy debates. In the situation of drought, however, combined with depleted aquifers, water conservation measures imposed on local citizens are not enough to mitigate water scarcity. The existing desalination plants in California and Florida provide a buffer for municipal water use and, if needed, could potentially help redirect water for municipal purposes towards agricultural production.

Current developments show that the total desalination capacity has been increasing in all three states, with Florida leading in brackish groundwater and river desalination, followed by brackish groundwater desalination in California and Texas (Fig. 1). Seawater desalination is very limited in all three states, despite their close proximity to the ocean. This can be explained with higher seawater desalination costs compared to brackish groundwater desalination. However, assuming the anticipated technological developments in the desalination sector and the resulting decrease in desalination costs, seawater desalination could prove to be a viable and cost-competitive solution in case brackish groundwater resources start to diminish.

Among the analyzed states, Florida is leading both in terms of the number of the operating plants and the total desalination capacity (in MGD), followed by California (Table 4). In all three states, RO is the prevailing desalination technology. The same trend applies to the final consumer of desalinated water, where in all three states the main beneficiaries are municipalities, while industry takes second place. In 2013, desalination plants in California, Florida, and Texas made up 45% of the total number of all operating desalination plants in the United States, while they provided 66% of the total desalinated water in the country. This proves the strength and maturity of the desalination sector in the analyzed states. Moreover, in terms of the technology, RO accounts for 68.5% of the total desalinated water in the country. The production of desalinated water for drinking purposes in the three states makes 77% of desalinated water for this customer group in the United States, while 53.5% is for power generation, and 31% for the industry sector [authors'

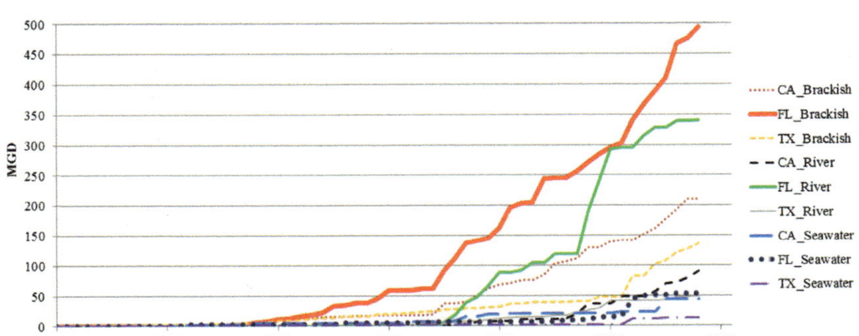


**FIGURE 1** Cumulative production capacity (MGD) by feed water source in California (CA), Florida (FL), and Texas (TX) in 1955–2013. *Authors' calculations and presentation based on data from GWI (2013).*

**TABLE 4 Operating Desalination Plants in California, Florida, and Texas in 2013**

| 2013 | California | Florida | Texas | % of Three States in United States Total |
|---|---|---|---|---|
| **# Of plants online** | 98 | 144 | 67 | 45 |
| **Total MGD** | 290.3 | 674.7 | 152.8 | 66.3 |
| **Technology (MGD)** | | | | |
| RO | 263.5 | 655 | 146.8 | 68.5 |
| MED | 2.9 | 2.1 | 5.9 | 37.6 |
| **User (MGD)** | | | | |
| Municipalities | 208.3 | 617.1 | 107.0 | 77.3 |
| Industry | 41.7 | 23.5 | 39.0 | 31.2 |
| Power generation | 21.9 | 10.7 | 4.6 | 53.5 |

*MED*, multieffect distillation; *MGD*, million gallons per day; *RO*, reverse osmosis.
Authors' calculations based on data from GWI (2013).

calculations based on GWI (2013)]. This comparison emphasizes that California, Florida, and Texas have been applying desalination as a supplemental water supply option because of the existing water scarcity. Desalination plants under construction and offline plants have not been accounted for in this analysis.

## 3.1 Experiences From California

California is the second largest state in the United States in terms of desalination capacity. This development can be attributed to a large number of low-capacity plants: 60 plants producing water for municipal purposes at a total capacity of 208 MGD. This makes 61% of all desalination plants and 71% of the total desalination in the entire state of California. The smallest desalination plant with its daily capacity of 0.02 MGD is located in San Luis Obispo County in Santa Margarita, while the largest plant in San Bernardino County in Yucaipa produces 12 MGD (GWI, 2013). Because of this large distribution and production dispersion, it is difficult to identify a desalination plant that could serve as a representative example of desalination processes happening in the state.

The Santa Cruz seawater RO desalination plant gained public and media attention when it was proposed by the City of Santa Cruz Water Department and Soquel Creek Water District in 2008. With its expected capacity of 2.5 MGD, the plant was envisioned to mitigate several water scarcity-related problems in the county: aquifer depletion, drought conditions, endangered environmental flows, and aquifers contamination because of saltwater intrusion caused by overpumping (SCWD2, 2014). The pilot operation of the desalination plant in Santa Cruz in 2008–2009 at the capacity of 0.07 MGD was successful, while the 2013 Environmental Impact Report for the project found only insignificant environmental impact on marine life (SCWD2, 2013; Cooley et al., 2013). The plant was also anticipated to be transformed from a desalination plant to a direct-to-potable water recycling facility. After years of planning, the Santa Cruz City Council put a halt to the project, one year before public vote because of strong and vocal opposition from a number of residents. The Santa Cruz plant constitutes a single example of the complexity in the process of designing a desalination plant, including environmental assessments and permits that can take up to several years in waiting time. This proves one of the discussed challenges of investment risks in the short or long term.

A successful example in California is the seawater RO desalination plant in Carlsbad with an anticipated operation start date in late 2015. The plant will be the largest seawater desalination plant in the Americas providing 54 MGD that can supply around 300,000 people per day. The plant will operate in coexistence with the Encina Power Station and provide up to 10% of the region's drinking water. It will alleviate demand for water from traditional sources (Colorado River, Northern California, groundwater aquifers, and local surface water) (Poseidon Water, 2015). The total capital costs for the Carlsbad plant amounted to approximately $1 billion, which makes it one of the most expensive plants in the United States. Total O&M costs (including pipeline distributing water to the community) are anticipated at $49–54 million annually. Despite high expenses, positive socioeconomic impacts of building the plant are expected as follows: 2500 direct and induced new jobs and economic output of $350 million over 2 years. Once in operation, the plant will have 18 full-time employees, support 500 direct, indirect, and induced jobs, and contribute $50 million in estimated annual spending to the county's economy. As the plant is close to completion, around 1200 decision-makers and stakeholders have sampled the desalinated water, with more than 99% of respondents rating the desalinated water taste as either "excellent" or "good," and 83.9% ranking it as "excellent" (Poseidon Water, 2015). This proves the effectiveness of this investment. However, also in this case the planning process took 12 years, while the state's permitting process took over 6 years. The San Diego County Water Authority and Poseidon agreed on a 30-year water purchase contract for the total water output of the plant (Poseidon Water, 2015). This example substantiates the need of curtailing administrative procedures to make desalination a less risky investment.

## 3.2 Experiences From Florida

Florida has been the leading state in brackish groundwater and river water desalination, and since 2006 in seawater desalination. The largest seawater desalination plant in the state (also the largest seawater plant in North America) is located in Tampa Bay and uses RO to produce 25 MGD (Tampa Bay Water, 2010). The plant is located next to Tampa Electric's (TECO) Big Bend Power Station, which withdraws and discharges up to 1.4 BGD of seawater from Tampa Bay, using it as cooling water for the power plant.

With the official start of the plant in March 2003 and several legal difficulties with construction deficiencies and revamping the plant, it started its production in a remodeled structure in spring 2007. The plant is a public-private partnership between American Water-ACCIONA (operating the plant), the Southwest Florida Water Management District (responsible for managing the public's water resources in 16 counties of West Central Florida), and Tampa Electric Company (leasing the 8.5-acre plant site to Tampa Bay Water and providing electricity and source water for the desalination plant).

The Tampa Bay desalination plant provides a valid example to discuss hands-on environmental issues and concerns of seawater desalination described as a risk/challenge component in the SWOT analysis. At its full capacity, the RO process produces around 19 MGD brine with its salinity twice as high as the feed water (seawater). Brine is returned to the Big Bend's cooling water stream and blended with up to 1.4 billion gallons of cooling water, which allows for achieving a blending ratio of up to 70:1. At the point of entering and mixing with bay water, brine salinity is on average only 1−1.5% higher than seawater in Tampa Bay. This slight increase is within Tampa Bay's normal seasonal salinity fluctuations. Moreover, to insure compliance with basic environmental permits, Tampa Bay Water's hydrobiological monitoring program collects water samples every 15 min near the desalination facility. Those measurements and other water quality monitoring since 2003 have shown no measurable salinity changes in Tampa Bay related to the plant production. The plant's alarm system will warn plant operators in case of deviations from the standard levels, while it will automatically shut down affected areas of the facility if monitored levels exceed predetermined parameters (Tampa Bay Water, 2010, n.d.).

Also, the desalination plant uses water from Tampa Electric's Big Bend power plant, which eliminates any potential risk of fish entrapment through the intake system. Cooperation with the power plant and the warm temperature of the power plant's cooling water combined with relatively low salinity of the Tampa Bay seawater is a benefit to optimizing the RO process and keeping costs down (Tampa Bay Water, 2010). Prior to the construction of the plant, several studies were conducted to address environmental concerns beforehand, such as cumulative impact analysis for Master Water Plan projects, US Geological Survey of the Big Bend Power Station area, as well as independent studies by Mote Marine Laboratory, Danish Hydraulic Institute, University of South Florida, Savannah Laboratory/STL Precision, Marinco Laboratory, and Hillsborough County. The studies were approved by the Florida Department of Environmental Protection (FDEP) and conducted in accordance with FDEP methods by an FDEP-approved laboratory. Each study concluded that the desalination plant would produce high-quality drinking water without any harm to Tampa Bay's water quality or marine life (Tampa Bay Water, n.d.).

Thus, experiences with the Tampa Bay seawater desalination plant provide examples of where environmental concerns related to desalination processes were successfully addressed. This can serve as a case study to inform the discussion about the commonly expressed environmental risks/challenges about desalination.

## 3.3 Experiences From Texas

Although the largest desalination plant in Texas is the Kay Bailey Hutchison plant in El Paso (25 MGD), this section will focus on a smaller plant—Southmost brackish groundwater RO desalination plant with its production capacity of 7.5 MGD and the expected lifetime of 50 years from the start of operation in 2004. The plant is located near the Gulf of Mexico and the Texas−Mexico border outside of Brownsville, Texas. It is owned and operated by the Southmost Regional Water Authority—a consortium of six partners including: Brownsville Public Utilities Board, City of Los Fresnos, Valley Municipal Utilities District No. 2, Town of Indian Lake, Brownsville Navigation District, and Laguna Madre Water District (Brownsville Public Utilities Board, n.d.; SRWA, 2014; Sturdivant et al., 2009). The plant represents a successful enterprise example of how desalination costs can be reduced, making desalination a viable and feasible solution.

The main reason for constructing the Southmost desalination plant was the rapid urban growth in the northern area of Brownsville, forcing the community to build either another water treatment facility or a desalination facility.

Because of the proximity to the Gulf, the plant is often misperceived and misclassified as a seawater desalination plant. The Southmost desalination plant utilizes brackish groundwater from the Gulf Coast aquifer (Sturdivant et al., 2009) with the approximate feed water salinity levels of 3500 ppm (parts per million). The desalination process reduces the salinity level down to 300–475 ppm (Sturdivant et al., 2009), which is below the maximum level (500 ppm) set by the US Environmental Protection Agency (EPA) for drinking water (Arroyo, 2005). The plant complies with the Texas Commission on Environmental Quality maximum discharge permit of 35,339 mg/L TDS per day (SRWA, 2014).

The actual production efficiency rate of the Southmost plant has varied over time because of operational and product-demand interruptions. In the first operation year (2004) production efficiency (measured as the ratio of the annual production and the design capacity production) accounted for only 13% of the total design capacity. Production efficiency increased up to 67.3% in 2007 and is anticipated to further increase up to 94%. The RO system of the plant operates at a very high recovery rate of 75%, meaning that three-quarters of the feed water is desalinated, while the remaining quarter is the waste product—brine (Sturdivant et al., 2009).

The Southmost desalination plant is an example of a facility that utilizes its geographical location to minimize desalination cost and thus the final prices for desalinated water. By using brackish groundwater with a lower salinity (compared to seawater), the facility is minimizing energy costs for the desalination process. At the same time, the facility disposes brine to the Gulf through a drainage ditch and ship channel extending to the Laguna Madre, thus reducing the total desalination costs.

The original construction costs of the plant amounted to $29 million, with the 2014 O&M costs of $3 million. The electricity costs made 23% of the O&M costs, while costs of chemicals accounted for 40% of the O&M costs. Because of the reduction of arsenic level in drinking water standards by the EPA in 2006, the Southmost desalination plant started the project of installing an additional pretreatment phase for the RO process, consisting of 12 MGD microfiltration membranes for arsenic and iron removal. In this way, the plant capacity has been expanded up to 11 MGD for an additional cost of $13 million. Despite its anticipated longevity, which assures the economic feasibility of this plant, regular capital replacement costs need to be considered as well. It has been estimated that wells/pumps need to be replaced every 3 years, which would create costs of $200,000, while the membranes are replaced every 6 years for $700,000 (Sturdivant et al., 2009).

Because of the proximity to the sea, seawater desalination has been considered as an alternative in Brownsville. A pilot study funded by the Texas Water Development Board with the demonstration capacity of 2.5 MGD found that while seawater desalination at the Brownsville Ship Channel is technically feasible, the anticipated full-scale capacity desalination plant (25 MGD) is not recommended at this time because of a lack of funds (NRS, 2008).

## 4. LESSONS LEARNED AND FUTURE PERSPECTIVES FOR DESALINATION IN THE UNITED STATES

Past and recent developments in desalination in the United States show that the desalination sector indicates both challenges and prospects. Since desalination gained its momentum in the 1990s and continued a steep upward development, in recent decades it has been described as a "drought-proof" water source.

The presented SWOT analysis shows a significant potential of desalination for mitigating water scarcity and initiating a debate about pricing water for its actual value as a resource. At the same time, several challenges still exist (eg, high desalination costs, environmental concerns, investment risks) that have been hindering a quick uptake of this technology. Solutions need to be found to provide desalinated water at a price level competitive with the water rates from traditional water sources. One of the solutions is a strategy of mixing desalinated water with traditional water sources that allows for keeping final water rates at an affordable and acceptable level. Another solution could include application of renewable energies, thus making desalination carbon neutral and partially eliminating environmental concerns. With the decreasing prices of renewable energies in the past decades, this solution seems feasible in the long term, while it might still be costly in some regions.

Despite the discussed obstacles, based on the data analysis and because of extreme weather events in many regions of the country, it can be anticipated that desalination will become a water supply technology of the future. This applies especially to the cases of the three analyzed states: California, Florida, and Texas, which have the fastest rates of population growth in the country. Consequently, water demand from all economic sectors in those states will increase in the future.

Currently, comprehensive economic analyses on desalination are still missing, mainly because of the complexity of the sector and missing knowledge and data on specific economic or environmental impacts. Moreover, regional differentiation of the plants and different desalination parameters in each single plant make it difficult to compare two desalination plants. More research and holistic approaches (eg, life cycle analysis, cost–benefit analysis) are needed to evaluate the complete picture of economic and environmental impacts of desalination processes.

## ACKNOWLEDGMENTS

This material is based on work supported by the National Science Foundation under Grant No. OIA-1301789. The authors thank Ms. Judy K. Adams—Water Treatment Manager at the Brownsville Public Utilities Board and the Southmost Regional Water Authority and Mr. Joe Saldivar for providing us with valuable information about the Southmost desalination plant.

## REFERENCES

Abdul-Wahab, S.A., Jupp, B.P., 2009. Levels of heavy metals in subtidal sediments in the vicinity of thermal power/desalination plants: a case study. Desalination 244 (1–3), 261–282.

Afgan, N.H., Darwish, M., Carvalho, M.G., 1999. Sustainability assessment of desalination plants for water production. Desalination 124 (1–3), 19–31.

Arroyo, J.A., 2005. Water Desalination. http://www.twdb.state.tx.us/publications/reports/GroundWaterReports/GWReports/R360AEPC/Ch15.pdf (05/20/2006).

Australian Government, 2008. Emerging Trends in Desalination: A Review. Waterlines Report Series 9, Australia.

BenJemaa, F., 2009. Water Use and Efficiency Branch. Water Recycling and Desalination. Logistics for Deploying Mobile Water Desalination Units. State of California—Department of Water Resources, Sacramento.

Borsani, R., Rebagliati, S., 2005. Fundamentals and costing of MSF desalination plants and comparison with other technologies. Desalination 182, 29–37.

Brownsville Public Utilities Board, n.d. Desalination. In: http://www.brownsvillepub.com/water/desalination.html (05/17/2006).

Bureau of Reclamation, 2012. Yuma Desalting Plant Pilot Run Final Report. US Department of the Interior, Yuma Area Office.

Colley, H., Ajami, N., 2012. Key Issues for Desalination in California – Cost and Financing. Pacific Institute, Oakland.

Cooley, H., Heberger, M., 2013. Key Issues for Seawater Desalination in California – Energy and Greenhouse Gas Emissions. Pacific Institute, Oakland.

Cooley, H., Gleick, P.H., Wolff, G., 2006. Desalination with a Grain of Salt – A California Perspective. Pacific Institute, Oakland.

Cooley, H., Ajami, N., Heberger, M., 2013. Key Issues in Seawater Desalination in California – Marine Impacts. Pacific Institute, Oakland.

Daniels, T., Daniels, K., 2003. The Environmental Planning Handbook for Sustainable Communities and Regions. APA/Planners Press, Washington DC.

Darwish, M.A., Al-Najem, N.M., 1987. Energy consumption and costs of different desalting systems. Desalination 64, 83–96.

Dawoud, M.A., Al Mulla, M.M., 2012. Environmental impacts of seawater desalination: Arabian Gulf case study. International Journal of Environment and Sustainability 1 (3), 22–37.

Drought Monitor, 2015. US Drought Monitor. http://droughtmonitor.unl.edu/ (03/21/2015).

Gasson, C., 2013. Desalination Market Update. Fourth Quarter Assessment. GWI, Webinar.

Ghaffour, N., Missimer, T.M., Amy, G.L., 2013. Technical review and evaluation of the economics of water desalination: current and future challenges for better water supply sustainability. Desalination 309, 197–207.

Greenlee, L.F., Lawler, D.F., Freeman, B.D., Marrot, B., Moulin, P., 2009. Reverse osmosis desalination: water sources, technology, and today's challenges. Water Resources 43, 2317–2348.

Gude, V.G., Nirmalakhandan, N., Deng, S., 2010. Renewable and sustainable approaches for desalination. Renewable and Sustainable Energy Reviews 14, 2641–2654.

GWI (Global Water Intelligence), 2013. Market Profile and Desalination Markets, 2009–2012 Yearbooks and GWI Website. http://www.desaldata.com/ (05/23/2014).

Karagiannis, I.C., Soldatos, P.G., 2008. Water desalination cost literature: review and assessment. Desalination 223, 448–456.

Koyuncu, I., Topacik, D., Turan, M., Celik, M.S., Sarikaya, H.Z., 2001. Application of the membrane technology to control ammonia in surface water. Water Science and Technology 1 (1), 117–124.

Lee, S., Lee, E., Ra, J., Lee, B., Kim, S., Choi, S.H., Kim, D.D., Cho, J., March 1, 2008. Characterization of marine organic matters and heavy metals with respect to desalination with RO and NF membranes. Desalination 221 (1–3), 244–252.

Mabrouk, A.N., Nafey, A.S., Fath, H.E.S., 2010. Steam, electricity and water costs evaluation of power-desalination co-generation plants. Desalination and Water Treatment 22 (1–3), 56–64.

Maurel, A., 2006. Seawater/Brackish Water Desalination and Other Non-conventional Processes for Water Supply, second ed. Lavoisier, Aarhus.

Mickley, M.C., 2001. Membrane Concentrate Disposal: Practices and Regulation. U.S. Department of the Interior, Bureau of Reclamation, Mickley & Associates.

Miller, J.E., 2003. Review of Water Resources and Desalination Technologies. Available from: Sandia National Laboratories (2003). http://www.prod.sandia.gov/cgi-bin/techlib/access-control.pl/2003/030800.pdf (05/09/2014).

Nair, M., Kumar, D., 2013. Water desalination and challenges: the Middle East perspective: a review. Desalination and Water Treatment 51 (10–12), 2030–2040.

NOAA, 2013. Climate of 2013 – April U.S. Standardized Precipitation Index. http://www.ncdc.noaa.gov/oa/climate/research/prelim/drought/spi.html (04/02/2015).

NRS, October 2008. Final Pilot Study Report Texas Seawater Desalination Demonstration Project.

Pankratz, T., 2008. In: MEDRC Workshop on Membrane Technology Used in Desalination and Wastewater Treatment for Reuse. Muscat, Oman (March 2008).

Pankratz, T., May 7, 2012. Permitting Morass Explained. Water Desalination Report.

Penate, B., García-Rodríguez, L., 2012. Current trends and future prospects in the design of seawater reverse osmosis desalination technology. Desalination 284, 1–8.

Poseidon Water, 2015. The Carlsbad Desalination Project. Enhancing Water Reliability for San Diego County. http://carlsbaddesal.com/ (03/27/2015).

Quteishat, K., 2009. Desalination and water affordability. In: SITeau International Conference, Casablanca, Morocco (January 2009).

Raluy, L., Serra, L., Uche, J., 2005. Life cycle assessment of desalination technologies integrated with renewable energies. Desalination 183, 81–93.

Reddy, K.V., Ghaffour, N., 2007. Overview of the cost of desalinated water and costing methodologies. Desalination 205, 340–353.

Romeril, M.G., April 1977. Heavy metal accumulation in the vicinity of a desalination plant. Marine Pollution Bulletin 8 (4), 84–87.

SCWD2, 2013. Proposed Regional Seawater Desalination Project Draft Environmental Impact Report. Santa Cruz, California: Santa Cruz and Soquel Creek Water Districts. http://www.scwd2desal.org/Page-EIR_Docs.php (03/12/2015).

SCWD2, June 2014. Desalination and Alternatives Water for a Thirsty County. 2013–2014 Santa Cruz County Grand Jury. White Paper.

Sethi, S., 2007. Desalination Product Water Recovery and Concentrate Volume Minimization (DRAFT). American Water Works Association Research Foundation Project #3030. Carollo Engineers, U.S. Department of the Interior, Bureau of Reclamation.

Sommariva, C., Hogg, H., Callister, K., 2003. Cost reduction and design lifetime increase in thermal desalination plants: thermodynamic and corrosion resistance combined analysis for heat exchange tubes material selection. Desalination 158, 17–21.

SRWA (Southmost Regional Water Authority), 2014. Southmost regional water authority. In: Presentation at Texas Alliance of Groundwater Districts (October 29, 2014).

Sturdivant, A.W., Rister, M.E., Rogers, C.S., Lacewell, R.D., Norris, J.W., Leal, J., Garza, J., Adams, J., September 2009. An Analysis of the Economic and Financial Life-Cycle Costs of Reverse-Osmosis Desalination in South Texas: A Case Study of the Southmost Facility. AgriLife Research and Extension. TR-295.

Tampa Bay Water, 2010. Tampa Bay Seawater Desalination Plant. TBW, Clearwater.

Tampa Bay Water, n.d. Tampa Bay Seawater Desalination Plant. In: http://www.tampabaywater.org/tampa-bay-seawater-desalination-plant.aspx (03/17/2015).

USGS, 2014. National Brackish Groundwater Assessment. http://water.usgs.gov/ogw/gwrp/brackishgw/study.html (03/12/2015).

Van der Bruggen, B., February 2003. Desalination by distillation and by reverse osmosis – trends towards the future. Membrane Technology 6–9.

Voutchkov, N., 2010. How Much Does Seawater Desalination Cost? Water Globe Consulting, Texas Innovative Water.

West Basin Municipal Water District, 2014. Ocean Water Desalination. http://www.westbasin.org/water-reliability-2020/ocean-water-desalination/about-desalination (04/02/2015).

Zhou, Y., Tol, R.S.J., 2004. Implications of desalination for water resources in China – an economic perspective. Desalination 164, 225–240.

Ziolkowska, J.R., 2015. Is desalination affordable? – regional cost and price analysis. Water Resources Management 29 (5), 1385–1397.

Ziolkowska, J.R., Reyes, R., 2016. Geospatial Analysis of Desalination in the US - An Interactive Tool for Socio-Economic Evaluations and Decision Support. Applied Geography 71, 115–122.

Chapter 3.1.4

# Water Trading Innovations: Reducing Agricultural Consumptive Use to Improve Adaptation to Scarcity

B. Colby
*University of Arizona, Tucson, AZ, United States*

## 1. INTRODUCTION AND BACKGROUND

Water trading programs now are emphasizing demonstrable reductions in agricultural consumptive use (CU), to assure flows for protected water-dependent habitats and species, for instream recreation, and to satisfy interjurisdictional agreements such as binational treaties and interstate compacts. Early water trading programs were aimed primarily at providing water for urban growth and were more concerned with preserving an overall water balance than with producing measurable improvements in stream flows, wetlands, and lake levels at specific locations and seasons. This chapter describes initiatives focused upon reducing agricultural CU to achieve specific surface flow changes, emphasizing electronic water trading initiatives and the role of baselines in measuring and monitoring changes in CU. The programs reviewed are located in the western United States and Australia.

The term "consumptive use water trading programs" or "programs," as used in this chapter, refers to water programs that implement voluntary agreements with farmers and/or agricultural districts specifically to reduce agricultural CU and provide measurable improvements in "wet water" available for other purposes. Such programs contrast with water trading that is focused on general reductions in irrigated acreage, rather than with real-time reductions in agricultural consumption to produce a desired improvement in surface flows at specific locations and seasons.

Potential participants in regional programs include nations, cities, farms, irrigation districts, special-purpose water districts, water brokers, investors, other private firms, public agencies, and nongovernmental organizations (NGOs). To motivate the necessary investments of money and time to create a CU trading program, potential participants must be facing significant positive incentives or high costs and uncertainties. Positive incentives to create CU trading programs come from attractive offers by NGOs and public agencies to pay farmers to reduce CU. Costs and uncertainties that motivate a

Competition for Water Resources. http://dx.doi.org/10.1016/B978-0-12-803237-4.00018-5

new program can come from legal mandates to assure specific stream flows to protect species, with noncompliance resulting in cutting off water to water users. Another common motivation is a treaty or compact-based surface water delivery obligation, with significant penalties for noncompliance.

Programs allow water users to acquire a more reliable water supply during dry years; provide a means for water users to temporarily use more water than their entitlement provides and incentivize reduced CU by allowing water users to lease the unconsumed portion of their entitlement. Water trading programs offer flexibility to water users, helping mitigate water shortage impacts on regional economies. Such programs reduce regional economic losses when water deliveries must be curtailed. They also help preserve local water user control and provide choices when drought, litigation, or other factors reduce access to water.

In regions where trading programs must achieve streamflow and lake-level mandates, changes in water use on farms must result in measurable increases in water availability at specific locations and seasons. Those paying for reduced CU to accomplish specific objectives require verification. Various programs have devised strategies, such as designated trading zones with different trading ratios and multiyear trading, to address variations in hydrologic connectivity and time lags between reduced water consumption on a farm and improved surface flows at the desired location. Differences in conveyance losses and aquifer characteristics can be incorporated into trading rules to ensure that decreased water use on a specific farm will improve surface flows at a target location and time period. A scientifically sound and stakeholder-accepted model of groundwater–surface water interactions is an important foundation for designing trading rules.

Differences in the economic value generated per acre-foot of water consumed, and in the ability to pay for water, are what drive water trading activity. In areas where water trading encompasses multiple types of water use, these differences are typically present. A city may have a high willingness to pay to avoid dry-year cutbacks in its water service, and an ability to pay stemming from its rate base. A state or federal agency charged with acquiring water for endangered fish presumably has a budget to acquire water for this purpose. And a public agency may be able to persuade farmers to provide water not only through offering an attractive price, but also through the implicit threat of costly and protracted litigation if habitat needs for protected species are not accommodated through voluntary water exchanges.

## 1.1 Ways of Structuring Consumptive Use Transactions

Transfers of water entitlements based on permanently retiring irrigated land generally are adequately handled by existing federal and state change of water use processes. Although every western US state has a formal procedure for changing the ownership, place, and/or purpose of use for a water right, these procedures can be slow and cumbersome. They are not well suited for circumstances in which water users need additional water temporarily and promptly. CU trading's real advantage is in cost effectively implementing temporary and seasonal idling of irrigated land to improve water supply conditions for those who pay for increased supply reliability. CU trading programs provide a way for water users to adapt quickly and cost effectively to changing water supply conditions and changing economic conditions.

Contingent contracts are a useful and valuable way to address water shortage risks. The acquirer (the party acquiring more reliable water supplies) pays an amount upfront for the assurance of water from a senior entitlement holder if a specified shortage event occurs. These are also called option contracts or dry-year reliability contracts. They allow trading of shortage risk across multiple years rather than only on a year-by-year basis. A contingent contract specifies the trigger that activates the exercise of the option, low reservoir elevations, for instance. Such contracts can extend over many years, but contract language typically limits the number of years (and the number of consecutive years) in which the contract can be exercised and irrigated land idled. These provisions are intended to prevent the contingent contract from becoming a means to permanently fallow irrigated land.

Another innovative way of structuring supply reliability transactions between holders of prior appropriation water rights is to implement what is effectively a "priority date exchange" between willing participants. The junior right holder pays the senior right holder to exercise the senior priority date for the time period of the agreement. The senior right holder accepts the increased risk that comes with the junior priority date in return for payment. This model has the advantage of leaving water right ownership unchanged, allowing for continued irrigation on the senior right holder's land in higher water supply years and providing for reversion back to the original priority dates at the end of the agreement. Legal expertise is necessary to create model contracts of this type that are consistent with applicable state and federal laws.

## 1.2 Allowable Activities to Create "Conserved" Water

There are a number of changes in farm water use that can be applied to create conserved water. Suspending irrigation for an entire year on a field that has a documented history of regular irrigation is the most straightforward method for reducing CU to create conserved water for other purposes. However, other approaches are being explored that interfere less with normal agricultural production.

Intentional deficit irrigation of crops like alfalfa is being closely studied in Colorado and other states as a means to conserve water for other uses. (In Colorado, this practice is referred to as "regulated deficit irrigation" to distinguish it from underirrigating crops because of water shortage conditions.) Regulated deficit irrigation has the advantage of having smaller impacts on farm net income and agriculturally linked jobs and businesses, compared to fallowing cropland. Farmers can reduce or halt irrigation during the season when alfalfa cuttings have lower protein content and earn lower prices. Farmers accept lower yields and revenues in return for a payment for the water saved. Reduced CU must be verified by a reliable method, such as advanced remote sensing imagery or crop evapotranspiration (ET) modeling (Jones and Colby, 2010).

Seasonal (rather than year-round) suspension of irrigation is another means to create saved water and this too has a smaller impact on farm net income and agriculturally linked jobs and businesses, as compared to year-round fallowing. As with deficit irrigation, reduced CU must be verified by a reliable method. Crop rotation considerations to build soil productivity or to comply with rules for federal farm programs may limit farmers' flexibility for seasonal irrigation suspension.

# 2. EFFECTIVE BASELINES AND MEASUREMENT PROTOCOLS

Any CU trading program must confront the challenge of measuring changes in agricultural CU attributable to the program's interventions. Ideally, the baseline will consider temperature, precipitation, wind, and other factors that determine CU. A baseline needs to provide program administrators a clear protocol for comparison and measurement of changes in water use attributable to the program, and is necessary to assess program effectiveness. The baseline for comparison should be the CU that would have occurred in the absence of program interventions during a specific time period.

Determining how much water would have been used had a field been irrigated instead of being fallowed because of enrollment in a water trading program is complex, as water use depends on many factors. These include irrigation management practices, temperature and rainfall patterns, soil characteristics, irrigation practices on neighboring farms, economic incentives to plant competing crops, and return flow patterns (Jones and Colby, 2012). These data are not available in many areas. Programs that operate in areas where these baseline data are lacking typically focus on changes in diversions instead of CU, or focus upon changes in the number of acres irrigated and fallowed. They also may use regional average cropping patterns and irrigation technologies and estimate "typical" ET or average water use.

A program that accounts solely for changes in diversions or changes in acres irrigated (and that pays farmers on the basis of reductions in diversions or acreage) will encounter problems such as detrimental effects on surface flows and downstream water rights, as well as on groundwater in storage. A farmer paid to cut back on the basis of reducing irrigated acres may change crop mix, irrigation technology, and water management on remaining acreage (acres not enrolled in the program) in ways that increase CU. For example, a farmer shifting fields from flood irrigation to sprinkler irrigation is likely to increase the CU portion of their water entitlement. For the farmer, this is a rational way to improve crop yields and farm profits. While improved on-farm irrigation efficiency can provide benefits at the farm and field scale, this is an unreliable policy tool for reducing agricultural CU on a broader geographic scale.

Trading program contracts that pay farmers based on changes in acres irrigated or on changes in diversion levels do not prevent increases in CU on acreage not enrolled in the program. Contract language needs to be crafted to effectively prohibit increased CU on the portion of a participating farmer's land (that is, a farmer with some fields enrolled in the program) that is not enrolled in an irrigation suspension program.

One problem encountered by land idling programs is "slippage," which occurs when one acre is retired from production through an idling program and another (not enrolled in the program) is returned to production (Leathers and Harrington, 2000). Slippage diminishes the effectiveness of a program in reducing overall CU. Another potential problem involves inadvertent agricultural consumptive water use by weeds growing on idled cropland. Additionally, deep-rooted crops such as alfalfa may consume groundwater when left unirrigated. Well-defined baselines are necessary to address these problems and identify the actual water "savings" from leaving some fields unirrigated. Also, if farmers are paid to practice intentional deficit irrigation (termed regulated deficit irrigation in Colorado), the program needs to account for the fact that deep-rooted crops such as alfalfa may consume groundwater when left unirrigated by surface water.

The potential effectiveness of programs to free up water for other purposes through reduced agricultural CU is compromised when the programs do not include a carefully defined baseline to use when calculating how much water is "saved" and made available for transfer. Baselines must be coupled with well-written contracts with farmers and with performance monitoring and enforcement to ensure compliance with the parameters of the program (Jones and Colby, 2012).

Since water rights typically are quantified as diversion quantities, a focus on CU accounting can be a difficult transition for farmers. If the benefits of participating in trades are attractive, participants can be required to "opt-in" to CU accounting to participate. Language suggested in recent Colorado water bank proposals offered farmers the option to revert back to standard diversion-based water right accounting when they are no longer participating in the water bank. This "reversion clause" alleviated some farmers' concerns about the focus on CU.

## 3. EXAMPLES OF CONSUMPTIVE USE ISSUES IN WATER TRADING PROGRAMS

Various programs, typically initiated by public agencies and NGOs, provide payments to farmers to reduce water consumption. Reduced CU is obtained through technology upgrades, shift in crops, adaptation of conservation irrigation practices, or fallowing land. The water conserved through these programs is then transferred to other uses, such as municipal uses or instream flows for environmental restoration. The following sections give several examples of agricultural water conservation program baselines that have been implemented in various areas of the United States and Australia.

### 3.1 Arizona and California

In the Lower Colorado River Basin (LCRB), water use within irrigation districts receiving water under contracts with Reclamation is tracked through one of the most sophisticated water accounting systems in the world, the Lower Colorado River Accounting System (LCRAS). Reclamation administers this system on behalf of the Secretary of the Interior. The US Supreme Court in Arizona v. California [547 US 150 (2006)] requires the Secretary to provide detailed records of diversions, return flows, and CU of water diverted from the mainstream of the Colorado River in the LCRB. LCRAS provides an excellent baseline from which to quantify changes in CU. However, such comprehensive water use measurement, monitoring, and accounting systems are lacking in most basins of the western United States and elsewhere in the world.

The California Department of Water Resources draws on weather data to determine accurate evapotranspiration of applied water (ETAW) values (Jones and Colby, 2012). The approach taken by the California Department of Water Resources provides a good example for designing protocols for assessing changes in CU. Farmers wishing to enroll irrigated land for fallowing must document 5 years of farm crop production including estimated ETAW. For irrigation districts proposing to suspend irrigation on cropland, the previous year's cropping patterns are used to establish ETAW, so long as the market for the crops in question has been stable and irrigation water application rates were normal over the past year.

The Imperial Irrigation District (IID) fallowing programs require extensive details to establish historical water use on a field: the canal, gate, and account number for the field, the crops grown for the last 3 years, and the field acreage (IID, 2010). Water conserved by fallowing is estimated based on a "water use history formula adopted by the IID Board of Directors," which is adjusted for recent water use trends (IID, 2010). IID's water use history formula is based on water deliveries in 8 of the last 10 years. IID drops the high and low years and calculates a rolling average of the remaining 8 years. They then examine the trend over the last 3 years. Using the average and the trend, IID predicts what the next year's water usage would have been. The calculation is also adjusted if the field was part of a previous fallowing program. Fields are allowed to participate up to 2 out of every 4 years. These measures are in addition to applying the LCRAS water accounting system described earlier, as IID is in the area governed by the system (Jones and Colby, 2012)

The Bureau of Reclamation (BOR) System Conservation Pilot Program, in LCRB of Arizona and California, was authorized in 2006 and reauthorized in 2008–2010. Its goal was to arrange and implement pilot agreements to provide supplemental water related to irrigation drainage water bypassed to the Cienega de Santa because of Yuma Desalting Plant operations, as well as to assess the effectiveness of land fallowing agreements in achieving system-wide conservation for the Colorado Basin (USBR, 2006a,b). The conservation savings would be achieved through voluntary agreements with irrigators to fallow fields in exchange for rental payments. Agreements reached and implemented with two irrigation districts include a pilot program in 2006 with Palo Verde Irrigation District (PVID) in partnership with Metropolitan Water District of Southern California (MWD), and agreements with Yuma Mesa Irrigation and Drainage District (YMIDD) in 2008 and 2009. The price of $170/acre-foot paid to PVID was based on an ongoing PVID–MWD agreement. The agreement set out to conserve 10,000 acre-feet of water from 2006 to 2007 through fallowing agreements with irrigators, in addition to savings from existing agreements between PVID and MWD (USBR, 2006a,b). In the BOR agreements with YMIDD, the price paid to YMIDD ranged from $90 to $120/acre-foot and approximately 10,000 acre-feet of water were acquired. In 2009, in recognition of the high administrative costs for enrolling and monitoring small fields, BOR created a 3-acre minimum for fields enrolled. In 2008, the YMIDD agreement fallowed 448 acres, in 2009 523 acres, and in 2010 529 acres. For all 3 years, a CU savings of 7 acre-feet water per acre land was assumed based on crop CU studies in this area (USBR, 2012).

## 3.2 Nebraska Consumptive Use Trading Initiatives

In 1969, the Nebraska Legislature created 24 Natural Resources Districts (NRDs), organized topographically, to manage and protect the state's natural resources (Nebraska's Natural Resources Districts, 2014). In 2004, Nebraska adopted Legislative Bill 962 requiring integrated management of surface water and groundwater. Various basins were declared overappropriated or designated as fully appropriated and the NRDs and the Nebraska Department of Natural Resources are responsible for developing Integrated Management Plans (IMPs) in fully appropriated basins where new uses can be allowed only through offsetting depletions to existing rights and uses. In overappropriated basins, the IMPs must replace or offset uses sufficient to return the basin to a fully appropriated (rather than overappropriated) status requiring uses exceeding fully appropriated status to be replaced or offset.

Nebraska entered into the Platte River Recovery Implementation Program (PRRIP) in 2006 with Wyoming and Colorado and the US Department of Interior. The program calls for no new depletions to US Fish and Wildlife Service "target flows" in the Platte River Basin and a return to the 1997 level of depletions. The PRRIP identified a goal of acquiring 10,000 acres of irrigated which was accomplished in 2014 land (CPNRD Board Meeting News Release, 2014). The PRRIP and LB 962 create strong mandates for controlling groundwater depletions and managing surface flow levels. Two NRD water trading programs developed to satisfy these mandates are described next.

The Central Platte Natural Resources District (CPNRD) in 2007 approved a Water Banking Policy to reduce the need to regulate irrigators while returning Platte River flows in its area to mandated levels. CPNRD acquires water rights from landowners through the bank. For every acre-foot river flow improvement from banked water, there is that much less regulation the CPNRD has to impose. The CPNRD board committed to pursue retirement of current uses to achieve the requirements before implementing new regulations. Nevertheless, regulation to reduce farm water use remains a viable option to meet the requirements of Nebraska law.

CPNRD has been acquiring permanent easements from willing sellers to retire irrigated acres (and other uses) and convert to uses that have lesser impacts on river flows. The program incentivizes retirement of lands that have a larger impact on river flows, since the fee structure is based upon a payment for each acre-foot of impact on the river. The higher the impact on the river, the higher the total payments. Water rights and uses deposited in the water bank by individuals are protected for 15 years. After that time, if unused, they revert to the CPNRD. Administrative fees for processing transactions through the water bank are calculated based on payments for the acre-feet transferred.

During the water bank's first year of operation, the CPNRD spent $2.25 million for purchasing water rights to get the overappropriated area back to a fully appropriated status. In 2011, the district adopted a new water pricing formula to incentivize landowners to convert irrigated acres to dryland production. In a 2012 effort to compete against private individuals looking to buy water rights, the CPNRD directors doubled the rate that they will pay for water rights to $8000 per acre-foot of depletion of the river; up from $3750. At the end of 2013, CPNRD had a balance of 2464 acre-feet of water rights available for offset in the overappropriated area (Board Meeting News Release, 2014). The CPNRD water bank subcommittee worked with Mammoth Trading, a private water banking entity, to establish an online trading system that incorporates the NRD's rules and regulations and matches buyers and sellers while calculating transferrable quantity and river flow effects.

In 2013–2014, Nebraska's Twin Platte Natural Resources District (TPNRD) created and implemented a mechanism "to help agricultural producers put groundwater to its best use by facilitating the transfer of certified irrigated acres." An online mechanism operated by a neutral market manager, Mammoth Trading, matches buyers and sellers anonymously and confidentially, simultaneously comparing bids and offers. The trading system ensures that transfers comply with TPNRD rules (such as flow lines, stream depletion factors, slope), manages the approval process with TPNRD, undertakes title searches, and works with farmers on finalizing paperwork and transferring funds. The first round of trading began in spring 2014 and several transactions have been fully approved and implemented. Each month, the market opens and bids are accepted throughout the month. At the end of the month, Mammoth Trading clears the next market cycle and determines the best possible matches, given the bids they have received that month.

The TPNRD also has undertaken a number of initiatives in collaboration with other NRDs, which highlight features potentially useful in designing new water bank programs. In 2012, four NRDs in western Nebraska bought nearly 20,000 acres of farmland in southern Lincoln County, the beginning of the N-CORPE Project (Nebraska Cooperative Republican Platte Enhancement Project, Nebraska Educational Telecommunications, 2014). Crops on over 16,000 acres of purchased lands are idled, the land is seeded to grass, and the irrigation water is sent to the South Platte and Republican rivers to help the NRDs and Nebraska meet legal obligations. The participating NRDs plan to pipe water from the farm's irrigation wells to the Platte and Republican Rivers. The project enhances stream flow with water that otherwise would have irrigated acres owned by N-CORPE in Lincoln County and is the largest grassland restoration project in Nebraska.

The four collaborating NRDs completed a pipeline in 2014 to pump saved groundwater into the Republican River (Knapp, 2013). The N-CORPE Project operates to satisfy Republican River Compact compliance and is credited with preventing shutdown on approximately 300,000 acres in the Republican Basin in 2014 and 2015. By providing the required volumes of water quicker than an irrigation shutdown would have, surface water was administered for compact compliance for a shorter amount of time. However, local farmers worry that the groundwater pumping will affect surface water for their irrigation canals and that future inflows will be depleted (Knapp, 2013). Representatives from Kansas have also raised objections, arguing that Nebraska has historically used too much water and needs to demonstrate that N-CORPE pumping water will not deplete future supplies (Knapp, 2013).

A 2015 compact compliance agreement approved by the US Supreme Court will decrease the amount of water that has to be pumped from N-CORPE. In 2015, the US Supreme Court released their decision on a suit filed by Kansas alleging Nebraska noncompliance with the Republican River Compact (Nebraska et al., 2015). At the time the court was considering the case, the N-CORPE Project was highlighted by Nebraska as a means to ensure Compact compliance. In 2016, the NRDs plan to use the N-CORPE Project to prevent a shortfall for 2016, though a smaller annual volume of pumping will be required than in 2014 and 2015 (Upper Republican NRD, 2015).

## 3.3 Australia

CU trading programs have been implemented in various parts of Australia, primarily in the Murray-Darling Basin (Water Find, 2009). The Northern Victorian Water Exchange utilizes auction methods and data indicate water traders indeed consider shifting supply and demand, as price does respond quickly to changing conditions (Bjornlund, 2003). The approval process streamlined to facilitate a relatively quick turnaround, as the banks are designed to make water resources available for immediate needs (Bjornlund, 2003).

In Australia's Murray-Darling Basin, the Australian government has designated $3.2 billion (approximately US$3.0 billion) under the "Restoring the Balance in the Murray-Darling Basin" program for acquiring water entitlements to meet environmental water needs (ADE, 2014). Australian researchers examined the market for short-term water leasing in the Murray-Darling Basin in Australia to better understand how irrigators bid for and offer water in the face of changing climatic conditions (Zuo et al., 2014). The researchers found evidence of price clustering on both sides of the marker. Buyers' prices tended to cluster around a central price because of increased water supply uncertainty (hotter and drier conditions). Sellers' offer prices cluster around a central value because of strategic behavior.

The Murray-Darling Basin Cap sets limits on diversions and measurement varies by state. Since most Murray River water diverted in South Australia is metered, the state has good data to manage diversions within the Caps (Flett et al., 2009). Queensland began a metering project in 2005 to have all unsupplemented diversions (ie, diversions not taken from regulated public storage) across the state metered. In New South Wales, diversions from unregulated streams are generally not metered although these diversions represent less than 4% of the state's long-term Diversion Cap. Unmetered use is estimated based on crop surveys and irrigation requirements (Flett et al., 2009; Jones and Colby, 2012).

## 4. REDUCING TRANSACTION COSTS—ONLINE WATER TRADING

Observers in the western United States sometimes lament a slow, stunted development of water trading mechanisms. Squillace notes: "Economists have long viewed water markets as an attractive solution for reallocating water to meet the demands of an evolving community of water users…The time for aggressively moving [water marketing] into the field is long overdue" (2013). Water trading can be stunted for a variety of reasons, including high transaction costs in the form of information costs, coordination costs, costs to implement a trade, and uncertainty surrounding a pending transfer's approval. Especially for temporary and intermittent transfers intended to produce results at specific locations and seasons, these costs can outweigh the benefits of trading.

Online trading helps provide the low transaction costs and timely approval that are essential for a trading program to be effective. This section examines online water trading in various locations.

Early electronic water trading programs predate widespread use of the internet and developed out of "bulletin board" trading schemes, utilized primarily by irrigation and water conservancy districts. These made information about potential buyers and sellers available to reduce search costs and promote water trading. Potential trading parties contact one another directly to negotiate trades. Even simple "bulletin boards" greatly reduce transaction costs of identifying and contacting potential trading partners.

In 1996, Westlands Water District, an irrigation district near Fresno, California, established what likely was the first electronic water marketing system, WaterLink. The system was funded by a grant from the Bureau of Reclamation and designed in a joint effort among the University of California Berkeley and Davis, the Natural Heritage Institute (a nonprofit conservation organization), farmers, and water district administrators. WaterLink allowed district water users to buy and sell water using their own computers, significantly reducing trading costs. WaterLink allowed users to post and read bids and asks, negotiate transactions, research price trends and trading volumes, schedule water deliveries, review rainfall summaries and water storage levels, and access their district water account balances (Olmstead et al., 1997). Westlands Water District reported significant cost savings from using WaterLink to facilitate the water transfer approval process and from the feature allowing farmers to order water deliveries electronically—the most commonly used WaterLink feature. In 1998, WaterLink was expanded to include 10 additional water districts in the San Luis & Delta-Mendota Water Authority on the west side of the San Joaquin Valley (Olmstead et al., 1997). Other trading systems have since superseded WaterLink.

Another bulletin board system has been operated by the Northern Colorado Water Conservancy District (NCWCD) since the 1950s (Howe and Weiner, 2002). The NCWCD manages water provided by the Colorado-Big Thompson Project (CBT). The CBT diverts water by tunnel across the Continental Divide from the headwaters of the Colorado River to the Front Range. The project supplements strained water supplies in the NCWCD. CBT water rights are divided into 310,000 shares, with the amount of water represented by a CBT share determined annually. CBT shares are freely marketable over the entire NCWCD, which includes the Front Range urban areas from Broomfield north to Fort Collins, as well as much of northeastern Colorado (Squillace, 2013). Shares for sale or rent are posted online at NCWCD's website along with owners' contact information, and trades are approved by the NCWCD's board of directors (Northern Water, 2014). This market is widely considered one of the most successful water markets in the world, with its frequent trading and low transaction costs. The majority of all permanent water rights transfers in the western United States over the past decade have involved CBT shares. Because little uncertainty exists about the prospects of a transfer approval, CBT shares are in high demand and command some of the highest prices in the western United States (Squillace, 2013).

Internet-based water trading systems facilitate trades across many geographical areas, even though the movement of the wet water is typically conducted locally. WaterBank.com is based out of Albuquerque, but operates throughout North America (WaterBank, 2014). Launched in 1999, the site lists a range of water assets from water rights to bulk water contracts and uses a two-stage auction process to promote trading of water for irrigation, municipal, and industrial uses. Waterbank.com acts as both a water rights broker and a merger and acquisition specialist, conducting water asset valuations and assisting in the legal transfer process (WaterBank, 2014).

Auctions are often a feature of online water trading. While auctions may be a one-time event, they contribute to more effective water trading by lowering search and negotiation costs and by expanding the population able to participate. In 2003, the San Antonio Water System (SAWS) purchased 10,000 acre-feet of pumping rights from Edwards Aquifer irrigators through an online reverse auction (Hartwell and Aylward, 2007). SAWS posted a maximum "reserve" price it was willing to pay and allowed bidders to submit and revise online bids. Nine bids at the reserve price were received, and the auction was considered successful despite the fact that the price was not lowered through competition (Eckhardt, 2003).

In 2004, the Deschutes River Conservancy operated an auction for permanent groundwater mitigation credits and temporary mitigation credits. The auction was conducted via a first-price English ascending-bid format, with bids taken by phone, fax, or online. The auction facilitated trading of mitigation credits with relatively low transaction costs (Hartwell and Aylward, 2007).

Many electronic bulletin boards and online water auctions evolved out of processes already occurring in water districts. However, recently companies and public agencies have looked to e-commerce to develop new markets for water or to expand existing markets to reach new participants. Fully online, business-to-business applications to facilitate water trading have developed, some allowing trades to be consummated online. Azurix, a subsidiary of Enron, in 2000 launched Water2Water.com, a site named one of the top 200 business-to-business sites by Forbes (Forbes, 2000). Water2Water was intended to be an online marketplace for buyers and sellers of water and water-related services, and it enabled auctions, reverse auctions, and bid/ask exchange solutions (Azurix Corp, 2000). However, this initiative did not progress past its pilot program, partially because of Enron's sale and breakup of Azurix (Smith and Lucchetti, 2000).

Australia has several markets that have become almost exclusively electronic. Though such markets now exist in the private sector, the initial push came from the public sector.

In 1998, Goulburn-Murray Water (G-MW), the largest regional water authority in Victoria, established the Northern Victorian Water Exchange (NVWE) to act as a clearinghouse for temporary water trading and to provide information about water prices and volumes traded. NVWE's success led G-MW in 2002 to introduce Watermove, a more sophisticated online platform covering a larger geographical area and facilitating permanent as well as temporary water trades. Watermove operated water exchanges in six regions, each of which was further divided into trading zones. Trading was restricted to water provided by specific water authorities for agricultural use, and trades between zones were limited. Sellers were considered on an ascending bid basis, while buyers were considered on a descending bid basis, within each trading zone. Once all buyers and sellers had been considered, a market equilibrium ("pool") price for the trading zone was calculated to maximize the volume of water traded, and trades were conducted at the pool price (Brooks and Harris, 2008). In 2008, Brooks and Harris conducted an empirical analysis of three trading zones within Watermove and concluded that Watermove generated substantial economic benefits by reallocating water from low- to high-valued uses and *promoting structural adjustment in the agricultural sector* by facilitating the exit of less efficient farmers. G-MW closed Watermove in 2012 because it had become a financial drain on the water authority (Adams, 2012a,b), operating at a $160,000 loss in 2011−2012 (Adams, 2012a,b). Watermove served the purpose of facilitating water trading at a time when other regional water trading services were not available.

The Victorian Water Register, established by the State of Victoria, improved on the model developed by Watermove. Implemented in Northern Victoria in 2007 and in Southern Victoria in 2008, the Victorian Water Register allows for online water share, allocation, and take and use license trading with near-instant approval and reduced fees. In addition to facilitating actual trades, the register provides information about water brokers and trading rules, allows users to check their current and past water allocations, and provides access to information about water availability and use, existing entitlements, and other water market information (Victorian Water Register, 2014).

In the private sector, the National Water Exchange commenced operations in 1994 in New South Wales, and today operates in major irrigation areas (National Water Exchange, 2014). The exchange uses a bulletin board approach via a broker intermediated electronic trading platform. National Water Exchange's online auction platform waterexchange.com.au was launched in 1998 and in 2002 it was changed to pure exchange-based trading with 24-h online trading services. In addition to its spot market, National Water Exchange began conducting trades in forward water contracts and groundwater in 2004 (National Water Exchange, 2014). Likewise, Waterfind Australia has carved out a niche providing an online spot market and forward water market for Australia's major irrigation regions. The company has received widespread recognition for its performance and provides brokerage services, weather information, and 24/7 online trading (Curran, 2014; Waterfind Australia, 2008).

Only recently in the United States have similar attempts been made in the private sector. As previously described, Mammoth Trading contracted with the CPNRD and the TPNRD in Nebraska to develop an online water trading system, which became operational in 2014. The program facilitates transfers of certified irrigated acres and Mammoth Trading acts as a neutral market manager (that is, the firm does not engage in speculation in the TPNRD market), matching buyers and sellers confidentially and anonymously. The market structure differs from the more traditional auction-style water exchanges;

Mammoth Trading simultaneously compares all bids and offers at the end of a specified bidding period, and then clears the market. This process takes into account all bids and offers as well as TPNRD rules, such as flow lines, stream depletion factors, slope, closeness to villages, etc., and splits the gains from transfers between buyers and sellers. No transfer is made if an individual's minimum selling price or maximum buying price (inclusive of all fees, such as title search, notarization of documents, Mammoth Trading's fee, and TPNRD fees) is not satisfied. Mammoth Trading also manages the regulatory transfer approval process with TPNRD on their customers' behalf, thereby lowering individual traders' transaction costs (TPNRD, 2014).

## 5. SUMMARY

Water trading programs continue to evolve where economic motivation for trading is strong. Legal mandates to protect species and habitat and to comply with surface water sharing compacts and treaties have spurred an emphasis on finer measurement and monitoring of reduced CU on irrigated land to produce temporally and spatially specific changes in stream flows. Examples of finer attention to measurement and monitoring are found in the western United States and Australia. Remote sensing is likely to play an increasingly important role in tracking reductions in CU attributable to water trading programs.

Online water trading is quickly becoming the norm in newer trading programs. This reduces transaction costs to farmers considering reducing their CU in exchange for a payment, as well as to those parties seeking water for instream needs. Seasonal and temporary transfers of water are particularly vulnerable to being rendered infeasible because of high transaction costs. Cost-effective verification of stream flow changes linked to trading and low transaction cost procedures to negotiate and implement trades have made temporary and intermittent arrangements to reduce CU feasible in many areas. This is particularly helpful to assuring flows for environmental needs and for recreation, uses in which water's value varies significantly with specific locations and seasons.

## ACRONYMS

**ACT** Australian Capital Territory
**ADE** Australia Department of the Environment
**ADWR** Arizona Department of Water Resources
**AMA** Active Management Area
**AWEP** Agricultural Water Enhancement Program
**BOR** Bureau of Reclamation
**CAGRD** Central Arizona Groundwater Replenishment District
**CAP** In Arizona: Central Arizona Project
**CAWCD** Central Arizona Conservation District
**CBT** Colorado-Big Thompson Project
**CDWR** California Department of Water Resources
**CFS** Cubic feet per second
**CIMIS** California Irrigation Management Information System
**CPNRD** Central Platte Natural Resources District
**CREP** Conservation Reserve Enhancement Program
**CRP** Conservation Reserve Program

**CUP+** Consumptive Use Program Plus
**CVP** Central Valley Project
**CVPIA** Central Valley Project Improvement Act
**DNR** Nebraska Department of Natural Resources
**DWR** Department of Water Resources
**ECP** Emergency Conservation Program
**EQIP** Environmental Quality Incentives Program
**ERS** Economic Research Service
**ESLT** Environmentally sustainable level of take
**ET(AW)** Evapotranspiration (of Applied Water)
**EWRI** Environmental and Water Resources Institute
**FAO** Food and Agriculture Organization of the United States
**FSA** Farm Service Agency
**GAO** U.S. Government Accountability Office
**GIS** Geographic Information System
**G-MW** Goulburn-Murray Water
**IDWR** Idaho Department of Water Resources
**IID** Imperial Irrigation District
**ISC** Interstate Stream Commission, New Mexico
**LCRAS** Lower Colorado River Accounting System
**MDBA** Murray-Darling Basin Authority
**MRGCD** Middle Rio Grande Conservancy District
**MWD** Metropolitan Water District of Southern California
**NASS** National Agricultural Statistics Service
**N-CORPE Project** Nebraska Cooperative Republican Platte Enhancement Project
**NCWCD** Northern Colorado Water Conservancy District
**NDVI** Normalized Difference Vegetation Index
**NOAA** National Oceanic and Atmospheric Administration
**NRCS** Natural Resources Conservation Service
**NRD** Natural Resources District
**NSW** New South Wales
**NVWE** Northern Victorian Water Exchange
**PRRIP** Platte River Recovery Implementation Program
**RCPP** Regional Conservation Partnership Program
**RGWCD** Rio Grande Water Conservation District
**SCP** System Conservation Program
**TNC** The Nature Conservancy
**TPNRD** Twin Platte Natural Resource District
**USDA** United States Department of Agriculture
**USGS** United States Geological Survey
**WTP** Water Transaction Program

## ACKNOWLEDGMENTS

The author appreciates the able assistance of research assistants and colleagues Lana Jones, Andrew Clarke, Dari Duval, and Tess Duffy. Support for this work came primarily from the US Bureau of Reclamation through the University of Arizona "Enhancing Water Supply Reliability" project and from the Climate Assessment for the Southwest (CLIMAS) program at the University of Arizona, through a National Oceanic and Atmospheric Administration's Climate Program Office grant. The author is solely responsible for the contents.

## REFERENCES


Adams, G., August 21, 2012a. G-MW's Watermove Closes. Country News. mmg.com.au.

Adams, G., October 9, 2012b. Loss for Watermove. Country News. mmg.com.au.

ADE, 2014. Restoring the Balance in the Murray-Darling Basin. https://www.environment.gov.au/resource/restoring-balance-murray-darling-basin.

Azurix Corp. Introduces Water2Water.com, December 28, 2000. Water and Wastes Digest.

Bjornlund, H., 2003. Efficient water market mechanisms to cope with water scarcity. The International Journal of Water Resources Development 19 (4), 553–567.

Brooks, R., Harris, E., 2008. Efficiency gains from water markets: empirical analysis of watermove in Australia. Agricultural Water Management 95 (4), 391–399.

CPNRD, June 26, 2014a. Board Meeting News Release. http://www.cpnrd.org/Board_Media2014.html.

CPNRD, September 25, 2014b. Board Meeting News Release. http://www.cpnrd.org/Board_Media2014.html.

Curran, R., June 25, 2014. How to Bet on the Price of Water. Fortune Magazine.

Eckhardt, G., 2000. The evolution of e-water markets. Global Water Intelligence 1 (11). http://www.edwardsaquifer.net/asr.html.

Eckhardt, G., 2003. SAWS Takes Online Bids for Edwards Water Rights. Edwards Aquifer News.

Flett, D., Baxter, P., Hillman, T., November 2009. Review of Cap Implementation 2008–09 Report of the Independent Audit Group. Report No. 45/09. Murray-Darling Basin Authority.

Forbes, July 17, 2000. Energy & Utilities New Postings.

Hartwell, R., Aylward, B., 2007. Auctions and the Reallocation of Water Rights in Central Oregon. Dechutes River Conservancy. River Papers. 1.

Howe and Weiner, 2002. Progress So Far. The Arkansas Water Bank Pilot Program. www.isse.ucar.edu/water.../Howe%20and%20Weiner,%20Arkansas%20Water%20Ban.ppt.

IID, 2010. Fallowing Programs. IID website. http://www.iid.com/index.aspx?page=190.

Jones, L., Colby, B.G., May 2010. Farmer Participation in Temporary Irrigation Forbearance: Portfolio Risk Management. Rural Connections, pp. 43–48.

Jones, L., Colby, B., 2012. Measuring, Monitoring, and Enforcing Temporary Water Transfers: Considerations, Case Examples, Innovations, and Costs, June 2012.

Knapp, F., July 16, 2013. Nebraska's Effort to Re-Water Republican River Making Waves. Net News. http://netnebraska.org/article/news/nebraskas-effort-re-water-republican-river-making-waves.

Leathers, N., Harrington, L.M.B., 2000. Effectiveness of conservation reserve programs and land 'slippage' in Southwestern Kansas. The Professional Geographer 52 (1), 83–93.

National Water Exchange, 2014. National Waterexchange Pty Ltd. Web. 9 July 2014.

Nebraska, K.V., et al., February 24, 2015. On Exceptions to Report of Special Master No. 126, Orig.

Northern Water, 2014. Northern Colorado Water Conservancy District, n.d. Web. 28 Jul.

"NRD History" Nebraska's Natural Resources Districts., 2014. http://nrdnet.org/about-nrds.php.

Olmstead, J., Sunding, D., Parker, D., Howitt, R., Zilberman, D., 1997. Water Marketing in the '90s. Entering the Electronic Age. Choices 12.3, 15–19.

Smith, R., Lucchetti, A., August 28, 2000. Rebecca Mark's exit Leaves Azurix Treading deep water. The Wall Street Journal.

Squillace, M., 2013. Water transfers for a changing climate. Natural Resources Journal 53 (1), 55–116.

Twin Platte Natural Resources District, 2014. Board of Directors Meeting Minutes – October 10, 2013. Subcommittee Meetings Report. Water Resources Subcommittee Twin Platte Natural Resources District. http://www.tpnrd.org/March2014.pdf.

Upper Republican NRD, http://www.urnrd.org/ncorpe-operations-cease-april-2015 and http://www.urnrd.org/new-imp-aid-water-users-compact-compliance-dec-2015.

USBR, August 15, 2006a. Agreement to Implement a Demonstration Program for System Conservation of Colorado River Water. United States Bureau of Reclamation. http://www.usbr.gov/lc/region/g4000/4200Rpts/DecreeRpt/2007/agreements/MWDSCA.pdf.

USBR, May 26, 2006b. Policy Establishing a Demonstration Program for System Conservation of Colorado River Water. United States Bureau of Reclamation. http://www.usbr.gov/lc/region/programs/demo/policy.pdf.

USBR, September 26, 2012. Review of Demonstration Program for System Conservation of Colorado River Water, Lower Colorado Region. United States Bureau of Reclamation (Presentation).

Victorian Water Register, 2014. Department of Environment and Primary Industries, the State of Victoria.

WaterBank, 2014. WestWater Resources.

Waterfind Australia, 2008. Waterfind Australia.

Water Find, 2009. Water Find National Water Brokers. Web. October 2009. http://www.waterfind.com.au/.

Zuo, A., Brooks, R., Wheeler, S.A., Harris, E., Bjornlund, H., 2014. Understanding irrigator bidding behavior in Australian water markets in response to uncertainty. Water 6, 3457–3477. http://dx.doi.org/10.3390/w6113457.

Chapter 3.1.5

# Groundwater Scarcity: Management Approaches and Recent Innovations

**Y. Kuwayama[1], R. Young[2], N. Brozović[3,a]**
[1]*Resources for the Future, Washington, DC, United States;* [2]*Mammoth Trading, Lincoln, NE, United States;* [3]*University of Nebraska-Lincoln, Lincoln, NE, United States*

## 1. INTRODUCTION

Groundwater resources are a major source of agricultural, potable, and industrial water. In the United States, approximately 83 billion gallons of groundwater are withdrawn per day, which is equivalent to a quarter of all daily freshwater withdrawals (Maupin et al., 2014). Irrigated agriculture is the largest category of groundwater withdrawals in the United States, with nearly 476,000 wells pumping 49 billion gallons daily (NGWA, 2015). At the same time, it is estimated that 39% of the US population relies on groundwater as a supply of drinking water, while 53% of freshwater used by industry, mining operations, and thermoelectric power generation is pumped from aquifers (Maupin et al., 2014; NGWA, 2015).

Despite the importance of groundwater for the economy and the well-being of individuals, groundwater withdrawals are often unmonitored and unregulated in the United States. For example, California, the state with the highest level of annual groundwater withdrawals (Maupin et al., 2014), follows the "reasonable use doctrine," which allows a user to pump an unlimited quantity of water as long as the water is put to a "beneficial use." While this doctrine prevents many wasteful uses of groundwater, it allows continued pumping even if water tables decline in the underlying aquifer or if flows are reduced in connected streams and rivers. In Texas, the third largest groundwater user by withdrawals (Maupin et al., 2014), the law governing groundwater use is known as the "rule of capture," which allows a person the right to pump whatever groundwater is available. Many states do not impose limits on the drilling of new groundwater wells.

---

a Yusuke Kuwayama is a Fellow at Resources for the Future. Richael Young is President and Co-Founder of Mammoth Trading and Program Associate at the Robert B. Daugherty Water for Food Institute at the University of Nebraska. Nicholas Brozović is Director of Policy at the Robert B. Daugherty Water for Food Institute, Associate Professor of Agricultural Economics at the University of Nebraska, and Co-Founder of Mammoth Trading.

**Competition for Water Resources. http://dx.doi.org/10.1016/B978-0-12-803237-4.00019-7**

However, symptoms of groundwater resource degradation such as aquifer depletion, stream depletion, well interference, and saltwater intrusion have led to rapidly changing groundwater use policies and management institutions in some regions. In 2014, after 3 years of severe drought, the California legislature passed the Sustainable Groundwater Management Act, which requires high- and medium-priority groundwater basins in California to develop groundwater management plans that achieve sustainability within 20 years of implementation. Texas has taken the general approach of allowing local governance of groundwater through the creation of groundwater conservation districts (GCDs). GCDs are authorized by the Texas Legislature and registered voters within proposed districts must confirm the creation of the GCD, appoint its directors, and set its tax rates in an election. Nebraska, the fourth largest groundwater user by withdrawals (Maupin et al., 2014), has implemented some of the most advanced groundwater regulations in several of its river basins, including moratoria on the drilling of new wells, the certification of acreage that can be irrigated with groundwater, annual volumetric limits on pumping with mandatory metering, and a framework for farmers to be able to transfer the right to pump groundwater.

In this chapter we describe some innovative policies, regulations, and initiatives that have been implemented in the United States as part of the rapidly shifting landscape in groundwater management. We take a broad interpretation of innovation that includes new policy instruments for managing groundwater as well as new technologies that can reduce the negative impacts of groundwater use cost effectively. Our focus is on the management of groundwater for irrigation, which is the single largest use of groundwater by withdrawals. The chapter starts with an overview of the motivations for groundwater management, focusing on (1) market failures that can lead to socially suboptimal levels of groundwater pumping and (2) equity considerations, both of which can provide an economic justification for a government role in management. In the following two sections, we describe the policy instruments and technologies that are available to address these market failures and provide examples of each. We then describe four case studies of existing or proposed water management institutions and initiatives that are innovative. The case studies are taken from the High Plains region of the United States, where a diversity of groundwater management programs and policies exists, but similar initiatives can be found in other regions of the country as well.

Some of the groundwater management systems that we will discuss are motivated by groundwater quality issues. However, this chapter will not address policies or technologies that directly affect groundwater protection and cleanup, such as those associated with the federal Safe Drinking Water Act, Resource Conservation and Recovery Act, and the Comprehensive Environmental Response, Compensation and Liability Act (the Superfund Act). This is not to say that groundwater quality is not an area of regulatory concern. For example, Craun et al. (2010) found that 54% of reported water-borne disease outbreaks between 1971 and 2006 were caused by the use of untreated groundwater or groundwater treatment deficiencies. We leave these issues surrounding the regulation of groundwater quality for a future study.

## 2. MOTIVATIONS FOR GROUNDWATER MANAGEMENT

There are many pathways through which unregulated groundwater pumping can lead to socially suboptimal outcomes. We address these pathways through the perspective of natural resource economics, which involves the identification of market failures in which the pursuit of self-interest on the part of individual groundwater users does not lead to an efficient result from a societal point of view. In the context of groundwater use, market failure most commonly results from externalities, principal–agent problems, imperfect information, high transaction costs, and the public good nature of aquifers. Understanding how these market failures arise in different agricultural systems and hydrologic contexts, in turn, can inform the assessment of the potential means of correction described in Section 3.

Although we discuss each market failure separately in the subsections that follow, it is important to note that these market failures can, and often do, occur simultaneously. For example, aquifer depletion, which can lead to lower water tables and thus increase pumping costs for users of the aquifer, can also lead to depletion of adjacent streams, which can affect aquatic ecosystems that depend on the stream and impinge on the rights of downstream surface water users. The presence of simultaneous market failures and the need to achieve multiple objectives can complicate the design of cost-effective groundwater management policies.

## 2.1 Aquifer Depletion

Aquifer depletion occurs when groundwater pumping extracts water out of the aquifer faster than it is replenished. Many aquifers in the United States have experienced significant depletion over the last century. For example, the Ogallala Aquifer, part of the High Plains Aquifer System and one of the largest aquifers in the world that underlies part of eight US states—Colorado, Kansas, Nebraska, New Mexico, Oklahoma, South Dakota, Texas, and Wyoming—has experienced an area-weighted, average water-level decline of 4.3 m from predevelopment to 2011 (McGuire, 2013). Similarly, satellite data reveal that California's Sacramento and San Joaquin River Basins are losing water at a rate of 31.0 mm per year; this translates to a total volume of 30.9 $km^3$ over the 2003–2010 study period, or nearly the capacity of Lake Mead (Famiglietti et al., 2011). In addition to reducing the total amount of groundwater stored in an aquifer, groundwater pumping can generate aquifer depletion impacts that are highly localized. For example, high recharge in the northern High Plains and lower recharge in the central and southern High Plains have resulted in focused depletion of 330 $km^3$ of groundwater, with about a third of that depletion occurring in 4% of the High Plains land area (Scanlon et al., 2012).

It is important to note that depleting an aquifer may actually be a socially optimal outcome. It may be optimal to deplete an aquifer over time if users have access to close substitutes that can serve as alternative sources of water or if the groundwater manager has a limited planning time horizon. However, in many cases, aquifer depletion is a socially suboptimal outcome that is being driven by one or more market failures. First, the rate of depletion of an aquifer may be suboptimally high if groundwater users or regulators do not account for scarcity rents when making pumping or groundwater allocation decisions. Scarcity rents are associated with aquifers for which withdrawals exceed recharge such that groundwater is mined over time until supplies are exhausted or the marginal cost of pumping additional water becomes prohibitive. They reflect the opportunity cost associated with the unavailability in the future of any unit of water used in the present. A "myopic" groundwater user or manager who ignores scarcity rents will make pumping decisions based on a marginal cost of groundwater use that is too low, leading to an extraction rate that is above the socially optimal level.

A related and often concurrent market failure is caused by the fact that aquifers are often shared by multiple users and groundwater is a nonexclusive resource. Such a scenario can lead to a dissipation of rents associated with the aquifer since a farmer may not expect to have more water in storage for the next year if he/she pumps less this year. This, in turn, can lead to competition among users to capture the groundwater reserves and over-exploitation of the aquifer, sometimes referred to in the literature as the "strategic" groundwater externality (see, eg, Negri, 1989). Several studies have attempted to quantify the significance of the strategic externality in determining the difference in outcomes for an aquifer pumped by competitive groundwater users versus being managed using optimal control trajectory that maximizes users' discounted welfare (see, eg, Gisser and Sanchez, 1980).

As noted by Koundouri (2004), the issue of groundwater users not facing scarcity rents (ie, myopic behavior) and the issue of competition (ie, the strategic externality) are often conflated in the literature. Furthermore, as noted by Provencher and Burt (1993), the market failures that drive aquifer depletion can exist even in the absence of well interference, which we discuss in the next subsection. To identify appropriate and feasible strategies for managing a particular aquifer, it is important to clarify which of these various market failures are occurring.

## 2.2 Well Interference

When groundwater is pumped from a well, the water levels in the aquifer around the well decrease. The amount of this decline becomes less significant with greater distance from the well, resulting in a cone-shaped depression in the water level radiating away from the well. In some cases, the cone of depression generated by pumping at a well can be large enough to lower water levels at nearby wells, a process known as well interference.

The degree of well interference that can occur within a set of wells is determined by the spatial distribution of the wells and physical properties of the aquifer. In a simple analytical solution, Theis (1935) showed that the magnitude and timing of the drawdown at a well caused by pumping at a neighboring well depend primarily on the rate and timing of pumping at the neighboring well, the distance between the two wells, aquifer storativity (the capacity of an aquifer to release groundwater from storage in response to gradients in the aquifer), and aquifer transmissivity (a measure of the rate at which water can be transmitted horizontally in an aquifer). According to hydrologic principles, the effect of drawdowns caused by multiple pumping wells is linearly additive, which implies that pumping by any number of nearby wells can contribute to drawdown at a well.

Well interference can negatively affect groundwater users because cones of depression increase the distance between the water table and the surface of the Earth (often referred to as the "pumping lift"), thus increasing pumping costs. As a result, well interference is a classic example of an externality because groundwater pumping by Farmer A can increase the pumping cost of neighboring Farmer B. Because the external cost on Farmer B is not accounted for in Farmer A's marginal cost of pumping, Farmer A is likely to pump too much water relative to the socially optimal level. Several economic studies have tried to empirically quantify the effect of well interference on irrigator behavior and ensuing exploitation of the aquifer, but isolating this effect from that of the myopic and strategic considerations described in Section 2.1 remains a challenge. Brozović et al. (2010) showed that accounting for well interference may be important in optimal groundwater management, particularly for large aquifers with surface areas of thousands of square miles.

## 2.3 Degradation of Groundwater Quality

While our chapter is not meant to address water quality management per se, some water quality outcomes are linked to groundwater withdrawals. Such a case would constitute an externality if the costs of negative impacts on water quality are borne by entities other than those who are pumping the groundwater. As a result, regulators and communities may wish to manage groundwater withdrawals to meet water quality goals. For example, as aquifer levels decline from chronic overdraft, natural and artificial pollutants can concentrate in the remaining groundwater, making it unsafe for irrigation or drinking without costly treatment. Exploitation of aquifers can also lead to increases in groundwater temperature and pressure that encourage the movement of previously immobilized toxic elements such as heavy metals and naturally occurring arsenic, leading to contamination of aquifers.

The most prominent example of a water quality problem that is caused by groundwater pumping is saltwater intrusion (sometimes referred to as seawater intrusion or saltwater encroachment), which is the movement of saline water into freshwater aquifers. Saltwater intrusion can occur near coastal areas as a result of groundwater pumping because extraction reduces the quantity of fresh groundwater in the aquifer, reducing its water pressure and allowing saltwater to flow further inland. Saltwater intrusion decreases freshwater storage in the aquifers, and, in extreme cases, can result in the abandonment of domestic supply or irrigation wells. Communities struggling with saltwater intrusion include the 2.5 million residents of Miami-Dade County, Florida, who rely on the Biscayne aquifer as their primary drinking water supply (Prinos et al., 2014), and users of the Central and West Coast Groundwater Basins in California, where potable water and recycled municipal wastewater have been injected into "barrier wells" since 1951 to build up a line of pressure that blocks the intrusion (Johnson, 2007).

## 2.4 Depletion of Surface Water

Groundwater and surface water are often connected, and groundwater discharge is the primary component of streamflow in many streams and rivers (Barlow and Leake, 2012). When a stream intersects an aquifer, if the height of the water table in the aquifer is greater than the height of the stream's surface, groundwater is discharged through saturated streambed and streambank sediments or permeable bedrock adjacent to the stream. Conversely, if the height of the stream surface is greater than the height of the aquifer's water table, streamflow seeps into the underlying groundwater system. As a result, groundwater pumping from aquifers can reduce the flow of surface water in nearby streams, rivers, and lakes—a process known as stream depletion. Stream depletion is subject to a time-dependent process known as diffusion, whereby pumping by users located far away from the stream will affect flows in nearby streams with a longer time lag than pumping by users located nearby (Glover and Balmer, 1954). Therefore over any finite time horizon, pumping by wells close to the stream will remove more water from the stream than pumping by wells that are further away.

As is the case with well interference and saltwater intrusion, stream depletion can become an externality if decisions regarding groundwater withdrawals take place without accounting for impacts on surface water users. Recent awareness of stream depletion in the United States has led to interstate conflict over the role of groundwater withdrawals in the fulfillment of compacts governing the distribution of water in transboundary rivers. Groundwater regulations have been implemented in response to claims filed in the US

Supreme Court by Texas against New Mexico over flows in the Pecos River, by Kansas against Colorado over the Arkansas River, and by Kansas against Nebraska and Colorado over the Republican River (Hathaway, 2011; Palazzo and Brozović, 2014). Interstate commissions that manage other transboundary rivers, such as the Delaware River, which flows through New York, Pennsylvania, New Jersey, and Delaware, have expressed concern over the role of groundwater withdrawals on streamflow (Delaware River Basin Commission, 2014). In addition, large-scale groundwater use in several parts of the country has degraded species habitats that depend on surface water. The ecological impacts of stream depletion have resulted in litigation between local stakeholders and federal agencies over species protected under the Endangered Species Act in Idaho (Hershler et al., 2014) and Texas (McCarl et al., 1999), as well as species that are economically and culturally important, like salmon (Fleckenstein et al., 2004; Nichols et al., 2014). Several studies have attempted to map out groundwater-dependent ecosystems using geospatial and remote sensing data, including in California (Howard and Merrifield, 2010), Texas (Gou et al., 2015), and Western Australia (Barron et al., 2014).

## 2.5 Land Subsidence

Aquifers are frequently associated with compressible layers of silt or clay. As groundwater is pumped out of these aquifers, forces that keep the silt or clay rigid are reduced and they consolidate. As a result, the total volume of the silts and clays is reduced, resulting in the lowering of the land surface. Although underground mining, drainage of organic soils, natural compaction, and thawing permafrost may also cause subsidence, more than 80% of the identified subsidence in the United States is caused by groundwater pumping (Galloway et al., 1999). Estimates of maximum subsidence amounts include a 1972 estimate of 28 ft in Mendota in California's San Joaquin Valley (Poland et al., 1975) and a 1977 estimate of 12.5 ft in Eloy, in South Central Arizona (Laney et al., 1978).

Land subsidence can impose external social costs through damages to buildings (Cooper, 2008) and land-surface infrastructure such as roads and aqueducts (California Department of Water Resources, 2015), the need to repair or replace steel casing for wells that are compressed or collapsed (Borchers et al., 2014), and increased risk of flooding (Dixon et al., 2006). In the Santa Clara Valley, the cost of replacing or repairing wells amounted to $756 million (in 2013 dollars) between 1960 and 1965 alone (Borchers et al., 2014). According to a National Research Council Report (1991), annual costs in the United States from flooding and structural damage caused by land subsidence exceeded $125 million. Sinkholes and other surface cavities may also be caused by subsidence and it may also cause changes in local hydrology.

In a similar problem, there is evidence that groundwater pumping, which represents a transfer of mass from land to the oceans, has contributed to sea level rise. A study by Konikow (2011) estimated that global groundwater depletion between 1900 and 2008 has caused sea levels to rise by 12.6 mm, equivalent to 6% of total sea level rise during the period, although more recent studies have estimated this effect to be smaller by accounting for the fact that not all groundwater extracted ends up in the oceans (Wada et al. 2016).

# 3. POLICY INSTRUMENTS FOR GROUNDWATER MANAGEMENT

In this section, we describe policy instruments that can help correct the market failures described in Section 2. In theory, it is possible to design a set of command-and-control regulations or market-based schemes that achieve the socially optimal outcome in the presence of any of the previously cited market failures. However, these first-best solutions are typically very complex, involving spatially and temporally differentiated treatment of groundwater users and a perfect understanding of the relevant economic and hydrologic systems. Therefore, in practice, these first-best solutions are unlikely to be implemented because of high transaction costs, monitoring costs, and informational requirements. Furthermore, regulators may face political opposition when proposing policies that treat firms differently based on their location or generate uncertainty in a user's ability to pump groundwater in the future. We thus focus on groundwater management policies and initiatives that are already in place or in the proposal stage. It is clear that none of these groundwater governance schemes correspond to the theoretical first-best solutions to the problems they seek to correct. However, because groundwater regulations are fairly limited in the United States, examining the current landscape of policies and initiatives can inform future management for unregulated aquifers.

## 3.1 Retirements and Buyouts

Retirement of irrigated agricultural lands involves long-term cessation of irrigation, and in some cases taking the land out of agricultural production altogether. Many land retirement programs have the primary goal of achieving load reduction of dissolved constituents and trace elements such as selenium that are present in subsurface drainage. A prominent example of such a program is the San Joaquin Valley Drainage Program, which sought to prevent excessive shallow groundwater conditions and the accumulation of salts in the root zones (SJVDIP, 1998).

One example of a land retirement program that has a groundwater focus is the Water Rights Transition Assistance Program, administered by the Kansas Department of Agriculture's Division of Conservation (DOC). The program started out as a 5-year pilot project in 2007 and has been extended until 2022 by the state legislature. The goal of the program is to reduce the demand on distressed aquifers and streams and return the overall level of water appropriation back into conformity with available water resources. This is accomplished via an incentive-based procedure in which participating landowners permanently retire their water rights in exchange for compensation by the state of Kansas. Mutually agreeable compensation is paid to a landowner in the form of a grant that is intended to aid willing sellers in the transition from irrigation to dryland farming. The amount of the compensation is largely determined by a fixed price point value determined annually by the DOC in conjunction with other agencies, and other relevant factors such as the seniority of the water right, its historic consumptive water use quantity, the distance of the water right within the targeted water supply, and a competitive bid price submitted by the owner. Through 2012 a total of $2,946,082 of state money was spent on the permanent retirement of 6169 acre-feet of annual water appropriation rights (Division of Conservation, 2012).

An example of a land retirement program that has benefited from funding from the US federal government is the Colorado Rio Grande Conservation Reserve Enhancement Program (CREP), which has been granted up to $109 million from the US Department of Agriculture (USDA) and is open to producers with irrigated acres in Subdistrict No. 1 of the Rio Grande Water Conservation District. The goal of the program is to conserve irrigation water and reduce groundwater withdraws from the Rio Grande Basin while also enhancing water quality, reducing erosion, improving wildlife habitat, and conserving energy. USDA irrigated rental payments for the Rio Grande CREP are equal to $175 per acre per year through the term of the contract, in addition to a cost-share of 50% of the total cost of revegetating a cover crop on the retired acres. Additional incentives and bonuses are provided by Subdistrict No. 1 (Rio Grande Water Conservation District, 2015). Similar federal−state partnerships exist in the Upper Arkansas River CREP (Division of Conservation, 2013) and the Republican River CREP.[1] These arrangements usually include a prior year irrigation requirement, which seeks to ensure that a producer is retiring irrigation that was actually taking place.

From an economic perspective, land retirement programs will be more cost effective if irrigation is retired from cropland where the environmental costs of groundwater irrigation are high relative to the value of agricultural production. As a result, spatial targeting of lands based on the contribution of irrigation to agricultural productivity and environmental damage may be able to approximate an optimal but more complicated groundwater allocation system. One drawback of land retirement is that it usually involves the full cessation of irrigation from a given land parcel. In cases in which partial reductions in irrigation would be sufficient to achieve the desired hydrologic improvement, this may impose additional costs to the program.

## 3.2 Moratoria on Well Drilling

Regulators can reduce the growth of groundwater extraction and irrigated acreage by prohibiting the drilling of new wells. In some cases these well-drilling moratoria serve as an immediate and temporary solution while the state or local water agency develops a more detailed, long-term policy. For example, in 2013, San Luis Obispo County in California issued Urgency Ordinance No. 3246, which established a moratorium on "new development dependent on a well" in the Paso Robles Groundwater Basin. This ordinance was established in response to groundwater monitoring wells that indicated 25-ft declines in water levels between 2011 and 2013 in some parts of the county. The ordinance also established moratoria on new or expanded irrigated crop production and the conversion of dry farm or grazing land to new irrigated crop production.

In 2014, Nebraska Legislative Bill 962 established a moratorium on the drilling of new wells with a capacity of over 50 gallons per minute and the growth of irrigated acreage in areas that are hydrologically connected to a basin that is considered fully appropriated by the Nebraska Department of Natural Resources (DNR).[2] Basins are determined to be fully appropriated if the long-term supply of hydrologically connected groundwater and surface water in the basin is not sufficient for supporting new uses without negatively affecting existing uses. This moratorium on wells and new acres will remain in place until an integrated management plan is developed or if it is lifted by the local Natural Resources District (NRD).

Moratoria on well drilling are appealing in that they impose a spatially uniform restriction that is relatively easy to monitor. However, moratoria do nothing to prevent pumping at wells that are already located in hydrologically sensitive areas, nor do they allow for the drilling of

1. See http://www.republicanriver.com/Programs/CREP/tabid/110/Default.aspx.
2. The Nebraska DNR defines a "hydrologically connected area" as one in which for a well drilled in the area, ten percent of the water pumped over 50 years would either come from the steam or would have gotten to the stream. The border of the hydrologically connected area is therefore often referred to as the "10 percent/50-year line."

new wells in areas in which additional groundwater use would have very little environmental impact. As a result, the cost effectiveness of moratoria is likely to be exceeded by some of the more flexible and targeted policy instruments described in this section.

## 3.3 Zoning and Setbacks

Zoning and setback regulations prohibit well drilling or groundwater pumping within spatially defined areas. These policies can be more targeted than moratoria in the sense that regulators can prohibit harmful activities in the most hydrologically sensitive or least agriculturally productive areas. For example, Nebraska has established so-called "Rapid Response Areas" in the Republican River Basin in areas with a streamflow depletion factor of 10% or more in a 2-year period. In years of reduced flow in the Republican River, pumping may be highly limited in these areas. While not intended for agricultural groundwater wells, another example of a setback regulation is Rule 17.36.323 of the Administrative Rules of Montana that requires all drinking water wells to be at least 100 ft from surface water or springs.

## 3.4 Groundwater Mitigation Programs

Some management programs require users to compensate the regulator for the negative impacts caused by their groundwater pumping. For example, in 2002, Oregon Revised Statute 537.746 and House Bill 3493 authorized the Oregon Water Resources Department to develop the Deschutes Groundwater Mitigation Program, which requires groundwater permit applicants to obtain "groundwater mitigation credits" prior to receiving permits. The stated goal of the program is to prevent injury to surface rights holders and to protect flows in rivers that are listed under the Oregon Scenic Waterways Program.[3] The mitigation program was authorized in response to a study showing that groundwater and surface water are directly linked in a portion of the Deschutes River Basin and removal of groundwater will ultimately diminish streamflow (Gannett et al., 2001). Mitigation requirements can be fulfilled by completing a mitigation project, such as instream transfers, storage releases, or aquifer recharge, or by obtaining mitigation credits from a credit holder or a mitigation bank in the Deschutes Ground Water Study Area (Water Resources Department, 2008).

The ability of a mitigation program to achieve cost effectiveness and environmental targets will depend on the specific design of the program. If mitigation is only required for new groundwater uses, the program will eliminate the net environmental impacts of that additional groundwater pumping and hydrologic conditions will remain at baseline levels. On the other hand, if the program requires mitigation for existing groundwater uses as well, the mitigation requirement will disincentivize existing groundwater use, and a net improvement in hydrologic conditions is possible.

## 3.5 Pumping Charges

A pumping charge is a fee that is charged to the irrigator per unit of groundwater pumped. Pumping charges can induce a first-best groundwater allocation if the fee accurately

3. See http://www.oregon.gov/oprd/RULES/pages/waterways.aspx.

reflects the social costs associated with the extraction of a unit of groundwater, including scarcity rents and externalities on third parties. As such, pumping charges are a form of the Pigouvian tax that reduces each user's groundwater pumping to the socially optimal level. As is the case with Pigouvian taxes in general, pumping charges do not require the regulator to know the cost functions of producers that use groundwater as an input. However, pumping charges do require metering of groundwater use, which increases monitoring costs, as well as costs associated with the assessment of the fees that may not be necessary with some of the simpler policies described in this section.

One example of an existing pumping charge is the Pajaro Valley Augmentation Charge in Pajaro Valley, California. It involves a per acre-foot fee ranging from $92 to $348 depending on location for the 2015–2016 fiscal year. Projects funded with the fees include conservation and distribution projects such as a water recycling program, purchase of Central Valley Project water, and pipeline construction. The fund will also support the implementation of the Agency's 2014 Basin Management Plan Update. The Coachella Valley Water District has a similar setup called the Replenishment Assessment Charge, which is meant to reduce overdraft of the East Whitewater River Subbasin and generate revenue to fund aquifer replenishment by charging entities that use over 25 acre-feet annually from the aquifer. The charge encourages use of surface water in wet years to allow for natural groundwater recharge and is a part of the Coachella Valley Water Management Plan. As of 2015, the charges are equivalent to $59 per acre-foot in the East Whitewater River Subbasin, $112 in the West Whitewater River Subbasin, and $112 in the Mission Creek Subbasin. The Skagit River Basin Stream Flow Enhancement/Groundwater Mitigation Program, which is still in the development phase, is administered by the Washington State Department of Ecology and the Upper Skagit Tribe. Its goals are to (1) create a managed recharge program to increase stream flows and offset decreased flow caused by groundwater use and (2) institute a mitigation fee program for property owners. Minimum streamflows were established in 2001 under the Skagit River Instream Flow Rule, and the fee-based mitigation program is designed to recover the costs of the recharge program.

Many management programs that assess a pumping charge use the revenue collected from the charges to invest in improvements in the groundwater basin. It is important to note that if a pumping charge is set equal to the marginal social cost of pumping a unit of groundwater, the pumping charge alone is sufficient to induce groundwater users to pump at the socially optimal level. As a result, use of revenues from fees in additional management activities that benefit the basin is not necessary for pumping charges to achieve a socially optimal outcome.

## 3.6 Tax Instruments

In some cases, regulators can use the local tax code to incentivize reductions in groundwater use. Nebraska Legislative Bills 701 and 862 allowed NRDs to impose a tax on irrigated land that is limited to $10 per acre. This land tax is intended to help the state comply with the Republican River Compact of 1942, which dictates the shares of surface water that can be withdrawn from the Republican River by Nebraska, Kansas, and Colorado. Land taxes are similar to pumping charges in that they reduce the incentive to irrigate, but they do so by requiring payment for additional acres irrigated as opposed to additional acre-feet pumped. To the extent that it is the volume of water pumped that is relevant for the health of hydrologic systems, land taxes operate through an indirect mechanism. However, regulations based on irrigated acreage may be almost as effective as regulations based on volume of water pumped if farmers tend to apply the same amount of water per acre to their crops regardless of their total irrigated acreage (Young, 2014).

Arkansas provides tax credits as a mechanism to incentivize the construction of impoundments for water storage, the conversion of groundwater use to surface water use, and the leveling of land to conserve water, all of which are meant to protect groundwater supplies. Impoundment construction allows for an income tax credit of 50% of the project cost, while conversion to surface water use allows for a credit of 10% of the project cost (or 50% in a critical groundwater area).

## 3.7 Pumping Quotas and Irrigated Acreage Certification

There are two approaches that regulators have taken to impose quantity restrictions on groundwater use. The first approach is a pumping quota, which is a limit on the volume of water that can be pumped by a user over a given period of time (in acre-feet per year, for example). The second approach is to restrict irrigated acreage. Chosen correctly, these quantity restrictions can achieve the socially optimal groundwater allocation in the presence of any of the market failures described in Section 2. As with well drilling moratoria, setbacks, and zoning, pumping quotas and irrigated acreage certification are examples of command-and-control regulations.

A prominent example of a pumping quota system was established by the Edwards Aquifer Authority Act in Texas, which implemented a cap of 572,000 acre-feet on total annual withdrawals from the aquifer and requires the authority to issue permits for groundwater pumping and administer metering of all wells (except wells that are categorized as being exempt). A well owner is granted an initial permit based on historical use of the water for beneficial purposes. Another example of a pumping quota system is the Harris Galveston Subsidence District, which is a special purpose district created by the Texas Legislature in 1975. The district was created to provide for the regulation of groundwater withdrawal throughout Harris and Galveston counties for the purpose of preventing land subsidence, which leads to increased flooding. All well owners are required to obtain a permit before a well is drilled or operated, and each permit specifies the amount of groundwater authorized to be withdrawn and the percentage of total water demand that may be met with groundwater. A groundwater meter is required for all permitted wells; owners must read their meters, record their readings in a log at least monthly, and report these readings either to the regional water supplier or to the districts.[4]

In irrigated acreage certification, the regulatory authority certifies and restricts the growth of irrigated acres. In such instances, there are no requirements to meter and report volumetric groundwater extractions. Irrigated acreage is therefore an indirect measurement of groundwater extraction where the regulatory authority estimates withdrawals by observing irrigated land area and crop type. Nebraska has implemented the certification of irrigated acreage in response to interstate litigation regarding the impact of groundwater use on transboundary flows in the Republican River. As of 2008 the Nebraska DNR maintains a record of over 10,000 active registered wells with certified irrigated acreage in the Nebraska portion of the Basin.

---

4. See http://hgsubsidence.org/wp-content/uploads/2014/11/RULES2013-09-11.pdf.

## 3.8 Markets for Groundwater and Irrigated Acreage

A pumping quota or irrigated acreage certificate, in effect, is equivalent to a right to pump groundwater. Regulators may elect to separate these rights from parcels of land and make them transferrable, thus allowing for a market for groundwater. In these markets, a transaction can take place if one user is willing to pay a larger sum of money for a unit of groundwater than another user is willing to sell it for. Both the buyer and the seller are better off after this transaction. For the buyer, the purchased groundwater is more valuable to him/her than what he/she paid for it; for the seller, the payment received exceeded the value that he/she placed on the sold groundwater. In theory, these trades will ultimately result in the optimal groundwater allocation. Markets also have the benefit of allowing more flexible movement of water to serve changing conditions and demands. Over time, they encourage users to evaluate conservation strategies because any water that is saved can be sold on the market. As a result, groundwater markets, and water markets in general, can help water-stressed regions better cope with decreased and uncertain water supplies in the face of climate change.

Note that in these markets, the good that is traded is not groundwater per se, but rather the right to pump a unit of groundwater (Brozović and Young, 2014). As a result, a trade only involves decreasing groundwater pumping at one well and increasing it at another, without the need for a water diversion or delivery system. This makes groundwater well suited for implementation of tradable permits since a trade carries no conveyance cost. Groundwater markets are also more likely to lead to a socially optimal allocation if monitoring of pumping is in place and enforcement is strong.

Several localized management institutions have allowed groundwater users to transfer groundwater pumping rights, either on a volumetric or areal basis. The transfers are sometimes subject to rules that reflect geophysical relationships for groundwater flow and extraction. For example, when the goal of management is to reduce stream depletion or seawater intrusion, management agencies typically implement trading ratios that reflect each well's expected damage relative to the one with which they trade. Trading rules may also include social or economic goals, such as prohibiting groundwater trading across county lines to prevent reductions of county revenues.

In the Edwards Aquifer, permit holders may lease or permanently sell a portion of their water rights to another party. In addition to agricultural users, the aquifer supports municipal and industrial demands, and this heterogeneity in users may create many opportunities for welfare gains from groundwater trading. The Edwards Aquifer Authority assists sellers by creating an online bulletin board for their offers of sale. Groundwater trading is also allowed in Nebraska's Upper Republican NRD, among other districts, and trades have taken place despite the lack of formal markets or water brokerages (Palazzo and Brozović, 2014).

## 3.9 Streamflow Augmentation

Streamflow augmentation is a strategy that is sometimes used when regulators are temporarily unable to meet flow requirements in a stream. The strategy involves pumping groundwater from an aquifer and discharging it into the stream or a tributary of the stream. Because the pumped groundwater needs to be conveyed to the stream or tributary via a pipe or channel, groundwater wells used for stream augmentation must be relatively close to the stream. One implication is that stream augmentation cannot serve as a long-term solution to chronic streamflow deficits because the portion of the aquifer from which the augmentation water is pumped is likely to be hydrologically connected to the stream, leading to stream depletion that negates the augmentation.

An example of a stream augmentation program is the Upper Republican Rock Creek Augmentation Project. In this project, Nebraska's Upper Republican NRD purchased 24 irrigated quarters in 2010, retiring 3261 acres in the process. The goal of the project is to comply with the Republican River Compact with Kansas and Colorado by increasing streamflow. In terms of infrastructure, the project involves 10 wells and a 20,000-ft pipeline that carries water to the outlet point, with a capacity of 15,000 acre-feet.

# 4. TECHNOLOGICAL INNOVATIONS TO SUPPORT GROUNDWATER MANAGEMENT

## 4.1 Hydrologic Modeling and Consumptive Use Models

There are several ways in which hydrologic modeling can aid groundwater management. First, it can provide a general sense of the availability of groundwater in a basin. Second, because monitoring of groundwater withdrawals is not a perfect indicator of the quantity of water that is actually consumed, hydrologic models can help estimate this latter effect. Finally, hydrologic modeling can help regulators quantify hydrologic processes that are otherwise difficult to measure. For example, while groundwater pumping can be readily measured with metering, stream depletion generally cannot be measured using in situ methods.

One example of a model that has been used for groundwater policy is the Republican River Compact Administration Groundwater Model, which was developed by a technical committee comprised of representatives from Colorado, Kansas, and Nebraska to administer the compact on the Republican River. Another example is the Cooperative Hydrology Study, which is a hydrologic study of groundwater and surface water resources in a portion of Nebraska's Platte River Basin. Among its goals the study is designed to assist NRDs in the basin with regulation and management of groundwater and provide Nebraska with the basis for groundwater and surface water policy.

## 4.2 Monitoring

### *4.2.1 Metering*

Metering of groundwater, meaning the measurement of the volumetric extraction at a well level, is fairly uncommon though growing in the United States. This information can provide dense, detailed information about spatial groundwater patterns and pumping that can be used for improved hydrologic modeling and targeted management strategies. Today, groundwater meters are generally recorded on an annual timescale, but as metering technology improves and becomes more affordable, it is conceivable that metering in the future will be real time. Real-time metering could greatly improve the ability to target groundwater management in both space and time.

Importantly, there are several regions where annual metering and reporting already exist and are mandatory. The Upper Republican NRD in southwest Nebraska has the densest and longest history of groundwater metering in the country, with district-wide metering since 1982. By 2004 the entire Nebraskan portion of the Republican River Basin was fully metered as a result of litigation over an interstate compact with Kansas and Colorado. The state of Kansas, through its Water Rights Information System, requires every well in the state to be metered and owners to annually report extraction volumes (Golden and Leatherman, 2011). There are several additional, though isolated, cases of metering that have emerged, typically as a result of federal or state litigation, such as the Mojave Basin in California and the Edwards Aquifer in Texas. Other metering initiatives are in a voluntary or precompliance setting, such as the South Platte NRD in Nebraska.

There are some concerns about the quality of monitoring data depending on the way in which it is reported. For example, where metering is mandatory, the management authorities in Nebraska use its own employees to read well meters each year. In Kansas, well reporting, while mandatory, is completed by the well owners. Monitoring and enforcement costs of metering can be quite high, which may include the meters themselves, staff time for the reading and recording, and necessary equipment and technological infrastructure.

### *4.2.2 Energy Records*

In cases in which comprehensive groundwater metering is not in place, tracking the energy consumed by groundwater wells may be a suitable alternative. However, while energy use is likely to be correlated with groundwater use, it is by no means a one-to-one relationship as the determinants of energy use efficiency (water extracted per energy used) vary wildly among farmers. Mieno and Brozović (2015) found that assuming uniformity in these determinants leads to measurement errors, and thus attenuation bias, in price elasticity estimation in previous studies. Nevertheless, an imperfect monitoring system based on energy records may still be preferable to an entirely unmonitored outcome, and further research is needed to quantify its benefits.

### *4.2.3 Measuring Irrigated Acreage*

The quantification and certification of irrigated acreage is a strategy to estimate and control groundwater use, but at much less cost to the regulatory authority and groundwater user. In such instances, there are no requirements to meter and report volumetric groundwater extractions. Instead, monitoring and enforcement are carried out through aerial photography and spot checks. Certification is oftentimes accompanied by restrictions on the growth of irrigated acres. Irrigated acreage is therefore a proxy to, rather than a direct measurement of, groundwater extraction, with varying degrees of accuracy depending on the region's mixture of stream depletion factors, crops, soils, and irrigation technologies, among other variables (Young, 2014). While this is an imperfect form of monitoring, it can approximate groundwater pumping at low cost. This methodology is especially useful if coupled with information about crop types or stream depletion factors.

## 5. EXISTING AND PROPOSED INNOVATIVE GROUNDWATER MANAGEMENT PROGRAMS

Advancements in the groundwater management arena are happening in every sector at every level: from cutting edge agricultural decision-support tools to redefining state laws, from technological improvements to unique public–private partnerships. This section discusses a few case studies of the pioneering work across the United States to improve groundwater productivity and management.

### 5.1 Smart Markets for Groundwater Trading

As previously mentioned, in the case of dynamic, spatially variable externalities, trading rules become necessarily complicated, and therefore time consuming to navigate.

Furthermore, there may be high search costs for users to find other interested parties with whom to trade. As a result, even in jurisdictions where groundwater trading is allowed, relatively few trades occur (Juchems et al., 2013).

Smart markets, or electronic clearinghouses, can reduce high transaction costs for trading. Smart markets pool interested buyers and sellers in a common market place, reducing the time and money spent searching for a party with which to trade. Further, smart markets, through tailored computer algorithms, consider the complex rules and regulations governing groundwater trades and automate the process of matching eligible parties, streamlining, and guaranteeing regulatory compliance. Because of the high granularity in groundwater movement and governance, most markets cover a fairly small geographic extent, meaning that the market for groundwater trading may be comparatively smaller or thinner than a typical surface water market. Nonetheless, not only are there emerging groundwater "markets," that is, the ability to trade groundwater rights, across the United States—including in Nebraska, California, Texas, Kansas, Washington, and more—there are also emerging smart markets, trading temporary and permanent groundwater rights, both on allocation and irrigated acreage bases.[5]

## 5.2 Changing Kansas Groundwater Management Law

The state of Kansas has a senior appropriative rights system for groundwater use, suggesting that in an overappropriated region, junior users would be the first to have groundwater pumping rights stripped. However, legislative actions have provided more flexibility to more equitably and efficiently mitigate quantity and quality concerns arising from groundwater use (Kansas Department of Agriculture, 2009). In particular, two management approaches have emerged, and one significant state policy has been repealed.

In 1978, the state gave authority to its chief engineer to implement an Intensive Groundwater Use Control Area (IGUCA). This law gave the chief engineer significant and sole discretion to impose restrictions, including well moratoria and reductions in withdrawals according to seniority. Today, only nine IGUCAs are in effect; there is generally strong local resistance to IGUCAs as restrictions can be costly and especially impair junior rights holders. For example, as ordered by the Wet Walnut Creek IGUCA, junior groundwater rights holders were allocated only 44% that of senior rights holders (Golden and Leatherman, 2011). As a result, in 2012 the state passed new legislation to permit its local management authorities, called Groundwater Management Districts, to initiate and implement with the chief engineer a Local Enhanced Management Agency (LEMA) (Kansas S.B. 310, 2012). Rather than the top-down approach of the IGUCAs, LEMAs instead offered a bottom-up approach that could be initiated and crafted by local groundwater users. In a LEMA, while the state cannot impose stricter controls than that which the region proposes, it retains authority to impose an IGUCA should the LEMA be ineffective. LEMAs, unlike IGUCAs, may suspend the priority rights system so that all users, regardless of seniority, share an equal burden. The first LEMA was ordered in Sheridan County in 2013, where locals wished to extend the life of their portion of the Ogallala Aquifer by reducing pumping in the following 5 years by 20% (Bossert, 2013).

5. Mammoth Trading, a company for which authors Nicholas Brozović and Richael Young are co-founders, designs and operates smart markets. Mammoth Trading currently operates several smart water markets across the United States.

In 2012 the state repealed its "use it or lose it" law, not uncommon to many other western states (Kansas H.B. 2451, 2012). Use it or lose it policies, which allow unused water rights to be recaptured by a state, were intended to provide states the flexibility to redistribute unused water rights to those who would make beneficial use of them. Instead, the policy has generated a powerful adverse incentive to waste and overuse limited water resources. Recognizing this, Kansas repealed the law, hoping to save surface water and groundwater alike for other users.

Several western states are struggling with antiquated laws concerning groundwater use and management, including California, Nevada, and Texas. The way that Kansas has, through policy and legislation, been adaptable and responsive to local concerns is a model to other states that are seemingly stuck with laws that predate a scientific understanding of groundwater. Kansas is still experimenting, and does not necessarily have a watertight set of policies. However, the state's willingness to innovate, learn from experience, and innovate again is promising.

## 5.3 Conjunctive Use and Managed Aquifer Recharge in Nebraska

Many states have evolved to manage groundwater and surface water conjunctively, acknowledging that the two resources are connected. In 2011, spurred by historic levels of streamflow in the North Platte River, public and private entities in Nebraska worked together to develop strategies that could turn flooding threats into opportunities to recharge the underlying Ogallala Aquifer. The Nebraska DNR, Twin Platte NRD, and several irrigation districts and private landowners developed an innovative solution to conjunctively manage the excess flows that were coming down the North Platte River. Flood waters would be diverted through irrigation canals and private reuse pits, thereby recharging the aquifer, reducing peak flood flows, and retiming flows that were stored in canals and shallow groundwater (Dimmitt, 2015). Given the short notice of the flood waters barreling down the North Platte River, the parties had little time, but managed to approve the paperwork and prepare the canals and pits for diversions. Since then, the process has been streamlined and repeated in 2013 and 2015, these times on the South Platte River (Dimmitt, 2015; Nebraska Department of Natural Resources, 2014).

## 5.4 Nebraska Water Balance Alliance

The Nebraska Water Balance Alliance (NEWBA), founded by agricultural producer Roric Paulman in 2011, is a producer-driven initiative to manage irrigation water most efficiently. To accomplish this, NEWBA combines state-of-the-art information technology, including telemetry, weather stations, pumping data and remote irrigation controls, and soils and soil moisture probes (NEWBA, 2013). Through a partnership of producers, equipment vendors, electricity companies, crop consultants, the University of Nebraska–Lincoln, NRDs, and the USDA NRCS, NEWBA is advancing field-level studies to identify effective water use strategies that can be quantified and replicated elsewhere (NEWBA, 2013). The partnership also seeks to demonstrate that regulation alone should not drive interest and action in conservation, but that the agricultural community should and can lead in water efficiency and sustainability. NEWBA recently announced plans to collaborate with the Texas Alliance for Water Conservation.

## 6. CONCLUSIONS

In this chapter we provided an overview of groundwater management policies, regulations, and initiatives that have been implemented in the United States. We considered various motivations for groundwater management, grounded in the need to resolve underlying market failures, and we described the policy instruments and technologies that are available to regulators.

Several overarching conclusions can be drawn from this chapter. First, groundwater use can be inefficient because of the presence of a variety of market failures, and in some cases regulators may be faced with multiple market failures in the same groundwater basin. This makes it less likely that an optimal groundwater allocation can be accommodated by existing infrastructure and institutional constraints. However, we find in some instances that second-best solutions are able to approximate the optimal allocation of groundwater, and in many cases are likely to lead to an improved outcome relative to a scenario with no regulation. Similarly, even if comprehensive metering is not currently in place in most groundwater basins, there are alternative options for regulators to start monitoring groundwater use, such as observing irrigated acreage and energy consumption.

Second, despite the fact that groundwater is generally unmetered and unregulated in the United States, many local institutions have gone ahead and adopted management strategies with varying degrees of sophistication. Some of the most innovative policies can be found in Nebraska and Texas, and while these policies are innovative in different ways, they are usually based on comprehensive metering of groundwater use and strict enforcement of permitted withdrawals. Water management institutions in states such as California that are seeking to develop groundwater regulations in previously unregulated regions can look toward this diverse existing landscape of groundwater management for inspiration.

Finally, while many groundwater management policies in the United States have been implemented in response to or under the threat of federal or state intervention, there are several instances of local groundwater users who have curtailed pumping voluntarily. In such cases, there is typically a strong sense of community that aspires to conserve natural resources for future generations and maintain a rural way of life.

## ACKNOWLEDGMENTS

The authors wish to thank Johanna Rayl for excellent research assistance. This research was funded in part by USDA NIFA Award number 2012-67003-23227 and by USDA NIFA Water CAP Award number 2015-68007-23133.

## REFERENCES

Barlow, P.M., Leake, S.A., 2012. Streamflow Depletion by Wells—Understanding and Managing the Effects of Groundwater Pumping on Streamflow. Circular 1376. U.S. Geological Survey, Reston, Virginia.

Barron, O.V., Emelyanova, I., Van Niel, T.G., Pollock, D., Hodgson, G., 2014. Mapping groundwater-dependent ecosystems using remote sensing measures of vegetation and moisture dynamics. Hydrological Processes 28 (2), 372–385.

Borchers, J.W., Grabert, V.K., Carpenter, M., Dalgish, B., Cannon, D., 2014. Land Subsidence from Groundwater Use in California. Luhdorff and Scalmanini Consulting Engineers, Woodland, California.

Bossert, W., February 7, 2013. An Overview of Kansas Groundwater Management District No. 4 (Personal interview).

Brozović, N., Young, R., 2014. Design and Implementation of Markets for Groundwater Pumping Rights. Water Markets for the 21st Century. Springer, Netherlands, pp. 283–303.

Brozović, N., Sunding, D.L., Zilberman, D., 2010. On the spatial nature of the groundwater pumping externality. Resource and Energy Economics 32 (2), 154–164.

California Department of Water Resources, 2015. NASA Report: Drought Causing Valley Land to Sink. Press Release.

Cooper, A.H., 2008. The classification, recording, databasing and use of information about building damage caused by subsidence and landslides. Quarterly Journal of Engineering Geology and Hydrogeology 41, 409–424.

Craun, G.F., Brunkard, J.M., Yoder, J.S., Roberts, V.A., Carpenter, J., Wade, T., Calderon, R.L., Roberts, J.M., Beach, M.J., Roy, S.L., 2010. Causes of outbreaks associated with drinking water in the United States from 1971 to 2006. Clinical Microbiology Reviews 23 (3), 507–528.

Delaware River Basin Commission, 2014. Water Resources Program, FY 2015–2017. West Trenton, New Jersey.

Dimmitt, A., September 24, 2015. Managed Aquifer Recharge Projects in the Twin Platte Natural Resources District (Telephone interview).

Division of Conservation, 2012. Water Right Transition Assistance Pilot Project Program: FY 2012 Legislative Report. Kansas Department of Agriculture, Manhattan, Kansas.

Division of Conservation, 2013. Upper Arkansas River Conservation Reserve Enhancement Program Performance Report. Kansas Department of Agriculture, Manhattan, Kansas.

Dixon, T.H., Amelung, F., Ferretti, A., Novali, F., Rocca, F., Dokka, R., Sella, G., Kim, S.-W., Wdowinski, S., Whitman, D., 2006. Space geodesy: subsidence and flooding in New Orleans. Nature 441, 587–588.

Famiglietti, J.S., Lo, M., Ho, S.L., Bethune, J., Anderson, K.J., Syed, T.H., Swenson, S.C., de Linage, C.R., Rodell, M., 2011. Satellites measure recent rates of groundwater depletion in California's Central Valley. Geophysical Research Letters 38 (3). http://dx.doi.org/10.1029/2010GL046442.

Fleckenstein, J., Anderson, M., Fogg, G., Mount, J., 2004. Managing surface water-groundwater to restore fall flows in the Cosumnes river. Journal of Water Resources Planning and Management 130 (4), 301–310.

Galloway, D., Jones, D.R., Ingebritsen, S.E., 1999. Land Subsidence in the United States. Circular 1182. U.S. Geological Survey, Reston, Virginia.

Gannett, M.W., Lite Jr., K.E., Morgan, D.S., Collins, C.A., 2001. Ground-Water Hydrology of the Upper Deschutes Basin, Oregon. Water-Resources Investigations Report 2000-4162. U.S. Geological Survey, Reston, Virginia.

Gisser, M., Sanchez, D.A., 1980. Competition versus optimal control in groundwater pumping. Water Resources Research 16 (4), 638–642.

Glover, R., Balmer, G., 1954. River depletion resulting from pumping a well near a river. American Geophysical Union Transactions 35 (3), 468–470.

Golden, B., Leatherman, J., 2011. Impact Analysis of the Walnut Creek Intensive Groundwater Use Control Area.

Gou, S., Gonzales, S., Miller, G.R., 2015. Mapping potential groundwater-dependent ecosystems for sustainable management. Groundwater 53 (1), 99–110.

Hathaway, D.L., 2011. Transboundary groundwater policy: developing approaches in the western and southwestern United States. Journal of the American Water Resources Association 47 (1), 103–113.

Hershler, R., Liu, H.-P., Howard, J., 2014. Springsnails: a new conservation focus in western North America. BioScience 64 (8), 693–700.

Howard, J., Merrifield, M., 2010. Mapping groundwater dependent ecosystems in California. PLoS One 5 (6). http://dx.doi.org/10.1371/journal.pone.0011249.

Johnson, T., 2007. Battling Seawater Intrusion in the Central & West Coast Basins. Technical Bulletin. In: Water Replenishment District of Southern California, vol. 13. Lakewood, California.

Juchems, E.M., Schoengold, K., Brozovic, N., 2013. Predicting groundwater trading participation in the upper republican natural resource district. In: 2013 Annual Meeting, August 4–6, 2013, Washington, DC. Agricultural and Applied Economics Association. No. 150511.

Kansas Department of Agriculture, 2009. Fact Sheet: Intensive Groundwater Use Control Areas.

Kansas House Bill 2451, 2012. Enacted.

Kansas Senate Bill 310, 2012. Enacted.

Konikow, L.F., 2011. Contribution of global groundwater depletion since 1900 to sea-level rise. Geophysical Research Letters 38 (17). http://dx.doi.org/10.1029/2011GL048604.

Koundouri, P., 2004. Potential for groundwater management: Gisser-Sanchez effect reconsidered. Water Resources Research 40 (6). http://dx.doi.org/10.1029/2003WR002164.

Laney, R.L., Raymond, R.H., Winikka, C.C., 1978. Maps Showing Water-Level Declines, Land Subsidence, and Earth Fissures in South-Central Arizona. Water-Resources Investigations Report 78–83. U.S. Geological Survey, Reston, Virginia.

Maupin, M.A., Kenny, J.F., Hutson, S.S., Lovelace, J.K., Barber, N.L., Linsey, K.S., 2014. Estimated Use of Water in the United States in 2010. Circular 1405. U.S. Geological Survey, Reston, Virginia.

McCarl, B.A., Dillon, C.R., Keplinger, K.O., Lynn Williams, R., 1999. Limiting pumping from the Edwards Aquifer: an economic investigation of proposals, water markets, and spring flow guarantees. Water Resources Research 35 (4), 1257–1268.

McGuire, V.L., 2013. Water-Level and Storage Changes in the High Plains Aquifer, Predevelopment to 2011 and 2009–11. Scientific Investigations Report 2012-5291. U.S. Geological Survey, Reston, Virginia.

Mieno, T., Brozović, N., 2016. Price elasticity of groundwater demand: Attenuation and amplification bias due to incomplete information, unpublished manuscript.

National Research Council, 1991. Mitigating Losses from Land Subsidence in the United States. The National Academies Press, Washington, DC.

Nebraska Department of Natural Resources, July 2014. The 2013 South Platte Flooding. Nebraska Resources Newsletter, pp. 1–5.

Nebraska Water Balance Alliance, 2013. Water for Generations, Summer Update 2013. Retrieved from: http://media.wix.com/ugd/1fbf66_01f9b7b8cdc40adc793f065129dc7c6f.pdf.

Negri, D.H., 1989. The common property aquifer as a differential game. Water Resources Research 25 (1), 9–15.

NGWA, 2015. Groundwater Use in the United States of America. National Ground Water Association, Westerville, Ohio.

Nichols, A.L., Willis, A.D., Jeffres, C.A., Deas, M.L., 2014. Water temperature patterns below large groundwater springs: management implications for Coho salmon in the Shasta River, California. River Research and Applications 30 (4), 442–455.

Palazzo, A., Brozović, N., 2014. The role of groundwater trading in spatial water management. Agricultural Water Management 145, 50–60.

Poland, J.F., Lofgren, B.E., Ireland, R.L., Pugh, R.G., 1975. Land Subsidence in the San Joaquin Valley, California as of 1972. Professional Paper 437-H. U.S. Geological Survey, Reston, Virginia.

Prinos, S.T., Wacker, M.A., Cunningham, K.J., Fitterman, D.V., 2014. Origins and Delineation of Saltwater Intrusion in the Biscayne Aquifer and Changes in the Distribution of Saltwater in Miami-Dade County, Florida. Scientific Investigations Report 2014–5025. U.S. Geological Survey, Reston, Virginia.

Provencher, B., Burt, O., 1993. The externalities associated with the common property exploitation of groundwater. Journal of Environmental Economics and Management 24 (2), 139–158.

Rio Grande Water Conservation District, 2015. Fact Sheet: Conservation Reserve Enhancement Program – Colorado Rio Grande. Rio Grande Water Conservation District, Alamosa, Colorado.

San Joaquin Valley Drainage Implementation Program, 1998. Drainage Management in the San Joaquin Valley: A Status Report. California Department of Water Resources, Sacramento, California.

Scanlon, B.R., Faunt, C.C., Longuevergne, L., Reedy, R.C., Alley, W.M., McGuire, V.L., McMahon, P.B., 2012. Groundwater depletion and sustainability of irrigation in the US High Plains and Central Valley. Proceedings of the National Academy of Sciences of the United States of America 109 (24), 9320–9325.

Theis, C.V., 1935. The relation between the lowering of the piezometric surface and the rate and duration of discharge of a well using groundwater storage. Transactions, American Geophysical Union 16 (2), 519–524.

Water Resources Department, 2008. Deschutes Ground Water Mitigation Program: Five-Year Program Evaluation Report. State of Oregon, Salem, OR.

Young, R., 2014. Tradable Permit Systems for a Spatially Heterogeneous Externality: A Microparameter Approach (Master's thesis). University of Illinois at Urbana-Champaign, Urbana, Illinois.

SUBPART 3.2

# Management Approaches in Europe

Chapter 3.2.1

# Wastewater Reuse to Cope With Water and Nutrient Scarcity in Agriculture—A Case Study for Braunschweig in Germany

P. Grundmann[1,2], O. Maaß[1,2]
[1]*Leibniz-Institute for Agricultural Engineering Potsdam-Bornim e.V., Potsdam, Germany;* [2]*Humboldt University of Berlin, Berlin, Germany*

## 1. INTRODUCTION

The main function of wastewater treatment is to avoid adverse effects on receiving surface water bodies from organic and inorganic pollutants as well as high inflows of nutrients contained in the wastewater. However, the recovery of nutrients and the reuse of wastewater are becoming more and more important, particularly in increasingly problematic situations involving water and nutrient scarcity in agriculture. This is reflected in an increasing interest in research and development on nutrient recovery from wastewater and reuse of wastewater in agriculture (Cordell et al., 2011; Maaß et al., 2014; Satorius et al., 2011; Singh et al., 2012; Zema et al., 2012). At the government level, the European Union's Urban Wastewater Treatment Directive emphasizes that "treated wastewater should be reused whenever appropriate" (EU, 1991, p. 4). Therefore options for reusing treated wastewater in agriculture are being investigated in Europe for reducing pressure on natural freshwater resources and limited nutrient reserves. Winpenny et al. (2010) and Aiello et al. (2007) ascribed a great potential for the adoption of wastewater reuse schemes, as several actors are simultaneously among the potential beneficiaries, including farmers, municipalities and their utilities, and rural and urban communities. However, some scholars have pointed out that to ensure long-term sustainability of wastewater reuse, it is also essential to take social, institutional, and organizational dimensions into account when evaluating reuse options (Pescod, 1992; Larsen, 2011). Some of the researched questions regarding wastewater reuse in agriculture include potential health hazards for farm workers and food consumers (Pedrero et al., 2010), soil salinization (Muyen et al., 2011), and the buildup of heavy metals and anthropogenic trace contaminants in soils and food crops (Khan et al., 2008; Mapanda et al., 2005; Pedersen et al., 2005; Toze, 2006). Precise management strategies, including the application of proper purification levels, periodic monitoring of soil and plant properties, as well as suitable irrigation, cultivation, and harvesting practices, are imperative

**Competition for Water Resources. http://dx.doi.org/10.1016/B978-0-12-803237-4.00020-3**

to minimize hazards to humans and the environment (Aiello et al., 2007; Muyen et al., 2011; Qadir et al., 2010; Rusan et al., 2007). The specific properties of agricultural wastewater reuse require strategies as well as appropriate governance and management models for integrating economic activities in the water sector and the agricultural sector. Learning from existing experiences should contribute to sharing knowledge about successful strategies, specifically for integrating wastewater reuse and agriculture.

This chapter aims to provide lessons learned from a case study on the activities and economic impacts of a wastewater reuse scheme in the vicinity of the city of Braunschweig in southeastern Lower-Saxony (Germany) that was developed for coping, among others, with water and nutrient scarcity in agriculture. We want to raise awareness and deliver information about the effects of wastewater reuse in agriculture for informed decision making by public bodies as well as involved enterprises, residents, and the farming community. The description draws from results from a detailed added-value analysis of the wastewater reuse scheme conducted by Maaß and Grundmann (2016). Added-value analyses contribute to understanding the net benefits and costs, as well as the increase in value generated through economic activities (Haller, 1997; Hamilton et al., 1991). The added-value is an indicator for assessing the effects of economic activities on local economies. The following presentation focuses on the monetary added-value effects of the wastewater reuse scheme as determined by Maaß and Grundmann (2016). An analysis of institutional economic aspects of the wastewater reuse scheme can be found in Villamayor-Tomas et al. (2015).

This chapter summarizes some selected findings and conclusions from the study conducted by Maaß and Grundmann (2016). For a more detailed and comprehensive description of the applied methodological approach and the results, we invite the reader to consult Maaß and Grundmann (2016).

## 2. CHARACTERIZATION OF THE WASTEWATER REUSE SCHEME IN BRAUNSCHWEIG

The case at hand was developed by the Abwasserverband Braunschweig, a wastewater association located in Braunschweig (Germany), from a number of wastewater treatment issues into a scheme using treated wastewater and sewage sludge for irrigation and fertilization of agricultural fields for energy plants, which are then used for biogas and energy generation (Lesjean and Remy, 2013).

The concept of the scheme is to use purified wastewater and sewage sludge for irrigation and nutrient provision for food and energy plants while using the energy plants for biogas generation and electricity and heat production, thus creating a water–nutrient–energy cycle (Fig. 1). The case of wastewater reuse in Braunschweig has been determined by its history, geographical features, and economic impacts; these will be described in the following sections.

### 2.1 History and Locational Characteristics

The wastewater reuse scheme started with the implementation of the first infiltration fields in 1895 for capturing wastewater from the nearby city of Braunschweig. With population growth in the nearby cities of Braunschweig and Gifhorn, the capacity of the infiltration fields was not sufficient anymore to capture and treat the increasing amounts of wastewater.

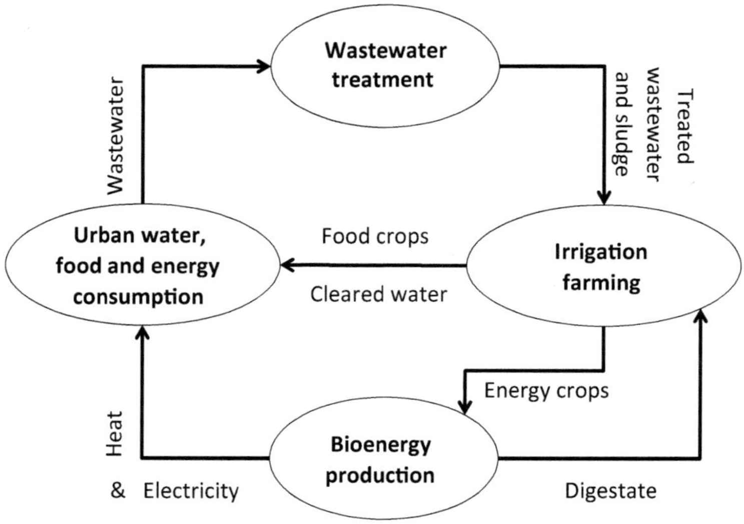


FIGURE 1 Concept of the wastewater-nutrient-energy cycle of the wastewater reuse scheme in Braunschweig (adapted from Hartmann et al., 2010).

This situation led to the creation of the Abwasserverband Braunschweig in 1954 with the mission of running a mechanical wastewater treatment plant that has been supplying treated wastewater for the irrigation of about 3000 ha of agricultural fields in the surrounding counties. The members of the association are the city of Braunschweig, the water association of the neighboring city of Gifhorn, and 430 owners of agricultural fields receiving the treated wastewater from the associations' treatment plant. The association has continuously refined the reuse scheme. The latest enhancements included an upgrade of the primary treatment facility to a full-biological treatment plant in 1992 and the commissioning of a sludge digester in 2000. Furthermore, a biogas plant, which is owned by the association, started operation in 2007.

Physical and natural conditions related to the wastewater reuse in agricultural production in Braunschweig are rather favorable. The agricultural fields selected for the application of treated wastewater have very sandy and light soils that require constant irrigation. Further, the sandy soils are generally poor in nutrients and have a limited water and nutrient retention capacity (Ternes et al., 2007). The structural characteristics of the soils allow for agricultural cultivation even in situations of excess water supply, as water percolates instead of causing waterlogging. Furthermore, the area experiences climatic water deficit from April to September each year (Ahlers and Eggers, 2004). A continuous additional supply of water and nutrients is therefore essential for crop production in the area.

## 2.2 Economic Activities

Fig. 2 displays the value chains related to different economic activities in the sectors of wastewater treatment, crop production, and bioenergy generation in the wastewater reuse scheme in Braunschweig. The outputs resulting from the primary and secondary treatment of wastewater, including secondary treated effluent and sewage sludge, are further

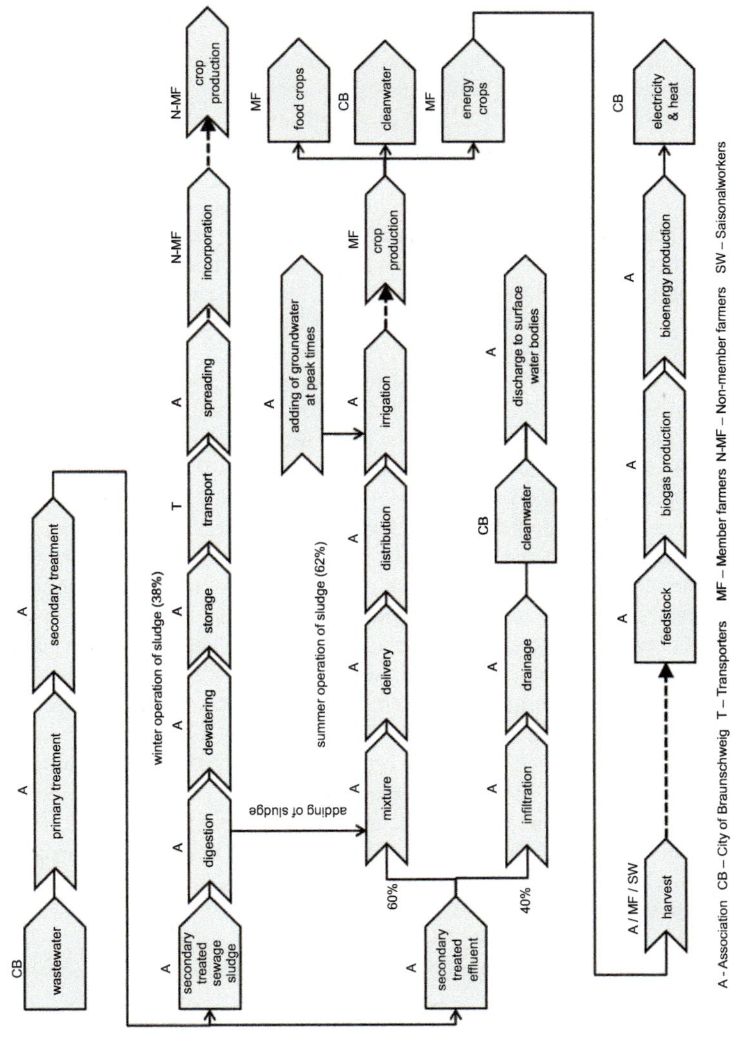


FIGURE 2 Value chains of the wastewater reuse scheme in Braunschweig (Maaß and Grundmann, 2016).

processed in the value chains of wastewater treatment and reused as inputs for crop production in the value chains of food and energy. The energy crops are feedstock inputs in the bioenergy value chain for the production of heat and electricity from biomass/biogas via anaerobic digestion. In this way, the material flows of the value chains are interconnected, including wastewater, biomass, and energy flows, based on the agricultural reuse of treated wastewater and sludge. The dotted lines in Fig. 2 indicate interconnections between the value chains in terms of material flows. The involved actors are indicated by the abbreviations above the single added-value steps and products.

Wastewater from Braunschweig and surrounding communities is delivered for primary purification to the wastewater treatment plant with a flow rate of 60,000 $m^3$ per day with a population equivalent of 350,000 people. The current treatment process includes mechanical treatment, biological phosphate removal in combination with nitrification and denitrification, and anaerobic stabilization of sludge (Ternes et al., 2007). In addition, a downstream system of irrigation and infiltration fields is used for the final treatment of the secondary effluent. The largest part of the effluent (60%) is used directly for irrigating about 2700 ha of agricultural fields of the member farmers. The remaining part (40%) is discharged to infiltration fields with a size of about 220 ha, located near the treatment plant (Maaß and Grundmann, 2016). The infiltration fields operate as part of an infrastructure for a natural treatment procedure using a meandering system and soil passage. After this process is completed, drained water is discharged to surface water bodies.

The sewage sludge produced in the described process is stabilized via anaerobic digestion and utilized in two different value chains. In the winter period the sewage sludge is dewatered and stored on-site before it is transported in the summer time to croplands (700 ha) outside of the wastewater scheme. The sludge is spread by the associations' employees and subsequently incorporated individually by the farmers into their soils.

During the vegetation period the sewage sludge is added to the effluent prior to irrigation. The mixed effluent–sewage–sludge is discharged to a gravity sewer system that brings the mixture to the irrigation fields. The mixture is then spread by the associations' staff on the agricultural fields of the member farmers. The irrigation system consists of four large pumping stations, a distribution net of about 125 km of pipes, 1000 discharge points, and 170 spray irrigation machineries. The association determines the watering schedule in coordination with the member farmers. In 2012, approximately 12 million $m^3$ of the secondary effluent were irrigated on the irrigation fields and approximately 8 million $m^3$ were discharged to the infiltration fields. In addition, about 604,000 $m^3$ of groundwater were used annually to meet the irrigation requirements at the peak times. The total quantity of sludge produced in 2012 was 4482 tons dry matter. Approximately 2770 tons dry matter of sludge were irrigated together with the effluent inside the association territory and 1712 tons dry matter were dewatered and applied as fertilizer outside the association territory (Maaß and Grundmann, 2016).

The legislative permission for reusing wastewater in the study area is given by the district government of Braunschweig (Bezirksregierung Braunschweig, 2001). High safety standards and a sophisticated monitoring system are applied to minimize hazards to the environment, inhabitants, and food consumers (Remy, 2012). The loading rates of heavy metals contained in the effluent–sludge mixture fall below the tolerable limits of the German Sewage Sludge Ordinance (LWKH, 2000). Furthermore, as shown by Ternes et al. (2007), residues of pharmaceuticals and personal care products can be degraded largely by top soil passage. Respecting precautionary hygienic restrictions, farmers in the association territory are not allowed to cultivate crops planted for direct consumption, for example, fruits or vegetables (Bezirksregierung Braunschweig, 2001), but instead the irrigation area is mainly cropped with maize, grain, and sugar beets.

The biogas plant, with a total capacity of 2.5 MW, is operated by the association that maintains quantity-based contracts, with 45 member farmers, for the delivery of maize to ensure a steady feedstock supply for the biogas plant. The machines for harvesting maize are provided by the association, whereas the operations are mainly executed by farmers themselves, as well as by seasonal workers. In return, the association purchases the maize at a preferential price.

## 2.3 Economic Impacts

Wastewater reuse in agriculture creates costs and benefits for the operators of the wastewater treatment plant as well as for farmers and bioenergy producers participating in the scheme in Braunschweig. The multiple benefits are paired by costs that cumulatively determine the profits of the equity providers. The added-value generated by the reuse scheme was assessed by analyzing the remunerations of all stakeholders participating in the different value chains (Maaß and Grundmann, 2016).

### *2.3.1 Costs and Benefits of the Wastewater Reuse Scheme*

The treatment and reuse of wastewater in the Braunschweig scheme entails costs related to wastewater treatment, irrigating wastewater and sewage sludge, spreading wastewater on infiltration fields, and spreading dewatered sludge on croplands. The total costs of the investigated activities are financed by contributions of the association members. The largest share is borne by the city of Braunschweig and the water association of Gifhorn. Landowners and member farmers contribute only to financing the costs of irrigation operations. For a more detailed analysis of the cost structure the reader is encouraged to consult Maaß and Grundmann (2016).

The operators of the wastewater treatment plant benefit from the savings of the wastewater fees because of avoided discharges of the effluent to the surface water bodies. Furthermore, the operators save the costs of incinerating sludge and the costs of dewatering the sludge quantities, which are irrigated together with the effluent on the irrigation fields (Maaß and Grundmann, 2016).

The benefits for member farmers participating in the wastewater scheme are mainly savings in irrigation costs because the supply of irrigation water and the irrigation operations are undertaken by the wastewater association. Accordingly, farmers save in investments and depreciations, capital, personnel, machine costs, and groundwater extraction (Maaß and Grundmann, 2016). A further benefit for farmers is reduced fertilization costs, since the nutrients in the irrigated wastewater and sludge substitute mineral fertilizers that otherwise would need to be purchased. Depending on the crop, farmers can reduce the application of mineral fertilizers by 40–66% (Maaß and Grundmann, 2016). This contributes to a reduction of variable costs for spreading mineral fertilizers, including related costs for machines, personnel, and capital. Maaß and Grundmann (2016) estimated that the reuse of wastewater and sludge, instead of groundwater and mineral fertilizers, results in a reduction of 32% in the total costs of crop production in the study area. Another benefit specific for farmers supplying energy maize as feedstock to the association's biogas plant comes from leaving the maize harvest with the association. However, the farmers also have to accept a lower price for the energy maize sold to the association's biogas plant compared to market prices for maize. The main benefit for the association as the operators of the biogas plant is purchasing the feedstock energy maize from the association's farmers at a price lower than the general market price (Maaß and Grundmann, 2016).

Benefits are also obtained by nonmember farmers from applying dewatered sludge with a high fertilization value as well as the financial compensation paid by the association to the farmers for incorporating sludge into soils (Maaß and Grundmann, 2016).

The total costs savings (ie, benefits) realized in the wastewater scheme in 2012 were estimated to be €3.3 million (Maaß and Grundmann, 2016). The shares in the total benefits from the reuse of treated wastewater and sludge were reported to be 48% for the member farmers, 40% for the operators of the wastewater treatment plant, 11% for the nonmember farmers for spreading dewatered sludge, and 1% for the operators of the biogas plant.

### 2.3.2 Added-Value Generation and Distribution

The total estimated added-value from wastewater treatment, crop production, and bioenergy generation in the Braunschweig scheme was €8.2 million in 2012 (Maaß and Grundmann, 2016). The value chains of wastewater treatment and reuse accounted for the greatest part of the added-value (53%). Food crop production in the association territory accounted for 28%, energy crop production (maize and rye) for 10%, and bioenergy production for 9% of the total added-value.

Maaß and Grundmann (2016) showed that the linkage of wastewater treatment and crop production resulted in a 50% increase of the added-value generated from crop production in the study area in 2012. The increase of the added-value is the result of connecting material flows of the value chains of wastewater treatment and crop production.

In addition to the equity providers, who participate in the different value chains, creditors, employees, and also the state obtain a share of the added-value in the form of interest payments, net wage payments and taxes, fees, and social contributions. Among stakeholders, the largest share of the added-value was captured by equity providers, including the operators of the biogas plant and farmers, who received altogether €2.7 million (or 33%) of the total added-value in terms of after-tax profits. Employees obtained €2.0 million (25%) in net wages. Creditors received €1.8 million (22%) in interest payments and the state received €1.6 million (20%) in taxes, fees, and social contributions of employers and employees (Maaß and Grundmann, 2016).

According to Maaß and Grundmann (2016) the amount of the added-value that remained in the communities in which the Abwasserverband Braunschweig is active (ie, the added-value benefiting the local economy) was 77% of the total added-value generated in the value chains composing the wastewater reuse scheme in 2012. The most important contribution to the added-value for the local economy comes from after-tax profits of equity providers (43%), followed by net wages of employees (33%), and interest payments to creditors (23%). About 95% of these remunerations remain in the local economy. In contrast, about 96% of the remunerations received by the state in the form of taxes, fees, and social contributions do not remain in the local economy.

### 2.3.3 Crowding Out Effects in Crop Production

Crowding out effects (resulting from the displacement of economic activities because of the reuse of wastewater) had an impact on the added-value from crop production mainly because of reduced on-farm labor inputs related to the irrigation and fertilization activities, which in turn decreased the wage payments to employees in agriculture and social contributions to the state (Maaß and Grundmann, 2016). Similarly, the reduction of capital use by farmers led to a decrease of interest payments received by creditors. The revenues of equity providers for feedstock for biogas production (ie, maize farmers) were reduced as well, because of the difference in the price paid by the association and the price on the region's market. Finally, the use of wastewater instead of groundwater decreased the farmer's payments of fees to the state for the extraction of groundwater. However, this effect was partially counteracted by the fees paid by the association for additionally using groundwater for irrigation at peak times (Maaß and Grundmann, 2016).

The crowding out effects and the benefits may have significant impacts on the distribution of the added-value from crop production among the stakeholders, as shown by Maaß and Grundmann (2016). Compared with the use of groundwater for irrigation, the remunerations of the creditors were reported to be 80% lower and the remunerations of the employees 33% lower when using wastewater for irrigation. Nevertheless, the remunerations of the equity providers (ie, farmers) were 88% higher. The authors attribute this net effect to the overall savings in food and energy crop production, which offset the partial decrease in the revenues of maize producers. The total remunerations of the state are reported to be 27% higher than in the case with groundwater use, mainly because of higher income tax payments by farmers resulting from profitability increases in crop production. This tax payment increase exceeds the decrease in social contributions of employers and employees and the reduction in groundwater extraction fee payments.

## 3. LESSONS LEARNED

The presented case study informs about economic impacts of agricultural wastewater reuse for local economic development. In general, the described reuse scheme guarantees the productivity of crop production despite the climatic water deficit in the area. In particular, the farmers benefit from the supply of inputs and services, such as the irrigation infrastructure and machines, as well as the nutrient-rich irrigation water and sludge. Furthermore, the reuse scheme provides an opportunity for the participating farmers to diversify their cropping practices, for example, by cultivating energy crops. Energy crop producers benefit further from business relations with the association that serves as a local buyer of their energy crops and also provides some agricultural services such as the harvest of energy maize. This way, the reuse scheme is advantageous for the development of bioenergy activities, which benefit from the local supply of energy crops. The operators of the wastewater treatment plant benefit from the farmers as customers of the wastewater and sludge, which otherwise would have to be further treated and disposed of without any subsequent reuse activities. The general population and communities benefit from the reuse scheme, as it provides clean water as well as electricity and heat from renewable energy at competitive costs (Maaß and Grundmann, 2016).

However, the wastewater reuse scheme in Braunschweig also imposes constraints on the actions of the involved actors, discourages them from engaging in certain economic activities (log-out effects), or makes them remain in less rewarding business activities (log-in effects) (Maaß and Grundmann, 2016). Member farmers of the Abwasserverband Braunschweig face restrictions regarding their choice of cultivation crops, which particularly prevents them from growing crops that are primarily consumed without previous processing. The wastewater association is allowed to prescribe crops and crop rotations to be cultivated if the cultivation plans of the member farmers do not guarantee a year-round acceptance of the wastewater and the nutrients contained in it. Because of the limitations for food crop production, farmers may miss the opportunity to achieve higher profits from producing high-value fruits and vegetables. Single farmers reported that in the case of Braunschweig, single producers of sugar beets and potatoes occasionally had to deal with problems when marketing their products, as not all customers and wholesalers are willing to accept products irrigated with treated wastewater and sludge. Some owners of agricultural land within the wastewater scheme are not allowed to freely opt out from the scheme, but need permission in case they want to leave it. Similarly, landowners are restricted from selling their land for purposes other than for agricultural production (eg, construction purposes). These restrictions for the landowners were implemented to ensure the size and the coherence of the irrigation area required for the continued existence of the reuse scheme.

The added-value generated in the wastewater scheme in Braunschweig remains mainly in the region where the wastewater association is located (Maaß and Grundmann, 2016). This coincides with findings from other authors who stated that local production, consumption, and disposal of goods and services are conducive to increasing the regional share of added-value (Bentzen et al., 1997; Hoffmann, 2009; Kimmich and Grundmann, 2008; Kosfeld and Gückelhorn, 2012; Marcouiller et al., 1996). The case of the wastewater reuse scheme in Braunschweig is particularly interesting, as the conversion of waste products from wastewater treatment into usable inputs for agriculture results in additional added-value that promotes the emergence of circular economies (Maaß and Grundmann, 2016), that are, economies based on local substances flow circulation for drawing added-value from restoring and reusing, repairing, refurbishing, and recycling existing materials and products (Bicket et al., 2014). Furthermore, the example shows how agricultural reuse of wastewater and sludge can replace groundwater and mineral fertilizers and therefore reduce the consumption of limited natural resources as well as the outflow of capital from communities and regions.

In the case of the wastewater scheme in Braunschweig, the operators of the wastewater treatment plant and farmers developed a variety of long-term commercial relationships with local suppliers, service providers, and traders. This has fostered economic development of upstream suppliers and thus may create additional added-value for the local economy. Furthermore, if spent in the locality, the additional income of the stakeholders can stimulate additional demand, which in turn can spur further local production. These examples constitute the so-called induced or multiplier effects. They may continue and contribute to a further increase in the added-value for local economies (Kosfeld and Gückelhorn, 2012).

Crowding out effects and changes in the distribution of the added-value among actors engaged in the wastewater scheme may also play an important role in gaining acceptance for wastewater reuse in agriculture. In the case of Braunschweig, some stakeholders experienced losses of added-value, for instance, farm employees in the value chains of crop production. However, the overall added-value gains in crop production outweighed these crowding out effects, ie, the net impact on the added-value was positive (Maaß and Grundmann, 2016). In addition, irrigation activities and other services generated extra payments of interests and wages in the value chains of wastewater treatment. In fact, Maaß and Grundmann (2016) found two major strengths of the wastewater reuse scheme in Braunschweig: (1) all involved actors have a financial benefit from their engagement in the scheme, and (2) the total share of the single stakeholders in the total added-value from the reuse scheme is distributed rather evenly. This is believed to have contributed to a general support and positive development of the wastewater reuse scheme in Braunschweig.

## 3.1 Opportunities for Optimization

Currently, the wastewater association in Braunschweig intends to enlarge the agricultural fields in the irrigation area to make possible the use of larger machinery and thereby increase the efficiency of technology and reduce operation costs (Abwasserverband Braunschweig, 2015).

Furthermore, the association is evaluating the feasibility of replacing short rotation coppices for protective hedges used to prevent wind erosion. Those measures are undertaken with the aim of gaining experience in cultivating short rotation coppices rather than to use biomass for energy production (Abwasserverband Braunschweig, 2015). As the maintenance of protective hedges is rather costly, the use of short rotation coppices instead of protective hedges may provide a future opportunity to save costs and gain additional revenues for financing costs of the reuse scheme.

Concerning irrigation practices, the crop production in the study may be enhanced by more needs-oriented irrigation practices including more frequent irrigation, but with fewer doses of water. However, because of awareness of the limited amount of wastewater available for irrigation at peak times and the corresponding increase in irrigation costs, the majority of farmers do not insist on changing present irrigation practices (Abwasserverband Braunschweig, 2015).

## 3.2 Future Challenges

Currently, the association copes with the possible changes in the institutional regulation of agricultural wastewater and sludge utilization (Abwasserverband Braunschweig, 2015). In particular, the possible new regulations of the planned amendment of the German Fertilizer Ordinance would challenge farmers in the association area.

The amendment of the Fertilizer Ordinance includes, among other new regulations, the creation of nutrient balances for nitrogen (N) and phosphorus (P) applications. These balances should prove that the nutrients supplied by fertilization correspond to the general nutrient requirements for crops and soils. The nitrogen contained in organic fertilizer (eg, sludge) is primarily bonded in the organic matter and therefore may not be accountable for fertilization. Since nutrient comparisons for nitrogen are based on the total level of N, farmers will face the problem of which proportions of nitrogen are accountable for plant nutrition and how to assess the necessarily arising excess amount of the total N fertilizer.

These issues also apply to irrigation of wastewater and sludge, as the organic bonded nitrogen contained in the sludge component will have to be fully included in the balance, although it is initially not fully available to plants. Accordingly, farmers have to replace the bounded organic nitrogen with mineral fertilizers to meet the nutrient requirements of the crops. The planned ceilings for spreading organic nitrogen with a limited permitted amount of excess N pose a serious challenge to farmers in the association area, since the supply of organic N through irrigation of wastewater and sludge is already high (Abwasserverband Braunschweig, 2015).

Further restrictions could arise from the planned expansion of the lock-up periods for fertilization. In case of longer lock-up periods, the association would have to significantly reduce the co-irrigation of sewage sludge as well as spreading dewatered sludge (Abwasserverband Braunschweig, 2015).

The Abwasserverband Braunschweig may even have to withdraw from reusing sludge in agriculture and seek alternative ways for sludge disposal in case the practice is prohibited in Germany in the future. The effects of a possible change in the legal framework for agricultural sludge utilization will significantly impact the future structure of the reuse scheme. On the one hand, a more restrictive regulation of agricultural sludge use would increase the necessity of dewatering and disposing larger sludge quantities via incineration. On the other hand, farmers could not make use of the nutrients contained in sludge and would have to replace the nutrients with imported mineral fertilizers. This will result in a decrease in benefits currently resulting from the reuse scheme as well as an increase in costs related to wastewater treatment, sludge disposal, and fertilization (Maaß and Grundmann, 2016).

### 3.3 Adaptation Strategies

To ensure the persistence of the reuse scheme and in preparation for possible future restrictions on agricultural sludge application, the association has considered and tested various adaptation strategies.

According to the ambitions of the German federal government to recover nutrients from sludge ashes, the produced sludge would need to be incinerated in mono-burning plants, in case of thermal sludge disposal. Since mono-burning plants have not been implemented in the study region yet, the association would need to seek cooperating partners to jointly build a mono-burning plant in the first place.

The strategies of the association to compensate farmers for the nutrients contained in sludge include stripping of ammonia for nitrogen recovery and precipitation of magnesium ammonium phosphate (struvite) from sludge liquors for phosphorus recovery (Abwasserverband Braunschweig, 2015). However, the nutrient quantities recovered by these techniques would not be sufficient to meet the full fertilization needs of the crops cultivated in the association area. Therefore techniques combinable with thermal sludge disposal options, enabling higher recovery rates (eg, through nutrient recovery from sludge ashes), may be more suitable for compensating crop producers for the missed nutrients supplied by agricultural sludge use.

In the case of a legal withdrawal from the agricultural sludge application and the resulting prohibition of co-irrigating sewage sludge with wastewater, the association may have an opportunity to apply other irrigation techniques such as linear or circle irrigation systems. These techniques have not been implemented so far because of technical problems related to the sludge component of the effluent–sludge mixture (Abwasserverband Braunschweig, 2015).

To increase the planning security for the development and implementation of such adaptation strategies, precise statements of the legislators about future regulations of agricultural wastewater and sludge utilization are imperative.

## 4. SUMMARY AND OUTLOOK

The reuse of treated wastewater and sludge in agriculture may be conducive to the development of regional value chains for providing water, crops, and bioenergy. Wastewater reuse in the analyzed case of Braunschweig in Germany brings about cost savings, a higher added-value from crop production, and a high share of added-value for the local economy (Maaß and Grundmann, 2016). In the value chain of wastewater treatment, the reuse of wastewater and sludge serves as a final treatment step and reduces the costs of wastewater treatment and sludge disposal. In the value chain of crop production, it reduces the costs of irrigation and fertilization (Maaß and Grundmann, 2016). Since it provides water from an alternative resource and a substitute for mineral fertilizers, it may also lessen the dependency of crop production on natural freshwater resources and mineral fertilizers in arid and nutrient-poor areas. In sum, agricultural reuse of wastewater and sludge can contribute to developing economic and substance cycles, thus enhancing competitiveness of regions and meeting the demands for a more sustainable use of limited resources (Maaß and Grundmann, 2016). However, agricultural reuse of wastewater and sludge also entails restrictions for economic activities and leads to crowding out effects and changes in the distribution of the added-value along the value chains (Maaß and Grundmann, 2016).

To compensate actors for possible disadvantages and limitations, complementary alternative economic opportunities, which increase their scope of possibilities, are important to allow actors to make up for missed economic potential and create acceptance for wastewater use (Maaß and Grundmann, 2016). In case of legal restrictions for food crop production as in the Braunschweig scheme, these opportunities can consist of combining wastewater reuse with bioenergy production. Besides the economic benefits for cities, operators of wastewater treatment facilities and farmers, acceptance of the buyers and consumers of agricultural products irrigated with wastewater is of equal importance. As buyers and consumers might reject purchasing crops irrigated with wastewater, it is vital to assure them that high-quality standards have been met, for example, through certification proofs. Furthermore, economic impacts and multiplier effects of agricultural reuse schemes for local economies can be enhanced by involving local-based employees, creditors, and suppliers, and by promoting the associated value chains (Maaß and Grundmann, 2016).

The future of the wastewater reuse scheme in Braunschweig in its present form depends highly on possible changes in the institutional regulation of agricultural sludge utilization as well as its capacity to secure benefits and ensure the safeguards provided by the wastewater reuse scheme. In case of potential future legal restrictions on agricultural sludge use, thermal options for the disposal of sludge may become necessary, which would require additional investments and operating costs for wastewater treatment. At the same time, farmers will need to substitute the nutrients supplied with sludge by acquiring mineral fertilizers from the market. This may significantly reduce the benefits of the reuse scheme, since the operators of the wastewater treatment plant and the farmers mainly take advantage of the cost savings in sludge disposal and fertilization (Maaß and Grundmann, 2016). In addition, higher application rates of commercial mineral fertilizers will likely increase the outflow of capital from the locality (region) and decrease the share of the added-value for the local economy.

## REFERENCES

Abwasserverband Braunschweig, 2015. Focus Group Discussion on the 28.04.2015 with T. Eggers, Dr G. Olfe, H. Ripke and B. Teiser from the Wastewater Association Braunschweig (Wendeburg/Braunschweig).

Ahlers, A., Eggers, T., 2004. Abwasserverband Braunschweig—50 Jahre erfolgreich tätig für Mensch und Umwelt durch Reinigung und landwirtschaftliche Verwertung kommunaler Abwässer. UK Verlag, Neubrück.

Aiello, R., Cirelli, G.L., Consoli, S., 2007. Effects of reclaimed wastewater irrigation on soil and tomato fruits: a case study in Sicily (Italy). Agricultural Water Management 93, 65–72.

Bentzen, J., Smith, V., Dilling-Hansen, M., 1997. Regional income effects and re-newable fuels. Increased usage of renewable energy sources in Danish rural areas and its impact on regional incomes. Energy Policy 25 (2), 185–191.

Bezirksregierung Braunschweig, 2001. Neufassung der wasserrechtlichen Erlaubnis zur Beregnung mit behandeltem Abwasser aus dem Klärwerk Steinhof für den Abwasserverband Braunschweig. Bezirksregierung Braunschweig, Braunschweig.

Bicket, M., Guilcher, S., Hestin, M., Hudson, C., Razzini, P., Tan, A., Withana, S., 2014. Scoping Study to Identify Potential Circular Economy Actions, Priority Sectors, Material Flows & Value Chains. Study prepared for the EU Commission, DG Environment. Publications Office of the European Union, Luxembourg. http://www.ieep.eu/assets/1410/Circular_economy_scoping_study_-_Final_report.pdf.

Cordell, D., Rosemarin, A., Schröder, J.J., Smit, A.L., 2011. Towards global phosphorus security: a systems framework for phosphorus recovery and reuse options. Chemosphere 84 (6), 747–758.

EU (European Union), 1991. Council Directive of 21 May 1991 concerning urban wastewater treatment (91/271/EEC). Official Journal of the European Communities. L135/40.

Haller, A., 1997. Wertschöpfungsrechnung. Ein Instrument zur Steigerung der Aussagefähigkeit von Unternehmensabschlüssen im internationalen Kontext. Schäffer-Poeschel Verlag, Stuttgart.

Hamilton, J.R., Whittlesey, N.K., Robison, M.H., Ellis, J., 1991. Economic impacts, value added, and benefits in regional project analysis. American Journal of Agricultural Economics 73 (2), 334–344.

Hartmann, A., Klein, D., Dockhorn, T., Dichtl, N., Sardet, C., Teiser, B., 2010. Long-term experience with water reuse in Braunschweig: Prospects and Risks. 15. Internationales Symposium Wasser, Abwasser, Abfall, Energie auf der IFAT ENTSORGA, 13–16. September 2010, Munich, Germany.

Hoffmann, D., 2009. Creation of regional added value by regional bioenergy resources. Renewable and Sustainable Energy Reviews 13 (24), 19–29.

Khan, S., Cao, Q., Zheng, Y.M., Huang, Y.Z., Zhu, Y.G., 2008. Health risks of heavy metals in contaminated soils and food crops irrigated with wastewater in Beijing, China. Environmental Pollution 152, 686–692.

Kimmich, C., Grundmann, P., 2008. Regional governance and economic impacts of decentral bioenergy value chains – the case of the bioenergy village Mauenheim. In: Schnitzer, H., Ulgiati, S. (Eds.), Proceedings of the 6th Biennial International Workshop Advances in Energy Studies: Towards a Holistic Approach Based on Science and Humanity. TU-Graz, Graz, pp. 564–573.

Kosfeld, R., Gückelhorn, F., 2012. Ökonomische Effekte erneuerbarer Energien auf regionaler Ebene. Raumforschung und Raumordnung 70, 437–449.

Larsen, T., 2011. Redesigning wastewater infrastructure to improve resource efficiency. Water Science and Technology 63 (11), 2535–2541.

Lesjean, B., Remy, C., 2013. Projekt CoDiGreen: Optimierung der Energie- und Nährstoffrückgewinnung in der Abwasserbehandlung. Kompetenzzentrum Wasser GmbH, Berlin.

LWKH, 2000. Gutachten zur Überprüfung der Tragbaren Belastung der landwirtschaftlichen Mitglieder des Abwasserverbandes Braunschweig. Landwirtschaftskammer Hannover, Hannover.

Maaß, O., Grundmann, P., 2016. Added-value from linking the value chains of wastewater treatment, crop production and bioenergy production: a case study on reusing wastewater and sludge in crop production in Braunschweig (Germany). Resources, Conservation and Recycling 107, 195–211.

Maaß, O., Grundmann, P., von Bock, Polach, C., 2014. Added-value from innovative value chains by establishing nutrient cycles via struvite. Resources, Conservation and Recycling 87, 126–136.

Mapanda, F., Mangwayana, E.N., Nyamangara, J., Giller, K.E., 2005. The effect of long-term irrigation using wastewater on heavy metal contents of soils under vegetables in Harare, Zimbabwe. Agriculture, Ecosystems and Environment 107, 151–165.

Marcouiller, D.W., Schreiner, D.F., Lewis, D.K., 1996. The impact of forest land use on regional value added. Review of Regional Studies 26 (2), 211–233.

Muyen, Z., Moore, G.A., Wrigley, R.J., 2011. Soil salinity and sodicity effects of wastewater irrigation in South East Australia. Agricultural Water Management 99, 33–41.

Pedersen, J.A., Soliman, M., Suffet, I.H., 2005. Human pharmaceuticals, hormones, and personal care product ingredients in runoff from agricultural fields irrigated with treated wastewater. Journal of Agricultural and Food Chemistry 53, 1625–1632.

Pedrero, F., Kalavrouziotis, I., Alarcóna, J.J., Koukoulakis, P., Asano, T., 2010. Use of treated municipal wastewater in irrigated agriculture—Review of some practices in Spain and Greece. Agricultural Water Management 97, 1233–1241.

Pescod, M.B., 1992. Wastewater Treatment and Use in Agriculture – FAO Irrigation and Drainage Paper. FAO, Rome. http://www.fao.org/docrep/t0551e/t0551e00.htm.

Qadir, M., Wichelns, D., Raschid-Sally, L., McCornick, P.G., Drechsel, P., Bahri, A., Minhas, P.S., 2010. The challenges of wastewater irrigation in developing countries. Agricultural Water Management 97, 561–568.

Remy, C., 2012. LCA Study of Braunschweig Wastewater Scheme (CoDiGreen Project Report). KZW, Berlin. http://kompetenz-wasser.de/fileadmin/user_upload/pdf/forschung/CoDiGreen/CoDiGreen_LCA_Braunschweig_final.pdf.

Rusan, M.J.M., Hinnawi, S., Rousan, L., 2007. Long term effect of wastewater irrigation of forage crops on soil and plant quality parameters. Desalination 215, 143–152.

Satorius, C., von Horn, J., Tettenborn, F., 2011. Phosphorus recovery from wastewater – state-of-the-art and future potential. In: International Conference on Nutrient Recovery and Management – Inside and Outside the Fence. International Water Association (IWA), Water Environment Federation (WEF), Miami, Florida, USA.

Singh, P.K., Deshbhratar, P.B., Ramteke, D.S., 2012. Effects of sewage wastewater irrigation on soil properties, crop yield and environment. Agricultural Water Management 103, 100–104.

Ternes, T.A., Bonerz, M., Herrmann, N., Teiser, B., Andersen, H.R., 2007. Irrigation of treated wastewater in Braunschweig, Germany: an option to remove pharmaceuticals and musk fragrances. Chemosphere 66 (5), 894–904.

Toze, S., 2006. Reuse of effluent water—benefits and risks. Agricultural Water Management 80, 147–159.

Villamayor-Tomas, S., Grundmann, P., Epstein, G., Evans, T., Kimmich, C., 2015. The water-energy-food security nexus through the lenses of the value chain and the institutional analysis and development frameworks. Water Alternatives 8 (1), 735–755.

Winpenny, J., Heinz, I., Koo-Oshima, S., 2010. The Wealth of Waste: The Economics of Wastewater Use in Agriculture. FAO Water Reports 35. FAO, Rome. http://www.fao.org/docrep/012/i1629e/i1629e.pdf.

Zema, D.A., Bombino, G., Andiloro, S., Zimbone, S.M., 2012. Irrigation of energy crops with urban wastewater: effects on biomass yields, soils and heating values. Agricultural Water Management 115, 55–65.

Chapter 3.2.2

# Avoiding Floods in Spring and Droughts in Summer—Water Regulation Strategies in Germany and Poland

C. Schleyer
*Alpen-Adria University Klagenfurt, Vienna, Austria*

## 1. INTRODUCTION

In many regions with naturally high water tables, various forms of drainage or reclamation infrastructure have been applied to provide water tables that allow for (efficient) agricultural land use. These reclamation infrastructures are hydromelioration systems that usually consist of networks of canals, ditches, and subsurface tiles that collect, transport, and dispose of gravitational water from soil. This agricultural land drainage eliminates or mitigates adverse effects of excess soil moisture on crops and cropping operations. Thus crop yields are usually higher, and costs and risks of farm management on drained soils are reduced (Spaling and Smit, 1995). In many reclaimed areas, agricultural land use had not been possible before the construction of the drainage works. Further, these systems of tertiary (very small), secondary (small), and primary (large) arterial drains are often equipped with weirs and pumping stations that enable irrigation by flood and even by infiltration. With these facilities, the water table can be regulated, excessive water runoff can be avoided, and the risk of droughts in summer can be reduced.

In many reclaimed areas, however, the long-standing and intensive arable farming of the drained land has led to an increasing, nearly irreversible degradation of soil (eg, bog subsidence). Soils run dry during arid periods, losing more and more fertile soil and organic matter because of wind erosion. Thus in the medium or long run, continued intensive drainage and arable farming are likely to make agricultural land use impossible. In addition, degradation combined with soil compaction from the use of heavy machinery has often resulted in the soil's almost complete inability to hold water. Thus a high level of water runoff will be experienced if not held in check by well-functioning and coordinated weirs (AVP, 1998).

However, drastic political, economic, and administrative changes after the breakdown of the socialist regimes in many Central and East European countries and other transition countries in 1990 have made such a sustainable governance of reclamation systems even

**Competition for Water Resources. http://dx.doi.org/10.1016/B978-0-12-803237-4.00021-5**

more difficult. More precisely, changes in effective property rights on land and reclamation infrastructure, a complete restructuring of water administrations, and the emergence of increasingly heterogeneous stakeholder objectives with respect to water management and land use have often worsened the situation and even added new problems. The visible consequences are droughts in the summer and waterlogged plots in the spring caused by neglecting the cleaning and maintenance of open or subsurface ditches and/or by dilapidated water management facilities operating in an uncoordinated or even unauthorized way.

This chapter aims to explore approaches to manage water and water (reclamation) infrastructure in two reclaimed agricultural areas in the transition countries of East Germany (Schraden) and northwest Poland (Pyrzyce). In particular, the role of institutional and organizational changes as well as of changes in stakeholder interests in managing water tables in space and time will be examined.

The chapter is based on an empirical material collected in: (1) the German Federal State of Brandenburg, in particular in the Schraden region, a fenland area in the south of this federal state, and (2) the Voivodship Zachodniopomorskie in northwest Poland, largely in the Powiat (district) Pyrzyce (Fig. 1). In the Schraden region, 12 qualitative, semi-structured interviews were conducted between July 2000 and February 2002. The 14 interviews in the Pyrzyce region took place between March 2005 and February 2006. In both regions, farmers, local environmentalists, and representatives of the agricultural, environmental, and reclamation administration at all subnational administrative levels as well as of the regional and local water associations were interviewed. Moreover, available planning materials, regional statistics, legal documents, and other (local) information available for both regions were consulted (Schleyer, 2012).

The chapter proceeds as follows: in Section 2, conceptual issues related to reclamation systems and reclamation infrastructure in particular are presented with a focus on issues of property rights and governance. In Sections 3 and 4, the German and Polish case studies are presented. Here the development and current state of reclamation systems, including the respective ecosystems and reclamation infrastructures, are introduced as well land use, stakeholder objectives, the distribution of property rights, and governance structures in place. A thorough explanation of those concepts further allows us to discuss critical issues potentially hindering a sustainable water management in the analyzed reclaimed areas. In Section 5, some conclusions are drawn.

## 2. RECLAMATION SYSTEMS: PROPERTY RIGHTS AND GOVERNANCE CHALLENGES

In this contribution, reclamation infrastructures are conceived as typical examples of rural infrastructures that provide various forms of public goods (Ostrom et al., 1993). Similarly, as flood protection measures, reclamation infrastructure also provides more spatially restricted (ie, local) public goods: "appropriate" levels of soil moisture and water tables conducive for agricultural land use. Farmers outside the reclaimed area are excluded from this local public good. Access to the benefits of the reclamation system is only possible if a farmer has, at least, use rights on the land within the reclaimed area. In turn, it is much more difficult, and often technically not feasible, to exclude farmers within the reclaimed area. Thus reclamation systems share the attribute of nonexcludability with many other physical rural infrastructures, such as irrigation systems and roads. Unlike the latter

FIGURE 1 Map of research regions. *Martin Mantel, Map © OpenStreetMap.*

examples, however, consumption by one farmer of the flow of services provided by the reclamation system usually does not subtract from the flow of services available to others. However, there may be exceptional circumstances when rivalry in consumption applies. For example, in case of heavy rain, small ditches may become "congested" and may need some time to transport all the water away into the primary arterial drains, thus resulting in temporary excess soil moistures. Still, even if the element of scarcity is largely "missing," since exclusion is difficult, farmers may have incentives to free-ride, which can lead to underinvestment in the provision and maintenance of reclamation infrastructures (Scheumann and Freisem, 2001). Because of the technical interdependence of the reclamation infrastructure, uncleaned, or insufficiently cleaned, segments of the network of ditches can severely hamper water runoff for larger parts of the reclaimed area. This can result in local spots of land with excessive soil moisture that are not restricted to the land in the immediate vicinity of a poorly cleaned ditch.

Some form of collective action and coordination between the farmers within the reclaimed area is necessary to ensure the provision of appropriate levels of this local public good. Despite theoretical skepticism about the ability of self-interested individuals to coordinate collective action for providing public goods (Olson, 1965), there is substantial empirical evidence that various forms of local self-organized resource governance systems are indeed able to reduce the probability of free-riding and to manage common resources sustainably (Ostrom, 1990). Indeed, there is a plethora of studies focusing on collective action within commonly managed irrigation systems as prime examples of common property regimes and on the determinants for success or failure of locally organized management of these systems (eg, Ostrom, 1992; Vermillion, 1999; Wade, 1988). However, there is a comparatively small body of research addressing collective action and property regimes within reclamation or drainage systems (eg, Scheumann and Freisem, 2001; Pant, 2002).

In this chapter, an institutional economics perspective is taken to analyze the determinants of water regulation in reclaimed areas. In most reclaimed areas, rights and duties related to the various system components are, to a varying degree, assigned to social actors like farmers, municipalities, regional and national state authorities, and water (user) associations. For example, in many reclaimed fenland areas, secondary ditches and weirs are the private property of the owners of the bordering land. This usually includes the responsibility to clean and maintain the ditches and to maintain and operate the weirs. Furthermore, farmers may own the reclaimed land privately; yet, often other actors lease it out to them. In turn, primary ditches and canals as well as the respective large weirs and pumping stations are often owned by the state, covering all related costs for cleaning, maintaining, and operating. Indeed, state ownership of primary canals and pumping stations within reclamation systems is rather common. It is often motivated by the fact that agricultural land drainage also contributes to flood protection of towns, villages and traffic infrastructure in the region; a task usually assigned to the state (Schleyer, 2004).

Suitable governance structures, that is, organizational solutions or coordination mechanisms, are necessary to make property rights effective, since individual land users within a reclaimed area may be tempted to neglect their duties to avoid the costs involved. Similarly, to some extent, neighboring farmers may be deprived of their benefits from the reclamation infrastructure if some farmers free-ride and do not stand up for their duties. Thus the suitability of the governance structures in place to guarantee the rights and duties of the various actors influences the likelihood both that an individual actor can actually realize the system's benefits and that they will be sanctioned for neglecting related duties. Overall, the distribution of de facto property rights on any components of the reclamation system and (the ability of) the governance structures in place to enforce the related rights and duties (de)motivate the economic decisions of individual actors to engage in collective action. This is because they determine the, admittedly subjective, actors' expectations about the actual costs and benefits of their decisions.

# 3. CASE STUDY SCHRADEN (GERMANY)[1]

## 3.1 Ecosystem, Reclamation System, Land Use, and Stakeholders

The Schraden fenland is a 15-kilometer-long section of the moderately sloped Breslau-Magdeburg glacial valley located 89 m above sea level in the west and rising to 94 m above sea level in the east; it covers approximately 11,400 ha. The first reclamation measures in this former wetland were carried out in the 14th century and consisted of small ditches equipped with weirs. Reclamation activities intensified in the second half of the 19th century, mainly for the extension of grassland farming in an area that was still dominated by inaccessible alder forests and swamps. Furthermore, it was intended to reduce the often-disastrous effects of seasonal floods on the villages and towns in the region, and to minimize health risks associated with extensive swamps, such as malaria. Hence, the major watercourses were straightened and diked, small water arms were backfilled, and extensive drainage works (ditches) were built. In the 1960s and 1970s, reclamation measures reached their peak in the German Democratic Republic (GDR) (Könker, 1993). Large drainage systems—mostly open ditches up to 3 m below ground, but also various forms of subsurface tiles—were installed to lower the groundwater table. This system of ditches and channels was also equipped with weirs and pumping stations to enable irrigation by flood and even by infiltration, if necessary. By 1997, there were 330 ditches and small channels totaling approximately 300 km in length and equipped with about 170 weirs, which were able to regulate the water table within the entire Schraden. By 1997, as much as 88% of the area was used as agricultural land, 78% of which was used for arable farming. Forests covered only about 3% of the total area (AVP, 1998, p. 33ff).

However, water management facilities in the Schraden are often degraded and operate in an uncoordinated way. Out of 108 weirs examined more closely in 1998, 42 were found to be out of order (AVP, 1998, appendix 10). The significance of this fact becomes apparent when one considers that the climatological water balance in the Schraden is negative (Landgraf, 2001) and the average annual rainfall is fairly low (573−631 mm from 1951 until 1980) (AVP, 1998, p. 20). As a consequence, plots with a relatively low (natural) groundwater table frequently suffer from drought periods in the summer, resulting not only in negative income effects for farmers, but also in negative environmental effects for the plants and animals that depend on a particular groundwater level. Since 1990, water-intensive crops such as potatoes have become increasingly replaced by maize and rapeseed as well as rye and barley (Hanspach and Kiβro, 2001). The reclamation infrastructure that was installed in the Schraden was designed to cover the entire area and to meet the needs of large agricultural firms farming very large plots. The relatively low number of weirs compared with the total length of ditches indicates that there was no need to regulate the water table for small plots. Indeed, there were only four large agricultural cooperatives farming agricultural land in the Schraden by 1976 (Hanspach and Kiβro, 2001).

After 1990, however, the interests of farmers, environmentalists, and other actors with respect to water management became significantly more diverse. First, the respective requirements primarily concerning the groundwater table of the newly restructured and reorganized agricultural firms have become quite heterogeneous and now greatly depend on farm size and location (eg, upstream or downstream), crop structure, and economic performance. Apart from a few part-time farmers with small plots, there are 13 different

1. Parts of this section are based on Schleyer (2004).

agricultural enterprises (one of which is a tree nursery) with various legal forms and ownership structures predominantly farming on leased land (AVP, 1998, p. 78; Hanspach and Kißro, 2001). Here farm size varies between 320 and 1870 ha (AVP, 1998, p. 102).

Second, interests regarding nature conservation have become much more prominent since 1990. These interests are predominantly represented by the respective environmental administrations, such as the Brandenburg Environmental Agency at the federal state level and the Lower Environmental Agency at the district level. Nongovernmental environmental associations, however, are still of lesser importance at the regional or local level. Nevertheless, it became clear during the interviews that these environmental interests are not homogeneous. Some environmentalists aim at bringing the degradation of the fenland to a halt or even to reverse the process by sustaining very high groundwater tables all year round. In contrast, other environmentalists might want to preserve the species and habitats typical for extensively used (wet) grassland with a comparatively high groundwater table during the winter, but would accept only moderately high water levels during the summer, thus allowing for extensive grassland farming. Hence the interests of farmers and environmentalists are not necessarily diametrically opposed, but conflicts in resource use do persist.

Other interest groups/sectors, such as forestry, industry, housing, construction, and transportation services, also demand "safe" groundwater tables to avoid flooding and other damages. Operators of gravel pits in this region, for example, require sufficient flood protection to make opencast mining possible. Yet, they also need a sufficiently high groundwater table to use floating excavators. In contrast, private and professional fishers prefer high water tables all year round. Thus agriculture and nature conservation are not the only divergent interests, but are clearly the dominant and most powerful ones.

## 3.2 Property Rights and Governance Structures

During the socialist times the formal ownership of land remained fragmented—almost as it was in the 1950s—while the agricultural firm structure underwent immense changes (Laschewski, 1998). As can be observed in the Schraden, these changes were often linked to comprehensive land consolidation measures and extensive reclamation measures. In other words, plots that have had little infrastructure were suddenly "enriched" by ditches and weirs. These assets of the newly built reclamation systems were regarded as collective property.

Shortly after the breakdown of the socialist regime and the subsequent reunification of Germany in 1990, collectivized land in the Schraden was restituted to legal owners, who received full property rights. This step revived the fragmented land ownership structure. Most of the new/old landowners quickly leased their land to the newly restructured and reorganized cooperatives; these are now joint stock companies, limited liability companies, or producer cooperatives. In 1994, the Brandenburg Water Act finally replaced the GDR Water Act, formally reorganizing responsibilities and rights for rivers, channels, and ditches and dividing them into two categories. Only those bodies of water belonging to the (new) first category were declared state property and thus the responsibility of the federal state of Brandenburg. In regard to the reclamation infrastructure in the Schraden, only a few of the former main ditches were classified as first category. The remaining infrastructure, that is, all open waters of the second category including the weirs, was to become legal property of the owners of the bordering properties.

After 1990, large agricultural cooperatives at the local level disintegrated into smaller and more focused enterprises with different legal forms and ownership structures (Laschewski, 1998). Consequently, the reclamation cooperatives—interfirm organizations of enterprises that began competing directly on the market—were soon dissolved without substitutes. To ensure the necessary water runoff, and to avoid damage by floods or a high groundwater table, the maintenance and cleaning duties of second category ditches were formally assigned to regional water associations in 1995. These water associations had been established shortly after the reunification. The Water Association Kleine Elster-Pulsnitz is responsible for the Schraden area and is supervised by the Brandenburg Environmental Agency. Membership in the associations is compulsory for all municipalities representing the landowners subject to property tax. The results of the interviews with farmers confirmed that the tenants pay the membership fee as an implicit part of the land rent. Other beneficiaries (eg, railway companies) can be voluntary members.

## 3.3 Critical Issues of Sustainable Water Management

In this section, some critical aspects are briefly presented that make sustainable water management in this reclamation system difficult, and sometimes even impossible.

### *3.3.1 Land Fragmentation and Leasehold*

After the restitution of land in the Schraden, the majority of owners decided to lease their land to the new agricultural enterprises instead of starting their own farming business or using the land for other purposes. According to the statements of interviewed tenants, most landowners do not know about the reclamation works on their land or are not aware of the related (legal) rights and duties. What is more, many owners no longer live in the region, are not involved in the farming business, and own only very small plots. There are also cases where the owners are not known, cannot be found, or ownership is legally disputed. In all cases, however, the owner of a section of the reclamation infrastructure, such as a weir, would have to explicitly agree to any maintenance or operating measures to be carried out. Otherwise the activity would be regarded as illegal. Yet, since recently most new or renewed lease contracts contain a clause transferring all rights and duties related to the reclamation works to the tenant for the time of the lease. Importantly, since 2004 the regional water association has also the legal right to maintain and operate at least those weirs on second-order ditches that are crucial for regulating the landscape water regime.

### *3.3.2 Financial Limitations*

Like other water associations in the federal state of Brandenburg, the activities of the regional Water Association Kleine Elster-Pulsnitz are financed solely by membership fees, as there are no regular subsidies from the state. In the interviews conducted, representatives of the Water Association stated that the available funds are only sufficient for the compulsory tasks of maintaining and cleaning ditches. Noncompulsory measures, such as maintaining or operating weirs, are only carried out occasionally and if auxiliary funds are available. Means to mitigate the problem include state support programs, which can be used for project-related maintenance tasks but not for basic operating costs. Some programs, however, require water associations to match this funding with up to 50% of their own funds. Thus scarce capital resources of water associations are yet again limiting their activities to maintain weirs.

### 3.3.3 Solidarity Principle

Another way to ease water associations' financial limitations would be to increase membership fees to make weir maintenance financially "profitable" as well. Interviews, however, indicate that contributors perceive the membership fees as too high for the region, and that they would veto a further increase. This is especially true since membership fees do not correspond to the actual distribution of benefits from the association's activities. While the Federal Water Associations Act allows for a differentiation that would better reflect the actual distribution of reclamation benefits, a solidarity principle was adopted for the Brandenburg Water Act that allows for the membership fees only to be proportionate to the land size.

### 3.3.4 Complete Restructuring

In 1994, the Brandenburg Water Act established a new administrative structure for water management and planning, which follows the example of the former West German states and emphasizes self-government at the communal level. In interviews with representatives from the Lower Water Agency in the Elbe-Elster district, restructuring was described as a drastic, complete, and bumpy process. Almost all relations among the various levels of the newly established water authorities had to be rebuilt from scratch. This process has not been completed yet and communication is still limited among the authorities to the absolute minimum. Interviewees described the process of long-standing and new civil servants acquiring competency with newly established laws and rules and exploring new space for maneuvering as time and energy consuming. The same is true for the relations among the water authorities and water users, water associations, interest groups, other administrative agencies, the municipalities, and the general public.

### 3.3.5 Transboundary Interrelations

The problems connected with the administrative restructuring were certainly made worse by the fact that this restructuring followed completely new political and administrative borders. What is more, one might suspect that this impedes or even hinders efficient and coordinated handling of a complex, transboundary biophysical system like a landscape water regime. However, one should not underestimate the implications of introducing or strengthening river basin management for existing institutional configurations and organizational structures. In fact, the water regime of the Schraden is greatly determined by the water inflow from the neighboring district of Oberspreewald-Lausitz as well as from the federal state of Saxony. Interviews with the Lower Water Agency, however, revealed that there are almost no joint activities, informational exchanges, or coordination meetings among the respective water authorities.

### 3.3.6 Organizational Subordination

The Lower Water Authority operates at the district level. Hence nearly all decisions related to water management are made from the viewpoint of the district as a whole. In other words, decision-making power is with the (political) head of the district administration. The Lower Water Agency, which is responsible for the practical work and professional input, is organizationally subordinate to the district's Lower Environmental Agency, which is in turn subordinate to the district's Department of Economics and Environment. With regard to statements made by the Lower Water Agency, this organizational structure is somewhat delicate in its conduciveness to executive decisions that are politically opportunistic rather than purely professional. This is also true in the choice of key issues such as improving the public wastewater disposal system and ensuring the public water supply that were perceived as being more urgent, thus largely reflecting political priorities.

### 3.4 Federal State of Brandenburg Program to Improve Landscape Water Regime

Problems caused by the already negative water balance in Brandenburg are exacerbated by climate change, including a series of extreme droughts in 2003, 2006, and 2008 that led to high income losses for farmers and negative effects on flora and fauna. Moreover, predictions for climate change in Brandenburg are even more sobering: more rainfall in winter, but less in summer; more high-precipitation events; increasing temperatures leading to higher evapotranspiration and reduced rate of groundwater recharge (−40%). Anticipating these developments, the federal state of Brandenburg initiated a program to increase landscape water retention in 2002. It was cofinanced by the European Union and the measures were carried out by the regional water associations. Between 2002 and 2006 about 350 projects were implemented (c. €70 million), while the total number of realized projects increased to 580 projects by 2010. Most of these projects were focused on technical measures like repairing or installing weirs or building rock ramps. Institutional deficits like the coordination of operating the weirs, however, were not addressed; and only 3% of the funds were directed to improvement of the monitoring system for groundwater tables, which is an important determinant of sustainable water management (Jörns et al., 2010).

## 4. CASE STUDY PYRZYCE (POLAND)[2]

### 4.1 Ecosystem, Reclamation Systems, Land Use, and Stakeholders

Pyrzyce is a flat agricultural region with a large number of small lakes, ponds, and rivers. It covers 726 $km^2$ and is situated about 50 km southeast of Szczecin, the main city in the most northwestern Polish Voivodship Zachodniopomorskie. Compared with other Polish Powiats (districts), the region is sparsely populated [57 inhabitants per $km^2$ in Pyrzyce vs. 124 on average in Poland (SOS, 2001)].

Before the beginning of the political and economic transition period after the collapse of the socialist regime, around 1990, state-owned agricultural firms (PGRs) were farming about 52% (30,000 ha) of agricultural land in the region; the rest (about 27,000 ha) was used by a very high number (c. 2000) of small family farms. In the PGRs, intensive livestock production (pigs and cattle) was the dominant form of agricultural production. Here the agricultural land was mainly used for intensive grassland farming and for growing maize, cereals, and sugar beet to ensure a sufficient supply of animal fodder. However, mixed farming (crops and livestock) was also characteristic for the family farms in the region. Soil quality in the Powiat Pyrzyce can be classified as good to very good.

2. Parts of this section are based on Schleyer (2009).

Reclamation activities in Pyrzyce started as early as in 1300. Under the reign of Friedrich the Great, drainage activities were intensified during the 17th/18th centuries by building many of today's primary canals and ditches. During socialist times, in particular since the 1960s, the network of secondary ditches and subsurface tiles was extended greatly. Reclamation infrastructures were built predominantly on state-owned land farmed by the PGRs, yet not exclusively so. About 90% of the land farmed by the PGRs was reclaimed, while this was only the case for about 50% of the land farmed by family farms. By 1990, there were about 900 km of canals, ditches, and subsurface tiles (of which 600 km were secondary ditches and subsurface tiles) that drained an area of about 30,000 ha of agricultural land; of which 20,000 ha were arable lands. This system of canals and ditches was also equipped with many weirs and nine pumping stations to enable irrigation by flood and even by infiltration, if necessary.

## 4.2 Property Rights and Governance Structures

Secondary ditches, subsurface tiles, and weirs on family farmland were private property of the owners of the (bordering) land, coming along with the legal responsibility to maintain and clean ditches and subsurface tiles on a regular basis and to maintain and operate weirs. Maintenance and cleaning of ditches and subsurface tiles is carried out to ensure their ability to drain water by gravity flow into a natural water system outside the reclaimed area. It usually includes an annual cut of grass, small trees, and other natural cover at the edges of the ditches as well as removal of sludge and sediments deposited by the water flow. To ensure the functionality of subsurface tiles it is particularly important to clean the outflows of the tiles in the secondary ditches. Cleaning and maintenance of secondary ditches as well as maintenance and operation of weirs at secondary ditches on family farmland were usually carried out by specialized state-owned enterprises assigned by the village water associations. Here farmers who were members of the respective water associations paid small fees according to the length of ditches and subsurface tiles on their land. This would effectively transfer their legal obligations to clean and maintain "their" secondary ditches to the water association. Alternatively, they could opt for carrying out the cleaning themselves under supervision of the water association, though this was rather exceptional. In the 1980s, all village water associations in a municipality (Gmina) merged to form larger local/Gmina water associations. Furthermore, in 1986, all six local water associations in the region which became the Powiat Pyrzyce in 1999 were organized under the roof of an umbrella association. Here a general manager (honorary post), an accountant, and a small technical team were supposed to coordinate the activities of the local water associations in the region as well as to collect the farmers' fees.

Unlike the secondary ditches and weirs, primary canals and ditches as well as primary weirs and pumping stations were a legal property of the state. Here building activities, cleaning and maintenance of the ditches, and maintenance and operation of the weirs and pumping stations were planned and organized by the respective state authorities at the Voivodship level and the respective regional branch office. Specialized state-owned reclamation enterprises or employees of the regional branch office carried out all these activities.

## 4.3 Critical Issues of Sustainable Water Management

Two main phases of water management can be distinguished here: (1) discontinuation of the cleaning and maintenance activities at secondary ditches and subsurface tiles of the local water associations in 1990; (2) revival of the local water associations and their resumption of cleaning activities at secondary ditches and subsurface tiles from about 2002 onward.

In 1990 the local water associations in Pyrzyce largely suspended cleaning and maintenance activities at secondary ditches and subsurface tiles. Two contextual factors were the main driving forces in this process: First, the liberalization of agricultural markets and the abolition of input-related subsidies, which resulted in a devalorization of regional agricultural production and land use, and thus in the abandonment of most grassland plots. Second, the inability of the state and communal authorities to effectively control and facilitate the activities of the local water associations.

The liberalization of food prices (1989) and prices for goods and services (1990) together with the abolition of input-related subsidies (1991) resulted in a systematic decline of agricultural production in Poland and severely worsened the agricultural terms of trade. In particular, there was a strong decline in cattle numbers and milk production, reflecting the shrinking demand for milk products as a consequence of the declining real incomes of major social groups in Poland and the abolition of food subsidies, which led to higher consumer prices (Mohr, 1997; OECD, 1995). In 1990/1991, for example, private family farms in Poland suffered an overall income loss of about 60%. However, the income losses from agriculture were partly compensated by a higher nonagricultural income whose share increased on average from 30% of total private farm household income in the late 1980s to over 50% in 1992 (OECD, 1995).

Despite the decline in agricultural production and the substantial income losses from agriculture, the structure, size, and number of family farms in Pyrzyce remained rather stable after 1990. Presently, there are more than 2200 family farms in the Powiat Pyrzyce, of which about 22% are classified as semisubsistence and subsistence farms. Many farms still practice mixed farming (crops and livestock), although a growing number of larger family farms specialize in crop farming. Cereals (wheat, rye, and barley) are the dominant crops, but triticale, potatoes, and sugar beet are also grown (Perez et al., 2005). The drastic decline of animal production was considered by all interviewed actors in Pyrzyce as the main reason that led to land devalorization. As a consequence, a significant share of the grassland in the region was either turned into arable land since crop farming had become more profitable or, more often, was simply abandoned (ie, not mowed anymore). In particular in the latter case, drainage of these large plots of land was hardly needed anymore. As a consequence, most farmers stopped paying their fees to the local water associations. Some of those farmers decided to fulfill their legal duties instead by cleaning their ditches on their own to avoid surcharges and to save costs. However, as was pointed out by the representatives of the agricultural and reclamation administrations in the region, these activities were carried out only occasionally and very poorly and superficially, since farmers lacked the relevant expertise and did not use appropriate cleaning machinery.

The interviews revealed, however, that the inactivity of the local water associations in the 1990s was only partly a result of the declining farmers' contributions because of changing market conditions. In 1990 the control and facilitation of the activities of the local water associations were transferred to the Gmina administrations, which had just become the basic units of local self-government in the context of the decentralization of the political and administrative structures in Poland. Yet, as was mentioned frequently by most of the interviewed actors, control and enforcement were limited or completely absent. Reportedly, this often resulted, among other things, in representatives of local water associations taking the family farmers' fees, but not carrying out any cleaning or any of sufficient quality. To some extent, this lack of effective control and support was because of limited financial resources and a shortage of competent personnel available at the communal level. Further, while the formal monitoring and sanctioning power now rested with the Gmina administrations, some ambiguities remained concerning the (new) distribution of, in particular financial, responsibilities for reclamation issues between Gminas, local water associations, and the state water administration.

At the same time, there was only very limited political interest in reclamation issues, in particular at higher governmental levels. According to the interviewed representatives of the state water administration at the Powiat and Voivodship levels, the state budget for maintaining and cleaning primary canals and ditches has been reduced drastically since 1990. Apart from general budget constraints on the part of the state, the reduced funding for reclamation also reflects a more critical perspective on the environmental effects of (excessive) drainage. Thus no new reclamation infrastructures have been built in the region since 1990. In Pyrzyce, however, because of specific natural soil conditions, drainage activities have not caused substantial negative environmental effects like soil degradation. Thus many primary canals and ditches were only cleaned every third year, or even less often than that. In 2005, for example, the budget of the regional branch of the state water administration in Pyrzyce was only sufficient to ensure professional cleaning and maintenance of 4–17 km of canals, that is, only up to 6% of all primary canals in the Powiat. At this point the representative of the regional branch of the state water administration noted that he had always tried to ensure adequate cleaning, at least in areas that had not been largely abandoned. By and large, however, cleaning activities on secondary ditches and subsurface tiles became largely ineffective since the water drained from the soils could not flow freely in the primary arterial drains. This effect was exacerbated by the almost nonexistent communication and coordination between local water associations and the state water administration.

The interviews highlighted another important determinant that contributed to the long duration (over one decade) of inactivity of the local water associations. This factor relates to a specific feature of drainage systems, that is, neglecting cleaning and maintenance activities does not necessarily cause flooding problems in the short run. Ditches and subsurface tiles only gradually lose their ability to transport excess water. Further, in the case of subsurface tiles, the need for maintenance is not easy to observe at all if there are no flooding events. Thus because of a succession of rather dry summers in the 1990s a decreasing ability of secondary ditches to reduce soil moisture was not perceived as a problem by most farmers. Only very few and local flooding events in the early 1990s could be remembered by the interviewed actors.

From about 2002 onward, more and more local water associations in Pyrzyce were revitalized and cleaning activities at secondary ditches and subsurface tiles were resumed. To a large extent this can be attributed to developments that mitigated the contextual factors that contributed to the poor performance of these associations during the 1990s and that again induced a process of revalorization of land and reclamation infrastructure.

Probably, most important was the administrative reform in Poland in 1999 that introduced Powiats or districts as intermediate units, between Gminas and Voivodships, of local self-government and administration. At the same time, the number of Voivodships in the country was reduced from 49 to 16. In the course of the revision of the national water law in 2001, administrative responsibility for secondary reclamation infrastructure and thus also the control and facilitation of the activities of local water associations was transferred to the head (Starosta) of the newly created and more potent Powiat administrations. All interviewed actors confirmed that in the Powiat Pyrzyce, the director of the Powiat Department of Environmental Protection, Forestry and Agriculture has indeed been very active in tackling the reclamation problems in the region. Among other things, he succeeded in convincing many farmers to revive the local water associations. In some Gminas, farmers elected new representatives and started to pay their fees again.

From the perspective of the Powiat administration, encouraging and supporting local water associations to organize cleaning activities at secondary infrastructures was by far the most convenient and efficient way to guarantee effective drainage. As was argued by the representative of the Powiat administration, the alternative would have been to enforce the obligations to clean and maintain ditches directly by individual farmers. Given the sheer number as well as limited and very heterogeneous capabilities and competences of individual family farmers in the region, this approach would not have guaranteed effective drainage. Starting in 2004, the director also initiated cooperation with the local employment center in the context of a state program to reintegrate long-term unemployed: the employment center pays the salaries for teams of workers that maintain and clean secondary ditches. In turn, the Powiat administration together with the respective local water associations coordinates and supervises the work and provides the technical equipment. Furthermore, cleaning activities at secondary ditches are closely coordinated with the regional state water administration and its respective activities at primary canals.

As stated by the interviewed actors, however, motivating and convincing farmers to pursue the reanimation of local water associations was (and still is) a difficult process. In particular, this was because the incentives for farmers to overcome the dysfunctionality of the reclamation infrastructure as well as their individual perceptions of the reliability or trustworthiness of the respective state authorities differed to some extent. The interviews revealed some characteristics of those farmers who pushed the process of reanimation by making the case for it in the village or Gmina assemblies. These would be the farmers who had continued or recommenced grassland farming after 1990 and who had suffered the most from excess soil moisture caused by the increasingly congested ditches over the years. But also farmers who had been less successful in compensating income losses by employment in the nonagricultural sectors and who did or could not switch to alternative crop production that is more tolerant to high water tables would support the reanimation of local water associations. In turn, farmers who have abandoned grassland farming completely and permanently, who have not suffered from flooding events, or who can make use of nonagricultural income alternatives tend to be more reluctant to engage in the reanimation process. Further, the revival of the local water associations has also been triggered by a succession of rainy summers (in particular in 1999 and 2002) that caused widespread flooding. The flooding was made worse by the fact that many secondary ditches and subsurface tiles had become totally dysfunctional (damaged, overgrown, and filled with thick sludge). In some Gminas, farmers whose plots had been flooded have also been very active in reviving their local water association.

As well as reinforced control by and support from the regional state authorities, two other contextual factors need to be mentioned. First, while the general agricultural terms of trade did not improve much until 2005 (EC, 2002), Poland's accession to the European Union in 2004 and integration into the Common Agricultural Policy substantially changed the market conditions and also the ways of generating farm income for the family farmers. In this context, the European Union premiums paid to farmers for the extensive use of grassland were often mentioned as a strong incentive to start using the formerly abandoned grassland plots (cutting the grass on those plots at least once per year is compulsory). Because of this revalorization of land, drainage regained importance since related farming operations are only possible if there is no excess soil moisture. Second, although the share of the state budget available for cleaning and maintaining primary canals was not increased, the state water administration representatives reported that several EU cofinanced programs had recently been initiated to undertake necessary activities to improve rural infrastructures. This would also allow for an intensification of cleaning activities at primary canals, which in turn would make individual contributions to keep secondary ditches clean more worthwhile.

Finally, the revised national water law of 2001 introduced a new method of determining the fees individual farmers have to pay to the local water associations. Unlike before, the fees are to be calculated based on the plot size that is actually benefiting from the system of secondary ditches and subsurface tiles. Thus the size of the whole agricultural area that is effectively drained by a respective ditch or subsurface tile will be taken as a basis, not the length of the directly adjacent ditches. This modified distribution of costs was perceived by the farmers as more "fair." For example, for 1 ha of drained land a farmer would have to pay about Zł32 (c. €8) per year. As was confirmed by the farmers, the absolute level of fees did not change much from what it had been before and was considered as rather low and, again, as "adequate."

## 5. CONCLUSIONS

The restitution of effective private property rights on land as well as the "privatization" of rights and obligations for secondary reclamation infrastructure has predominantly shaped the processes of revalorization in East Germany. In the Polish case, however, the distribution of formal property rights and obligations among social actors with respect to the various elements of the reclamation system as well as the property right regimes themselves did not change over the analyzed time period: land use has been governed privately by family farmers; the secondary reclamation infrastructure has been largely governed collectively by local water associations (albeit very poorly during the 1990s); while the primary reclamation infrastructure has been managed by the state in a hierarchical way. Thus, in the Polish case, it cannot be argued that the existence of different property regimes per se and the actual distribution of formal property rights among social actors caused or contributed much to the decline and revitalization of the local water associations.

Another trigger of revalorization that features prominently in many other reclamation systems in postsocialist and non-postsocialist countries is the growth of environmental concerns caused by negative environmental effects like soil degradation, of (excessive) reclamation activities (eg, FAO, 1995; Turnock, 1998). In Pyrzyce, however, because of specific natural soil conditions, reclamation activities have not caused any negative environmental effects. In addition, nonproduction-related rural property objects, such as landscape amenities, biodiversity, etc., have so far played only a very limited role in the region and accordingly are low valued.

In general, cooperative forms of local governance are badly needed for water management systems in the Central and Eastern European transition countries. Here, as in the Schraden and, largely, the Pyrzyce case, the reclamation infrastructure had been designed to serve large collective farms whose production goals were determined by the central planning system. After the breakdown of the socialist regimes, in some countries property rights on land were privatized or restituted (ie, made effective), thus resulting in or revitalizing highly fragmented land ownership. The transition process has also resulted in substantially smaller agricultural production units with heterogeneous interests, different production portfolios, and economic potential. Not surprisingly, the technical infrastructure cannot be adapted easily, if at all, to such structural changes. The problem of low excludability—common with local public goods and common-pool resources—is consequently aggravated. Thus the incentives of individual (small) landowners or farmers to maintain and operate their parts of the infrastructure are very weak since *private* investments would most certainly result in *public* benefits. This is also true in the case of clearly defined private property rights. As Penov shows for Bulgarian irrigation systems (Penov, 2004; Theesfeld, 2004) and Busmanis (2001) for drainage systems in Latvia, no appropriate forms of (local) governance have yet been developed to deal with these problems. Thus local irrigation canals and local drainage ditches have not been maintained for years and are now dilapidated.

## REFERENCES


AVP, 1998. Agrarstrukturelle Vorplanung Wasserbewirtschaftungskonzept Schraden (Entwurf). Planungsbüro für Natur und Siedlung Lindenau/PL3 Planungsbüro für ländliche Regionalentwicklung Berlin. Im Auftrag des Ministeriums für Ernährung, Landwirtschaft und Forsten vertreten durch das Amt für Flurneuordnung und ländliche Entwicklung Luckau.

Busmanis, P., Zobena, A., Grinfelde, I., Dzalbe, I., 2001. Privatisation and soil drainage. In: Gatzweiler, F., Hagedorn, K., Judis, R. (Eds.), Sustainable Agriculture in Central and Eastern European Countries: The Environmental Effects of Transition and Needs for Change. Proceedings of the ACE Phare Seminar. Shaker Verlag, Aachen, pp. 117–121.

European Commission, EC, 2002. Agricultural Situation in the Candidate Countries: Country Report on Poland. European Commission, Brussels.

Food and Agriculture Organization, FAO, 1995. Environmental Impact Assessment of Irrigation and Drainage Projects. FAO, Rome. Irrigation and Drainage Paper 53.

Hanspach, D., Kißro, R., 2001. Gebietsstruktur und Landnutzung. In: Grundmann (Ed.), Der Schraden: Eine landeskundliche Bestandsaufnahme im Raum Elsterwerda, Lauchhammer und Ortrand. Böhlau Verlag GmbH & Cie, Köln, pp. 48–60.

Jörns, S., Dammann, A., Hämmerling, R., Kappel, T., Schöfer, J., Zeckel, C., 2010. Umsetzung der Richtlinie Landschaftswasserhaushalt in Brandenburg. In: Kaiser, K., Libra, J., Merz, B., Bens, O., Hüttl, R.F. (Eds.), Aktuelle Probleme im Wasserhaushalt von Nordostdeutschland: Trends, Ursachen, Lösungen. Scientific Technical Report STR10/10. Deutsches GeoForschungs Zentrum, Potsdam, pp. 82–85.

Könker, H., 1993. Neue Aufgaben für das Meliorationswesen in den neuen Bundesländern. Zeitschrift für Kulturtechnik und Landentwicklung 34, 147–156.

Landgraf, L., 2001. Chancen für den Landschaftswasserhaushalt in Brandenburg. In: Proceedings of the Berlin Interdisziplinäre Fachtagung Nachhaltige Entwicklung in der Wasserwirtschaft – Konzepte, Planung und Entscheidungsfindung, June 27–28, 2001. Berlin. Schriftenreihe Wasserforschung, vol. 7. Wasserforschung e.V. Interdisziplinärer Forschungsverbund, Berlin, pp. 205–223.

Laschewski, L., 1998. Von der LPG zur Agrargenossenschaft: Untersuchungen zur Transformation genossenschaftlich organisierter Agrarunternehmen in Ostdeutschland. Sigma, Berlin.

Mohr, E., 1997. Landwirtschaft in Polen – Strukturdefizite behindern volle Ausnutzung des Produktionspotentials. ifo Schnelldienst 24. ifo Institut für Wirtschaftsforschung. Abteilung Agrarwirtschaft und Agrarpolitik, München.

Organization for Economic Co-operation and Development, OECD, 1995. Review of Agricultural Policies: Poland. OECD, Paris.

Olson, M., 1965. The Logic of Collective Action. Harvard University Press, Cambridge, MA.

Ostrom, E., Schroeder, L., Wynne, S., 1993. Institutional Incentives and Sustainable Development: Infrastructure Policies in Perspective. Westview Press, Boulder, CO.

Ostrom, E., 1990. Governing the Commons. The Evolution of Institutions for Collective Action. Cambridge University Press, Cambridge.

Ostrom, E., 1992. Crafting Institutions for Self-Governing Irrigation Systems. Institute for Contemporary Studies, San Francisco, CA.

Pant, N., 2002. Drainage Institutions in Western Europe: England, the Netherlands, France and Germany. CWP Research Papers, No. 3. Irrigation and Water Engineering Group, Wageningen, The Netherlands.

Penov, I., 2004. The Use of Irrigation Water in Bulgaria's Plovdiv Region During Transition. Environmental Management 34 (2), 304–313.

Perez, I., Wery, J., Heckelei, T., Bergez, J.E., Leenhardt, D., Thenard, V., Rapidel, B., Bazile, D., Coulibaly, M., Majewski, E., Bigot, G., Josien, E., 2005. Report on the Major Characteristics of Scenarios and Agricultural Systems to Be Studied in Test Case 1' PD6.1.1. Seamless Integrated Project, EU 6th Framework Programme (contract no. 010036–2). Available at: www.SEAMLESS-IP.org.

Scheumann, W., Freisem, C., 2001. The Forgotten Factor: Drainage. Its Role for Sustainable Agriculture. GDI (German Development Institute, Bonn (Discussion Paper).

Schleyer, C., 2004. Economic and ecological transformation processes in East German water management regimes: the role of property rights and governance structures. Environmental Management 34 (2), 281–290.

Schleyer, C., 2009. Re-valorisation of property objects and collective action: the case of reclamation systems in Northwest Poland. International Journal of Agricultural Resources, Governance and Ecology 8 (1), 74–89.

Schleyer, C., 2012. Institutioneller Wandel von Meliorationssystemen. Eine vergleichende Studie in Ostdeutschland und Polen. (Beckmann, V., Hagedorn, K. (Eds.), Institutional Change in Agriculture and Natural Resources), vol. 39. Shaker, Aachen.

Statistical Office of Szczecin, SOS, 2001. Statistical Yearbook–Zachodniopomorskie Voivodship: Part II. Statistical Office of Szczecin, Szczecin.

Spaling, H., Smit, B., 1995. A conceptual model of cumulative environmental effects of agricultural land drainage. Agriculture, Ecosystems and Environment 53, 99–108.

Theesfeld, I., 2004. Constraints on collective action in a transitional economy: the case of Bulgaria's irrigation sector. World Development 32, 251–271.

Turnock, D., 1998. Privatization in Rural Eastern Europe. The Process of Restitution and Restructuring. Edward Elgar, Cheltenham, UK, Northampton, MA, USA.

Vermillion, D.L., 1999. Property rights and collective action in the devolution of irrigation systems management. In: Proceedings of the International Conference on Collective Action, Property Rights and Devolution of Natural Resource Management – Exchange of Knowledge and Implications for Policy. June 21–25, 1999. Puerto Azul, The Philippines, pp. 183–219.

Wade, R., 1988. Village Republics: Economic Conditions for Collective Action in South India. ICS Press, Oakland, CA.

Chapter 3.2.3

# Water Storage and Conjunctive Water Use

A. Sahuquillo[1,2]

[1]*Polytechnic University of Valencia, Valencia, Spain;* [2]*Royal Academy of Sciences, Madrid, Spain*

## 1. INTRODUCTION

During the last 50 years, groundwater pumping increased in all developed countries, and contributed to the Green Revolution in many developing countries. The estimated groundwater use in India, Pakistan, and Bangladesh is about of 300 km$^3$. In northern China 15 million hectares are irrigated with 3.5 million wells. Significant increases in groundwater use occurred in Mexico and Australia, while spontaneous and unplanned groundwater use caused an excessive depletion in some aquifers (Mukherji and Shah, 2005). Hydrogeology is a flourishing scientific discipline; although it is a relatively recent science. It has been incorporated in the curriculum of most technical universities only recently, although there are lapses between the introduction of a new technique and its later application. Most hydraulic engineers have limited experience in groundwater studies and most managers do not have any formal education in hydrogeology even though an appeal to increase those studies is frequently heard among engineers and geologists. With these considerations, in addition to political motivations, authorities tend to favor larger and more spectacular engineering projects (the so-called *monument syndrome*). Hence groundwater use has not been considered by most water decision makers and water engineers, who are not aware of those advances (Custodio and Llamas, 2003). In addition, the implementation of aquifer management is more difficult in countries where surface water has historically dominated water supply developments (Coe, 1990).

Groundwater is usually a significant hydrological component of watersheds. Although varying largely between basins, average drainage from aquifers in continental areas is in the order of 30% of stream flow and is essential for sustaining stream flow during a dry period; the so-called "base flow." Significant volumes of groundwater are stored in aquifers as large as hundreds to thousands of times their annual recharge. The volume of the aquifer storage provided by a relatively small fluctuation in water levels of unconfined aquifers considerably exceeds available or economically feasible surface storage. This allows both the use of groundwater in storage during dry seasons or droughts and the use of the subsurface space for storing surface or subsurface water. In fact, groundwater has traditionally been used worldwide to back up supply in times of shortage or drought and that embodies a certain type of conjunctive use.

**Competition for Water Resources. http://dx.doi.org/10.1016/B978-0-12-803237-4.00022-7**

An unquestionable argument in favor of the joint consideration of ground and surface water is the fact that, to a greater or lesser extent, both water sources are hydraulically connected and have a mutual effect on each other and on other components of the hydrological cycle. Groundwater is integral to the hydrologic system, and plays a vital role in the functioning of ecosystems and biological habitats. Groundwater pumping has been causing a reduction of river or spring flow and other externalities such as land subsidence, or destruction of wetlands and aquatic habitats. In many cases, gaining rivers have been transformed into losing or ephemeral reaches because of the lowering of groundwater levels. Building a dam can increase the recharge to downstream underlying aquifers in losing reaches of the river channel if water stored in the dam during flood periods is released afterward to the channel more slowly. On the other hand, recharge from losing streams to underlying aquifers can decrease as a result of upstream water diversions. Groundwater recharge can be increased by filling losing surface reservoirs or by return flow irrigation. Excessive return flow irrigation and canal losses in arid areas have produced drainage and salinity problems caused by the increase of water levels and groundwater evaporation. Because groundwater and surface water are usually hydraulically connected, contamination in one water source can migrate to the other water source. These effects can create physical, geotechnical, environmental, legal, and economic problems. In many of these situations conjunctive use can bring out positive effects and limit the negative ones.

Aquifers can perform complementary functions of storage, water distribution, and treatment, which are classical components of a surface system. The existence of aquifers over a large basin area adds distribution and conveyance advantages to the benefits of water storage; the water distribution role is directly related to the storage function. Moreover, the passage of contaminated water through soil and the nonsaturated zone or long-term residence in a groundwater aquifer can improve water quality by mixing it with the innate water and eliminating pathogenic microbes and other contaminants.

Table 1 shows the differences between classical water resources systems and conjunctive use systems. Aquifers have widely performed water storage and distribution

**TABLE 1** Components of a Water Resources System

| Classical System | Conjunctive Use System |
|---|---|
| **Storage** | |
| Dams | Dams and aquifers |
| **Transport** | |
| Rivers, canals, and aqueducts | |
| Water transport in aquifers is irrelevant in most cases | |
| **Distribution** | |
| Irrigation and urban nets | Includes aquifers |
| **Water treatment** | |
| Classic treatment for drinking water and wastewater | Includes soil treatment in artificial recharge |

functions through conjunctive use systems. They have also been used for water treatment, but their capability of transporting water is negligible compared to the flows that rivers, canals, or aqueducts can convey.

In Israel, because of large volumes of groundwater stored in aquifers, coastal and carbonate aquifers were overpumped well above their recharge, to defer building of the more costly National Water Carrier of Israel. In other cases, aquifers have allowed, through unplanned overdraft, to develop primary economic activities, which have been the origin of a further economic growth, for example, in California and southeastern Spain.

Likewise, groundwater has other advantages such as its adaptability to a progressive increase in water demand, in addition to mitigating the impacts of droughts, and alleviating drainage problems in irrigated zones of some arid areas. Another virtue of groundwater in conjunctive use schemes is its role in offsetting uncertainty surrounding surface flow, hydrological parameters, or water demand. Groundwater pumping can be augmented in dry or drought periods, or lowered if surface flows are above the expected or forecasted flow levels. Accordingly, groundwater management can be changed if the knowledge of the aquifer improved along the operation of the system. The strongest argument in favor of conjunctive use is the fact that including aquifers in the water management system provides alternatives not only for augmenting the number of storage and distribution options, but above all because their functionality is complementary with the ones of a surface system.

Those alternatives are usually far more economical than those involving building big dams or canals. Groundwater cost is usually much lower, has shorter execution time, and investments can be applied according to the increased demand. In most cases, conjunctive use can prove to be a cheaper solution than dependence on either surface water or groundwater. The suitability of conjunctive water use is not restricted to arid or water-scarce areas. If mutual influence of surface water and groundwater is considered, conjunctive use is advisable in most areas, including cases where scarcity or pollution problems exist (Sahuquillo, 1985, 2000, 2002, 2005).

A system with surface and subsurface components that is properly maintained and flexible, subject to changing water demands and hydrologic variability, can provide economic, functional, and environmental benefits to ecosystems and the region. In recent years, structural solutions have been questioned, while consensus can be reached on the need for better management of the existing options rather than expensive investment in construction of new water infrastructure, like dams. In many countries, building new dams is not a valid option any more. The most favorable and less controversial dams have already been built, keeping pace with growing environmental conscience. Conjunctive use alternatives should be considered for development of water resources or expanding the existing ones.

Global warming in many regions causes negative impacts on water availability, such as reduced average river flows, increased flow variability with more floods and droughts, shifts in the spring dates of mountain snow melting, and a general need for water storage to meet growing water demands. The functions of water storage provided by conjunctive use allow the following: increase in water availability without augmenting surface storage, inclusion of aquifer as an additional water source, advantages of aquifer storage and distribution characteristics, increase of water supply guarantee, drought mitigation, decrease of aquifers' overexploitation, and in some arid areas reduction of drainage problems caused by irrigation return flows.

As with most human activities, the practice of conjunctive use is subject to many political, social, and economic factors. The advantages obtained by implementing conjunctive use depend on physical factors, but the rules and institutions might foster or limit the benefits of this system. Water rights and institutional and social aspects implicate possible incentives or problems related to conjunctive water use (Coe, 1990).

## 2. METHODS OF CONJUNCTIVE USE

There are two ways of using the storage provided by aquifers. The most intuitive one is through augmenting aquifer recharge. The second one is through alternate conjunctive use (ACU). In ACU the target yield is obtained in dry years through increased pumping; when more water than on average is available in streams or surface storage, more surface water can be used allowing more groundwater to remain in storage. In this way, subsurface storage is provided by the differences between the extremes of the aquifer water levels, these being high at the end of wet periods and low at the end of dry periods. Both possibilities of artificial recharge and alternate use are not exclusive. In fact, there are many sites where both approaches are applied, although in most cases one predominates.

### 2.1 Artificial Recharge

The usual practices of artificial recharge are through injection wells or infiltration ponds to store surface flows or nonused, surplus water that otherwise would be lost. Aquifers are also recharged by infiltration in leaky canals, losing rivers or reservoirs, and irrigation return flows, sometimes called *passive recharge*. Unintended, and unwelcomed, aquifer recharge from pervious reservoirs in some Mediterranean karstic areas in Spain became advantageous when conjunctive use was established. The Algar dam north of Valencia has been purposely built with this aim, and other dams have been suggested.

Numerous studies have been published since 1900 on artificial recharge, although references dating back even to the late 18th century about urban water supply in southern France can be found (Custodio, 1986). Europe has a long tradition of artificial recharge in alluvial aquifers to complete the treatment of heavily polluted river water. In many countries of Northern, Eastern, and Central Europe, aquifers are widely used benefiting from the ability of the soil and unsaturated zone to retain and absorb heavy metals and other organic pollutants. In most cases, the main function of an aquifer is water treatment and not water storage per se. In Germany, more than 500 million $m^3$/year of water are recharged annually for urban water supply. In southern California, sewage-treated effluent has been recharged in some aquifers after having passed advanced treatment. In Israel, treated sewage water from the metropolitan area of Tel Aviv is recharged in the sand dunes to be pumped at a later time for accepted irrigation uses. In Holland, coastal dunes are used as storage and for additional water treatment. Since 1940, around 100 million $m^3$/year of water from the Meuse and Rhine rivers have been recharged annually for urban use. In France, 20 million $m^3$/year are recharged from the Seine River into the chalk aquifer near Paris every year (Custodio, 1986).

The method known as aquifer storage and recovery (ASR) was first employed in the state of Florida in the United States; its use is predominantly for urban water supply. It consists of underground storage of treated water during periods of low demand and its recovery for potable water uses during periods of high demand. The recharge operation is carried out with dual-purpose wells that inject water into the aquifer and pump it at a later time (Pyne, 1989). The chalk aquifer in the Lee Valley is recharged through wells during the winter with treated water from the Rivers Thames and Lee in the United Kingdom, and pumped in summer. The same temporary storage function of treated potable water is used in Barcelona (Spain). Up to 20 million $m^3$/year are recharged by 12 dual-purpose wells (probably contemporary to the first Floridian ASR wells) to be stored in the Llobregat Delta Aquifer when the water tanks of the raw water treatment plant are full. In the delta of the Besós River north of Barcelona, two pioneering ASR dual recharge and pumping wells of 1 m diameter were built in 1953, surrounded by 16 small diameter wells to help clean the injection wells. Also for the water supply of the metropolitan area of Barcelona, artificial recharge is made in the alluvial aquifer in three basins of the lower Llobregat River and by scarifying the river bed (UK Groundwater Forum, 1998; Custodio et al., 1969). Though other sites have been tested for artificial recharge of varying duration, there are no other places in Spain where artificial recharge is conducted regularly.

It is in the western states of the United States where most artificial recharge for conjunctive use happens, mostly in spreading basins. California is the region where experience in groundwater and conjunctive use is broader. The state accounts for about 18% of the total groundwater withdrawals in the United States. With its 3000 million $m^3$/year of recharged water, the state probably has much more artificial recharge than the rest of the world. One of the first examples of conjunctive use in California were the diversions and spreading of stream flow east of Los Angeles, in the channels of San Antonio Creek in the 1890s and San Gabriel River in the early 1900s. Another early case, including passive recharge, is Vaquero Dam built by the US Bureau of Reclamation where storm flows are released to percolate downstream in the riverbed with infiltration up to 750,000 $m^3$/day. Those practices improve groundwater quality and eliminate overdraft and the risk of seawater intrusion. In 1913, the Los Angeles Aqueduct was built, from the Owens Valley (just east of Sierra Nevada) to San Fernando Valley. Water was applied for direct use in the City of Los Angeles, with the excess water recharged in the valley basins. In 1940, the aqueduct was extended north into the Mono Basin. Later, wells in Owens Valley were incorporated into the aqueduct. This allowed conjunctive use at both ends of the original aqueduct, Owens and San Fernando Valleys (Fig. 1) with a total length of 550 km (Coe, 1990). The introduction of the deep-well turbine pump in 1910 in the then rich farming area of the Santa Clara Valley, south of San Francisco Bay, led to an explosive increase in groundwater pumping that caused a significant drop in water levels in the aquifer. In 1931, 0.9 m of subsidence was identified in the City of San Jose. Later on, subsidence reached up to 4 m and was halted during the 1950s with artificial recharge of local and imported water from the Hetch Hetchy Aqueduct. The water supply infrastructure includes several reservoirs and 18 recharge facilities (Fig. 1) (Kretsinger and Narasimhan, 2006).

In 1948, the Orange County Water District (OCWD) imported water from the Colorado River for artificial recharge to remediate groundwater level decay and seawater intrusion. In 1971, the OCWD implemented two seawater control intrusion barriers: the Alamitos Barrier with 28 injection wells using imported potable water and the Talbert Barrier (since 1975) using reclaimed water from an advanced treatment facility called Water Factory 21 (Kretsinger and Narasimhan, 2006).

In California, groundwater overdraft reached 5 $km^3$/year in 1955, causing subsidence in many areas; up to 9 m in Los Banos in Tulare Basin in southern Central Valley. Because of

**FIGURE 1** California's major water projects. *After the California Department of Water Resources.*

artificial recharge, overdraft was reduced in 1982 to values estimated between 2.5 and 3.1 km$^3$ (State Water Resouces, 1982). The majority of California's projects do recharge through spreading basins rather than through injection wells. The Central Valley Project (CVP) is a federal project built by the Bureau of Reclamation, the largest water purveyor in California. Today, the CVP manages more than 8 km$^3$ of water annually, about 90% of which is for irrigation. It was authorized in 1937 and began operating in 1951. The State Water Project is the main component of the 1957 California Water Plan. It includes an integrated system of dams and canals parallel to the CVP, aimed at an efficient transfer of large volumes of water from the more humid north to the driest and more populated south (Fig. 1). Communities in the Sacramento Valley take some of the water, but most is pumped into the California Aqueduct (Fig. 1). The construction of this project started in 1961. In 1966, water began to flow south through the California Aqueduct. Five years later, water arrived in southern California and the Department of Water Resources signed contracts with 30 agencies throughout the state for permanent water services (Hanak et al., 2011; State of California Department of Water Resources, 1957). The Tulare Basin is one of California's most productive agricultural regions in the Central Valley. Water consumption is high (around 16 km$^3$/year) importing annually between 4 and 7 km$^3$ water from the CVP and

State Water Project. Despite the water imports, overdraft continues in the basin (Harou and Lund, 2008).

The 2009 California Water Plan update estimated that 11 $km^3$ of new underground storage would be needed to increase water availability by 0.6 $km^3$. This implies both storage reoperation and recharge in the provided space. More aggressive alternative estimates indicate that there is a potential of increasing water availability up to 2.4 $km^3$ using about 24 $km^3$ of new storage. An institutional issue raised in California is the question of local versus regional groundwater management. Local agencies have traditionally constructed and operated water projects to develop and deliver water to meet local needs. This has included the diversion of storm runoff and delivery of imported water for artificial recharge. However, in recent years in southern California it has become clear that certain increased efficiencies and advantages could be achieved if several local agencies jointly operated and developed regional water management strategies. As California's drought is in its third year, it seems that more storage is needed in the state and that surface water and groundwater storage should be considered and analyzed as part of a large system with a large number of operation, conveyance, and storage alternatives (Coe, 1990; Lund et al., 2014).

In Arizona, before 1980, when the Groundwater Management Act (GMA) was passed, no mechanisms existed to secure the rights of water users who wanted to recharge water in an aquifer. After GMA was approved artificial recharge increased rapidly. Accordingly, between 1990 and 1991 around 8 million $m^3$ of water were recharged, 290 million $m^3$ until 1992, and 590 million $m^3$ until 1997. In Arizona, conjunctive use is driven by the state, which greatly differs from conjunctive use in California, which is carried out by organizations created by individual users. In 1996, the state established the Arizona Water Banking Authority to store Arizona's unused water diverted from the Colorado River through the Central Arizona Project (CAP) (Heikkila et al., 2001). Water supply for the Phoenix metropolitan area has two sources of surface water, one from the Salt-Verde Rivers watershed and the other from the CAP and the storage of the aquifer in the alluvial Salt-Verde River Valley (Sahuquillo, 2005).

In Texas, conjunctive management of surface and groundwater has long been recognized as a potential strategy for increasing water availability of limited water resources. But institutional constraints, including water law considerations, currently severely limit conjunctive use (Wurbs, 1987).

The development of water marketing and groundwater banking have been important tools to reduce water scarcity in drought years as they allow for a temporary movement of water from areas of relative abundance to areas of critical needs. Water markets started in many US western states. They were launched in California in the early 1990s and by the early 2000s the annual volume of water sales was in the range of 2 $km^3$. There was much initial political resistance to water marketing and concerns within source regions about local economic harm from transfers. Transactions have leveled off since the early 2000s. An active water market system has been developed within the state of California, enabling temporary and longer-term reallocation of water from lower-value (mainly agricultural) activities to higher-value activities in farming and urban sectors and to the environment.

The state played a major role in launching the market, through legislation in the early 1980s and the establishment of drought water banks in some areas during the early 1990s. The state and federal governments have been major purchasers of environmental water (water needed to maintain endangered fish species in the Sacramento–San Joaquin

Delta). Opportunities for market development are still considerable because many acres of farmland are still planted with low-value crops. Expanding underground storage can be much more cost effective than building new surface storage to stretch available water supplies. The still large discrepancies in crop values and water use suggest a potential for a better water market system in the future. A well-functioning water market is important for encouraging water demand reduction and reallocation of water from lower to higher value uses (Hanak et al., 2011).

## 2.2 Alternate Conjunctive Use

The ACU aims at satisfying water demand alternately from surface water or aquifers, depending on water levels in each system. The difference between the water volumes stored in an aquifer at the end of a dry period and the end of a wet one is the storage provided by the unaffected aquifer. Making use of ACU, in dry periods, groundwater recharge decreases, pumping increases, and groundwater levels drop. During wet periods, rain recharge increases and so do irrigation return flows, passive recharge from ephemeral rivers and losses from reservoirs or canals, and finally pumping decreases so that groundwater levels rise more than in the unaffected case. It is an induced storage that occurs without resorting to artificial recharge. Demand increases can be met by augmenting pumping in dry periods and use of surface water in wet periods if unused surface water is available. Similarly, for a fixed demand, reliability is augmented with more pumping without changing surface storage. Both water availability and groundwater levels in storage are higher when surface water is used in wet periods. Overexploited aquifers can be alleviated through ACU with additional existing or projected surface water storage. In many cases, new connecting infrastructure needs to be built or enlarged. An important aspect that needs to be stressed is that ACU allows for a more efficient use of surface water without the need of artificial recharge.

### *2.2.1 Alternate Conjunctive Use in the Central Valley of California*

The previously mentioned 1957 California Water Plan proposed a large-scale ACU for the Central Valley that is the first and larger planned scheme of this type. In 1957, its total projected storage, adding existing and new dams, amounted to 24 km$^3$ and the planed subsurface storage was 37 km$^3$, as shown in Fig. 2. By using this subsurface storage more surface water can be supplied without having to resort to artificial recharge. The plan anticipated some supplementary use of artificial recharge (State of California Department of Water Resources, 1957). However, it was not managed with a simple and uniform operating rule, as originally planned. Local agencies built and operated their projects directly or through agreements with water rights holders, and instead of continuing with the ACU concept, they preferably used artificial recharge. They have the right to store water underground and to recapture it. More than 20 types of local districts or agencies have statutory authority to provide water for beneficial uses in California (Hanak et al., 2011).

The idea of using ACU, which was proposed in the 1957 California Water Plan for the entire Central Valley, was anticipated for the Upper San Joaquin Valley in the 1930 Water Plan. ACU, passive recharge, and excess irrigation were proposed without artificial recharge and were estimated to amount to more than 20 km$^3$ of water that could economically be stored in the southern San Joaquin Valley aquifer, a goal that could not be reached with surface reservoirs (Wurbs, 1987).

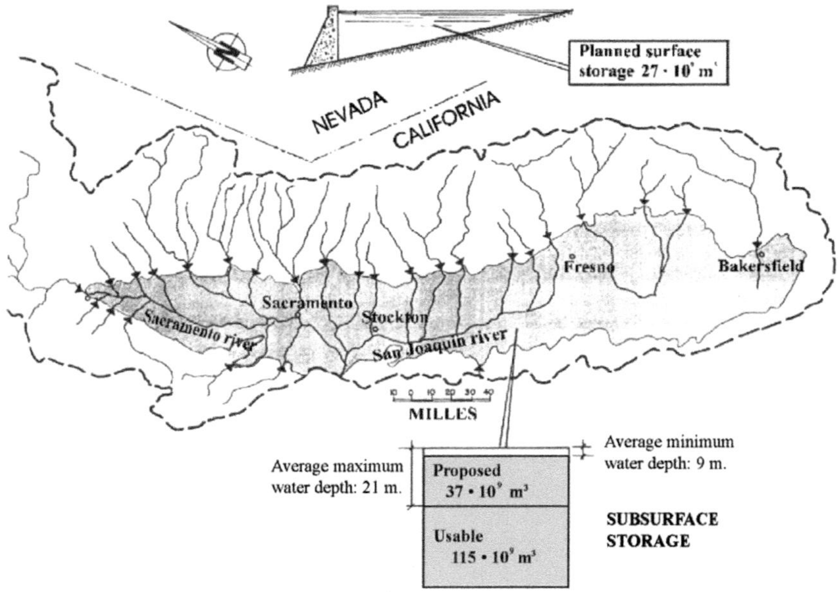


FIGURE 2 Conjunctive use in the Central Valley, California, USA. *Modified from State of California Department of Water Resources, 1957. Bulletin No 3. The California Water Plan.*

After the 1980s, in many instances, excess surface water that could not be stored was applied to meet existing demands. This strategy paradoxically has been termed "in-lieu recharge" because, instead of pumping groundwater, it is applied directly and the volume of water in storage increases because it is not pumped off. To a certain extent this can be considered as an occasional ACU. Simulations in this regard prove that significant quantities of new water can be generated in California at considerably lower costs than building new dams. This possibility could be amplified by analyzing operation alternatives of dam storage. Consideration of weather forecast can also improve system operation and water availability (Jaquette, 1981). Further advances in climate modeling could improve hydrologic forecast and increase efficiency of the anticipated operation of conjunctive use systems.

### 2.2.2 The Mijares River-Plana de Castellón Aquifer Scheme

Spain is the driest country in Europe, with a large number of conjunctive water use operations. Most of them use the ACU concept, except for the ones mentioned before that rely on artificial recharge. Historically, in many Mediterranean basins, besides fields traditionally irrigated with surface water, more areas were irrigated in wet years and after the rapid increase of aquifer exploitation in the 1960s, they were integrated smoothly into the existing water systems. Thus more surface water was used during wet years and more groundwater was pumped during dry years. In many cases, the schemes were proposed by individual users. In other cases, canals have been built by the water administration in areas partly irrigated with groundwater. When diverted surface water is insufficient, ACU can be applied.

The Mijares basin on the Mediterranean coast of Spain, 60 km north of Valencia, has three surface reservoirs. Two of them built in karstified limestone are connected with the Plana de Castellón aquifer. They have important leaks, of the order of 45 million $m^3$/year, which recharge the aquifer. The Mijares River losses recharge the aquifer with a similar volume. About one-third of the irrigated surface is supplied alternately depending on surface water flowing in the river and stored in reservoirs. The difference between high and low values of aquifer storage can reach over 700 $hm^3$, around four times the existing surface storage (Figs. 3 and 4). This allows for the use of a large percentage of the average annual river flow. The scheme and its operating rules were proposed by individual users and legally approved in 1973. Also in the early 20th century the Palancia River farmers established a similar ACU scheme. ACU strategy is also applied in several areas of the Júcar Basin Agency, as the Canal Júcar-Túria, the Canal Campos del Túria, and the Marina Baja (Sahuquillo, 2002, 2005).

The Júcar Basin Water Agency drilled and installed 65 large capacity wells near the main canals in La Plana de Valencia aquifer at the end of the 1991–1995 drought period.

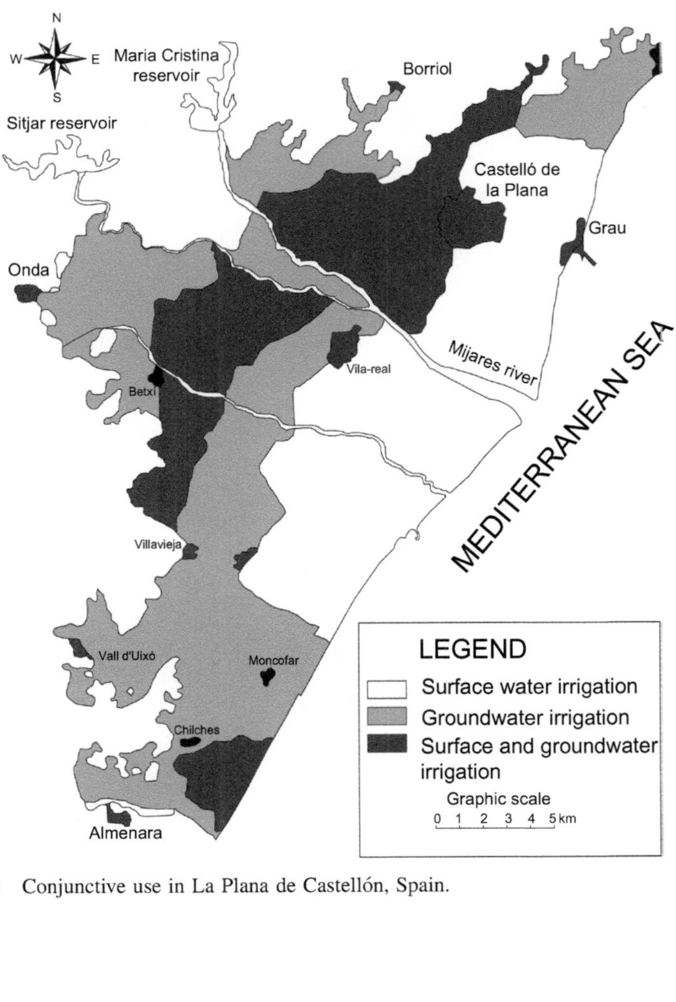


**FIGURE 3** Conjunctive use in La Plana de Castellón, Spain.

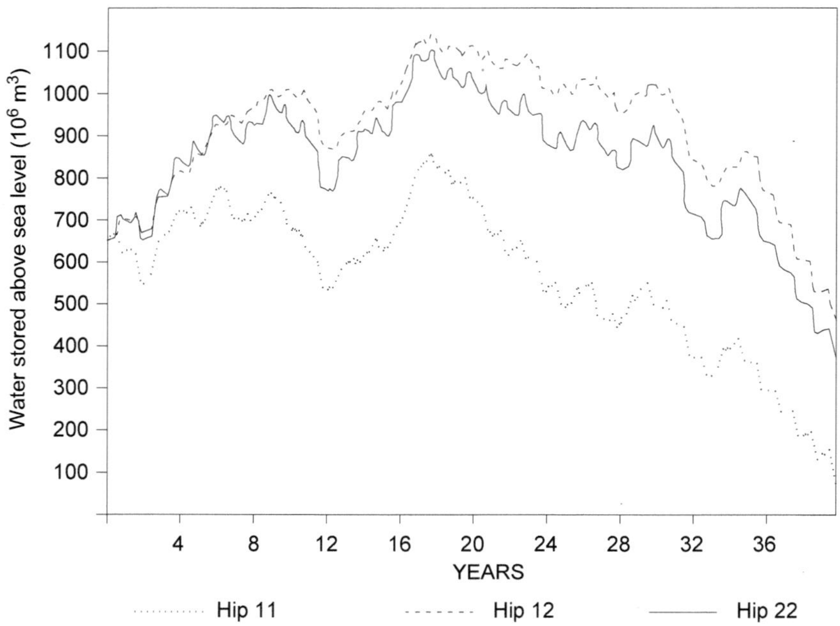


**FIGURE 4** La Plana de Castellón aquifer. Change in storage for different use alternatives.

They scarcely began to run as the drought ended, but during the 2007 drought, 50 $hm^3$ of water were pumped from those wells. Similarly, 50 $hm^3$ were pumped in 2007 from wells made by the Segura River Agency in the Vega Media y Vega Baja alluvial aquifers.

To increase the capacity of the water supply system of the Madrid metropolitan area, pumping from wells, up to 500 m deep, in dry years has been preferred to building new dams. This provides more security in drought years. Additional increments of pumping from the existing 4 $m^3$/s have been foreseen and the use of artificial recharge to reduce the depression cone caused by the low permeability of the aquifer is analyzed. Thus preparedness for drought increases the guaranty of water supplies (Sahuquillo, 2002).

Another example is the Campo de Cartagena that has traditionally been irrigated with very saline water from overexploited aquifers. With the arrival of imported water from the Tajo Basin to the Segura Basin, a significant rise in groundwater levels was noticed, as well as a decrease in groundwater salinity because of pumping reduction and return flow irrigation, depending on the variability of imported water flows. High groundwater levels have produced drainage problems in some areas that have been solved with drainage. Small desalination plants exist in areas where salinity exceeds 4 g/L. In 2001, around 5 $hm^3$ of desalinated water were incorporated into the irrigation canals, while brine is injected in a well 750 m deep.

### *2.2.3 Stream-Aquifer Systems*

The ACU concept has been applied to alluvial and other small aquifers in conjunction with connected rivers. The mutual influence between rivers and aquifers is more rapid than in larger aquifers, lasting from several months to a few years to produce a significant decrease in river flows. The delay depends on the distance from the pumped well to the river, the aquifer geometry and diffusivity (ratio of transmissivity to aquifer storativity), and the aquifer–river connection, the last one being of foremost importance. Pumping during dry intervals, when more water is needed, increases water availability by an amount equal to the pumped quantities minimized by the effect of pumping from the river flow. A part of the effect of pumping from river flows subsequently carries on over wet periods, when river flows are higher and demands are lower. After several seasons the stream depletion is about equal to average annual rates of pumping. The further a well is away from the stream, the smaller is the aquifer diffusivity, and the less the river–aquifer connection, the less the annual fluctuations of pumping will affect stream depletion. Significant gains for irrigation were obtained from pumping the alluvial aquifers in the Arkansas and South Platte Rivers in the state of Colorado in the United States. The result of an economic model showed that conjunctive use in the South Platte doubled the net benefit for farmers and that having more wells to increase pumping in dry years acts as an insurance against droughts (Young and Bredehoeft, 1972; Bredehoeft and Young, 1983).

In Colorado, a conflict emerged in the 1960s between surface water and groundwater users when pumping of the aquifer considerably affected river flows. Conjunctive management of both resources solved the conflict. Colorado's prior appropriation doctrine allocates water rights based on time priority for both surface and groundwater. Since surface water was used before groundwater, most surface rights have priority to most groundwater rights, which are considered tributaries. As pumping tributary groundwater depresses surface water flows, this impedes the use of groundwater. To solve this conflict, Colorado introduced tributary groundwater use in the prior appropriation rule considering the effect of groundwater pumping from river flows and obligating groundwater users to release pumped water to the river flow as due compensation. This practice was called "river augmentation." Temporary plans of augmentation exist in both Arkansas and South Platte. A farmer who wishes to continue or increase his/her pumping must receive approval from the State Engineer with estimates of the water amounts he/she will pump and in addition the amount of water that he/she will discharge to the river to satisfy prior rights (Heikkila et al., 2001).

In the United Kingdom a very efficient use is being made of aquifer–river systems. Aquifers are located mainly in consolidated rocks, limestone, sandstone, and chalk. Aquifers are generally small and their storage is less than in alluvial deposits, meaning that pumping has a relatively fast effect on river flows. Groundwater is pumped and piped into certain rivers during dry periods to maintain adequate flows in them to meet supply and environmental requirements. These schemes have also been termed "river augmentation" and are used systematically in a very efficient way in water planning in England and Wales (Downing et al., 1974). The approach to development on a regional scale in England and Wales has only been possible since the Water Resources Act became law in 1963. This placed the management of water resources in the hands of 29 river authorities. Adequate river flows for amenity benefits have to be considered. One result of the Act has been that the use is now controlled by a licensing system. The Shropshire Groundwater Scheme (SGS) is the largest conjunctive use scheme in the United Kingdom to cope with increasing demand for water supply from the River Severn. The SGS is being constructed in a phased manner so that the capacity of the scheme is only increased when required. Then, it will pump from up to 68 sites (81 wells). This is estimated to increase flows in the River Severn by 225,000 $m^3$/day with a net gain of 68% (Shepley et al., 2009).

Alluvial aquifers are also used in many wet areas of the eastern United States. The alluvial formation of the Hunt-Annaquatucket Rivers has been analyzed with an aquifer model to determine if it could increase its exploitation. Modeling results show that it is possible to increase the amount of the current withdrawal from the aquifer by as much as 50% by modifying current withdrawal schedules and the number and configuration of wells (Barlow et al., 2003).

### 2.2.4 Use of Karstic Springs

In Spain, several karstic springs have been regulated to augment water availability for irrigation and urban water supply. In some cases the best or unique physical possibility has been to locate wells near the spring, in the proximity of existing canals, or aqueducts used to transport the captured spring flow. In such cases, the influence of pumping from wells is immediate: the spring dries out and all demanded water must be pumped. This approach implies that supply can be augmented well over the natural flow of the spring during the irrigating season, or for urban or industrial needs. In this way, large changes of flow, common in most karstic springs, can be adjusted to water demands. The use of the aquifer as a subsurface reservoir is very clear when the spring dries out. In many cases, very high flows have been obtained from wells (up to 1.2 $m^3/s$ in two wells in the Los Santos River spring in Valencia, Spain). In the Deifontes spring near Granada, in southern Spain, water flow of 2.25 $m^3/s$ was provided by five 100 m deep wells. In other cases, the spring is a component of more complex schemes. An example is the Marina Baja water supply scheme, consisting of two dams, two aquifers (one of them being their outlet of the El Algar spring), and treated water reused for irrigation. Alternate use of groundwater and surface water and the regulation of the El Algar spring by wells solved the acute water supply problem in the late 1970s caused by a frequently visited tourist area near Alicante, on the Mediterranean coast of Spain. The two wells near the spring can pump up to 400 L/s each, while the wells are used exclusively during dry periods. The underground storage provided by the aquifer during the severe drought of 1990–1996 was estimated to be in the order of 40 million $m^3$, three times the existing surface storage (Sahuquillo, 2002).

### 2.2.5 Alleviation of Land Drainage and Salinization

In many irrigation projects, aquifer recharge has increased because of water losses from unlined canals and distribution systems in addition to the infiltration surplus of applied water. The increments in the aquifer recharge augment the potential for groundwater development, and in some zones they also produce drainage and salinity problems caused by rising groundwater levels. This is a customary problem of large surface irrigation projects in arid countries. The Planning Commission of the Government of India has recognized problems of waterlogging caused by an average water table rise exceeding 1 m/year in several schemes. The Commission suggested, in addition to enhancing water use efficiency, an increase in groundwater use with canal water to augment supplies and prevent land salinization (Sondi et al., 1989).

Waterlogging and salinity problems created in the Punjab plain in Pakistan had the same origin of surface water infiltration along the irrigation system of the Indus River and its tributaries. Irrigation started to be intensively developed in the late 19th century under British colonial rule. In the middle of the last century, 25,000 ha had to be abandoned every year and in 1960 2 million hectares of a total 14 million hectares of irrigated land were abandoned. The irrigated area is dominated by 43 big canals with a total length of 65,000 km, in addition to secondary and tertiary canals, the biggest 15 of them have a capacity between 280 and 600 $m^3/s$. They are fed by several big dams, among others Mangla dam and Tarbela dam with 5.5 $km^3$ and 10.6 $km^3$ of storage, respectively. Most canals are unlined, experiencing significant losses that feed the huge aquifer below. In the past 80–100 years, water levels in this aquifer rose between 20 and 30 m, and up to 60 m in some places. The problem has been intensively studied since the 1960s. The water resources group at Harvard University proposed to drill 32,000 high-capacity wells to pump 70,000 million cubic meters per year to lower the water levels. The idea was to take out the pumped salty water to the sea through lined canals and use fresh groundwater jointly with surface water to increase irrigation. A public tube-well development called Salinity Control and Reclamation Projects (SCARP) was initiated by the Pakistan government. As drainage projects do not have an immediate economic profit, most groundwater pumped from wells was freshwater used to increase irrigation. Pumping of saline water and lining of canals to avoid infiltration of salt water was not addressed in that project. When brackish water was pumped from a well, it was blended with surface water to irrigate crops. Thus the salt storage of the aquifer increased instead of decreasing. In some areas, pumping and mixing water with different salinity levels has erratically raised water salinity. Nevertheless, improvements of drainage and decrease of soil salinity were important developments. Another important aspect not considered in early plans was the capacity of the private sector to fund development of groundwater and drilling of deep high-capacity wells, an idea that was triggered by SCARP performance. In the end, groundwater exploitation has enabled many farmers to supplement their irrigation requirements (Burke and Moench, 2000; Task Committee on Water Conservation, 1981).

The target in heavily irrigated arid areas in the developing world is to use existing aquifers, recharged by irrigation return flows and by surface water infiltrated in the conveyance and distribution channels, jointly with surface water, while maintaining groundwater levels at a certain depth below the ground level, to prevent salinity and drainage problems. Equally important is the control of migration and dispersion of more saline groundwater bodies. In this way, groundwater quality can be maintained in addition to augmenting the total water availability. Extensive hydrogeological analysis and monitoring are needed in addition to the long-term simulations of groundwater flow and salinity to be able to better understand and cope with water shortages.

The same drainage and salinity problem exists in Egypt, northern China, and the Asiatic countries of the former Union of Soviet Socialist Republics (USSR). In the former USSR countries, the use of the estimated 25,000 million $m^3$ of water drained from irrigated lands has been suggested, jointly with surface water. It has been estimated that one-fifth of the irrigated land in the United States, and one-third in the world, is plagued by salt accumulation. Losses at canals and distribution systems can be alleviated by lining the conductions. However, if lost water feeds usable aquifers and conjunctive use is practiced, it can be more convenient to leave canals unlined, unless drainage problems exist and water losses create excessively high aquifer levels. In addition to falling groundwater levels, water management practices should also consider the problems of rising groundwater levels. Conjunctive use of surface water and groundwater resources can both solve the problem of water shortages and improve water use efficiency and regional environment of irrigated areas (Task Committee on Water Conservation, 1981; Van Steenbergen and Oliemans, 2002; Kats, 1975).

In large plains, such as the Indo-Ganges, with large irrigation projects, important aquifers exist that have given rise to spontaneous developments of conjunctive use. In the case of the Indus aquifer, it was stimulated by the SCARP project. In other cases the stimulus was produced, in addition to increasing water availability, to remedy problems in the design and operation of irrigation projects. Some of the problems addressed are: inadequately maintained canals unable to sustain design flows throughout the system, poorly administered canals, allowing unauthorized or excessive off-takes and oversized channels in relation to surface water availability, or water flows tied to rigid canal-water delivery schedules and unable to respond to crop needs. Any combination of these circumstances leads to inadequate irrigation water service levels throughout much of the canal system and especially toward the tail sections. At the head-ends of the canal, irrigation causes water tables to rise, resulting in waterlogged areas, whereas at the tail-ends, salinity problems are increasing because of excessive use of bad quality groundwater for irrigation. Thus private water well drilling usually proliferates. Also sociopolitical and institutional impediments to an adequate management of water resources in the area play a role, like sociopolitical dominance of farmers in areas excessively endowed with surface water, who refuse to reduce canal intakes to allow a greater proportion of available flow to reach small areas. Other problems are: inadequate understanding of conjunctive use and the potential of groundwater use by water and irrigation engineers, split responsibility for surface water and groundwater management among different organizations and agencies, or else water resources management being mainly addressed by surface water-oriented agencies, because of the historical relationship with the development of irrigated agriculture (Foster et al., 2010; Foster and Van Steenbergen, 2010).

The Ganges River basin has serious water problems because of topographic limitations to build large surface storage, contrary to the Indus River. The basin is almost 800,000 km$^2$ large and accommodates a 10th of the world population, mainly working in agriculture, with very low per capita income. In 1975, pumping water from the aquifer along the Ganges River and canals was proposed, inducing recharge that would provide 60 km$^3$ of water annually. The total annual Ganges flow is estimated at almost 300 km$^3$ in the monsoon season from July to October and in the order of 70 km$^3$ during the other 8 months of the year (Revelle and Lakshminarayana, 1975). Similar proposals have been suggested with several different relative scenarios for wells and canals with the aim of increasing induced aquifer recharge (Chaturvedi and Srivastava, 1979; Khan et al., 2014). In the vast Ganges aquifer with a rainfall ranging from 2 m/year in the east to 1 m/year in the west, groundwater exploitation has grown in the past decades, while aquifer recharge has been increased with channel water. An important increase in aquifer pumping seems a viable option, although it has only been simulated with schematic and very simplified models.

## 3. ANALYSIS OF CONJUNCTIVE USE SYSTEMS

It is important for the design and operation of conjunctive use resource systems to adequately evaluate their performance. Performance evaluation is required for convincing stakeholders (governments, water agencies, and other public administrations and users). The institutional dimension of conjunctive use management is much more complex than when surface water or groundwater alone is the main water supply source. Models have become an essential tool for analyzing complex systems. These models are applied to simulate historical conditions and to evaluate future water resources management strategies. The analysis of conjunctive use alternatives has to include streams, reservoirs, canals, aquifers, and flow interchanges between groundwater and surface water, as well as water supply facilities for different instream and offstream uses. Consequently, the analysis needs to be conducted at the basin or regional scale.

Similarly, the assessment of surface and groundwater resources needs to be conducted with comprehensive models including both water sources simultaneously. Groundwater models and surface hydrology models are usually calibrated by reproducing the observed values of surface flow and piezometric heads at some locations. Model calibration should be oriented to capture in the best possible way the interactions between both subsystems. For conjunctive use modeling this is more important than achieving a better tuning to other responses.

Effective conjunctive use is a matter of proper resource management. Therefore operating rules are important components of the water management alternatives. The same set of structural facilities can produce very different yields, depending on the system operation. Thus it needs to be explicitly incorporated in the analysis and be realistic enough to be applied in real life. There are many political, historical, sociological, and cultural factors that may impede the application to the real world of otherwise perfect operating rules. We can have surface and groundwater models that have successfully replicated the historical behavior of the system, but future river flows and groundwater recharge are unpredictable and uncertain in the mid and long term. It seems probable that the variability of river flows will increase in the coming decades because of global warming. This could cause more intense floods and droughts. The latter has two important consequences: (1) it will be necessary to have a greater water storage capacity, where conjunctive use can be useful and (2) it will be vital to conduct more simulations of the behavior of each system, with different operating rules, for dry and wet time series and periods as well as for more intense droughts.

## 3.1 Methods of Analysis

This section will address how to analyze and evaluate conjunctive water use and system interactions between surface and subsurface components. The purpose is to present, in a very simplified way, what types of models and tools exist to simulate all the components, with emphasis on the interactions between groundwater and surface water. The simplest among the *lumped* (aggregated) models to reproduce the flow between a river and an aquifer is the deposit model. Next in complexity is the *unicellular linear model* in which the aquifer–river flow is equal to the volume stored above the riverbed multiplied by a factor of discharge. These models do not consider the spatial location of pumping and only calculate the total river flow into the aquifer. *Pluricellular models* (based on the few analytical solutions for simple geometries of the groundwater flow equation for homogeneous and isotropic aquifers) are more elaborate and consider pumping locations. Those models are composed of a few virtual cells. Each virtual cell corresponds to one of the infinite terms of the analytical solution. All cells behave as the unicellular model, each one has a discharge coefficient, and each different external action is partitioned between all the cells of the pluricellular model (Sahuquillo, 1983a,b, 1993). Partition coefficients are different for each distinct external action over the aquifer; pumping in A, B, or C, artificial recharge M, N, or P, etc., and must be previously calculated. In all cases, only a few terms (up to five in most cases) of the solution are needed for a sufficient approximation. The link to the surface model is very simple and straightforward and this is an interesting option when a distributed model is not available or for preliminary or screening modeling (Sahuquillo, 1993; Pulido-Velázquez et al., 2005).

For more detailed and exact results, the classical finite differences or finite elements numerical models can be used to solve a large algebraic linear system of equations in sequential time steps providing water heads in all the cells of the aquifer. However, its simultaneous simulation with the surface model is very cumbersome, even for analyzing a few management alternatives of several decades of duration. An important advance has been the use of the superposition strategy using influence functions when an aquifer behaves linearly (Maddock, 1971; Morel-Seytoux et al., 1973). However, for large models and modeling periods and many alternatives, it is needed to handle and store many influence functions and consider all the previous stresses, which in practice makes the models more difficult and their use more costly. To apply influence functions, the influence of all previous external actions has to be calculated and all previous external actions and each influence function have to be kept stored. Simple cases, like the river augmentation cases, river aquifer systems, or other not too complex systems, have been modeled successfully with lumped or distributed models. For more complex cases or major integrated systems, more comprehensive models are needed.

The *eigenvalue method* (an explicit solution of the groundwater flow equation in linear problems) is a more appropriate option for simulations of large time periods and many alternatives. This approach solves explicitly the space discrete flow equation obtaining modal orthogonal components through very simple explicit state equations in the function of time. The method allows the following to be explicitly obtained: river–aquifer flow interchange in selected zones, groundwater heads in selected points, and groundwater volumes in different areas of the aquifer or flows between two zones of an aquifer. The complete solution of the eigenproblem is the most demanding task, but it must be done only once (Sahuquillo, 1983a,b, 2013; Andreu and Sahuquillo, 1987). To reduce the computational load, the simulation can be simplified with appropriate truncation using only dominant modes of the components, at the cost of a small error. Efficient codes have been developed to get the modal components computed very effectively (Alvarez Villa, 2014).

AQUATOOL is a decision support system developed at the Polytechnic University of Valencia (UPV), Spain, to optimize and simulate complex systems, including conjunctive use. This model has been applied for many Spanish basins and in other countries (eg, Brazil and Morocco). It can handle several tens of dams, aquifers, and water demand areas, including rivers, canals, aqueducts, and aquifer–river interactions. It includes legal priorities of use and environmental restrictions, restrictions in conduits, and return flows to groundwater and surface water. It can tackle most common nonlinear situations. AQUATOOL is a continuing project at the UPV and therefore it is continuously upgraded and expanded. It is a model for simulating water resources systems, creating a flow network that is optimized monthly by an out-of-kilter algorithm. The system is based on operating rules optimizing dam and aquifer storage, and which can be refined in the simulation process. The model can be used in a water-resource system to screen, design, and manage alternatives, check and refine the screened alternatives by means of the simulation module, and conduct sensitivity analysis by comparing the results after changes in the design or in the operating rules. Furthermore, once an alternative is implemented, the models can be used as an aid in the operation of water resource systems. It has been designed to help decision makers to analyze complex systems and answer specific questions, facilitating the use of a set of models and databases in an interactive way and in a user-friendly control framework (Andreu et al., 1996). It can include deposit, unicellular, pluricellular, and eigenvalue models.

Another model of a similar kind is CALVIN (optimization model used to analyze different conjunctive use alternatives in California) developed at the University of California at Davis. The model determines monthly system operations and provides shadow prices and marginal benefits and costs of the operation of the system of water resources. It includes environmental restrictions and restrictions in conduits as well. It enables changes in water allocation through water markets, allowing the transfer of water from agricultural to urban uses and changes in crops and fallow. It takes into account savings in urban use, irrigation efficiency, reuse, and desalination. It includes aquifers as deposits and deep percolation from conveyance losses and rainfall. Stream–aquifer and interbasin interactions are preprocessed as a fixed time series of groundwater inflows and thus they are not dynamically represented in CALVIN. The software is based on an optimization solver for the water resource system called HEC-PRM, developed specifically to examine the economic operation of large water resource systems. The results of several studies indicate the physical possibility of an interconnected, complex, and diverse system to adapt to dramatic changes in climate and population, although at significant costs. The model has been applied to all California, but can be used for any complex system of water (Jenkins et al., 2004).

CALVIN formulates economic optimization and evaluation, which makes it a plausible tool considering water markets, surface storage, and groundwater banking, something that AQUATOOL does not provide. AQUATOOL instead may include aquifers and aquifer–river connections with the same accuracy as provided by distributed aquifer models at the expense of a moderate increase in the computational cost. Seemingly, both models can benefit from the other's advantages.

The operation of several multipurpose reservoirs is one of the most complicated and difficult tasks of water management. Different solutions have been suggested in such cases, depending on particular historical and socioeconomic conditions. The problem is even more complicated if interactions between surface water and groundwater and climate change are considered. So far in California, water agencies using their aquifers and surface reservoirs have managed their own and imported water resources independently. However, as discussed earlier, it appears clear that significant gains could be obtained if water resources of various agencies were managed jointly. At the beginning of the century, the CalSim II model was developed for the Central Valley in California that only deals with surface water. The model was criticized for missing groundwater representation and possible limitations in the model's applicability to planning, policy, and operational problems under future water management and hydrologic conditions. Integrated basin management models, which handle groundwater and surface water simultaneously, are needed to improve the performance of California's water management systems. In 2013 a groundwater model of the Central Valley aquifer and a conjunctive model of surface and groundwater (C2VSim) was developed. Improvements in the operation of the northern California reservoirs could contribute to enlarge the assignment of aquifers and reservoirs, and are currently under intense scrutiny (Lund et al., 2014).

## 4. CONCLUSIONS AND RECOMMENDATIONS

Conjunctive use of surface and groundwater provides possibilities of significant functional and economic improvements in water management and increasing water availability in general. Groundwater can provide an additional water supply as well as a means for water storage, distribution, and treatment, which can be combined advantageously with surface water sources. The yield of surface water projects can be augmented if groundwater is included and operated in the system jointly with surface components without a need to augment surface storage capacity. Some of the valuable aspects that can be expected from conjunctive use of surface water and groundwater, in addition to the increase in yields and economic gains, are: alleviation of drainage and salinity problems, alleviation of aquifer overexploitation, alleviation of seawater intrusion, and higher water supply reliability and smaller infrastructures. The enumerated possibilities lead on to the need for a fully integrating groundwater management system in all planning and management stages. Improvements of many schemes can be achieved, but adequate legal, institutional, and social changes would be needed in many cases. On the other hand, the analysis for implementing conjunctive use has to be performed carefully. This in turn implies the need for more complex models in view of the need to simulate surface water, groundwater, and river aquifer flow interchange for multiple scenarios and alternatives during longer modeling periods.

After analyzing multiple cases, the question arises: which methods are most adequate for evaluating artificial recharge or ACU? The answer will depend on economic, social, political, and engineering conditions, to mention just a few. It seems inappropriate to project long channels of interbasin transfers (as those in the CVP, or Los Angeles Aqueduct and others in California, or the CAP in Arizona) for the transport of highly variable flows. Rather, its storage through artificial recharge in receiving areas seems more plausible. By contrast, in river–aquifer cases, or in basins where water is conducted largely through a losing river channel or unlined channels (like the Castellón la Plana aquifer or the Indus River), ACU appears to be more adequate. However, environmental, social, historical, legal, and hydrological issues exist, which have largely influenced the decisions taken in each situation.

## ACKNOWLEDGMENTS

The author wishes to thank Professor Emilio Custodio for his constructive comments and suggestions that helped improve this manuscript. Thanks are given to the Department of Water Resources of California for permission to use Fig. 1.

## REFERENCES

Alvarez Villa, O.D., 2014. Simulación eficiente de la relaciones río-acuífero en sistemas de utilización conjunta mediante técnicas de reducción de modelos lineales invariables en el tiempo.

Andreu, J.Y., Sahuquillo, A., 1987. Efficient aquifer simulation in complex systems. Journal of the Water Resources Planning and Management Division, American Society of Civil Engineers 113 (1), 110–129.

Andreu, J., Capilla, J., Sanchis, E., 1996. Generalized decision support systems for wáter resources planning and management including conjunctive water use. Journal of Hydrology 117, 269–291.

Barlow, P.M., Ahlfeld, D.P., M.ASCE, Dickerman, D.C., 2003. Conjunctive-management models for sustained yield of stream-aquifer systems. Journal of the Water Resources Planning and Management Division, American Society of Civil Engineers 129, 35–48.

Bredehoeft, J.D., Young, R.A., 1983. Conjunctive use of groundwater and surface water for irrigated agriculture: risk aversion. Water Resources Research 19, 1111–1121.

Burke, J.J., Moench, M.H., 2000. Groundwater and Society. Resources, Tensions and Opportunities. United Nations Department and Social Affairs and Institute for Social and Environmental Transition, 170 pp.

Chaturvedi, Srivastava, 1979. Induced groundwater recharge in the Ganges basin. Water Resources Research 15-5, 1156–1166.

Coe, J.J., 1990. Conjunctive use advantages, constraints, and examples. Journal of Irrigation and Drainage Engineering, American Society of Civil Engineers 116, 427–443.

Custodio, E., Llamas, M.R., 2003. Intensive use of groundwater: introductory considerations. In: Llamas, M.R., Custodio, E. (Eds.), Intensive Use of Groundwater. Challenges and Opportunities. A.A. Balkema Publishers.

Custodio, E., Isamat, F.J., Miralles, J.M., 1969. Twenty Five Years of Groundwater Recharge in Barcelona (Spain). In Int. Symposium on Artificial Groundwáter Recharge. Dortmund. Paper 10. DVWK. Bull.11. Verlag Paul Parey Hamburg-Berlin, pp. 171–192.

Custodio, E., 1986. Recarga artificial de acuíferos. Boletin de Informaciones y Estudios n° 45 Servicio Geológico. Ministerio de Obras Públicas y Urbanismo, 149 pp.

Downing, R.A., Oakes, D.B., Wilkinson, W.B., Wright, C.E., 1974. Regional development of groundwater resources in combination with surface water. Journal of Hydrology 22, 174–177.

Foster, S., Van Steenbergen, F., 2010. Conjunctive groundwater use: a 'lost opportunity' for water management in the developing world. Hydrogeology Journal 2011 (19), 959–962.

Foster, S., Steenbergen, F., van Zuleta, J., Garduno, H., 2010. Conjunctive Use of Groundwater and Surface Water: From Spontaneous Coping Strategy to Adaptive Resource Management. GW-MATE Strategic Overview Series 2. World Bank, Washington, DC. www.worldbank.org/gwmate.

Hanak, E., Lund, J., Dinar, A., Gray, B., Howitt, R., Mount, J., Moyle, P., "Buzz" Thompson, B., 2011. Managing California's Water. From Conflict to Reconciliation. Public Policy Institute of California, 483 pp.

Harou, J.J., Lund, J.R., 2008. Ending groundwater overdraft in hydrologic-economic systems. Hydrogeology Journal 2008 (16), 1039–1055.

Heikkila, T., Blomquist, W., Schlager, E., 2001. Institutions and conjunctive water management among three western states. In: Tenth Biennial Symposium. Artificial Recharge of Groundwater. Tucson, Arizona. June 2001.

Jaquette, D.J., 1981. Efficient water use in California: conjunctive management of ground and surface reservoirs. In: Beard, L.R. (Ed.), Water for Survival. Journal of Hydrology 51, 187–204.

Jenkins, M.W., Lund, J.R., Howitt, R.E., Draper, A.J., Msangi, S.M., Tanaka, S.K., Ritzema, R.S., Marques, G.F., 2004. Optimization of California's water system: results & insights. Journal of Water Resources Planning and Management 130 (4), 271–280.

Kats, D.M., 1975. Combined Use of Groundwater and Surface Water for Irrigation. Soviet Hydrology. Selected papers. Published by the American Geophysical Union, International Water Resources Ass. & American Water Resources Association, pp. 190–194.

Khan, M.R., Voss, C.I., Yu, W., Michael, H.A., 2014. Water resources management in the Ganges basin: a comparison of three strategies for conjunctive use of groundwater and surface water. Water Resources Management 28, 1235–1250.

Kretsinger, G., Narasimhan, T.N., 2006. California's evolution toward integrated regional water management: a long-term view. Hydrogeology Journal 14, 407–423.

Lund, J., Munévar, A., Taghavi, A., Hall, M., Saracino, A., 2014. Integrating Storage in California's Changing Water System. University of California at Davis.

Maddock, T.D., 1971. Algebraic Technological functions from a simulation model. Water Resources Research 8 (2), 129–134.

Morel-Seytoux, H.J., Young, R.A., Radosevich, G.E., 1973. Systematic Design of Legal Regulation for Optimal Surface Groundwater Usage – Phase 1. Colorado State University, Environmental Resources Centre report no. 53, 81 pp.

Mukherji, A., Shah, T., 2005. Socio-ecology of groundwater irrigation in South Asia: an overview of issues and evidence. In: Sahuquillo, A., Capilla, J., Martinez-Cortina, L., Sanchez-Vila, X. (Eds.), Groundwater Intensive Use. A.A. Balkema Publishers.

Pulido-Velázquez, M., Sahuquillo, A., Ochoa, J.C., Pulido-Velázquez, D., 2005. Modeling of stream-aquifer interaction: the embedded multireservoir model. Journal of Hydrology 313 (3–4), 166–181.

Pyne, R.D.G., 1989. Aquifer storage recovery: a new water supply and ground water recharge alternative. In: Johnson, A.I., Finlayson, D.J. (Eds.), Proceedings of the International Symposium on Artificial Recharge of Ground Water. American Society of Civil Engineers, New York, pp. 107–121.

Revelle, R., Lakshminarayana, V., 1975. The Ganges water machine. Science 188, 611–616.

Sahuquillo, A., 1983a. Modelos pluricelulares englobados. In: Varela, M. (Ed.), En Utilización conjunta de aguas superficiales y subterráneas. Servicio Geológico de Obras Públicas y Universidad Politácnica de Valencia 1983.

Sahuquillo, A., 1983b. An eigenvalue numerical technique for solving unsteady groundwater models continuously in time. Water Resources Research 12, 87–93.

Sahuquillo, A., 1985. Groundwater in water resources planning: conjunctive use. Water International 10 (2), 57–64.

Sahuquillo, A., 1993. Simulación de Acuíferos en modelos de gestión. In: Andreu, J. (Ed.), Conceptos y métodos para la planificación hidrológica. CIMNE, Barcelona. Spain.

Sahuquillo, A., 2000. La utilización conjunta de las aguas superficiales y subterráneas en las sequías. Real Academia de Ciencias Excactas, Físicas y Naturales. Monográfico "Aguas Subterráneas y Sequías" 94 (2) (In Spanish).

Sahuquillo, A., 2002. Conjunctive use of surface water and groundwater. In: Silveira, L. (Ed.), UNESCO Encyclopedia of Life-Support Systems. EOLSS, Cambridge, UK.

Sahuquillo, A., 2005. Strategies for the conjunctive use of surface and groundwater. In: Andreu, J., Rossi, G., Vagliasindi, F., Vela, A. (Eds.), A. Sahuquillo in Drought Management and Planning for Water Resources. Taylor and Francis. http://www.amazon.es/Drought-Management-Planning-Water-Resources/dp/1566706726/ref=sr_1_1/276-1601614-1966150?s=foreign-books&ie=UTF8&qid=1433868035&sr=1-1.

Sahuquillo, A., 2013. Potential and merits of the eigenvalues method to simulate the conjunctive use of groundwater and surface water. British Journal of Environment and Climate Change 3 (3), 480–498. SCIENCEDOMAIN International.

Shepley, M.G., Streetly, M., Voyce, K., Bamford, F., 2009. Management of stream compensation for a large conjunctive use scheme. Water and Environment Journal 23 (2009), 263–271.

Sondi, S.K., Rao, N.H., Sarma, P.B.S., 1989. Assesment of groundwater potential for conjunctive water use in a large irrigation project in India. Journal of Hydrology 107, 283–296.

State of California Department of Water Resources, 1957. Bulletin No 3. The California Water Plan.

State of California Division of Water Resources, 1930. Bulletin 25. Report to Legislature of 1931 on State Water Plan 1930.

State Water Resources Control Board and the Department of Water Resources, 1982. State of California Policies and Goals for California Water Management. The Next 20 Years, Joint Report.

Task Committee on Water Conservation, 1981. Perspectives on Water Conservation. Journal of the Water Resources Planning and Management Division. ASCE 107(1). Water Resources Planning Committee. Water Resources Planning and Management Division.

UK Groundwater Forum, 1998. In: Downing, R.A. (Ed.), Groundwater Our Hidden Asset. Wallingford, UK.

Van Steenbergen, F., Oliemans, W., 2002. A review of policies in groundwater management in Pakistan 1950–2000. Water Policy 4 (4), 323–344.

Wurbs, R.A., 1987. Reservoir management in Texas. Journal of Water Resources Planning and Management 113 (1), 130–131.

Young, R.A., Bredehoeft, J.D., 1972. Digital computer simulation for solving management problems of conjunctive groundwater and surface water systems. Water Resources Research 8, 533–556.

Chapter 3.2.4

# Product, Process, and Organizational Innovations in Water Management

B. Ziolkowski[1], J.R. Ziolkowska[2]

[1]*Rzeszow University of Technology, Rzeszow, Poland;* [2]*The University of Oklahoma, Norman, OK, United States*

## 1. INTRODUCTION

Effective and innovative water management is highly relevant for sustainable development in each economy. Because of decreasing water availability and growing water demand at the same time, it is challenging to provide sufficient water supply at affordable rates, while protecting water resources from depletion. Furthermore, extreme droughts can cause spikes in food prices, and consequently famine and death in extreme cases like, for example, in Ethiopia in 1975 (analyzed extensively by Amartya Sen, the 1998 Nobel Prize winner in Economics) (Marsily, 2007).

While water quantity has been among the main issues of scientific, social, and policy debates for a long time, low water quality and water-borne diseases have also become a concerning issue. This has been substantiated by changing publication trends regarding water quantity and water quality. Historical publication records in Elsevier databases show that between 1838 and 2016 the number of international publications addressing water quality has exceeded publication numbers on water quantity by 28%.[1] A spike in publications on water quality and quantity started in 2007 and 2008, when almost half of all papers published in Elsevier journals focused on those two issues.

While in the past decades innovations were prevailing in the field of water quantity to ensure a proper and adequate water supply, innovations in water quality started to emerge in the most recent decade. One of the reasons for this development was the extent of global diseases and limited human abilities with their treatment and cure.

1. The statements are based on the results of the authors' bibliographic survey in Elsevier data bases conducted on July, 18, 2015 which included both "water quality" (as a search term in the titles, abstracts, and keywords) and water quantity (embracing the following categories: water quantity, water sources, and water resources).

**Competition for Water Resources. http://dx.doi.org/10.1016/B978-0-12-803237-4.00023-9**

One of the examples related to water quality and missing innovations in this area in the past is diarrhea (and other "fecal–oral" diseases like cholera, typhoid, and dysentery). According to the World Health Organization, prior to 2008 approximately 88% of those diseases originated from impure water, inadequate sanitation, or insufficient hygiene (Prüss-Üstün et al., 2008). Only in the last decade, dirty water and poor sanitation caused annual deaths of 1.4–1.8 million children (nearly 5000 children every day) (UNESCO, 2009, 2012). Since the beginning of the 21st century around 3.5 million people have been dying every single year because of inadequate water supply, sanitation, and hygiene, mainly in the developing countries (UNESCO, 2012; WHO, 2008).

Neither water quality nor water quantity in Europe has been as acute as in the developing countries of Africa, South America, and Asia. However, despite well-established laws and regulations in place determining water resource management and environmental protection (Martins et al., 2013), recent droughts significantly affected water availability and water withdrawals in some European countries. An indicator of water supply per river basin in 1995 and anticipated in 2025 shows that the countries most affected (or expected to be most affected) by water scarcity[2] were France, Spain, Italy, Greece, the United Kingdom, and the Netherlands (UNEP, 2008). At the same time, an opposite trend and the most excessive water use was recorded in Malta, Bulgaria, Germany, Italy, Poland, Spain, and Ukraine (UNEP, 2008).

Given the multitude of water management innovations established and successfully applied in many EU countries it might seem surprising that no effective solutions have been found yet that would help with mitigating and adapting to extreme weather events (here: droughts) at a larger scale. The reason for this is a clash of high innovativeness in the water sector on the one hand, and existing water problems on the other, which can be interpreted as an example of ineffective water management practices. Those contrary tendencies are common in many countries around the world, where concerted collaborations of the water industry organizations to improve water management efficiency are missing, while strategic visions are very rare (Matheson, 2013). Moreover, missing knowledge about existing innovations constitutes one of the main barriers for improving water management practices and strategies at the regional scale and transferring them to the national level.

The main aim of this chapter is to present various types of the most current water management innovations applied in different countries and/or enterprises in Europe. It is important to emphasize that different water management innovations are dispersed across different European countries. Thus the landscape of water management innovations and practices is unique for each single country. The examples of water management innovations presented here are not geographically limited and could theoretically be applied in all regions around the globe. However, socioeconomic, environmental, technological, political, and juridical conditions in each country will be determinant for the final implementation of specific water management innovation practices. The presented examples represent three types of innovations: product, process, and organizational innovations. The next sections define the nature of innovations. They also present selected examples of innovation approaches used in water management in Europe as well as the results of representative research studies conducted in this field.

2. Water scarcity occurs when the annual water availability is less or equal to 1000 $m^3$ per capita.

## 2. NATURE AND TYPES OF INNOVATIONS

The very first notion of innovations was created more than 100 years ago. According to Unger (2005), the term "innovation" was defined for the first time in the 19th century by A.F. Riedel (an economist) (Riedel, 1839) and G. De Tarde (a sociologist) (De Tarde, 1993). In the following years, the Oslo Manual (Guidelines for Collecting and Interpreting Innovation Data) defined an innovation as "a new or significantly improved product (good or service), or process, a new marketing method, or a new organizational method in business practices, workplace organization or external relations" that has been implemented in practice/on the market to improve efficiency or effectiveness (OECD, 2005).

According to the theory of J.A. Schumpeter (1883–1950), innovations are attributed to the so-called "creative destruction" that is the main element in the theory of economic development. The term "creative destruction" was defined as a process of creating new values on the basis of the older ones that ultimately needed to exit the market. Thus between the 19th and 20th centuries, innovations were understood as new combinations of production factors (Matusiak, 2011; Schumpeter, 1939). Further, according to Schumpeter's theory, knowledge about structural market changes resulting from the implementation of innovations should be used as a base for designing appropriate adjustment programs. These programs ought to support the process of market transformation and, where possible, reduce or eliminate negative factors impacting organizations, individuals, and society (Schumpeter, 1995). The real destruction of many enterprises (Schumpeter, 1995) after the introduction of innovations on the market became a core element for a new theory of the economic growth [next to other new theories created at the same time by, for instance, J. Schmookler and K.H. Oppenländer (Matczewski, 2005)]. All those theories generated a potential for new technologies and thus creation of an additional market value. Currently, innovation theory is still in the development stage. It distinguishes among different types of innovations and their nature, subject to different criteria (Table 1).

The introduced interpretations present the nature of the respective innovation types, while they do not describe all aspects of the terms. The most common innovation typology in Europe relates to product, process, marketing, and organizational innovations (OECD, 2005). It is based on the following guidelines for collecting and interpreting innovation data (OECD, 2005):

1. *Product innovation* is a good or service that is new or significantly improved with respect to its characteristics or intended uses. A significant improvement includes technical specifications, components and materials, software, user friendliness, easiness of use, or other functional characteristics, for example, speed and performance.
2. *Process innovation* is a production process or delivery method that is new or significantly improved. A significant improvement in this case includes changes in techniques, equipment, or software [eg, new production line, radio-frequency identification (RFID) or global positioning system (GPS), new reservation systems or techniques for project management, information, and communication technology], which help improve efficiency or quality of an ancillary support activity.
3. *Marketing innovation* is defined as a marketing method that is new or significantly improved. It relates to changes in the product design, packaging, placement, promotion or pricing, which are all aimed at satisfying customers' needs or opening up new markets.

**TABLE 1 Types and Nature of Innovations**

| No. | Criterion | Types of Innovations | Nature of Innovations |
|---|---|---|---|
| 1 | Form of innovation | 1. Product innovation<br>2. Process innovation<br>3. Marketing innovation<br>4. Organizational innovation | 1. Product innovation is a new or significantly improved good or service<br>2. Process innovation is a new or significantly improved production process or delivery method<br>3. Marketing innovation is defined as a new or significantly improved marketing method<br>4. Organizational innovation is a new organizational method (OECD, 2005) |
| 2 | Orientation for law | 1. Codified innovation<br>2. Uncodified innovation | 1. "Codified innovation refers to all technological capabilities listed, codified, and legally identified through patents and concessions" (Lecerf, 2012)<br>2. "Uncodified innovation is the kind of innovation that happens as services and products are adapted to the needs and preferences of a new set of customers" (Saporito, 2008) |
| 3 | Place where innovations are created | 1. Exo-innovation<br>2. Endo-innovation | 1. Exo-innovation is an innovation that overthrows established practices in a field<br>2. Endo-innovation is an innovation that relies on and conforms to established boundaries (Rehn, 2006) |
| 4 | Distance to customer | 1. Intra-innovation<br>2. Market innovation | 1. Intra-innovation is the own innovation capacity of an organization (Malinsky, 2010)<br>2. Market innovation is improving the mix of target markets/customers and how these are served (Axel, 1999a,b) |
| 5 | Originality | 1. Original innovation<br>2. Modified innovation | 1. Original innovation is an innovation created for the first time, which has never been proposed before (Yixin, 2011)<br>2. Modified innovation is a slight improvement in technology and process, which creates new ways of producing or servicing (Joshi, 2010) |
| 6 | Dimension of change | 1. Radical (revolutionary) innovation<br>2. Incremental (evolutionary) innovation | 1. Radical innovation is a revolutionary change (eg, new process, product), which requires changes in the technology or organization (Ravn and Petersen, 2005)<br>2. Incremental innovation is a gradual and often small change in a company or a sector (Søndergaard et al., 1997) |

**TABLE 1 Types and Nature of Innovations—cont'd**

| No. | Criterion | Types of Innovations | Nature of Innovations |
|---|---|---|---|
| 7 | Grade of autonomy | 1. Autonomous innovation<br>2. Systemic innovation | 1. Autonomous innovation can be implemented independently from other innovations, that is, without the need to completely redesign other innovations<br>2. Systemic innovation operates only with complementary innovations because it is dependent on them (Chesbrough and Teece, 1996) |
| 8 | Grade of materialization | 1. Hardware innovation<br>2. Software innovation<br>3. Mixed innovation | 1. Hardware innovation is created mainly in applied and life sciences and has a material form<br>2. Software innovation is created mainly in social sciences and has a nonmaterial form<br>3. Mixed innovation is the combination of both<br>(Zajączkowski, 2003) |
| 9 | Costs of creation | 1. Cheap innovation<br>2. Expensive innovation | 1. Cheap innovation does not require many resources for research and development<br>2. Expensive innovation requires many resources in the creation phases of research and prototyping (Zajączkowski, 2003) |
| 10 | Field of impact | 1. Nano-innovation<br>2. Bio-innovation<br>3. Genetic innovation<br>4. Social innovation<br>5. Eco-innovation/ sustainable innovation<br>6. Effectition | 1. Nano-innovation is a novelty or improvement in nanotechnology<br>2. Bio-innovation relates to new or improved biotechnological solutions<br>3. Genetic innovation is a novelty in genetics<br>4. Social innovation is a new solution of social problems (Phills et al., 2008)<br>5. Eco-innovation is a new or modified process, technique, practice, system, or product, which allows avoiding or reducing environmental harm (Kemp et al., 2004)<br>6. Effectition is an innovation with no ("zero") negative environmental impact. Because humans are a part of the economy and society spheres, the 100% effectiveness in reducing environmental harm equals a 100% reduction in social and economic harm (Ziółkowski, 2012; Ziolkowski and Ziolkowska, 2014) |

Authors' presentation based on Zajączkowski, M., 2003.

4. *Organizational innovation* is an organizational method that is new for entrepreneurial practices, workplace organization, or external relations. In contrast to a process innovation, an organizational innovation relates primarily to people and aims at reducing administrative or transaction costs, improving workplace satisfaction, and gaining access to new tradable assets such as knowledge to improve entrepreneurial performance. Organizational innovations are reflected, for instance, in the following:
   a. Enterprise practices: improvement in learning within an organization (eg, databases of best practices), practices for employee development (eg, internal education and training systems), management systems for general production or supply operations (eg, supply chain management systems, business reengineering, lean production, and quality management systems);
   b. Workplace organization: methods for delegating decision-making to employees, integration of different business activities, organizational models for employees' empowerment, decentralization of group activities, management control, establishment of formal or informal work teams, and integration of engineering and development with production; and
   c. External relations: ways of organizing relations with other bodies (eg, establishment of new types of cooperation with customers or R&D, new methods of integration with suppliers, and outsourcing).

The contemporary science trends reveal many different definitions and classifications of the aforementioned innovation types. For instance, product innovations are identified with extraction innovations, process innovation are used as a synonym for resource innovations, organizational innovations are compared with strategic innovations or system innovations, while process innovations can be defined as an element of system innovations. According to some researchers (eg, Murphy and Gouldson, 2000; Bernauer et al., 2006), all organizational innovations support the implementation of technical innovations, that is, process innovations and product innovations. Moreover, it is a very common scientific practice to describe different types of innovations by means of the term "technologies." Thus product innovations or marketing innovations are also called technologies.

Another relevant aspect of the innovation theory is the fact that many innovations could be simultaneously attributed to more than just one innovation type (eg, product and process innovations) (OECD, 2005). This phenomenon is also confirmed by the numerous definitions and classifications of innovations. Therefore, in many cases it is challenging to specify and clearly define a single form of innovation in place (Freier, 2003), because a product innovation for one enterprise can mean a process innovation for another enterprise. Thus from the theoretical and practical point of view, the novelty of innovations at the enterprise, market, and global levels is a decisive factor for determining the innovation form. When a technology is introduced into a company for the very first time, it is called innovation in that specific company, although it is not necessarily an innovation for other market players that have been using the same solution for more than one year (OECD, 2005).

The process of innovation development has occurred in several stages, which are called "innovation waves," phases of Industrial Revolution or Industrial Revolutions. According to Hargroves and Smith (2006), the first innovation wave was triggered by a swift uptake of heavy industry around 1785. After the age of iron manufacturing, water power, textiles, mechanization, and commerce, the second industrial revolution occurred around the year

1800 and was triggered by a proliferation of steam power, trains, and steel and cotton production. During the third wave, around the year 1900, the dominant innovations resulted from the discovery and a wide use of electricity, chemicals, and the internal combustion engine (the era of automobile production and electrification of cities). The fourth innovation wave started around 1945 and embraced the mass production of semiconductors and electronics, the rise of petrochemicals, the aviation and space race, as well as new energy sources. During the fifth industrial revolution the main determinants of growth were: information and communication technologies (ICT) with computers, digital networks, and bio- and nanotechnologies. The present times are classified as the sixth wave of innovations and are represented and voiced with the following concepts: sustainability, radical resource productivity, whole system design, biomimicry, green chemistry, industrial ecology, renewable energy, green nanotechnology, etc. (Hargroves and Smith, 2006; Lovins, 2008).

The enumerated innovations have been successfully applied in many fields and disciplines. However, considering current water problems, it can be deduced that innovations in water management have been lagging behind. At the same time, scientific and geopolitical interest in water management innovations has been steadily growing. The main reason for this development is the global population growth and extreme weather events, like droughts, heat waves or floods around the world. Despite the growing need for innovations in water management, many common innovations are not disseminated as widely as needed in many regions of the world, which can be explained by missing knowledge about existing innovations and their positive effects.

To provide a comprehensive picture of innovations and their successful applications in the field of water management, the following section will focus on three innovation types in the European Union.

# 3. INNOVATIONS IN WATER MANAGEMENT IN EUROPE

## 3.1 Legal and Organizational Background of Water Management

In the European Union, innovations in the water sector started to expand in the late 1970s as a result of legal regulations related to environmental problems. Numerous European Commission treaties stipulated the creation of water policies, regulations, and laws (Dworak et al., 2007), and served as a jump start for innovative water management practices. Table 2 presents three regulatory periods in EU water policy creation.

The regulations mentioned in Table 2—combined with thousands of additional water policy Acts—have significantly impacted water innovations in the business and administration sectors in European countries and generated an added value since the end of the 1970s. It is difficult to assess the strength or character of this impact because research is missing that would provide specific measurement methods to evaluate the significance or success of all regulations in place. Simultaneously, the areas subordinated to these regulations determined the creation and direction of the respective innovation types. The most relevant and widespread organizational and institutional measures and frameworks for respecting water laws in Europe and creating innovative solutions include, among others: the European Innovation Partnership on Water, Water supply and sanitation Technology Platform (WssTP), European Water Conferences,

**TABLE 2** Regulations Constituting Water Policy in the European Union

| First Period (1973–1989) |
|---|
| Environmental Action Programs in 1973 |
| Surface Water Directive (75/440/EEC) |
| Bathing Water Directive (76/160/EEC) |
| Dangerous Substances Directive (76/464/EEC) |
| Fish Water Directive (78/659/EEC) |
| Shellfish Water Directive (79/923/EEC) |
| Drinking Water Directive (80/778/EEC) |
| **Second Period (1990–1996)** |
| Urban Wastewater Treatment Directives (91/271/EEC and 98/15/EEC) |
| Nitrate Directive (91/676/EEC) |
| Directive Concerning Integrated Pollution Prevention and Control (96/61/EC) |
| Guideline to Control the Dangers in the Event of Major Accidents (96/82/EEC, the so-called Seveso II Directive) |
| **Third Period (1996 until present)** |
| A Communication of the European Commission on the Water Policy of the Community in 1996 |
| Water Framework Directive (2000/60/EC) |
| Groundwater Directive (2006/118/EC) |
| Floods Directive (2007/60/EC) |
| Marine Strategy Framework Directive (2008/56/EC) |
| Directive on Environmental Quality Standards in the Field of Water Policy (2008/105/EC) |

Authors' presentation based on Dworak, T., Kampa, E., de Roo, C., Alvarez, C., Bäck, S., Benito, P., 2007.

EUREKA Cluster for water, European water programs and strategies, and the Water Information System for Europe. All these initiatives foster water-related innovations and thus provide useful knowledge about a wide spectrum of different solutions for effective water management.

An example worth discussing in this context is the WssTP (also called the European Technology Platform for Water), initiated by the European Commission in 2004. The activities encompassed by this platform to promote water-related innovations were brought to special attention in 2014, when the WssTP SME awards (for small and medium-sized enterprises) were established for the EU member countries and EU associated countries.

The spectrum of water-related innovations evaluated in this competition was determined in the following WssTP Working Groups and areas (WssTP SME Awards, 2015):

1. *Agriculture and Irrigation*, which aims at advanced technologies for a rational use of water and agrochemicals in precision agriculture (including sensors for soil water content and evapotranspiration measurements, machinery, and agricultural runoff management).
2. *Bathing Water*, focused on the European Bathing Water Directive implementation by research initiatives and demonstration projects.
3. *Eco-systems Services*, promoting protection and restoration of ecosystems, ecosystem services, provision of Payments for Ecosystem Services, EU Green Infrastructure Communication, and alternative wastewater solutions.
4. *Emerging Compounds*, aiming at information exchange in the field of technological developments and in the context of emerging compounds or building joint research and innovation agendas to facilitate international consortia.
5. *Fin4WatComp*, which stands for Financing for EU Competitiveness and aims at supporting actions for internationalization of innovations in the water sector.
6. *Green Infrastructure*, promoting green infrastructure innovations for local, regional, river basin, and coastal environments.
7. *Hydro-climatic Extremes*, addressing management and predictions of hydroclimatic extreme events.
8. *Membrane Technologies*, supporting creation and promotion of breakthrough water membrane inventions.
9. *Resource Recovery*, supporting projects for sustainable resource (nutrients) recovery from water.
10. *Shale Gas*, aiming at new knowledge delivery and exchange about the best innovations to avoid water contamination during shale gas mining and to identify future research needs.
11. *Techwatch* (Emerging Technologies), focused on identifying promising innovations in all WssTP Working Groups to accelerate their commercialization.
12. *Urban Water Pollution*, oriented toward challenges of water pollution sources in urban environments from the industry sector, agriculture, or wastewater treatment.
13. *Water Beyond Europe*, with its main objectives of creating networks of key stakeholders in the water sector and working out common action plans with shared vision and goals facing the call of Horizon 2020 (EU Framework Program for Research and Innovation).
14. *Water and ICT*, promoting applications of ICT in the European water sector and creating worldwide reference standards in this domain.
15. *Water and Industry*, focused on creating ideas/topics, strategies, and system approaches on behalf of the European water industry (including water supply/use/reuse/recycling, wastewater treatment, and resource recovery from water) to be considered for future program calls.
16. *Water—Energy—Food—Biodiversity Nexus*, oriented toward the development of green economy innovations in the field of water—energy—food—biodiversity, mainly by means of open platforms for modeling, information exchange, or joint activities supporting a smooth research and innovation uptake.

The next sections depict three selected types of water management innovations: product innovations, process innovations, and organizational innovations. Several examples were used, based on the trade or manufacturers' information data to present a selective overview of the scope of water management innovations in Europe. As shown in the following, some innovations resemble solutions from the field of effectitions that do not generate any harm for the economic, social, and natural environment.

## 3.2 Product Innovations in Water Management

As described previously, the way in which technologies are applied determines the type of innovations. Some technologies viewed as product innovations in some places can be understood as process innovations in other places. Similarly, the application areas of new technologies as well as the geographical location determine whether some solutions are acknowledged as innovations in the first place. While some technologies in Europe may be new, they may be commonly applied in the United States or other countries in the world and vice versa.

The prevailing trend in the field of water management innovations regards water savings. Water metering is only one of many examples that have spread quickly in many countries in recent years. Among the known product innovations for water metering, the Swiss Amphiro Smart Water Meters represent an interesting example of innovation progress occurring in water management (Amphiro AG, 2014, 2015). The Amphiro meters are mounted at the outlet of water installation systems and embrace four functionally independent, but visually quite similar technologies under the following trade names: amphiro a1, amphiro a2, amphiro b1, and OEM amphiro (Picture 1).

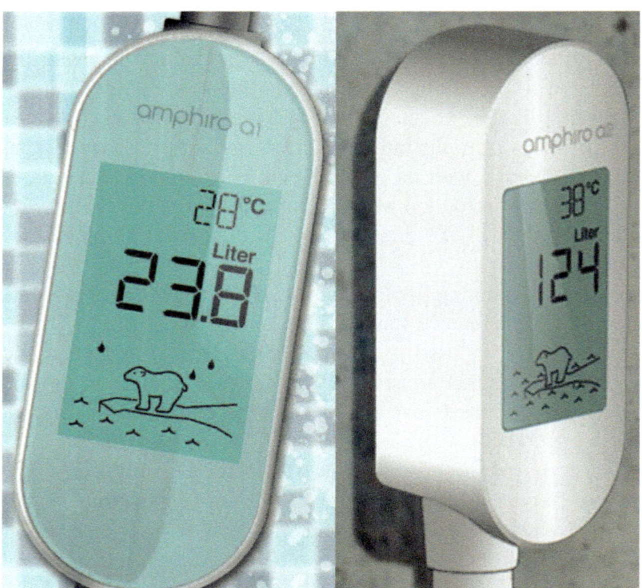


PICTURE 1 Amphiro a1 and amphiro a2. *Adapted from Amphiro AG, 2014 and Amphiro AG, 2015.*

The meters use a set of microprocessor units and sensors, powered by energy produced by water-flow driven generators, to measure the temperature and water consumption levels. The Amphiro meters allow for combining information on people's behaviors with information on energy production and metering. Thus information can be generated about individual user's energy and water consumption for different household activities (eg, taking a shower, laundry, lawn watering) or at different times during the day and night. The information is displayed electronically for all components separately on a display screen. In addition to the information on energy used for water heating and the actual amount of water consumed during a shower time, information on the shower energy efficiency class can also be measured and displayed on a scale from A (the highest) to G (the lowest). The collected information can be manually computed into Amphiro's online portal and compared over a timeline or with other households' information. The collected data can also be transferred via the Bluetooth 4.0 technology to mobile phones and to the Internet (amphiro b1). According to the producer, the use of amphiro a1 in a typical household saves 440 kWh of energy per year (from the 2000 kWh used for heating water) and 8500 L of drinking and wastewater (Amphiro AG, 2014). On the other hand, the OEM amphiro uses different modules that can be integrated with water systems produced directly by faucet manufacturers. The single modules differ with regard to their functionality levels in water monitoring. Compared to the aforementioned amphiro components, the OEM solutions can additionally monitor water quality (eg, for presence of *Legionella bacterium*) and transfer this information through Bluetooth 4.0 or WLAN to help with disease prevention and control. This product innovation could be especially useful in hotel management and in enterprises launching their energy or water services on the regional markets (Amphiro AG, 2013).

Another example of product innovations in the field of water management relates to water retention and collection in lakes, ponds, drinking water reservoirs, irrigation reservoirs, water treatment plants, cooling towers, and swimming pools. These tanks are exposed to a constant risk of a biological invasion by different algae species, with green and blue-green algae (toxic cyanobacteria) creating the biggest problem. Algae toxins distort the taste and smell of water, and they can also cause diseases and damage industrial clarifiers or sand filters. A very effective product innovation to address those issues was created based on a cooperation agreement of several universities and SME partners from the European countries under the leadership of a Dutch enterprise specialized in the development of the ultrasound technology. The outcome of this cooperation was the MPC-Buoy combining an innovative device and technology for monitoring water quality as well as for predicting and controlling algae blooms (LG Sonic, 2015). The devices are assembled on a buoy-like platform and powered by solar panels (Picture 2).

In the first stage, water quality monitoring is conducted to detect the presence of green algae (chlorophyll) and blue-green algae (phycocyanin) as well as the pH, total suspended solids (TSS),[3] dissolved oxygen, and temperature. This information is transmitted instantly to a web application by means of GPS, wireless computer networking (Wi-Fi), or 3G network. Based on this information the LG Sonic e-line (ultrasound transmitters mounted on the platform) is activated to treat algae. The

3. Total suspended solids (TSS) are solid materials suspended in water and impacting its quality. TSS are reported in milligrams per liter of water (mg/L) (for more information, consult Murphy, 2015).

PICTURE 2 MPC-Buoy. *Adapted from Dutch Water Sector, 2014.*

information is also used in a next step to predict future algae blooms. The algae controlling process is conducted by means of the chameleon technology—ultrasonic technology using 12 treatment programs for different types of algae by modulating amplitude, frequency, signal duration, and the wave form of the ultrasound. Therefore, there is no risk of algae resistance to a single ultrasonic parameter. Ultrasound damages the internal areas of algae cells without breaking the cell walls, which prevents the release of algal toxins into water. The additional benefits of the MPC-Buoy technology include, for instance: high process efficiency (elimination of 90% of algae blooms), no application of chemicals, and safety for ecosystems (LG Sonic, 2015; LG Sound, 2015).

Another product innovation in water quality management is the Colifast technology introduced by a Norwegian biotechnology company. The technology was built to reduce the time necessary to evaluate microbial water quality. Water quality and safety is commonly tested by means of microbiological indicators (eg, presence of *Escherichia coli* bacteria). In the industry sector, water analyses are often performed retrospectively after water has already been supplied to the end users. The lack of instantly available knowledge on water quality might be detrimental for decision-making processes (European Commission, 2015a). The Colifast method provides information on water quality within 6–14 h, which is significantly faster than traditional methods that require 1–3 days (Colifast, 2015a). The technology comes in two different innovation categories: Colifast ALARM (At-line Automated Remote Monitor) and Colifast At-Line Monitor (CALM) detecting the total count of coliforms, thermos-tolerant coliforms, or *E. coli* samples collected at the programmed intervals (Picture 3). Colifast ALARM is used for drinking water and groundwater testing, whereas CALM is designed for surface water, wastewater, and industrial use water (Colifast, 2015a).

The technology is based on the detection of a fluorescence signal in the analyzed water samples. A higher signal indicates a high amount of different bacteria strains and thus a lower water quality. Colifast consists of three components: incubator reaction chamber, a flow injection pump system, and a detector system including a spectrometer and wavelength emitters. The entire system allows for bacterial growth and measures the amount of the *E. coli* enzyme indicating the concentration of a fluorescent product.

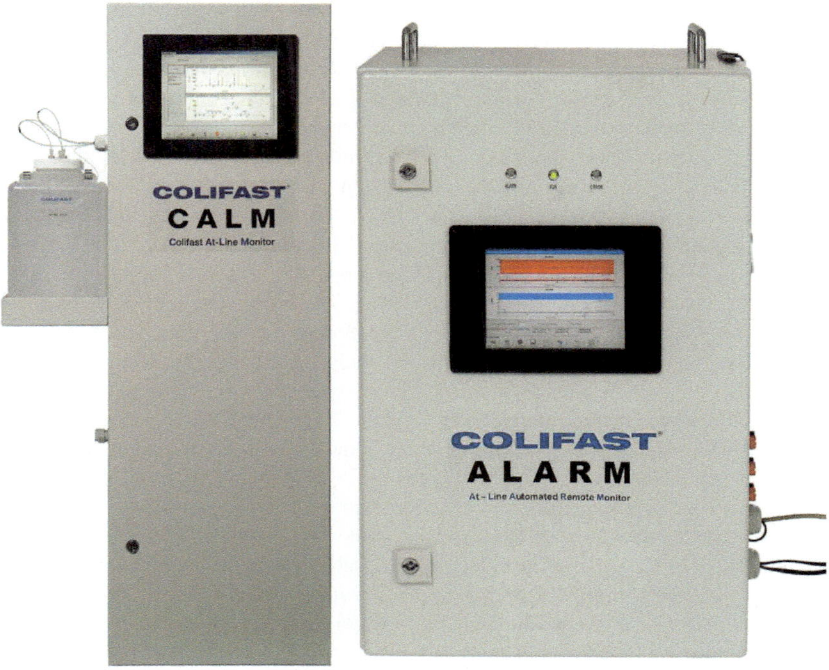


PICTURE 3 Colifast ALARM and Colifast CALM. *Adapted from Colifast, 2015b,c.*

The application of these innovations does not require either microbiological skills or laboratory facilities, while the only operational activity is limited to refilling reagents every 14–20 days. The results of the tests can be sent automatically via LAN, digital signals or the mobile phone network (text message). Because of the online enumeration of microorganisms by means of remote sampling and warning messages, the process of water quality analyses can be reduced to the possible minimum, which can bring about significant cost savings in the laboratory infrastructure, service provision, and transportation. The technology is beneficial not only for water testing, but also for the environment because of its low carbon footprint (Colifast, 2015a).

## 3.3 Process Innovations in Water Management

In some circumstances, very complex process innovations consisting of many different processes are also called system innovations. In most cases, however, process innovations are applied separately as single modules.

One of the process and product innovations worth mentioning here relates to wastewater treatment. OxyMem technology (also called membrane aerated biofilm reactor—MABR) is a mobile modular wastewater plant that is characterized by high technological innovativeness, technological efficiency (around 75% reduction in energy costs), and high effectiveness (perfect mobility) (OxyMem Technology, 2015; OxyMem Smart Aeration, 2015). The mobility feature of this product and process innovation allows for providing services in different areas and environmental conditions. The OxyMem reactors can be connected with each other, thus increasing the wastewater treatment capacity. The reduction in energy costs of biological wastewater treatment results from the smart aeration based on the membrane aerated biofilm and closed in a transport container. The membranes within the reactor are formed by tiny permeable silicone tubes mounted next to each other, similar to whalebone. They fill up the entire container, which allows for 95% of transfer efficiency in the aeration process (OxyMem Technology, 2015). The oxygen used for aeration of wastewater is transferred through those silicone tubes that are covered with a biofilm containing wastewater treating bacteria. This technology, quite different from the traditional wastewater treatment technologies, assures that oxygen is not released into the atmosphere, which also eliminates the unnecessary energy consumption. According to OxyMem (OxyMem Technology, 2015; OxyMem Technology-Brochure, 2015), additional efficiency parameters include: reduction in the sludge production by about 40–50%, no need to recycle sludge, resilience to changes in the wastewater influent, ability to manage the treatment capacity during increased or variable wastewater loading periods (especially in the industry), reduction in operating hours by about 50%, significant space savings after uplifting/elevating the reactor above the ground, fast and easy installation, and scalability to service any municipal or industrial requirement.

Another interesting process innovation in water management is the eutectic freeze crystallization (EFC) process invented at the Delft University of Technology in the Netherlands. While its initial purpose was to provide energy efficient water treatment, nowadays it is also used for desalination, clean water reclamation, and salt production that can further be used for industrial purposes. The material input in the process of freeze crystallization is aqueous inorganic waste stream. During the entire process performed in the vast range of temperatures between +50 and −50°C two different product streams are generated separately and simultaneously: very clean water in the form of ice and high purity salt. Depending on the process scheme, different sorts of salts can be mixed at the output stage or crystallized separately, which guarantees their high purity. The perfect salt purity corresponds to the complete salt crystallization, which provides the zero liquid discharge process.

Three operational EFC schemes have been applied in Europe so far:

- Single stage EFC based on one crystallizer working at a constant temperature and used for ice and salt crystallization;
- Multistage EFC combining several crystallizers located next to each other and operating at different temperatures; and
- EFC combined with pretreatment and/or after-treatment processes, which requires another auxiliary process (chemical treatment, evaporation, ion exchange, or osmosis) before the EFC process can be performed.

The capacity of a large-scale EFC installation ranges from 1 to 5 tons of crystallized ice and salt per hour. In comparison to the triple stage evaporation system, the EFC generates about 50% savings in total energy costs. The EFC process can also be used to treat wastewater from different sources. Currently, the following sectors are dominantly applying this technology: mining, petroleum and chemical industries, waste deposition and incineration, recycling industry, and agriculture. Moreover, the EFC process does not require the application of any chemicals, while the problem of wastewater deposition generates simultaneous benefits: salt and fully usable clean water (EFC Separations, 2015).

## 3.4 Organizational Innovations in Water Management

Organizational innovations presented in this section play a decisive role in implementing product and process novelties. Organizational innovations in the field of water management have probably been the fastest growing innovations in recent years. For many organizational innovations, implementation costs are among the most hindering factors. In many cases, organizational innovations do not generate direct costs, for example, when new project teams are created, new internal strategies are conceptualized, or quality circles[4] are developed. However, some other innovation types (eg, formal management systems like ISO standards) constitute a considerable cost component in an enterprise or organization budget, in spite of additional benefits they generate.

An example of a beneficial yet costly organizational innovation is Eco-Management and Audit Scheme (EMAS). This voluntary environmental management system was established by the European Union in 1993 and updated in 2001 and later in 2009. The core elements of EMAS are environmental management systems requirements under EN ISO 14001:2004, based on the Regulation of the European Parliament and the Council on the voluntary participation by organizations. The objective of this regulation is "to promote continuous improvements in the environmental performance of organizations by the establishment and implementation of environmental management systems by organizations, the systematic, objective and periodic evaluation of the performance of such systems, the provision of information on environmental performance, an open dialogue with the public and other interested parties and the active involvement of employees in organizations and appropriate training" (EC, 2009). Thus an overall improvement in resource management (including water) is relevant for many organizations registered in the EMAS records. Years of experience and implementation have shown that this voluntary organizational innovation has been successful also in the field of water management. Because of the implementation of EMAS rules, many enterprises have reduced operational water costs and were further awarded with the EU EMAS Award distinguishing innovative and environment-friendly actions (European Commission, 2015b). Some of these enterprises include:

- Regional Centre for Water and Wastewater Management Co. (Poland)
- Landeskrankenanstalten-Betriebsgesellschaft—KABEG (Austria)
- HR Björkmans Entrémattor AB (Sweden)
- Aeropuerto de Menorca (Spain)
- Abwasserverband Anzbach Laabental (Austria)
- Riechey Freizeitanlagen GmbH & Co. KG (Germany)
- Neumarkter Lammsbräu Gebr. Ehrnsperger e. K. (Germany)
- Lafarge Cement (United Kingdom)
- Municipality of Ravenna (Italy)

The new standard ISO 14046:2014 "Environmental management—Water footprint—Principles, requirements and guidelines" incorporated by EMAS is a promising next step to ensure a basis for improvement and creation of an effective water management system across the European Union. EMAS is both an organizational innovation generating benefits

4. Quality circles are employees groups meeting regularly who strive for solutions of quality problems in their organizations. They are often created as a part of quality management systems in organizations.

in water management and a powerful trigger of future organizational innovations next to product and process innovations.

Improvement of water management by enterprises and entities registered in EMAS is frequently determining adaptation of product, process, or organizational innovations by unregistered organizations. For example, because of the EMAS membership, Comune di Tavarnelle Val di Pesa (Italy) formulated and administered the following organizational water management innovations over time (European Commission, 2015c):

- Formal requirement to install water saving technologies in new buildings;
- Plans to reduce agricultural soil erosion caused by water runoffs;
- A formal principle considering water issues in decision-making processes on capital investment and procurement;
- A new cooperation and education-related practice oriented toward hospitality provided by the enterprise.

Another example of a company implementing organizational water management innovations is the Abwasserverband Anzbach Laabental (Austria) microorganization that was awarded by EMAS with the European Eco-Management and Audit Scheme Award (European Commission, 2012). As an outcome of EMAS membership, the company adopted a new environmental policy and defined a set of rules for its clients regarding waste disposal (directly related to wastewater cleaning processes). One of them is a ban on allowing harmful substances, such as chemicals, oils, medicines, solid materials, etc. to enter sewage installation systems (Abwasserverband Anzbach Laabental, 2012). Another organizational innovation for effective water management was introduced with regard to future business relations. The company created a new practice of monitoring and classifying suppliers according to their certified compliance with the EMAS regulations (the monitoring process was applied only to the suppliers generating at least 80% of their income from the cooperation with Abwasserverband) (Abwasserverband Anzbach Laabental, 2012).

Organizational innovations in water management are very often both a topic and the final outcome of many European projects. One of the examples is the WaterBee Smart Irrigation Systems Demonstration Action conducted under the leadership of Ireland. The system targets mainly growers, vineyards, golf clubs, public authorities, and landscape managers. It aims at economic and environmental optimization of any water-related operations and actions. The WaterBee system was constructed on the basis of field irrigation sensors networked through the Internet and tested in 14 reference sites in Estonia, Malta, Italy, Spain, Sweden, and the United Kingdom. The system is integrating numerous interactive solutions of information and communication technologies, such as wireless meteorological stations, soil moisture models, software systems for recommendation of irrigation activities, and computer and mobile devices operability. According to WaterBee (2015), the benefits of the system include:

- Instant and complete information on crop and field conditions;
- Comprehensive information and control of irrigation requirements;
- Irrigation based on scientific (weather forecasts and soil conditions) and not intuitive assumptions;
- No risk of overirrigation;
- Better economic efficiency per acre through greater yields and higher quality crops;
- User-friendly software for smartphones;
- Support of real-time crop irrigation decisions and scheduling irrigation tasks;

- No software and hardware maintenance costs, with full operational service;
- Lower operating costs of agricultural activities;
- Reduction of pollution from nutrient leaching; and
- Ability to integrate the WaterBee system with other farm management systems.

The adaptability of this innovation to customers' needs (including business models) and the learning ability of the system are anticipated to guarantee a higher quality of the irrigated areas (WaterBee, 2015). Although WaterBee is a new and innovative system in Europe, similar systems have been operating in the United States. For instance, the Oklahoma Mesonet has been providing valuable weather information to farmers, policymakers, water managers, emergency responders, media, and K-12 teachers for over 20 years.

The examples provided earlier (and other examples from practice not mentioned here) have shown that organizational innovations are generally (and especially technologically) not as complex as product or process innovations, although they might provide similarly valuable outcomes.

## 4. CONCLUSIONS

In recent decades, effective water management became a relevant issue at the global scale. In Europe, water management has faced two problems: (1) many innovative technologies applied in the past have not been sufficient enough to solve critical water problems and crises, and (2) knowledge on those innovations has not been sufficiently disseminated.

Nowadays, the most important innovations are technologies incorporating ICT, especially RFID, Wi-Fi, and GPS. They allow for a quick information transfer and create a basis for successful water management at different scales, which has been proven by many successful examples in the water sector in the European Union. Fast dissemination of effective water management innovations could be facilitated by the creation of an effective and transparent water policy in the European Union. Positive impacts could be anticipated as soon as knowledge on existing and available water management innovations became easily accessible. Consequently, the policy-induced actions (eg, creation of open access databases on water management innovations) could lead to broader knowledge and innovation dissemination. Finally, as a result, market maturity would increase faster, which again would establish new innovations as common and readily available market items (as it was the case with wristwatches, mobile phones, personal computers, laptops, and most recently even with solar panels and very silent micro wind turbines).

The examples of product, process, and organizational innovations in water management presented in this chapter are only a small selection of the latest achievements in research and development in this field. Those examples represent in many cases the first proactive steps toward converting waste (eg, wastewater) into valuable resources according to the "cradle-to-cradle" concept. For this reason, all innovations described in this chapter can be called eco-innovations, that is, innovations delivering better environmental outcomes than the solutions already applied in practice. A few of them can be classified as effectitions—innovations with no negative environmental impact. The implementation of effectitions in the form of product, process, or organizational innovations in water management could generate a significant progress in the field.

Finally, new product, process, and organizational innovations can be expected to appear on the market in the next few years as a result of both population growth, extreme weather events (eg, droughts, floods), and advancements in the technological state-of-the-art in the water management sector. For example, vehicle engines fueled by water are the next technological challenge and a prospective innovation of the future. An innovation of this kind could create a new market niche for water purification and rain harvesting technologies that could be further utilized in the transport sector and other sectors.

The presented examples prove a clear trend in the development of water management innovations in Europe. It is apparent that this development will continue in the future or even accelerate, depending on the available technology, water demand, and human needs to combat water shortages.

## REFERENCES


Abwasserverband Anzbach Laabental, 2012. Umwelterklärung 2012. Abwasserverband Anzbach Laabental, Markersdorf.

Amphiro AG, 2013. Amphiro Smart Water Meters. Modules, Zurich. http://amphiro.com/ (retrieved on 01.04.15.).

Amphiro AG, 2014. Amphiro Smart Water Meters. Amphiro a1, Zurich. http://amphiro.com/ (retrieved on 01.04.15.).

Amphiro AG, 2015. a2, http://amphiro.com/products/a2/, http://amphiro.com/wp-content/uploads/2013/04/amphiro-a2-intro-image.jpg (retrieved on 01.04.15.).

Axel, J., 1999a. Successful market innovation. European Journal of Innovation Management 2 (1), 6–11.

Axel, J., 1999b. Using market vision to steer innovation. Technovation 19 (4), 203–207.

Bernauer, T., Engels, S., Kammerer, D., Seijas, J., 2006. Explaining Green Innovation. CIS Working Paper No 17.

Chesbrough, H.W., Teece, D.J., January–February 1996. When Virtual Is Virtuous? Organizing for Innovation. Harvard Business Review.

Colifast, 2015a. http://www.colifast.no/ (retrieved on 07.04.15.).

Colifast, 2015b. Colifast CALM – On-line Microbial Analysis. http://www.colifast.no/products/calm/. http://www.colifast.no/wp-content/uploads/2015/03/Calm_cutout.png (retrieved on 21.07.15.).

Colifast, 2015c. Colifast ALARM – Rapid, On-line Microbial Water Analysis. http://www.colifast.no/products/alarm/. http://www.colifast.no/wp-content/uploads/2015/01/ALARM_etv.png (retrieved on 21.07.15.).

De Marsily, G., 2007. An overview of the world's water resources problems in 2050. Ecohydrology & Hydrobiology 7 (2), 147–155.

De Tarde, G., 1993. Les Lois de L'imitation. réimpression 1890. Kimé Editeur, Paris.

Dutch Water Sector, 2014. LG Sound Successfully Demonstrated Its New MPC-Buoy to Monitor Algae in Skrzyneckie Male Lake Poznan, Poland. http://www.dutchwatersector.com/news-events/news/10643-lg-sound-successfully-demonstrated-its-new-mpc-buoy-to-monitor-algae-in.html. http://www.dutchwatersector.com/uploads/2014/05/dws-lgsound-pmc-buoy-poznan-lake-650px.jpg (retrieved on 15.07.15.).

Dworak, T., Kampa, E., de Roo, C., Alvarez, C., Bäck, S., Benito, P., 2007. Policy Department, Simplification of European Water Policies, Study for the European Parliament's Committee on Environment. Public Health and Food Safety. IP/A/ENVI/FWC/2006-172/C1/SC5, Brussels.

EC, 2009. Regulation (EC) No 1221/2009 of the European Parliament and of the Council of 25 November 2009 on the Voluntary Participation by Organisations in a Community Eco-management and Audit Scheme (EMAS), Repealing Regulation (EC) No 761/2001 and Commission Decisions 2001/681/EC and 2006/193/EC.

EFC Separations. http://efc.nl (retrieved on 7.04.15.).

European Commission, 2012. Environment: Commission Recognises Innovative Water Management Solutions. http://europa.eu/rapid/press-release_IP-12-1294_en.htm. Brussels (retrieved on 19.04.15.).

European Commission, 2015a. CORDIS, Projects and Results, Demonstration of a Rapid Microbial Monitor for Operations and Quality Decision-Making in the Water Industries. http://cordis.europa.eu/project/rcn/69967_en.html (retrieved on 08.04.15.).

European Commission, 2015b. http://ec.europa.eu/environment/emas/emasawards/ (retrieved on 21.07.15.).

European Commission, 2015c. EU EMAS Award Winners 2012. http://ec.europa.eu/environment/emas/emasawards/winners_2012.htm (retrieved on 21.07.15.).

Freier, I., 2003. Environmental management from an ecological modernisation and innovation perspective. In: DRUID Academy Winter 2003 PhD Conference, 16–18 January, Aalborg.

Hargroves, K.C., Smith, M.H. (Eds.), 2006. The Natural Advantage of Nations, Business Opportunities, Innovation and Governance in the 21st Century. Earthscan, London, Sterling, VA.

Joshi, M., 2010. Seizing the opportunity for innovation: a case study from India. International Journal of Technology Marketing 5 (1), 90–104.

Kemp, R., Andersen, M.M., Butter, M., 2004. Background Report about Strategies for Eco-innovation. Report for VROM, zaaknummer 5060.04.0041, Final version, 22 May.

Lecerf, M.-A., 2012. Internationalization and innovation: the effects of a strategy mix on the economic performance of French SMEs. International Business Research 5 (6), 2–13.

Lovins, H., 2008. Rethinking production. In: State of the World 2008. Innovations for a Sustainable Economy. Worldwatch Institute.

Malinsky, E., March 2, 2010. Innovating Innovation: It's Time for the Canadian Government to Show Real Leadership in Innovation. Centre for Social Innovation. http://socialinnovation.ca/blog/innovating-innovation-it%E2%80%99s-time-for-canadian-government-show-real-leadership-innovation (retrieved on 10.07.15).

Martins, G., Brito, A.G., Nogueira, R., Ureña, M., Fernández, D., Luque, F.J., Alcácer, C., 2013. Water resources management in southern Europe: clues for a research and innovation based regional hypercluster. Journal of Environmental Management 119, 76–84.

Matczewski, A., 2005. Raport końcowy z realizacji Pilotażowego Projektu Foresight w polu badawczym Zdrowie i życie, Panel główny Zdrowie i Życie. Warszawa.

Matheson, J., 2013. How to get real innovation flowing in the water industry. Journal AWWA 105 (5), 108–117.

Matusiak, K.B.R., 2011. Innowacje I Transfer Technologii. Słownik Pojeć. Wydanie III, zaktualizowane. Polska Agencja Rozwoju Przedsiebiorczości, Warszawa.

Murphy, J., Gouldson, A., 2000. Integrating environment and economy through ecological modernisation? an assessment of the impact of environmental policy on industrial innovation. Geoforum 31 (1), 33–44.

Murphy, S., 2015. General Information on Total Suspended Solids. City of Boulder/USGS Water Quality Monitoring. BASIN. http://bcn.boulder.co.us/basin/data/NEW/info/TSS.html (retrieved on 15.07.15.).

OECD, 2005. Oslo Manual. OECD, Paris.

OxyMem Smart Aeration, 2015. http://www.oxymem.com/ (retrieved on 01.04.15.).

OxyMem Technology, 2015. http://www.waterinnovationeurope.eu/wp-content/uploads/OxyMem_1pager_June2014-Wsstp.pdf (retrieved on 29.03.15.).

OxyMem Technology-Brochure, 2015. http://www.waterinnovationeurope.eu/wp-content/uploads/OxyMem_1pager_June2014-Wsstp.pdf (retrieved on 29.03.15.).

Phills Jr., J.A., Deiglmeier, K., Miller, D.T., 2008. Rediscovering social innovation. Stanford Social Innovation Review 6 (4), 34–43.

Prüss-Üstün, A., Bos, R., Gore, F., Bartram, J., 2008. Safer Water, Better Health: Costs, Benefits and Sustainability of Interventions to Protect and Promote Health. World Health Organization, Geneva.

Ravn, L., Petersen, J.W., 2005. Cluster Development in Hungary – A Study of the Possibilities for the Automotive Industry and the Wood and Furniture Industry to Enhance Regional Sustainable Development. Department of Environment, Technology and Social Studies, Roskilde University, Roskilde.

Rehn, A., 2006. Anachronism and innovation: a case of hybrid economies in the early 19th century. Management & Organizational History 1 (1), 71–86.

Riedel, A.F., 1839. Nationalekonomie Oder Volkswirtschaft? (2 volumes) F. H. Morin, Berlin.

Saporito, F., April 3, 2008. Uncodified Knowledge: The Middle East's Unique Innovation Opportunity. Executive Magazine.

Schumpeter, J.A., 1939. Business Cycles. A Theoretical, Historical and Statistical Analysis of the Capitalist Process. McGraw-Hill Book Company, New York, London.

Schumpeter, J.A., 1995. Kapitalizm, Socjalizm, Demokracja. PWN, Warszawa.

Søndergaard, B., Hansen, O.E., Kerndrup, S., 1997. Renere produktion i et innovationsperspektiv. In: Holm, J., Kjærgård, B., Pedersen, K. (Eds.), Miljøregulering – Tværfaglige Studier. Roskilde Universitetsforlag, Frederiksberg, pp. 293–317.

Sonic, L.G., 2015. Leading in Ultrasonic Algae Control. http://www.lgsonic.com/lg-sonic-technology/ (retrieved on 01.04.15.).

Sound, L.G., 2015. Ultrasonic Algae Control. http://www.waterinnovationeurope.eu/wp-content/uploads/Speeddate-sessions-Company-description.pdf (retrieved on 01.04.15.).

UNEP, 2008. Vital Water Graphics – An Overview of the State of the World's Fresh and Marine Waters, second ed. UNEP, Nairobi, Kenya.

UNESCO, 2009. The United Nations World Water Development Report 3, Water in a Changing World, Facts and Figures. UNESCO Publishing, Paris, London.

UNESCO, 2012. Managing Water under Uncertainty and Risk. The United Nations World Water Development Report, 4(1). UNESCO Publishing, Paris.

Unger, B., 2005. Problems of measuring innovative performance. In: Caper, S., van Waarden, F. (Eds.), Innovation and Institutions. A Multidisciplinary Review of the Study of Innovation System. Edward Elgar, Cheltenham, Northhampton, pp. 19–50.

WaterBee (2015). http://waterbee.iris.cat/ (retrieved on 07.04.15.).

WHO, 2008. The Global Burden of Disease. WHO, Geneva.

WssTP SME Awards (2015). http://www.waterinnovationeurope.eu/awards/wsstp-sme-awards/ (retrieved on 23.03.15.).

Yixin, L., 2011. On the connotation and evaluation of original innovation, IACEED2010. Energy Procedia 5, 643–646.

Zajączkowski, M., 2003. Podstawy Innowacji I Ochrony Własności Intelektualnej. Economicus, Szczecin.

Ziolkowski, B., Ziolkowska, J., 2014. Effectitions – a new trend in sustainable innovations. International Journal of Business and Management Studies 3 (3), 473–479.

Ziółkowski, B., 2012. Effectitions as a new instrument in sustainable development policy – the conceptual approach. Management 16 (2), 282–295.

Chapter 3.3

# Comparison of Water Management Institutions and Approaches in the United States and Europe—What Can We Learn From Each Other?

M. Pulido-Velazquez[1], F.A. Ward[2]

[1]*Research Institute of Water and Environmental Engineering (IIAMA), Universitat Politecnica de València, Valencia, Spain;* [2]*New Mexico State University, Las Cruces, NM, United States*

## 1. INTRODUCTION: HISTORY AND STATUS OF WATER POLICY IN THE EUROPEAN UNION AND THE UNITED STATES

### 1.1 EU Water Policy and Institutions

#### *1.1.1 Evolution of Water Policy in the European Union*

In the European Union the role of the development of environmental policies is shared between the European Commission (EC), which sets minimum standards, and its Member States, which might require a stricter protection. Member States are obliged to transpose and implement the minimum standards approved at the EU level usually as directives. The water sector was one of the first sectors to be covered by the EC environmental policy legislation in the 1970s.

The earliest legislative approach (1975–80) established directives and decisions providing environmental quality standards for specific types of water, including surface water, fish water, bathing water, and drinking water directives, and establishing emission limit values for specific water uses addressed by the dangerous substances directive. The second approach to water legislation, seen in 1980–91, tackled new issues and approaches including Nitrates and Urban Wastewater Directives. It also completed the dangerous substances directive through daughter directives on specific substances. However, all these directives had been mostly developed piecemeal to address specific problems (Chave, 2001). The presence of diverse regulatory instruments as well as growing environmental

Competition for Water Resources. http://dx.doi.org/10.1016/B978-0-12-803237-4.00024-0

problems derived from the intensification of economic water uses required more coherent action from the European Union.

The 1990s saw a worldwide movement toward an ecosystem approach. Within Europe this led to the draft of the EU Directive on the Ecological Quality of Surface Waters and many countries adopted monitoring schemes and environmental quality standards—an embryo of the Water Framework Directive (WFD) (Hering et al., 2010). The WFD had a precedent in the 1972 US Clean Water Act (CWA), with clear parallels in terms of objectives, implementation, and ecological approaches (Hoornbeek, 2004). Both the WFD and the CWA shifted the decision-making authority from the federal or supranational level toward the regional level—decentralized policy implementation (Whitford and Clark, 2007; Young, 2011).

### 1.1.2 The EU Water Framework Directive

After years of negotiation among the European Community, the WFD was finally approved at the end of 2000 and Member States were required to transpose it into their national laws by December 2003. The WFD introduced a remarkable change in community water legislation, incorporating new standards and criteria, institutions (river basin districts, river basin authorities, river basin plans), and planning processes for managing Europe's waters. For the first time, the primary focus was the good status of all water bodies, including inland water bodies, transitional waters, and coastal waters.

To this end, the Directive obliges Member States to prevent further deterioration and enhance and restore the status of aquatic ecosystems. The Directive binds together the previously fragmented environmental legislation adopting an "integrated ecosystem approach": it requires the combination of natural and social sciences for addressing environmental problems, incorporating ecosystem-based criteria within a formal planning process at the basin scale. River basin management plans will have to be approved, defining the necessary measures to be implemented for achieving specific environmental objectives in a basin. In this way, water policy moved from the protection of particular waters of special interest, such as nature areas, bathing water, and drinking water, to the protection and use of all waters based on the consideration of the links of the hydrology and ecology of the entire natural cycle of each river basin (Barreira, 2006).

The *essential objectives* of the WFD are:

- Prevention of further deterioration of water bodies
- Protection and enhancement of aquatic ecosystems and associated wetlands
- Achievement of a good "water status" for all waters within a certain time scale
- Promotion of sustainable water use
- Mitigation of the effects of floods and droughts

The *most important and innovative features* of the WFD are:

- Management and planning at the river basin scale (river basin management plans for each river basin district)
- Application to all waters: inland surface waters, ground waters, and transitional (estuaries) and coastal waters
- Combined approach for controlling water pollution: emission limit values plus water quality objectives

- Application of economic principles (eg, the polluter pays), methods and tools (eg, cost-effectiveness analysis), and instruments (eg, water pricing), ensuring that the users bear the real cost of providing and using water (cost recovery principle)
- Public involvement in decision making on water management

The WFD requires that each Member State establishes a program of measures for achieving the environmental objectives for each river basin district. For surface water bodies, the objective is to reach a good water status, initially by 2015, and then with timeframes of 6 years, based on a good chemical and ecological status. For groundwater bodies, the objective is to reach a good groundwater status, requiring both a good quantitative and chemical status. The Groundwater Directive (Directive, 2006) includes further criteria for assessing the groundwater chemical status and identifying upward trends on pollutant concentrations and baselines for trend reversals.

The WFD clearly integrates economics into water management and policy making, and economics has a decisive role in the development of new river basin management plans. The most cost-effective program of measures should be selected to meet the WFD environmental objectives. Cost-effectiveness analysis will need to be expanded to cost–benefit analysis (CBA) to determine if the cost of the more cost-efficient program of measures is likely to be disproportionally expensive (cost exceeding benefit), which might require the application of nonmarket economic valuation techniques (Brouwer, 2008; Hanley et al., 2006; Pulido-Velazquez et al., 2009).

Member States should apply the principle of cost recovery of water services, including not only financial costs, but also environmental and resource costs. Estimation of financial costs is the most straightforward. However, the definition and method of assessing resource and environmental costs for the purposes of the Directive remains controversial, and can cause difficulties in implementing this ambitious approach (Brouwer et al., 2009; Heinz et al., 2007). The Directive also requires the implementation of monitoring programs for a comprehensive overview of the water status and protected areas.

## 1.2 US Water Policy and Institutions

### *1.2.1 Evolution of Water Policy in the United States*

In spite of the importance to living standards in the United States to provide water of the right quantity, quality, timing, and location, a nationally consistent water policy in the United States is fragmented. There is no nationally formulated or implemented program of monitoring and research to develop a better understanding of the best ways to use science to inform policy in the United States. Moreover, the United States depends heavily on groundwater pumping in many parts of the country for urban and domestic supply. No national program currently exists to monitor or act on the results of the monitoring, nor are there programs to guide science to inform policy actions to govern those supplies (Lawford, 2003). The United States has nothing like the European WFD throughout the 50 states to unify water management plans, actions, performance, and review.

Despite this fragmentation, there has been a considerable development for many years of a growing set of unifying economic principles to guide water programs and policies, known as CBA. The political and economic motivations for the use of CBA to guide water resource development and later policy decisions originated in the United States as an outgrowth of a Congressional legislation known as the Flood Control Act of 1936

(Ward, 2012). The motivation for the development of CBA to guide federal water resources policy was driven by the need to assure that taxpayer dollars with a high opportunity cost, by being removed through taxation of the private sector, would not be used for low-valued uses financed by federal water development projects. It came from the need to conduct economic performance assessments of expensive federal water projects. CBA was established to compare total economic gains and losses resulting from a proposed federal water policy.

After many years of improvement that produced a fair amount of intellectual heft, CBA can be viewed as something of a gold standard to provide decision makers with a comparison of the impacts of two or more water policy options using methods that are grounded in well-established economic principles. Net economic efficiency, measured as the difference between added benefits and added costs, can be used to inform water managers and the public of the economic impacts of water programs to address peace, development, health, the environment, climate, and poverty. When confronted by scarce taxpayer dollars and scarce water, CBA can inform policy choices by summarizing tradeoffs involved in applying, designing, reviewing, or implementing a wide range of water programs (Ward, 2012).

### *1.2.2 US Water Quality and Environmental Policy and Institutions*

The desire for a better environment became a widespread movement in the United States at the end of the 1960s, marked by a generation of writers and environmental activists (Allit, 2015). When a critical mass of these young people were old enough to vote, this movement encouraged political leaders to enact an innovative body of federal environmental legislation in the early 1970s, including the National Environmental Policy Act (1970), the Endangered Species Act (1973), and the creation of the Environmental Protection Agency (EPA). The EPA's Office of Water was created under the mission of ensuring drinking water safety, and restoring and maintaining oceans, watersheds, and their aquatic ecosystems.

The CWA passed in 1972, based on precedent from the Federal Water Pollution Control Act (1948), was the first major US law to address water pollution, and established the regulation of pollutant discharges into US waters, regulating quality standards for surface waters. The CWA set a new national goal "to restore and maintain the chemical, physical, and biological integrity of the Nation's waters," with interim goals that all waters be fishable and swimmable where possible. The Act established a new federal−state partnership, where federal guidelines, objectives, and limits were set under the authority of the US EPA, states, territories, and authorized tribes that would largely administer and enforce the CWA programs, with federal technical and financial assistance. The Act also gave citizens an important role to play in protecting and restoring waters (EPA, 2015). Core programs under the CWA are: establishing water quality standards; identifying polluted waters and developing plans to restore them; permitting discharges of pollutants from point sources through the National Pollutant Discharge Elimination System; addressing diffuse, nonpoint sources of pollution; and protecting wetlands, coastal waters through the National Estuary Program, and large aquatic ecosystems. Although initially the CWA programs were primarily directed at point source pollution, in 1987 that was changed by a new federal program that provides money to states, tribes, and territories for the development of programs to reduce pollution from unregulated diffuse sources such as agriculture. EPA grants were used to identify waters affected by nonpoint sources, help stakeholders implement best management practices to reduce runoff, and monitor and evaluate progress to restore waters (US Environmental Protection Agency, 2015).

The Safe Drinking Water Act was originally passed by Congress in 1974 with the aim to protect public health by regulating the nation's public drinking water supply. Amended in 1986 and 1996, the law requires actions to protect drinking water and its sources: rivers, lakes, reservoirs, springs, and groundwater wells.

### *1.2.3 Integrated Watershed Management*

Integrated river basin management has been an ideal to guide water management for many years in the western world. The European Union responded to this ideal in 2000 by establishing the WFD as described earlier. Despite the fragmentation of federal water policy across the numerous federal agencies, US water policy has been guided in many cases by turning to a consistent use of economic principles to guide water management, potentially at a river basin scale.

A 2008 work described three periods in US history in which CBA was used in progressively more sophisticated ways to inform debates on the most desired policies for implementing integrated river basin management:

- River Basin Development, 1933–65
- River Basin Commissions, 1965–80
- The Watershed Movement, 1980–2007

Nationwide efforts toward comprehensive, integrated watershed management in the United States have seen a long history. This history has evolved from large-scale river basin commissions motivated by economic challenges of the Great Depression to the more recent regional watershed partnerships organized around smaller more localized watersheds (Schlager and Blomquist, 2008).

### *1.2.4 The Cost–Benefit Framework: US Water Policy Guidance*

By 1950, in an attempt to deal with the requirements of the Flood Control Act, an interagency water resources committee produced the "Green Book," the first attempt to define principles, standards, and procedures for measuring the benefits and costs of federal water projects as a guide to water program design. The Green Book was updated in 1962 (United States Senate, 1962), then later in 1979 and 1983. The updates continue. The principles, standards, and procedures governing the use of CBA for federal water and environmental programs continue to see vigorous debate in the Obama administration (Ward, 2012; US Council on Environmental Quality, 2013). As can be imagined, the scope for the use of CBA in the United States has widened since its inception in the 1930s. Current challenges are to inform the design of policies and programs to promote economic development, advance peace, address climate, eradicate poverty, protect the environment, and improve health (van Grinsven et al., 2010; Ward, 2012).

Since the mid-1990s, several kinds of legislation proposed by the US Congress would have expanded the use of CBA to underpin the regulatory process. Because of differences among the planning, development, regulatory, and review missions of agencies who do conduct or would be asked to conduct CBA to guide policy choices, the proposed legislation has generated much debate, discussion, and controversy. Greater use of CBA to inform water policy decisions in the United States would significantly change how government agencies develop and implement certain types of regulations (Ward, 2012). In the United States, CBA has nearly 100 years of application as a method to support economic review of water programs (Prest and Turvey, 1965; Ward, 2012). CBA is a rational and systematic decision-support tool that draws on nearly a century of developments in theoretical and empirical economics beginning with the foundational work from 1920 (Pigou, 1920).

In 2015 in the United States, important water policy debates often surround more questions of how intensively and to what end CBA should be practiced than whether or not it should be used. CBA has for many years seen widespread use by numerous US federal water and environmental agencies to support the review of regulatory activities as well as supporting the on-the-ground implementation of proposed programs or review of actual ones. The question of how economic principles can be used to identify better ways to mitigate damages caused by extreme weather events and climate change will likely continue to receive widespread attention in the United States as well as internationally (Ward, 2012).

In the United States, the question of the economic value of programs that could limit damage from floods, drought, and other extreme water events continues to receive much attention. In addition, debates continue to surround decisions on water conservation, development, allocation, transboundary water allocation, and environmental improvements inside and outside the United States. Debates are especially intense where there is competition for water between food production, energy generation, key ecological assets, and urban water uses. Excess water, inadequate water, water at the wrong time, the wrong place, of the wrong quality (van Grinsven et al., 2010), or for the wrong use all present a challenge for which policy interventions are needed (Ward, 2012).

## 2. WATER GOVERNANCE AND MANAGEMENT

### 2.1 EU Principles and Institutions

The WFD recognizes the value of integrated planning and comprehensive development at the basin scale, considered as the unit for water management planning across the European Union. Member States have to identify river basins within their territory and assign them to River Basin Districts (RBD), which constitute the spatial unit for all planning tools under the WFD, from river basin management plans and programs of measures to monitoring activities.

The Member States shall designate a competent authority for each of its RBDs. In some EU countries river basin organizations to manage their water resources have already existed. Among them, Spain has the greatest experience in this regard. The first river basin agencies ("Confederaciones Hidrográficas") were created back in 1926 to promote, construct, and manage hydraulic works in cooperation with the central government. User participation was already considered in the development of the national and river basin plans, with a focus on water allocation and hydraulic infrastructure (Bhat and Blomquist, 2004; Del Moral and Saurí, 1999). In France, the Agences de l'eau covered the six main river basins of France, but with no role in discharge or abstraction control or infrastructure (Chave, 2001). In a federal state like Germany, however, the organizational structures of the competent local and regional authorities were not based on the river basins but rather on the local administrative boundaries (Chave, 2001).

Transboundary rivers are widespread in Europe. The WFD requires that river basins covering the territory of more than one Member State should be assigned to international river basin districts. Some large international river basins have already been subject to international conventions and protocols (eg, the Rhine, passing through 10 countries, and the Danube, the largest river in the European Union, spanning 17 countries). The WFD has significantly improved transboundary river management in Europe, from an exchange of information to a joint problem diagnosis and joint decisions on management measures (EC, 2012). Joint river basin management plans in large transboundary basins have been prepared. However, the level of management and intervention required by the European Union still needs significant work and input to improve collaboration and coordination.

The first task of the RBD authorities is to collect information about the basin. The WFD requires an analysis of each RBD including identification of water bodies, a review of the pressures and impacts of human activities, an economic analysis of water use, and a register of areas requiring special protection. River basin management plans have to be developed and implemented, including the previously mentioned information and the program of measures to achieve the environmental objectives in the water bodies.

Public participation in water management is another key element introduced by the WFD. The public needs access to information, which has been secured and defined by specific consultation phases. Public participation in the implementation of the WFD constitutes a major opportunity to achieve better informed decisions and more effective implementation of measures, and to reduce uncertainty in the implementation process (De Stefano, 2010; Newig et al., 2005). Despite the emphasis given in the Directive to the role of public participation, some authors have pointed out the need to improve the actual practices of the participatory process, improving working capacity of water stakeholders through timely and effective provision of adequate information, as well as by training and financial support for their active participation in decision-making processes (De Stefano, 2010).

The Directive also incorporates obligations and control mechanisms for encouraging compliance and enforcement. Penalties can be imposed for breaching the measures to be implemented and failing to meet the objectives. A strict timetable for the monitoring progress and achievement of the objectives, as well as for submitting data and reports to the Commission, has been established in the Directive.

## 2.2 US Principles and Institutions

The motivation for US federal economic guidelines for water resource project evaluation springs from the Flood Control Act of 1936, in which the American Congress directed that flood control projects should be undertaken "if the benefits to whomsoever they may accrue are in excess of the estimated costs…" The application of this mandate in the United States has been far wider than to flood control projects. The normative branch of economics that had developed prior to 1936, and which has since been further developed, is known as "welfare economics." It was the intellectual foundation for implementing these guidelines that launched this branch of economic thought. The opportunity afforded by the Congress to apply this normative theory to the field of water and related land resource use and development was welcomed vigorously by academic economists and continues to be embraced to this day. A few limitations reflecting unfinished business as of 2015 include:

- Weak capacity to measure the economic value of environmental improvements
- Unresolved debates over the proper way to measure "secondary benefits"

- Weak treatment of optimization models as a method to estimate economic values of water resource programs constrained by scarce water
- Weak distinction between prospective (forward-looking) and retrospective (backward-looking) CBA and the correct way to use either to inform policy debates
- The need to make a sharper distinction between objectives and constraints of water resources planning

### 2.2.1 International Treaties

The United States has signed onto several international water-related treaties that are important institutions for governing its water and political relations over transboundary waters shared with Mexico and Canada. Two important treaties with Mexico include the US Mexico Treaty of 1906 addressing allocation of flows by the United States to Mexico of the Rio Grande and a 1944 treaty between the United States and Mexico relating to the utilization of the waters of the Colorado and Tijuana Rivers, and of the Rio Grande.

Under the 1906 treaty, the United States obligated itself to delivery 60,000 acre-feet per year. However, that treaty includes a special qualification that in case of extraordinary drought or serious accident to the irrigation system in the United States, the amount of water delivered to Mexico shall be diminished in the same proportion as the water delivered to US lands under the irrigation system in the United States downstream of Elephant Butte Dam.

The 1944 treaty, sometimes labeled the Treaty of the Utilization of Waters of the Colorado and Tijuana Rivers and of the Rio Grande, directs Mexico to deliver water to the United States from six tributaries that feed the Rio Grande, in exchange for water delivered to Mexico from the Colorado River. The Mexican government obligated itself to release 1,750,000 acre-feet of water every 5 years. Mexico has had a long history of difficulties in meeting these requirements (Robinson et al., 2010), made more difficult in periods of severe and sustained drought.

Water issue linkage is a strategy for enhancing cooperation and joint management of transboundary waters (Fischhendler et al., 2004). Such a linkage can have both short-term and long-term implications for guiding water policy. Treaties operate by constraining the options available to the treaty parties. These constraining implications may be redressed by mechanisms that will allow the parties to adapt the linkage to new conditions and by tactics that reduce the political cost of a linkage strategy. Fischhendler et al. examine the negotiation process in connection with the Colorado and Rio Grande, two US—Mexico transboundary water sources. One important long-term implication identified by the authors has been the difficulty of Mexico to adapt the linkage in response to the 10-year drought along the Rio Grande, as manifest in the inability of Mexico to meet its water obligations to the United States along the Rio Grande. This has resulted in a controversy between Mexico and the United States, as well as between the Mexican federal government and several Mexican border states. Flexibility provisions are included when linkages involving water resources are advanced, so that needed adjustments are implemented without requiring a renegotiation of the treaty (Fischhendler et al., 2004).

The use of "minutes" provides flexibility that can be added to existing transboundary water treaties, addressing important management issues that arise after signing. For example, on November 20, 2012, Minute 319 of the International Boundary and Water Commission was signed by the commissioners of Mexico and the United States. This minute establishes measures for binational water management until 2017. This agreement aims at finding cooperative mechanisms that address water shortages in the Colorado

Basin. The agreement for the most part addresses storage of Mexican water by the Hoover Dam in the United States, which, under the rule of the 1944 Water Treaty described previously, would be managed for joint and general benefit by both countries. The more recent minute addressed anticipated climate change impacts such as extreme drought and the consequent reduction of water supply available for both countries (Sanchez and Cortez-Lara, 2015).

### 2.2.2 Interstate Compacts

Within the United States, water allocation agreements among states (interstate compacts) have provided an innovative and lasting mechanism for resolving transboundary conflicts of water supplies (McCormick, 1994). McCormick described 22 compacts allocating the water of rivers among states in the western United States. His examination provides guidance for drafters of future compacts worldwide, a motivation that could see growing attention in the face of potential climate change.

Percentage allocations can equitably apportion risk, but can conflict with principles of prior appropriation, important in the American West. Guarantees of minimum flows should be approached with caution, to avoid any upstream deliverer guaranteeing natural flows over which it has no control. Disputes can be anticipated, and a dispute resolution mechanism designed in advance is needed. To reach sustainability, a compact will be comprehensive in scope, encompassing groundwater as well as surface use, normal water supply years, flood years, and dry years. The Rio Grande Compact of 1938 has been one of the more flexible interstate compacts developed in the United States, having both kinds of flexibility (Hill, 1974; Ward et al., 2007).

## 3. POLICY INNOVATIONS FOR ADAPTING TO WATER SUPPLY VARIABILITY

### 3.1 Integrated Water Resources Management

Integrated Water Resources Management (IWRM) has been defined as a process that "promotes the coordinated development and management of water, land and related resources, in order to maximize the resultant economic and social welfare in an equitable manner without compromising the sustainability of vital ecosystems" (Agarwal et al., 2000).

IWRM is an old idea, dating back to storage reservoirs and aqueducts built thousands of years ago (Evans, 1997). One version of IWRM was established in the United States in the 1930s, as an outgrowth of multiple use river basin development spurred by US federal financing of flood control and reclamation projects. Despite the long history of this high level of ambition, several participants in the professional water community started noticing by the 1980s that achieving IWRM is a tall order, fraught with financial, scientific, and political challenges. This recognition grew during the 1990s, when growing numbers began to recognize that water challenges are multidimensional, compounded by numerous sectors and several competing interests, each with their own political and economic agendas. While widely recognized as a desirable goal, an important question asks how can this be achieved in practice, including a demonstration that IWRM is more economically efficient than piecemeal policy (Biswas, 2004). A central challenge of IWRM is to convince those with authority, votes, and financing that it can be implemented, and to demonstrate when it has been implemented. IWRM to date has a weak record in terms of its implementation when faced by those who would critically assess it. There remains little agreement on basic issues, including what dimensions of the water problem should be integrated, how, by whom, and with what resources and financing (Biswas, 2004).

The notion of an entire river basin as a management unit has passed through several stages and it continues to evolve. It was compatible with the notion of full control of highly fluctuating supplies and demands for water in a river basin with multipurpose US dam building the 1930s–60s. It later reemerged to face growing challenges associated with water quality and pollution. Later still it grew into the 1990s notion of IWRM, complemented with watershed and ecosystem management approaches. A 2009 work reviewed the evolution of the concept of a river basin and how it has been associated with various branches of thinking embraced by various special interests to add strength to their ambitions (Molle, 2009).

### *3.1.1 EU Applications*

In the European Union, the WFD has attached considerable importance to the use of IWRM to guide water planning by Member States. IWRM has been embraced by the WFD since 2000. IWRM has served as a guiding principle underpinning the development of River Basin Management Plans, which are expected to have central importance as a tool in widespread use for future planning and managing water resources in the European Union.

### *3.1.2 US Applications*

In the United States, IWRM is difficult to achieve. Each municipality, state, and other political organization likes independence in the financing and implementation of water management plans. Shively and Mueller (2010) reexamined a potential for a successful implementation of IWRM in the United States with application to a river basin in Montana. That investigation examined an organization unique in the American West, namely, Montana's Clark Fork River Basin Task Force. They assessed the task force's activities since its formation in 2001 to answer a central question: are the activities and contributions of this group functioning effectively to promote a better-integrated approach to water resources management in that region? In reviewing the physical, interactive, and protocol underlying planning policy components of IWRM, the authors concluded that the task force has contributed to the constructive development of the state's water resources management making progress toward achieving the framework goals, though numerous unresolved constraints remain (Shively and Mueller, 2010).

## 4. TECHNICAL INNOVATIONS FOR ADAPTING TO WATER SUPPLY VARIABILITY

### 4.1 Cost Recovery and Pricing

There is a considerable disagreement about what the principle of water as an economic good means for policy choices. According to Savenije and van der Zaag (2002), water pricing should serve the mission of financial sustainability through cost recovery, so that water systems are sustainably and financially supported by water users. When water pricing is used as a policy instrument, sufficient attention should be given to justice goals, through measures such as rising block tariffs, sometimes known as tiered pricing. An important challenge along these lines is the definition of a water price that signals the opportunity cost of water use, promotes full cost recovery, but that also protects ecological values as well as access to safe water for those who cannot pay the full cost of supply.

Pricing water at a level that ensures financial sustainability, including tiered pricing, secures the efficiency mission of providing a clear signal to the users that water should be used carefully. Nevertheless, an important target of water pricing is cost recovery. A central proposition of economic research conducted over many years is that pricing water at its marginal cost can facilitate the movement of water from economic sectors that produce low economic values at the margin, for example, irrigation, to sectors that produce a higher marginal economic value such as city water use (Savenije and van der Zaag, 2002; Pulido-Velazquez et al., 2013).

A pricing policy is efficient, according to economic theory, if the prices charged correspond to the marginal cost of water. Therefore it must take into account supply costs, opportunity costs, and externalities (Rogers et al., 2002). Measuring the opportunity costs of scarce water is difficult: since water markets are usually absent or ineffective, scarcity values are not reflected in the water prices. Given that opportunity cost depends on the alternative uses, an integrated basin-wide approach is needed to simultaneously account for all major competing water uses in the basin (Rogers et al., 2002; Pulido-Velazquez et al., 2013).

Integrated hydrologic and economic optimization models at the basin scale provide a framework for policy design, implementation, and evaluation in water-stressed basins (Ward and Pulido-Velazquez, 2008b; Pulido-Velazquez et al., 2008). One US investigation (Ward and Pulido-Velazquez, 2008b) developed a basin-scale framework to identify hydrologic and economic impacts of alternative water pricing programs that comply with environmental regulations for protecting water quality. Significant issues were examined that challenge those who develop integrated hydroeconomic models of river basins. These challenges included linking the physical aspects of the water resource with economic values, integration of spatial and temporal scales, as well as quantity−quality relationships. In that investigation, economic efficiency was defined and measured for each of two urban water pricing arrangements that were compatible with urban water quality protection measures. Alternative measures of equity were assessed in both temporal and spatial dimensions. Sustainability was assessed physically for protecting the water supply as well as financially for long-term revenue viability. The approach is shown practically from results of a dynamic nonlinear programming optimization model of water use in North America's Rio Grande Basin. The model optimizes the net present value of the basin's total economic benefits subject to constraints on equity, sustainability, hydrology, and institutions. Results indicated that two-tiered pricing of urban water supply offers a great potential to perform well in meeting the aims of efficiency, equity, and sustainability (Ward and Pulido-Velazquez, 2008b).

### *4.1.1 Cost Recovery Through Pricing: The European Union*

The main challenge of the European WFD is to define a framework for the protection of inland surface waters, transitional and coastal waters, and groundwater, which should protect the aquatic ecosystems, promote sustainable water use, guarantee reduction of pollution, and mitigate the impacts of droughts and floods. The WFD aims to achieve full cost recovery while being compatible with the Polluter Pays Principle[1], as part of a package supporting an adequate and sustainable water use. The WFD has paid special attention to three different costs associated with water use: resource costs, financial costs, and environmental costs. In article 9 (WFD) water pricing policies are proposed with a double role: as an adequate incentive to use water resources efficiently and as an instrument to recover the costs of water services.

The recent Blueprint to Safeguard Europe's Water Resources (EC, 2012) has the goal to guarantee that enough good quality water is available for people's needs, the economy, and the environment. To achieve this goal, the initiative has proposed the following targets: good water status, water efficiency, and resilience to adapt to extreme events. Because of rising water scarcity and stress, water efficiency measures should be taken to save water resources and, in many cases, to save energy too. Pricing is considered in the Blueprint as a key instrument for improving the efficiency of water use in Europe. Pricing is a powerful awareness-raising tool for consumers, while it also combines environmental and economic benefits stimulating innovations in water management. The Blueprint proposed to enforce water pricing/cost-recovery obligations under the WFD, including metering when relevant, and to make water pricing/cost recovery an ex-ante condition under the Rural Development and Cohesion policy funds.

### *4.1.2 Cost Recovery Through Pricing: A US Example*

A requirement to recover costs through water pricing is common in the United States, especially for urban water utilities. However, in some cases, innovative incentive pricing programs have a potential to promote economically efficient water use patterns and provide a revenue source to compensate for environmental damages (Ward and Pulido-Velazquez, 2009; Pulido-Velazquez et al., 2013; Macian-Sorribes et al., 2015). However, risks remain that incentive pricing has the potential to impose disproportionate costs and worsen poverty where high water prices are levied for basic human needs. Ward and Pulido-Velazquez (2009) presented an analysis of a two-tiered water pricing system that sets a low price for subsistence needs, while charging a higher price equal to marginal cost, including environmental cost, for discretionary uses beyond basic human needs, such as irrigation of outdoor landscape. That work used data on water sources and uses from the Rio Grande Basin of North America to develop a mathematical programming model. Compared to the Law of the River and marginal cost pricing, the two-tiered pricing approach performs well in regard to water efficiency and acceptably well in regard to sustainability and equity, specified in the model as evaluation criteria. The findings provided a template for designing water pricing plans that could promote economically and environmentally efficient water use programs, while also addressing other important policy goals (Ward and Pulido-Velazquez, 2009).

## 4.2 Market Transfers of Water

A water market enables the short-term, long-term, or permanent transfer of the rights to use water in exchange for compensation, either monetary or in kind. The capacity to transfer these rights or use adds flexibility to a community's water supply, helping to successfully

1. Polluter Pays Principle (PPP) is an environmental policy principle requiring that the costs of pollution should be borne by those who cause it.

adapt to temporary drought conditions and to address longer-term changes in the water demand patterns. This is an especially common occurrence for the case where growing demands are seen by expanding urban and manufacturing uses. Market transfers of water can move water from lower to higher-valued uses, and have been applied to both surface and aquifer uses of water (Ramirez et al., 2009). One version of groundwater banking can involve the planned storage of surface water in aquifers with room for more water in wet runoff years, for later recovery and use when the inevitable dry year occurs (Dillon, 2005). This practice is termed aquifer storage and recovery.

### 4.2.1 Innovations: Optimization Frameworks

Optimization approaches can offer considerable insights into economic values gained by allowing market forces to guide the movement of water from low to higher economically valued uses (Salman et al., 2014). Such approaches view scarce water as a resource that can be allocated or reallocated to increase the value of an economic objective while respecting constraints on total supply or institutions that are required for social acceptability (Maass et al., 1962).

Booker et al. (2005) examined market-related management options for the Rio Grande. In that basin, water is overallocated, demands are growing, and river flows and uses are vulnerable to drought and climate change, a condition that has occurred periodically for years. The basin has passed through many periodic drought periods. As of 2015, the basin is in its third year in a row of severe drought, in which irrigation and municipal water diversions have been severely curtailed, numerous stream diversions threaten endangered species, and reservoir storage volumes have returned to historic lows.

An ongoing and continuing challenge in the Rio Grande Basin is the development of programs that efficiently and justly allocate the basin's water resources among competing demands as well as across political and institutional jurisdictions. The authors described the development and application of a basin-wide mathematical programming model that economically optimizes water allocations and water use levels for the upper part of the Rio Grande Basin. Results of the model inform policies that wish to examine institutional innovations that could reduce damages caused by drought. Compared to existing institutions at the time of the work, future drought damages could be reduced by one-fifth to one-third per year through interstate various kinds of water markets that would permit or encourage water transfers across water management jurisdictions. Results revealed numerous economic tradeoffs among water uses, regions, climate futures, and drought control strategies (Booker et al., 2005).

### 4.2.2 Innovations: Transferable Rights

Since the late 1960s, water supply development projects used to meet growing economic and population demands have been reduced, and the interest of public providers in demand management measures increased importantly. An example of water innovations is market-supported transfers of water and water rights (Chong and Sunding, 2006; Ramirez et al., 2009) such as water banking. Water demands have also expanded in the face of increased urban growth, changes in irrigation water use patterns, and growing debates over climate and environmental quality protection needs.

Water rights systems based on historical use rights or connections to the land can lead to an economically inefficient allocation of water resources. Such water right systems can also give rise to other economically poor outcomes, including food insecurity (Salman et al., 2014) as well as overuse of land and weak incentives to convert to water conservation measures (Ward and King, 1998). Water can be locked into existing patterns of irrigated agriculture and into existing irrigation technologies (Ward and Pulido-Velazquez, 2008a). Water trading based on transferable water rights has been proposed by many scholars (Chong and Sunding, 2006) in the United States as well as the European Union as a mitigating measure to address these challenges. Water trading helps equalize the marginal economic values faced by various numerous competing water users, thereby supplying information and informing debates about the value of water in alternative uses, and creating compatible economic incentives. Putting water markets into practice introduces real-world complications of transaction costs and third-party externalities.

### 4.2.3 EU Example

The political context is very relevant for water markets in Europe, since most of water supply management and control is in public hands. In the Blueprint to Safeguard Europe's Water Resources (EC, 2012), the EC stated that "water trading is another instrument, used mostly outside the EU, which could help to improve water efficiency and overcome water stress, if a sustainable overall cap for water use is implemented. Water trading entails relatively significant administrative costs and, in principle, only makes sense among water users in a defined river basin. Although it would not be helpful to set up such a system at EU level, the Commission proposes developing CIS (common implementation strategy) guidance to help the development of water trading in the Member States that choose to employ it."

Informal water markets have been common in Spanish Mediterranean agricultural areas (Ostrom, 1990; Garrido, 2011). The Reformed Water Law (Law 46/1999) has incorporated formal water markets into the Spanish legal and regulatory framework, including spot water markets (formal lease contracts of water rights) and water banks (water exchange centers), allowing water rights holders to temporarily trade their water rights. Besides the potential of this measure, water markets have been operative only during drought periods under extreme scarcity situations, and the narrowness of the market suggests that there are some barriers hampering their effective functioning (Rey et al., 2014; Palomo-Hierro et al., 2015).

### 4.2.4 US Example

Transferable water rights are common in certain parts of the United States, for example, in New Mexico (De Mouche et al., 2011) and parts of California (Israel and Lund, 1995). Many studies have been conducted on the economic performance of such transfers. Knapp et al. (2003) found that water transfers from agricultural to urban and environmental uses will likely become more widespread worldwide in the future. Numerous irrigation regions currently rely on groundwater aquifers that can be tapped when surface supplies are low or unreliable. Surface water transfers to locations out of the basin of origin are likely to put upward pressure on groundwater pumping from regional aquifers, while also reducing aquifer recharge. The authors conducted an empirical analysis in southern California. They found that opening up water markets can increase the quantity of water exported over time as well as reduce economic benefits in the basin when property rights are set up under a common property arrangement (eg, in nonregulated aquifers).

## 5. CONCLUSIONS: WHAT CAN WE LEARN FROM EACH OTHER?

US water planners can profit from many lessons on European water policy innovations. Especially important are lessons from the WFD's goals, steps to achieve them, actual achievements, shortcomings, and remaining challenges. Despite a number of road bumps along the way, the WFD has had a remarkable influence on promoting greater consistency and unity in water program planning and policy implementation among the EU Member States.

Many US water planners are aware that the WFD changed water management in all EU Member States assigning aquatic ecology and water quality as the foundation of water management decisions. Hering et al. (2010) reviewed the successes and challenges encountered with the implementation of the WFD over the past 10 years and provided recommendations to further improve the implementation process. Their study examined three challenges, all of which could provide valuable experiences for US water policymakers:

- Consistent development of assessment methods, including reference conditions, typologies, and calibration
- Design and use of assessment systems in monitoring programs
- Outcomes produced by river basin management plans, including the design, monitoring, and success of restoration measures

The development of assessment methods of the WFD has been a transparent process and resulted in improved and more standardized tools for evaluating water bodies across the European Union. Despite that success, an important lesson is that the process has been more time consuming, and because of complex interactions between hydrology and economics, methods were found to be more complex than European planners had initially anticipated back in 2000. Remaining challenges regard Europe specifically, including estimating uncertainty of assessments for biological quality improvements.

US planners could benefit from the knowledge generated in Europe and monitoring data as a result of implementing the WFD. Remaining challenges include the need for improved data accessibility and the establishment of an EU-wide central monitoring system for reference sites (Hering et al., 2010).

Another important experience from Europe is that the WFD river basin management plans evaluate management decisions based on the response of aquatic resources to environmental stress. It is important to know that in contrast to the effects of degradation, biological response to restoration efforts is poorly understood and hard to predict and manage. The short time scale of the programs of measures in the WFD, to secure good water quality and good ecological status for all surface waters within a 6-year planning horizon, was found to be too ambitious to be realistic. If theoretically US water planners wished to initiate a similar directive as the WFD, the enumerated challenges, including long-term monitoring of restoration measures to better learn about ecosystems and their recovery patterns, would need to be considered. Also establishing priorities for measures according to the real potential for restoration of ecological watershed status is relevant (Hering et al., 2010).

The integrated, cross-sectoral and cross-jurisdictional river basin management approach adopted in Europe by the WFD has proven to be necessary if water resource management is to be efficient (Cao and Warford, 2006). River basin-based water institutions with a clear legal status and responsibilities for strategic water-related issues are essential for integrated water quality, quantity, and land planning (Cao and Warford, 2006). Basin-based planning and management for environmental purposes is a key challenge introduced by the WFD. Water management in the United States similarly indicates a clear transition from a decentralized approach toward a more integrated, comprehensive watershed management.

In terms of water pricing, the developments in Europe and the United States are focused on introducing a stricter financial discipline with more cost-recovery tariffs, recognizing that subsidies can be counterproductive. Irrigation, water, and sewerage pricing continues to be subsidized by governments, which could stimulate excessive and inefficient water use. Water abstraction fees do not even reflect the financial cost (they ideally should reflect the long run marginal cost of augmenting public supplies), and are far from reflecting the environmental and opportunity costs, despite the principles stated in the EU WFD. Besides their potential and increasing role, the use of pricing and taxation to create incentives for efficient water consumption is a long-term process both in Europe and the United States (Cao and Warford, 2006; Lago et al., 2015). Although informal water markets have been actively applied for a long time in Europe (eg, for irrigated agriculture in Mediterranean Spain), the United States (especially the western states) has more than 30 years of experience with formal water markets and water banking as well. Europe can certainly learn from this experience to improve the existing framework to satisfy the demand for water trading (Delacámara et al., 2015). This would require in-depth investigations of the conditions under which this instrument could serve the objectives of EU water policies (Rinaudo et al., 2015).

Finally, private sector participation in water resource management and water supply has a long tradition both in the United States and in Europe, even though the countries differ in terms of water management circumstances and contexts. Public–private partnerships have proven as effective measures to reduce drought and flood risks. Both European and United States governments have also recognized the importance of public participation in water resource management, implementing legislative measures to facilitate this participation.

## REFERENCES


Agarwal, A., Angeles, M.S.D., Bhatia, R., Chéret, I., Davila-Poblete, S., Falkenmark, M., Villarreal, F.G., Jønch-Clausen, T., Kadi, M.A., Kindler, J., Rees, J., Roberts, P., Rogers, P., Solanes, M., Wright, A., 2000. Integrated Water Resources Management. Technical Advisory Committee Background Papers, 4. Global Water Partnership, Stockholm, Sweden.

Allit, P., 2015. A Climate of Crisis. America in the Age of Environmentalism. Penguin History American Life. Penguin Press.

Barreira, A., 2006. Water governance at the European Union. Journal of Contemporary Water Research & Education 135 (1), 80–85.

Bhat, A., Blomquist, W., 2004. Policy, politics, and water management in the Guadalquivir river basin, Spain. Water Resources Research 40, W08S07. http://dx.doi.org/10.1029/2003WR002726.

Biswas, A.K., 2004. Integrated water resources management: a reassessment – a water forum contribution. Water International 29 (2), 248–256.

Booker, J.F., Michelsen, A.M., Ward, F.A., 2005. Economic impact of alternative policy responses to prolonged and severe drought in the Rio Grande basin. Water Resources Research 41 (2).

Brouwer, R., Barton, D., Bateman, I., Brander, L., Georgiou, S., Martín-Ortega, J., Pulido-Velazquez, M., Schaafsma, M., Wagtendonk, A., 2009. Economic Valuation of Environmental and Resource Costs and Benefits in the Water Framework Directive: Technical Guidelines for Practitioners. Institute for Environmental Studies, VU University, Amsterdam, the Netherlands.

Brouwer, R., 2008. The role of stated preference methods in the water framework directive to assess disproportionate costs. Journal of Environmental Planning and Management 51 (5), 597–614.

Cao, Y.S., Warford, J., 2006. Evolution of Integrated Approaches to Water Resource Management in Europe and the United States: Some Lessons from Experience. China addressing water scarcity background paper series; no. 2. World Bank, Washington, DC. http://documents.worldbank.org/curated/en/2006/04/7149364/evolution-integrated-approaches-water-resource-management-europe-united-states-some-lessons-experience.

Chave, P.A., 2001. The EU Water Framework Directive: An Introduction. IWA publishing.

Chong, H., Sunding, D., 2006. Water markets and trading, annual review of environment and resources. Annual Review of Environment and Resources 239–264.

De Mouche, L., Landfair, S., Ward, F.A., 2011. Water right prices in the Rio Grande: analysis and policy implications. Water Resources Development 27 (2), 291–314.

De Stefano, L., 2010. Facing the water framework directive challenges: a baseline of stakeholder participation in the European Union. Journal of Environmental Management 91, 1332–1340.

Del Moral, L., Saurí, D., 1999. Changing course: water policy in Spain. Environment 41, 12–36.

Delacámara, G., Gómez, C.M., Maestu, J., 2015. Water trading opportunities and challenges in Europe. In: Burnett, K., Howitt, R., Roumasset, J., Wada, C. (Eds.), Routledge Handbook of Water Economics and Institutions, ISBN 978-0-415-72856-0, pp. 279–293 (Chapter 17).

Dillon, P., 2005. Future management of aquifer recharge. Hydrogeology Journal 13 (1), 313–316.

EC, 2006. Directive 2006/118/Ec of the European Parliament and of the Council of 12 December 2006 on the protection of groundwater against pollution and deterioration. Official Journal of the European Union, L, 372.

EC, 2012. Report from the Commission to the European Parliament and the Council on the Implementation of the Water Framework Directive (2000/60/Ec) River Basin Management Plans. Com (2012), 670.

Evans, H.B., 1997. Water Distribution in Ancient Rome: The Evidence of Frontinus. University of Michigan Press.

Fischhendler, I., Feitelson, E., Eaton, D., 2004. The short-term and long-term ramifications of linkages involving natural resources: the US-Mexico transboundary water case. Environment and Planning C-Government and Policy 22 (5), 633–650.

Garrido, S., 2011. Governing scarcity. Water markets, equity and efficiency in pre-1950s eastern Spain. International Journal of the Commons 5 (2), 513–534. http://doi.org/10.18352/ijc.274.

van Grinsven, H.J.M., Rabl, A., de Kok, T.M., 2010. Estimation of incidence and social cost of colon cancer due to nitrate in drinking water in the EU: a tentative cost-benefit assessment. Environmental health 9 (1), 58.

Hanley, N., Wright, R.E., Alvarez-Farizo, B., 2006. Estimating the economic value of improvements in river ecology using choice experiments: an application to the water framework directive. Journal of Environmental Management 78 (2), 183–193.

Heinz, I., Pulido-Velazquez, M., Lund, J., Andreu, J., 2007. Hydro-economic modeling in river basin management: implications and applications for the European water framework directive. Water Resources Management 21 (7), 1103–1125.

Hering, D., Borja, A., Carstensen, J., Carvalho, L., Elliott, M., Feld, C.K., Heiskanen, A.S., Johnson, R.K., Moe, J., Pont, D., et al., 2010. The European water framework directive at the age of 10: a critical review of the achievements with recommendations for the future. Science of the Total Environment 408 (19), 4007–4019.

Hill, R.A., 1974. Development of Rio-Grande-Compact-of-1938. Natural Resources Journal 14 (2), 163–200.

Hoornbeek, J.A., 2004. Policy-making institutions and water policy outputs in the European Union and the United States: a comparative analysis. Journal of European Public Policy 11 (3), 461–496.

Israel, M., Lund, J.R., 1995. Recent California water transfers: implications for water management. Natural Resources Journal 35, 1.

Knapp, K.C., Weinberg, M., Howitt, R., Posnikoff, J.F., 2003. Water transfers, agriculture, and groundwater management: a dynamic economic analysis. Journal of Environmental Management 67 (4), 291–301.

Lago, M., Mysiak, J., Gómez, C.M., Delacámara, G., Maziotis, A., 2015. Use of Economic Instruments in Water Policy: Insights from International Experience. Global Issues in Water Policy. Springer, The Netherlands.

Lawford, R., 2003. Water: Science, Policy, and Management: Challenges and Opportunities, 16. American Geophysical Union.

Maass, A., Hufschmidt, M.M., Dorfman, R., Thomas Jr., H.A., Marglin, S.A., Fair, G.M., 1962. Design of Water-Resource Systems. Harvard University Press Cambridge, Massachusetts.

Macián-Sorribes, C., Pulido-Velazquez, M., Tilmant, A., 2015. Definition of efficient scarcity-based water pricing policies through stochastic programming. Hydrology and Earth System Sciences 19, 3925–3935. http://dx.doi.org/10.5194/hess-19-3925-2015.

McCormick, Z.L., 1994. Interstate water allocation compacts in the Western United-States – some suggestions. Water Resources Bulletin 30 (3), 385–395.

Molle, F., 2009. River-basin planning and management: the social life of a concept. Geoforum 40 (3), 484–494.

Newig, J., Pahl-Wostl, C., Sigel, K., 2005. The role of public participation in managing uncertainty in the implementation of the water framework directive. European Environment 15, 333–343. http://dx.doi.org/10.1002/eet.398.

Ostrom, E., 1990. Governing the Commons. The Evolution of Institutions for Collective Action. Cambridge University Press, Nueva York.

Palomo-Hierro, S., Gomez-Limon, J.A., Riesgo, L., 2015. Water markets in Spain: performance and challenges. Water 7, 652–678.

Pigou, A.C., 1920. The Economics of Welfare, fourth ed. Macnillam, London.

Prest, A.R., Turvey, R., 1965. Cost-benefit analysis: a survey. The Economic Journal 683–735.

Pulido-Velazquez, M., Andreu, J., Sahuquillo, A., Pulido-Velazquez, D., 2008. Hydro-economic river basin modelling: the application of a holistic surface-groundwater model to assess opportunity costs of water use in Spain. Ecological Economics 66 (1), 51–65. http://dx.doi.org/10.1016/j.ecolecon.2007.12.016.

Pulido-Velazquez, M., Hernandez-Sancho, F., Ferrer-Polo, F.J., Latorre, J.I., 2009. Economic tools for selecting a programme of measures to meet the Wfd standards. In: Application to the Serpis River Basin, Spain, IWA Conference on Water Economics.

Pulido-Velazquez, M., Alvarez-Mendiola, E., Andreu, J., 2013. Design of efficient water pricing policies integrating basinwide resource opportunity costs. Journal of Water Resources Planning and Management 139 (5), 583–592. ISSN:0733-9496.

Ramirez, O.A., Ward, F.A., Al-Tabini, R., Phillips, R., 2009. Efficient water conservation in agriculture for growing urban water demands in Jordan. Water Policy 13 (1), 102.

Rey, D., Garrido, A., Calatrava, J., 2014. The water markets in Spain: moving towards 21st century mechanisms and approaches with 20th century regulations. In: Easter, W., Huang, Q. (Eds.), Water Markets for the 21st Century: What Have We Learned? Global Issues in Water Policy, Springer, The Netherlands.

Rinaudo, J.D., Berbel, J., Bouscasse, H., Calatrava, J., Duponteil, A., Giannocarro, G., Garrido, A., Figureau, C., Hérivaux, C., Kervarec, F., Montginoul, M., Raggi, M., Rey, D., Rinaudo, J.-D., Vernier, M., Viaggi, D., Zavalloni, M., 2015. Water Cap & Trade (Water Markets Scenarios for Southern Europe: New Solutions for Coping with Increasing Water Scarcity and Drought Risks). Final project report. Available at: https://sites.google.com/site/capandtraderesearch/final-report.

Robinson, J.R.C., Michelsen, A.M., Gollehon, N.R., 2010. Mitigating water shortages in a multiple risk environment. Water Policy 12 (1), 114–128.

Rogers, P., de Silva, R., Bhatia, R., 2002. Water is an economic good: how to use prices to promote equity, efficiency and sustainability. Water Policy 4, 1–17.

Salman, D.A., Amer, S.A., Ward, F.A., 2014. Water appropriation systems for adapting to water shortages in Iraq. Journal of the American Water Resources Association 50 (5), 1208–1225.

Sanchez, V., Cortez-Lara, A.A., 2015. Minute 319 of the international boundary and water commission between the US and Mexico: Colorado river binational water management implications. International Journal of Water Resources Development 31 (1), 17–27.

Savenije, H.H.G., van der Zaag, P., 2002. Water as an economic good and demand management – Paradigms with Pitfalls. Water International 27 (1), 98–104.

Schlager, E., Blomquist, W.A., 2008. Embracing Watershed Politics. University Press of Colorado Boulder.

Shively, D.D., Mueller, G., 2010. Montana's Clark Fork River basin task force: a vehicle for integrated water resources management? Environmental Management 46 (5), 671–684.

United States Senate, 1962. Policies, Standards and Procedures in the Formulation, Evaluation and Review of Plans for Use and Development of Water and Related Land Resources. Senate Document, 97.

US Council on Environmental Quality, 2013. Principles and Requirements for Federal Investments in Water Resources. https://www.Whitehouse.Gov/Administration/Eop/Ceq/Initiatives/Pandg.

US Environmental Protection Agency, 2015. 40th Year Celebration. http://water.epa.gov/action/cleanwater40/.

Ward, F.A., King, J.P., 1998. Reducing institutional barriers to water conservation. Water Policy 1 (4), 411–420.

Ward, F.A., Pulido-Velazquez, M., 2008a. Water conservation in irrigation can increase water use. Proceedings of the National Academy of Sciences 105 (47), 18215–18220.

Ward, F.A., Pulido-Velazquez, M., 2008b. Efficiency, equity, and sustainability in a water quantity-quality optimization model in the Rio Grande basin. Ecological Economics 66 (1), 23–37.

Ward, F.A., Pulido-Velazquez, M., 2009. Incentive pricing and cost recovery at the basin scale. Journal of Environmental Management 90 (1), 293–313.

Ward, F.A., Michelsen, A.M., DeMouche, L., 2007. Barriers to water conservation in the Rio Grande basin. Journal of the American Water Resources Association 43 (1), 237–253.

Ward, F.A., 2012. Cost-benefit and water resources policy: a survey. Water Policy 14 (2), 250–280.

Whitford, A.B., Clark, B.Y., 2007. Designing property rights for water: mediating market, government, and corporation failures. Policy Sciences 40 (4), 335–351.

Young, C., 2011. Shaping water policy: political influences in the European Union and United States. Claremont-UC Undergraduate Research Conference on the European Union 2011. http://dx.doi.org/10.5642/urceu.201101.12. Article 12.

PART 4

# The Future of Water: Prospects and Challenges for Water Management in the 21st Century

A. Tarhule
*The University of Oklahoma, Norman, OK, United States*

## 1. INTRODUCTION

Water is the driver of nature and its availability and quality often constitute a limit on economic development and human welfare. Because of this primacy, the task of providing water in sufficient quantity and acceptable quality to the world's human population while assuring its availability for future needs constitutes one of the pre-eminent challenges of the 21st century. As the chapters in this book show, the challenges are numerous and widely cited. Globally, around 750 million people, most living in developing countries, lack access to safe drinking water, 2.5 billion lack adequate sanitation, and 842,000 people still die annually because of diarrhea from contaminated water sources [World Health Organization/UNICEF Joint Monitoring Program (WHO/JMP), 2015]. These challenges are real and tragic, and ongoing efforts, such as the Strategic Development Goals of the United Nations Development Program, must be sustained until they are resolved.

Looking ahead, we can expect that the water challenges of the future will be vastly different from those of the past and present in a number of important ways. Notably, both the quantity and quality of water are increasingly impacted by human actions either directly through activities like river channel modifications, diversions and storages, land use land cover alterations, interbasin water transfers, as well as point and nonpoint source pollution discharges or indirectly through anthropogenic climate change. Without concerted effort, we can expect that the scale, intensity, and impacts of these activities will continue to increase in tandem with rising population and increasing economic development.

**Competition for Water Resources. http://dx.doi.org/10.1016/B978-0-12-803237-4.00025-2**

A number of salient challenges are illustrative of the scope and nature of the water-related problems of the future. In this chapter, I highlight three of them, namely:

- Should we attempt to satisfy demand for water or curtail it?
- What is the economic status and value of water?
- What is the risk of future megadroughts caused by climate change?

These issues have not been selected because they are the most critical or important. Rather, I have attempted to integrate seemingly disparate facts and information to showcase historical patterns and trajectories, and to highlight how complex and interconnected the factors involved are. To the extent possible, an attempt is made to identify possible solutions and pathways forward.

## 2. SHOULD WE ATTEMPT TO SATISFY DEMAND FOR WATER OR CURTAIL IT?

### 2.1 Historical Context and Developments

The world's water challenge can be summarized fairly succinctly: supplies are finite, demand is rising, and competing needs are becoming more complex to reconcile. Confronted with this reality, water resources managers the world over have asked themselves exactly the same question: should they try to increase supply to meet demand or curtail demand to the level of available supplies? Until a few decades ago such a question would have been considered preposterous. From the time of the Romans and perhaps predating that, the accepted paradigm in water resources management has been decidedly supply oriented, characterized by reliance on engineered structures and solutions. Over time, this approach produced some of the largest, most iconic, and ambitious engineering projects of the modern era including the Hoover, Glen Canyon, Grand Coulee dams and the Colorado River Aqueduct in the United States, China's south—north water transfer project and the Three Gorges Dam, Libya's Great Man-Made River, the Almendra dam in Spain, and the Itaipu dam at the border between Brazil and Paraguay, to name just a few.

To be completely fair this outcome has not been simply a matter of engineering hubris; many of these projects were critically needed to meet municipal and domestic water supply and sanitation, flood control, irrigation to support agricultural productivity, electricity generation, and to facilitate the settlement and economic viability of entire regions like the western United States. Big dams and reservoirs in particular made sense because unlike other valuable commodities such as energy (electricity), it is cheaper to store water than it is to transport it over long distances. Thus in many locations, dams were not only the cheapest option; with the constraints and status of knowledge at the time, they were also the most obvious and logical option for underpinning rapid economic and energy transformation, as well as ensuring food security. So closely did dams become associated with economic development (and perhaps a symbol of mastery over nature) that they adorned the currencies of many developing countries.

It is important to note that supply-side water resources management has been phenomenally successful. Improved water supply and sanitation enabled humanity to lay the foundation for our modern civilization by eradicating diseases and epidemics such as typhoid, diarrhea, supporting the growth of megacities, industrialization, and generally raising life expectancy and higher standard of living. Not surprisingly, resources for these projects were generally available both in Europe and the United States, the populations welcomed them and there were hardly any concerns about their environmental or other impacts.

Today, the situation is diametrically different. Certainly in Europe and North America, dams and other mega water projects have become among the most controversial of all projects, perhaps ranking among nuclear and waste disposal facilities. Critics blame dams, for example, for a long list of environmental, ecological, and social ills including irreversible biodiversity loss, disruptions of aquatic habitats and upsetting ecological balance, permanent destruction and loss of cultural landmarks, disruptions of livelihoods of thousands of often vulnerable and marginalized groups, and the promotion of diseases, among numerous others. The pendulum has swung so far in the opposite direction that today a Google search using the term "science of dam removal" returns many more websites than the term "dam construction." Awareness of these adverse effects came about partially from accumulated effects and information gained from several decades of dam operations. Riding public awareness and angst about Earth's fragility and dwindling resources, environmental movements grew steadily in numbers, economic leverage, and gained intellectual and political legitimacy in many countries. In Europe, for example, the pan-European coalition of Green Parties won 50 seats in the EU Parliament (nearly 7%) during the 2014 elections. In North America, although having a long history and constituting strong lobby groups, Green movements have not been able to transmute into effective political parties. Instead, progress has been much more rapid on the academic front. In just a few years, hundreds of new "green" academic degree programs have been established in North American universities with nearly all experiencing explosive student enrollments. Indeed, it has been observed that the current generation of students is the "Green Generation." An increasingly environmentally educated and conscious electorate, combined with political and economic clout, has led to opposition to many large-scale projects perceived to portend risks to water resources. In the United States, the battle over the Keystone pipeline project is a case in point. Finally, the decline of the glory days of mega water projects can be attributed at least in part to the fact that most of the obvious and readily developed water sources have already been exhausted so that new developments must compete with other high-value uses.

As enthusiasm for supply-side water solutions waned and opposition mounted, financing for mega water infrastructure projects from sources like the World Bank all but dried up and engineers shifted attention to improving water use efficiency. In nearly all respects, this too has been a spectacular success. By shifting from gravity to pressure irrigation technology, for example, irrigation water use in the United States declined by 30% from its peak in 1980 (approximately 207.23 $km^3$ or 168 million acre-feet) (Donnelly and Cooley, 2015). Similarly, following the lead of the state of Massachusetts, which first mandated low flush toilets in 1988, President George H.W. Bush signed the Energy Policy Act in law in 1992, which mandated that all toilets in the United States should have a maximum flush capacity of 1.6 gallons per flush (6 L) as opposed to the previous industry standard of 3.5 gallons (13.2 L) per flush. Toilets were specifically targeted because they account for approximately 30% of residential indoor water use, the largest single source of water use in the home. In fact, the Environmental Protection Agency estimates that by replacing all full flush toilets with the new low flush toilets, the United States would save nearly 640 billion gallons of water per year (about 2000 gallons, or 7.57 $m^3$, per year for every person in the United States!). Following a rocky introduction in 1994 (due to poor performance), low flush toilets have now improved to the point where many use 20% less water than the mandated standard while providing equal or superior performance. The success of this federally-led initiative has since spread into other residential water appliances including shower heads, faucets, and dishwashers. Indeed, the United States Geological Survey data on water use trends in the United States show that per capita and total water use have been declining across all sectors in the Unites States since about 1980 (Donnelly and Cooley, 2015). The numbers are actually quite impressive: per capita water use has declined by nearly 40% since 1980 and both agricultural and municipal water uses have declined by approximately 20% during the same period. The decrease in the municipal water use is especially noteworthy because the urban population in the United States increased nearly 50%, from approximately 167 million in 1980 to about 249 million between 1980 and 2010 (United States Census Bureau, 2015).

Other countries have also experimented with mandatory demand management policies. In England the government has set a target of reducing per capita domestic water consumption from 150 to 130 L by 2020 (Defra, 2008). In support of the strategy, Ofwat, which regulates water and sanitation in England and Wales, has set a voluntary reduction of 1% in domestic water demand with plans to introduce other target reductions in the future.

## 2.2 Water Soft Path: An Emerging Paradigm

The type of water efficiency strategies and technologies just described constitute what has been termed traditional demand management, which focuses on ways to reduce the amount of water used to accomplish a specific task (Brandes et al., 2011). This approach, while commendable and generally recognized as a necessary first step (low-hanging fruit) in demand management, has nevertheless been criticized as inadequate. The major criticism is that it represents an anthropocentric view focused on short-term cost effectiveness as opposed to an ecosystem perspective, focused on long-term ecological sustainability. Motivating such criticism is a strong and growing belief in some sectors that "current water management practices are simply unsustainable and cannot continue to deliver the benefits they have in the past" (Brandes et al., 2011). These critiques argue that to achieve the desired goal of balancing water use and ecological sustainability a new comprehensive and integrated paradigm is needed.

Termed "water soft path" (WSP), the new approach goes beyond the goal of water use efficiency to a focus on changing how some tasks are performed so that "the use of water is reduced much further or eliminated entirely" (Brandes et al., 2011, p. 8). Where traditional demand management promotes low flush toilets, the WSP approach questions their use as opposed to waterless or fully integrated resource recovery systems in large buildings and/or at the neighborhood scale. Where traditional demand management advocates more efficient sprinklers, WSP questions why portable water is used to irrigate lawns when recycled water would serve just as well. To qualify as a WSP system, four fundamental principles need to be satisfied. First, water is treated as a service rather than an end in itself. Second, ecological sustainability is a fundamental criterion and desired end state. Third, the quality of water delivered is matched to the intended use. Fourth, planning proceeds from the future back to the present.

There is no question that this approach represents a revolutionary departure from prevailing practice. Is this the future of water resources management? Or will it end up as a niche initiative to be implemented only at selected locations or times when unique sets of circumstances prevail?

WSP proponents freely concede the nearly daunting barriers to a widespread adoption and implementation of the principles and practices they advocate. They recognize that it will take major changes in human behavior and perceptions at both the scale of individuals and societies as well as political/institutional/policy changes. Numerous theories have shown and explained why it is not easy to change the behavior of individuals and organizations. One such theory is path dependency. As summarized by Sambu and Tarhule (2013), path dependency theory states that context-dependent decisions by or about institutions (individuals) create frameworks (arrangements, policies, or practices) that can become self-reinforcing through a system of feedbacks or returns. Because individuals, society, and institutions invest time, money, skills, or expectations in these established ways of doing things, a considerable amount of inertia builds up against alternatives, making change difficult. Despite some criticisms (Kay, 2003) the path dependency theory has been used for illuminating the reasons and dynamics of reform failures as well as clarifying events and institutional behavior that are otherwise difficult to explain (North, 2005; Prado and Trebilcock, 2009).

Behavioral change with respect to a common pool resource also suffers elements of the tragedy of the commons inasmuch as the cost of the actions and decisions of individuals are externalized while individual benefits are maximized. Experience and history suggest that changing societal perceptions and attitudes often requires perception of public crisis, and risk and success are not guaranteed even then. Certainly, past antismoking campaigns, automobile seatbelt use, and ongoing efforts to regulate firearm use in the United States all attest to the difficulty of the challenge. A big part of the problem is that these issues are complex and multifaceted. In the case of water, for example, strengthening belief in the principle of water as a human right complicates efforts to restrict access to water, even to developments and user groups deemed illegal.

In the United States a number of other emerging dynamics are also worth monitoring. Notably, there appears to be increasing erosion of trust and confidence in public institutions and science among some segments of the society. The best publicized of these issues is climate change where there is a wide (and growing!) gap between the confidence that scientists have in their data and conclusions versus public belief and perception of climate change. Another example is the effort to either replace evolution with creationism or to include both concepts in school curricula. Declining public support for science also manifests, for example, in efforts and agitations to cut funding for various types of research often based on what appears to be political or perceptual arguments. Although neither necessarily widespread nor directly related to water, these dynamics bear watching because large-scale attitudinal and institutional transformations in water resources management will almost certainly require public support and political action. That task will be much harder if the public becomes less trusting of science and institutions.

Such misconceptions are already evident. In what Jordan et al. (2011) term the myth of overabundance, people find it difficult to believe that water is or could be in short supply. To the extent that they acknowledge the existence of the problem, most also believe that simple conservation measures will address the shortage. There is also pervasive belief that reductions in water supply and use will lead to loss of jobs and increase operational costs for industry and economic decline. Countering such attitudes and beliefs will be critical to the success of WSP.

Jordan et al. (2011) list other barriers to the implementation of WSP, such as: organization and management, financial, data and information, and policy and governance.

The authors also offered thoughtful suggestions for how some of these barriers could be overcome, including multistakeholder involvement in planning and decision making, as well as a carrot-and-stick approach to monitoring and enforcement.

The dynamics just discussed are welcome and mark a natural progression that began with making sure a majority of people in Europe and North America had access to water to now ensuring that the water is used efficiently. To succeed, continuing public education and sustained stakeholder engagement at all levels will be key. Water resources managers and policymakers will need to be attentive to opportunities that offer favorable entry points for transformative change. These might include, for example, opportunities to introduce new technology, replace existing infrastructure, or redesign new urban layouts with human and ecosystem water use needs as a guiding philosophy. Ultimately, it will take more than moral arguments or environmental concerns for the new paradigm to become widely accepted; people will have to derive tangible economic benefits with the innovations it introduces and the changes they are compelled—or choose—to make.

## 3. THE ECONOMIC STATUS AND VALUE OF WATER

### 3.1 Historical Context and Developments

The economic conception of water, as well as the associated question of whether water is unique or can be treated as a regular economic commodity, has been debated for millennia, testifying to the complexity of the issue. Adam Smith addressed the economic value of water in his treatise on the paradox of water and diamonds, distinguishing between value in use (ie, utility of a product) and value in exchange (ie, the purchasing power of a commodity). Water, he noted, has a high value in use, but limited value in exchange. Diamonds, on the other hand, have a high value in exchange, but limited value in use. This distinction underscores a realization reached by many people across the times, namely, that the market price of an item does not reflect the true value…of some priceless commodities (paraphrased from Hanemann, 2005).

While this realization had been around a long time, a different school of thought held that commodities that lacked economic value tended to be used inefficiently (in an economic sense), and that the free market is the best mechanism for the determination of the true value of a commodity. During the International Conference on Water and the Environment in Dublin in 1992, proponents of this thinking argued, "past failure to recognize the economic value of water has led to wasteful and environmentally damaging uses of the resource. Managing water as an economic good is an important way of achieving efficient and equitable use, and of encouraging conservation and protection of water resources." In fact, among the guiding principles that emerged from the Dublin Accord was a declaration that "Water has an economic value in all its competing uses and should be recognized as an economic good." Perhaps reflecting the background intellectual contestations, the same declaration recognized the human rights to water when it held, "…it is vital to recognize first the basic right of all human beings to have access to clean water and sanitation at an affordable price."

Exposing divergent value propositions, the economic value philosophy is unimpressed with the so-called unique status of water. As noted by Baumann and Boland (1998), water is no different from any other economic good. It is no more a necessity than food, clothing, or housing, all of which obey the normal laws of economics (quoted in Hanemann, 2005,

p. 70). Other researchers have contended since the declaration that the question is poorly phrased. "The question is not whether water is an economic good...rather, the question is whether it is purely a private good that can reasonably be left to free market forces, or a public good that requires some amount of extra-market management to effectively and efficiently serve social objectives" (Perry and Seckler, 1997, p. 1) and ecosystem needs. Ecosystems require water in much the same way that humans do. Indeed, it has been estimated that ecosystems require 75% of all available freshwater compared to 25% needed for human use (Falkenmark and Rockstrom, 2004). Without adequate water, the ecosystem can fall into disequilibrium or even collapse, imperiling its ability to deliver a range of services as well as the livelihoods and welfare of all lifeforms that depend on it. Yet, ecosystem water needs have not always been a priority for water resources projects or managers. Thus, beginning from an entirely different perspective, this position arrives at approximately the same conclusion reached by the opposing viewpoint, namely, without adequate valuation, water and ecosystem resources "may be implicitly undervalued and decisions regarding their use and stewardship may not accurately reflect their true value to society" (Green Facts, 2015). The point of divergence between these two positions is in the manner in which "value" is understood and interpreted. The one position refers to market valuation and the other to nonmarket valuation.

As pointed out by Hanemann (2005, p.66), "the history of non-market valuation in the USA is closely intertwined with water projects." The Flood Control Act of 1936 specifically instructs the Army Corps of Engineers to "involve itself in flood control provided that the benefits to whosoever they may accrue are in excess of the costs" although the methodological basis for such valuation did not exist at the time. Over the next several decades, economists came to realize (see Maler, 1971,1974) that economic valuation theory could be extended to anything from which people derive satisfaction, not just market commodities. Accordingly, nonmarket valuation evolved from a largely qualitative concept to rigorous mathematical treatments by the late 1960s. With the introduction of the National Environmental Policy Act in 1969, federal agencies were required to carry out environmental impact analysis associated with proposed federal projects. In 1979 the US Water Resources Council officially endorsed and recommended an approach, contingent valuation, for water resources valuation. Thus, by 1980, environmental impact analysis was a prerequisite for all water resources projects in the United States and the methods for conducting such analysis, including nonmarket components, had been developed and endorsed by the largest water resources organization in the country.

Today, nonmarket valuation is an accepted, even required, practice in many development projects. With the growth of green academic programs, the techniques for environmental impact analysis continue to grow in sophistication and use. Looking ahead, it appears likely that scholars and policymakers will continue to grapple with the challenge of how to place a value on water that reflects its true economic worth without undermining its utility as a social good or even human right.

## 4. COMPETITION FOR WATER BETWEEN ECOSYSTEMS AND HUMANS

During the 1990s a number of scholars and policymakers promoted the idea that water scarcity would be a major cause of wars in the 21st century (Bulloch, 1995; Shiva, 2002). At core, the view was essentially anthropocentric; the ecosystem, when mentioned at all, was only of peripheral interest. It is too early to dismiss the notion that water indeed could be a cause of war during this century. To date, however, research by Aaron Wolf (Wolf, 1999) and others (eg, Alam, 2002) appears to suggest that the opposite may well be the more likely scenario. That is, rather than a cause of war, the need to find sustainable solutions to diminishing water scarcity and other issues seems to be encouraging greater cooperation among nations and stakeholders, not more enmity.

But some variation of this argument persists: might competition for water between humans *and ecosystems* become a flashpoint in the coming decades (Van Emmerik et al., 2014). The issues involved are indeed complex and border on questions of economics, sustainability, and even morality and self-preservation. In his critically acclaimed book *Water Follies*, Robert Glennon (2002) used a wide range of case studies to illustrate the nature and scope of the emerging water conflicts. One of these is the case of the Upper San Pedro River in Upper Arizona. During the 19th century, the river was rich in wet and marshlands and the riparian habitats supported hundreds of species of both breeding and migrating birds. From an environmental and ecosystem perspective, the San Pedro River was a real gem in an otherwise desolate and unforgiving landscape. However, between 1970 and 2010, the population of Cochise County, within the Upper San Pedro Basin, more than doubled from around 62,000 to over 130,000. During the same period, the population of the city of Sierra Vista, within the basin, grew from just under 7000 to nearly 44,000. Over a 16-year period, the city approved 4700 new single family construction permits. Significantly, the average cost per house went up from $64,000 in 1997 to over $206,000 in 2012. Thus, not only was there a tremendous increase in new housing construction, housing units were also getting larger, all of which combined to put enormous pressure on water resources and the physical environment.

How could such development and pace of growth be sustained while simultaneously preserving the San Pedro River as a natural resource that delivers valuable ecosystem services? Unfortunately, many stakeholders approach this question as though it were a zero sum game, setting up a false proposition that argues for one *or* the other. This mindset must change to one in which stakeholders learn to recognize and appreciate the value of competing needs. The goal should be to find win–win solutions that increase the net value of water to the community as a whole as opposed to the ruination of one sector to support another.

The case of the competition for water between blueberry growers and wild Atlantic salmon in Maine, USA, described by Glennon (2002) is also illustrative. Like its Pacific cousin, the Atlantic salmon also faces threats caused by numerous hydroelectric power generation dams, clear-cut logging with the attendant river channel siltation, and overfishing. More recently, however, the most serious threat to wild salmon has come from the extraction and diversion of water to irrigate blueberry fields. Both industries are important to Maine's economy as well as its sense of heritage and identity. The state accounts for nearly 20% of the US domestic salmon production although much of that is increasingly farmed. In 2010, the salmon industry contributed $76 million to Maine's economy, while blueberries accounted for $250 million (Gabe et al., 2011). Clearly, it is in Maine's interest to ensure the continued viability and sustainability of *both* industries rather than undermine one for the other. The challenge is that planning is not always about the equitable allocation of a scarce resource, but rather about which sectors or stakeholders can exert the larger influence on the decision-making process.

Yet, human kind must find a way to strike a balance between the needs of the ecosystem and humans. In this regard, the emerging science of sociohydrologic modeling

(Van Emmerik et al., 2014; Sivapalan et al., 2012), provides promising frameworks and analytic tools. The goal of social hydrology is "...understanding, interpretation, and prediction of the flows and stocks in the human-modified water cycle at multiple scales, with explicit inclusion of the two-way feedbacks between human and water systems" (Van Emmerik et al., 2014, p. 4240). By providing improved understanding of how humans interact with the environment, it is hoped that tools and mechanisms for mediating the competition for water between the two would be developed. Indeed, in the last several years there has been increased focus on development of what is broadly termed "coupled human natural systems" (CHNS). New degree programs as well as initiatives at major funding agencies, such as the US National Science Foundation, focusing on CHNS continue to be introduced. This trend is to be supported and encouraged because the need for mediation and intervention is likely to become more intense as human demographic and economic pressures continue to intrude and make mounting demands on ecosystems and water resources.

## 5. FUTURE CLIMATE VARIABILITY AND CHANGE AND THE RISK OF MEGADROUGHTS

No discussion of future water risks or challenges could be complete without a consideration of the possible role of climate change and climate variability. With respect to water resources, an aspect of the climate change discussion that has received comparatively little attention until recently is the risk of multidecadal or so-called megadroughts. This kind of drought is of special interest for a number of reasons. First, they occur very rarely. Since the beginning of modern observational hydroclimatic records, the best and arguably only example of multidecadal drought anywhere in the world has been the Sahelian drought in sub-Saharan Africa, which began in the late 1960s and persisted into the 1990s. Certainly in Europe and North America, there have been no droughts of comparable duration during the last 100 years. As a result of their infrequency, water resources management systems are not designed for coping with megadroughts.

We know, however, that such droughts do occur purely from natural climatic variability. As far back as 1929, A.E. Douglass described a 24-year drought (1276–99) based on tree-ring reconstructions. This drought may have been responsible for the abandonment of the Anasazi cliff dwellings across the Colorado Plateau (Cook et al., 2009). In fact, numerous authors (eg, Douglass, 1929; Douglass, 1935; Stine, 1994; Stahle et al., 2000; Pederson et al. 2005; Cook et al., 2009) have since described or confirmed occurrences of megadroughts during the Medieval Climate Anomaly (MCA aka Medieval Warm Period), which occurred approximately between 900 and 1300 AD, and was characterized by warmer than average temperatures. Based on detailed analysis, Cook et al. (2009) confirmed that two megadroughts occurred in the California–Nevada region during the MCA. The first drought persisted for over 200 years and the second for over 140 years. The two droughts were separated by a pluvial period of about 47 years. Similar analysis showed that in the Mississippi Valley, a 46-year drought occurred between AD 940 and 985; a 148-year drought occurred between AD 1100 and 1247; and a 61-year drought occurred between AD 1340 and 1400. The last two droughts appear to have contributed to the collapse of the Cahokia culture in the Mississippi Valley (Cook et al., 2009). The causes of these droughts are complex, but appear to be linked to the elevated temperatures (and therefore increased evaporative demand), and cool La Niña-like sea surface temperatures in the tropical pacific El Niño-Southern Oscillation region.

This historical review is significant because virtually all General Circulation Models agree that temperatures during the latter part of this century will be higher than they were during the MCA. This begs the question: how likely is it that megadroughts of the kind experienced during the last MCA might occur again? Alternatively, how likely is it that droughts longer than any we have witnessed during the period of observational records might occur as a result of the unprecedented temperatures projected during the latter part of this century? Associated with that, how could megadroughts impact water resources in the affected areas?

To investigate the possible effects of human activities and policy on future gas emissions and climate change, the Intergovernmental Panel on Climate Change relies on a set of scenarios called the Representative Concentration Pathways (RCP). The baseline greenhouse gas concentration scenario, representing no mitigation of current emissions trends, is termed RCP8.5 (Riahi et al., 2011) in which temperature increase is expected to be in the range of 2.6−4.8°C during 2081−2100 relative to the baseline. RCP4.5 is a moderate scenario or gas concentration pathway in which emissions peak around 2040 and then decline as a result of mitigation actions. The expected average global temperature range (2080−2100) under this scenario is 1.8°C with a likely range of 1.1−2.6°C. While the exact magnitude of temperature increase during the MCA is subject to some uncertainty and controversy, most researchers believe it was likely 0.1−0.2°C below the 1960−90 reference period or an order of magnitude less than even the moderate RCP4.5 scenario.

Using results of empirical drought reconstructions and soil moisture indices from 17 state-of-the-art general circulation models, Cook et al. (2015) showed that under RCP8.5 there is a greater than 80% chance of multidecadal drought during the 2050−99 period in the Central Plains and Southwest of the United States. As a comparison, the risk of a multidecadal drought during the observational period was less than 12% for the same region. In terms of the severity of the droughts, the models suggest that the level of aridity might exceed that of the megadroughts of the MCA. The results are similar for the moderate RCP4.5. Notably, these conclusions are robust across a strong majority of the models investigated.

The conclusion that emerges is that there is a higher likelihood of drought in the latter part of this century because of higher temperatures. To see what implications this might have on water resources and associated sectors, we only have to look at the recent droughts in California and the southwestern United States. Since 2000, the January level of Lake Mead has fallen by more than 30.48 m (100 ft), from >365.76 m to 335.28 m above mean sea level, (msl) (ie, <1200 ft to <1100 ft above msl). Such levels have not been reached since the dam began filling in 1937. As a result, more than 17.27 $km^3$ (14 million acre-feet) of water has been lost. Should the lake level drop below 327.66 m above msl (1075 ft), a water shortage declaration will take effect and Nevada and Arizona will see reduced water allocations. California will not be affected because it has priority or senior rights to the water. Such a scenario is completely uncharted territory and its likely effects remain largely unknown. In California, which in 2015 entered its fourth consecutive year of drought, an economic analysis of the drought's impact on agriculture showed that surface water is in deficit by nearly 10.73 $km^3$ (8.7 million acre-feet) or −48%, necessitating a 72% increase on groundwater pumping (Howitt et al., 2015). Even so, more than 540,000 acres of agricultural land (45%) has been idled, 21,000 jobs have been lost, and overall economic impact is likely to top $2.7 billion.

So far, California has shown remarkable resilience to the drought probably because of the variety of sources from which the state derives its water. But the drought is only in its fourth year. What could happen if it persists for another four years? Or, how will water resources management cope if future droughts last 10 or more years as projected by climate models? There are no easy answers. First, the rarity of megadroughts and their lack of precedence means that policymakers and economic planners are likely to be reticent about tying up scarce resources against the possibility of such a drought. It is easy to see how this kind of thinking works today with respect to snow removal in cities that receive snow infrequently. In many instances, city managers prefer to take the economic hit of shutting down the city for a few days than bear the cost of maintaining snow plow fleets that might never get used in an entire year. Hurricane Katrina provides another example. In that instance, retention walls were built to withstand a Category Three hurricane on the logic that the risk of a Category Five hurricane was low. Such precedence suggests that the outlook for making the kinds of investments necessary for dealing with a multidecadal drought is not promising.

Second, policymakers concerned about making such investments will have a convenient rationale, namely: the cone of uncertainty around a climate outlook so far out is very large. Indeed, given current rates of environmental, technological, and human transformations, we can expect that the whole world will be radically different in 2080 and maybe unrecognizable to today's planners. Of course, no one expects investments in megadrought resilient water infrastructure for 2080 to begin now. But we must be aware of them and continuously update the climate projections and our own expected responses and mitigation strategies. On the other hand, as the current California drought shows vividly, you do not really need a multidecadal drought to wreck serious havoc on socioecological and agroeconomic systems. Indeed, as the population continues to grow and economic activities expand, each future drought or water scarcity episode costs more in terms of human and economic impacts. This means a drought of comparable magnitude occurring two decades into the future is likely to be more disruptive than the current one.

## 6. CONCLUSIONS

Water is life for humans and ecosystems alike. Our demand for water has increased over time and will continue to do so in the foreseeable future. But the supply of water is finite and human activities frequently serve to further degrade the quality of this already dwindling resource. Also, the degree of interdependence between humans and ecosystems and the vital importance of water on ecosystem health is not necessarily fully appreciated or embraced by all. Even in instances where such understanding exists, striking the right balance of water allocation and conservation measures among the competing demands can be challenging. For North America and Europe, where the basic question of access to water has been conquered in the main, developing the principles and frameworks for resolving the kinds of issues highlighted in this chapter is likely to constitute an important part of the dialogue on the future of water.

Indeed, one could identify a nearly endless list of challenges related to the future of water all deserving of discourse and urgent attention. The few identified here are illustrative only and used to frame the theme of this chapter. They underscore the need for strengthening the resilience of water management infrastructure and systems, identifying causal pathways and linkages among dynamics that impact water resources, and developing comprehensive monitoring systems of relevant drivers of change, including land use, land cover, climate, and human behavior. It is important also to note that changes in water resources management take a long time to implement. Major engineering projects, such as dams, routinely require more than 20 years from conception to completion, many considerably longer. Changes in societal behavior such as advocated by the WSP approach might require an entire generation of concerted awareness campaigns and promotions to take effect and that is just at the level of policymakers and managers. The point is that whatever the water resources approach of the future will be, we must give ourselves plenty of running room to start. In that sense, this book and the issues and conversations it stimulates are highly timely and needed.

## REFERENCES

Alam, U.Z., 2002. Questioning the water wars rationale: a case study of the Indus Waters Treaty. The Geographical Journal 341–353.

Baumann, D.D., Boland, J.J., 1998. The case for managing urban water. In: Baumann, D.D., Boland, J.J., Hanemann, M.W. (Eds.), Urban Water Management and Planning. McGraw Hill, pp. 1–30.

Brandes, O.M., Brooks, D.B., Gurman, S., 2011. Why a water soft path, and why now. In: Brooks, D.B., Brandes, O.M., Gurman, S. (Eds.), Making the Most of the Water We Have. earthscan, Washington, DC, p. 273.

Bulloch, J., 1995. Water Wars: Coming Conflicts in the Middle East. Gollancz.

Cook, E.R., Seager, R., Heim Jr., R.R., Vose, R.S., Herweijer, C., Woodhouse, C., 2009. Megadroughts in North America: placing IPCC projections of hydroclimatic change in a long-term paleoclimate context. Journal of Quaternary Science 25, 48–61.

Cook, B.I., Ault, T.R., Smerdon, J.E., 2015. Unprecedented 21st century drought risk in the American Southwest and Central Plains. Science Advances 1 (1), e1400082.

Defra, 2008. Future of Water: The Governments Water Strategy for England. Retrieved from Department of Food, Environment and Rural Affairs: www.defra.gov.uk/Environment/Water/Strategy.

Donnelly, K., Cooley, H., 2015. Water Use Trends in the United States. Pacific Institute.

Douglass, A.E., 1929. The secret of the Southwest solved with talkative tree rings. National Geographic 736–770.

Douglass, A.E., 1935. Dating Pueblo Bonito and other ruins of the Southwest. National Geographic Society Contributed Technical Papers. Pueblo Bonito Series 1–74.

Falkenmark, M., Rockstrom, J., 2004. Balancing Water for Humans and Nature: The New Approach in Ecohydrology. Routledge.

Gabe, T.M., McConnon Jr., J.C., Kersbergen, R., 2011. Economic contribution of Maine's food industry. Maine Policy Review 36–45.

Glennon, R., 2002. Water Follies: Groundwater Pumping and the Fate of America's Freshwaters. Island Press.

Green Facts, 2015. Non-Market Value. Retrieved from Green Facts. Org: http://www.greenfacts.org/glossary/mno/non-market-value.html.

Hanemann, M.M., 2005. The Economic Conception of Water. U.C. Berkeley: Department of Agricultural and Resource Economics, Berkeley, California. Working Paper No 1005.

Howitt, R., McEwan, D., Medellin-Azuara, J., Lund, J., Sumner, D., 2015. Economic Analysis of the 2015 Drought for California Agriculture. Davis: Center for Watershed Science. University of California, Davis.

Jordan, S., Stevens, C., Brooks, D.B., 2011. Removing institutional barriers to water soft paths: challenges and opportunties. In: Brooks, D.B., Brandes, O.M., Gurman, S. (Eds.), Making the Most of the Water We Have: The Soft Path Approach to Water Management. earthscan, Washington, DC, p. 273.

Kay, A., 2003. Path dependency and the CAP. Journal of European Public Policy 10 (3), 405–420.

Maler, K.G., 1971. A method of estimating social benefits in pollution control. Swedish Journal of Economics 121–133.

Maler, K.G., 1974. Environmental Economics: A Theoretical Enquiry. Johns Hopkins University Press for Resources for the Future.

North, D., 2005. Understanding the Process of Economic Change. Princeton University Press, Princeton, NJ.

Pederson, D.C., Peteet, D.M., Kudyla, D., Guilderson, D., 2005. Medieval Warming, Little Ice Age, and European Impact on the Environment during the Last Millennium in the Lower Hudson Valley, New York, USA. Quaternary Research, pp. 238–249.

Perry, C.J., Seckler, D., 1997. Water as an Economic Good: A Solution, or a Problem? International Irrigation Management Institute, Colombo, Sri Lanka.

Prado, M., Trebilcock, M.J., 2009. Path dependence, development and the dynamics of institutional reform. University of Toronto Law Journal 341–379.

Riahi, K., Rao, S., Krey, V., Cho, C., Chirkov, V., Fischer, G., et al., 2011. RCP 8.5-A scenario of comparatively high greenhouse gas emissions. Climatic Change 33–57.

Sambu, D.K., Tarhule, A., 2013. Institutional water reforms in Kenya: an analytical review. Water Policy 777–793.

Shiva, V., 2002. Water Wars: Privatization, Pollution and Profit. South End Press, Cambridge, MA.

Sivapalan, M., Savenije, H.G., Bloschl, G., 2012. Socio-hydrology: a new science of people and water. Hydrologic Processes 1270–1276.

Stahle, D.W., Cook, E.R., Cleaveland, M.K., Therrell, M.D., Meko, D.M., Grissino-Mayer, H.D., et al., 2000. Tree-ring data document 16th century megadrought over North America. EOS, Transactions of the American Geophysical Union 121–125.

Stine, S., 1994. Extreme and persistent drought in California and patagonia during mediaeval time. Nature 546–549.

United States Census Bureau, 2015. Population Projections. Retrieved from United States Census Bureau: http://www.census.gov/topics/population/population-projections.html.

Van Emmerik, T.M., Li, Z., Sivapalan, M., Pande, S., Kandasamy, J., Savanije, H.G., et al., 2014. Socio-hydrologic modeling to understand and mediate the competition for water between agriculture development and environmental health: Murrumbidgee River basin, Australia. Earth System Science 4239–4259.

Wolf, A., 1999. Water wars and water reality: Conflict and cooperation along international waterways. In: Lonergan, S. (Ed.), Environmental change, adaptation and human security. Kluwer Academic, Dordrecht, Netherlands, pp. 251–265.

World Health Organization/UNICEF Joint Monitoring Program, 2015. Progress on Drinking Water and Sanitation, 2015 Update and MDG Assessment. WHO.

# Index

'*Note*: Page numbers followed by "f" indicate figures and "t" indicate tables.'

## N



## O



## P



## R



## S